THE 18TH INTERNATIONAL CONFERENCE ON ENGINEERING DESIGN

15th-18th August 2011
Technical University of Denmark (DTU)
Copenhagen, Denmark

Organised By
Section for Engineering Design and Product Development
Department of Management Engineering at DTU
and the Design Society

Proceedings Volume DS68-4
IMPACTING SOCIETY THROUGH ENGINEERING DESIGN
VOLUME 4: PRODUCT AND SYSTEMS DESIGN

Edited By
Steve Culley
Ben Hicks
Tim McAloone
Thomas Howard
Udo Lindemann

Published by the Design Society

First published in 2011

This publication is copyright under the Berne Convention and the International Copyright Convention. All rights reserved. Apart from any fair dealing for the purpose of private study, research, criticism or review, as permitted under the Copyright, Designs and Patents Act of 1988, no part of this publication may be reproduced, stored in a retrieval system, or transmitted in any form or by any means, electronic, electrical, chemical, mechanical, photocopying, recording or otherwise, without the prior permission of the copyright owners. Unlicensed multiple copying of the contents of this publication is illegal. Inquiries should be addressed to the Design Society.

© 2011 The Design Society, unless otherwise stated.

The Design Society is a charitable body, registered in Scotland, number SC 031694

ISBN 978-1-904670-24-7, 522 pages.

Printed by Lightning Source, Inc., La Vergne, TN, USA
and by Lightning Source UK Ltd., Milton Keynes, UK

The publishers are not responsible for any statement made in this publication. Data, discussion, and conclusions developed by authors are for information only and are not intended for use without independent substantiating investigation on the part of potential users. Opinions expressed are those of the Author and not necessarily those of the Design Society.

Preface by the Programme Chair

Welcome to the ICED11 Proceedings!

A large team of individuals working together are responsible for the selection of the 416 papers being presented here at ICED11, the establishment of the themes, sessions and the creation of the podium and discussion activities. This team consists of the programme chairs, the theme chairs and their assistants, and last but by no means least the reviewers. In particular, the written comments of the reviewers have been critical to the programme team in making their final choices and grouping papers into the conference themes.

The papers have been collated into a multiple range of formats: a book of abstracts, a memory stick of full proceedings and ten volumes of proceedings, available via a print-on-demand supplier. These have been numbered against both Design Society and ISSN referencing. This will enable more extensive access, referencing and citation in the future.

For ICED11 there is no difference in prestige between the papers in podium and discussion sessions. All have passed the ICED quality threshold and papers in the discussion sessions have been selected and grouped to stimulate fundamental and exciting debate. To facilitate this we have put in place a new 5 x 5 format – 5 slides in 5 minutes. We are also introducing a number of techniques to encourage audience participation and for the first time we are trialling live minutes – so as to provide a record of the debates. Further to this, recorded interviews with all the theme chairs will be undertaken on the final day of the conference so that there is also a persistent summary of each theme following the conference. The records from the discussion sessions will be made available at the conference and also uploaded to the DS website alongside the interviews with the theme chairs.

So we hope that you enjoy the programme and participate fully in what is arguably the Premier engineering design research conference in the world. We also hope that you find time to enjoy Copenhagen, catch up with old friends and make some new ones.

Steve Culley
Programme Chair

Ben Hicks
Assistant Programme Chair

Preface by the Conference Chair

With these, the proceedings of the eighteenth International Conference on Engineering Design, ICED11, it can safely be stated that engineering design research is firmly established as a strong research discipline. In its 30 year history this is the first time that an ICED conference has returned to the same city, "Wonderful Copenhagen". In the Danish official 'Year of Design' the city is the perfect conference location, hosting the highest quality design, ranging from industrial design, through stunning architecture, to a dynamic engineering design industry, which has extensively backed the ICED11 conference.

As design researchers, design practice is our research object and industrial companies are our research laboratories. Based on our observations, discussions and participation in design activities, we gather knowledge and insights and crystallise these into both academic models and practical methods. Our customers are students, in training as the product developers and innovators of the future, and industrialists, engaging with us to get insight into tools and methods, which fit to their practices and empower them to meet the challenges of global competition.

The theme we have chosen for ICED11 is Impacting Society through Engineering Design. Design has a central role in bringing engineering and technology to practical use. Each of the 416 papers in these proceedings provides its own contribution to the ICED11 theme. We're delighted to see the variety and the quality of contributions that our colleagues from the design research community have contributed to ICED11.

We have taken great care to create a conference showing leading edge research into engineering design and product development practice and to provide a lively backdrop for knowledge exchange and research discussion.

Our goals with ICED11 have been to place particular emphasis on industry participation, provocative and relevant keynote speeches, maximum time for debate and discussion, and space to go in to depth, via the SIG workshops. And all this with a Danish flavour, which we hope you find welcoming, fun and "hyggelig"!

Welcome to ICED11!

Tim McAloone
Conference Chair

Tom Howard
Assistant Conference Chair

Preface by the Design Society President

The International Conference on Engineering Design (ICED) is the flagship conference of the Design Society, a Society created in 2001, on the foundations laid by Workshop Design-Konstruktion (WDK), to promote the development of the understanding of all aspects of design. The ICED conferences were inaugurated in 1981 in Rome, and with one extra conference have been held every two years since then, in fifteen countries of the world. Thirteen conferences were held under the auspices of WDK, and this is the fifth organised by the Design Society. It is also the first time that ICED has returned to a city in which the conference has been held before – Copenhagen was the second host city in 1983, and this revisit is most appropriate in view of the leading role that the city and its technical university have played in design research and practice over many years.

The 2011 Conference continues the format established in 2009, with a programme made up of plenary sessions, podium presentations, discussion sessions with focused debate and workshops led by the Design Society's Special Interest Groups. We hope that this varied programme, combined with extensive opportunities for networking, will provide an exciting opportunity for researchers and practitioners to learn about the latest developments in engineering design.

Organising an international conference takes an enormous amount of work, and I would like to express the thanks of the Society to the great team that has worked over many months to ensure the success of the Conference. Especially I would like to thank Tim McAloone, Tom Howard and colleagues at the Technical University of Denmark for their great work in the Organising Committee, and Steve Culley, Ben Hicks and the Programme Committee for bringing together such an excellent programme. Of course, their work would be in vain without the fantastic contributions of the authors, reviewers, theme chairs and session chairs, and the thanks of the Society are due to all of them.

Chris McMahon
Design Society President

ICED11 Design Society Programme Committee

Stephen Culley - Programme Chair
University of Bath, United Kingdom

Ben Hicks - Assistant Programme Chair
University of Bath, United Kingdom

Thomas Howard - Organising Committee Representative
Technical University of Denmark, Denmark

Udo Lindemann - Design Society Representative
Technical University of Munich, Germany

Tim McAloone - Organising Committee Representative
Technical University of Denmark, Denmark

Margareta Norell - Design Society Representative
KTH Royal Institute of Technology, Sweden

Martin Grimheden – Design Society Representative
KTH Royal Institute of Technology, Sweden

ICED11 Organising Team

Tim McAloone – Technical University of Denmark, Denmark
Thomas Howard – Technical University of Denmark, Denmark
Sofiane Achiche – Technical University of Denmark, Denmark
Saeema Ahmed-Kristensen – Technical University of Denmark, Denmark
Mogens Myrup Andreasen – Technical University of Denmark, Denmark
Per Boelskifte – Technical University of Denmark, Denmark
Hans Peter Lomholt Bruun – Technical University of Denmark, Denmark
Georg Christensen – Technical University of Denmark, Denmark
Claus Thorp Hansen – Technical University of Denmark, Denmark
Anja Maier – Technical University of Denmark, Denmark
Krestine Mougaard – Technical University of Denmark, Denmark
Birna S. Colbe Månsson – Technical University of Denmark, Denmark
John Restrepo – Technical University of Denmark, Denmark
Hamish McAlpine – University of Bath, United Kingdom

ICED11 Theme Chairs

John Clarkson and David Wynn, Design Processes
Yoram Reich, Design Theory and Research Methodology
Marco Cantamessa, Design Organisation and Management
Udo Lindemann, Product and Systems Design
Andy Dong, Design Methods and Tools
Johan Malmkvist, Lars Almefelt and Andreas Dagman, Design for X, Design to X
Wei Chen and Harrison Kim, Design Information and Knowledge
Petra Badke-Schaub, Human Behaviour in Design
Bill Ion, Design Education

ICED11 Scientific Committee

Aakjaer Jensen, Thomas – Universe Foundation, Denmark
Achiche, Sofiane – Technical University of Denmark, Denmark
Adams, Robin – Purdue University, United States
Ahm, Thorkild – IPU, Denmark
Albers, Albert – Karlsruher Institute of Technology (KIT), Germany
Almefelt, Lars – Chalmers, Sweden
Anderl, Reiner – Technische Universitat Darmstadt, Germany
Andersson, Kjell – KTH Royal Institute of Technology, Sweden
Andrade, Ronaldo – Universidade Federal do Rio de Janiero, Brazil
Andreasen, Mogens Myrup – Technical University of Denmark, Denmark
Antonsson, Erik – California Institute of Technology, United States
Aoussat, Ameziane – Ecole Nationale Superieure des Arts et Metiers, France
Arai, Eiji – Osaka University, Japan
Arlbjorn, Jan – University of Southern Denmark, Denmark
Aurisicchio, Marco – Imperial College London, United Kingdom
Badke-Schaub, Petra – TU DELFT, Netherlands
Bathelt, Jens – Eidgenoessische Technische Hochschule Zuerich, Switzerland
Ben-Ahmed, Walid – RENAULT, France
Bey, Niki – DTU Management Engineering, Denmark
Bhamra, Tracy – Loughborough University, United Kingdom
Bigand, Michel – Ecole Centrale de Lille, France
Binder, Thomas – the Danish Design School, Denmark
Birkhofer, Herbert – Darmstadt University of Technology, Germany
Bjärnemo, Robert – Lund University, Sweden
Björk, Eva-Stina – NHV Nordic School of Public Health, Sweden
Blanco, Eric – Grenoble INP, France
Blessing, Lucienne T.M. – University of Luxembourg, Luxembourg
Blount, Gordon – Coventry University, United Kingdom
Boelskifte, Per – DTU - Technical University of Denmark, Denmark
Bohemia, Erik – Northumbria University, United Kingdom
Bojcetic, Nenad – University of Zagreb, Croatia
Boks, Casper – Norwegian University of Science and Technology, Norway
Bolognini, Francesca – University of Cambridge, United Kingdom
Bonjour, Eric – FEMTO-ST Institute, France
Booker, Julian David – University of Bristol, United Kingdom
Borg, Jonathan C. – University of Malta, Malta
Boujut, Jean-François – Grenoble Institute of Technology, France
Bouwhuis, Dominic G – University of Technology Eindhoven, Netherlands
Bracewell, Robert Henry – Cambridge University, United Kingdom
Brandt, Eva – The Danish Design School, Denmark
Brissaud, Daniel – Universite de Grenoble, France
Broberg, Ole – Technical University of Denmark, Denmark
Brown, David C. – WPI, United States
Bruun, Hans Peter Lomholt – DTU, Denmark
Buch, Anders – DTU-Management, Denmark
Burchardt, Carsten – Siemens Industry Software GmbH & Co, KG, Germany
Burvill, Colin Reginald – The University of Melbourne, Australia
Bylund, Nicklas – Sandvik Coromant, Sweden
Cagan, Jonathan – Carnegie Mellon University, United States
Caillaud, Emmanuel – Universite de Strasbourg, France
Campbell, Matthew Ira – University of Texas at Austin, United States
Cantamessa, Marco – Politecnico di Torino, Italy
Cardoso, Carlos Coimbra – Delft University of Technology, Netherlands
Casakin, Hernan – Ariel University Center, Israel
Cash, Philip – University of Bath, United Kingdom
Chakrabarti, Amaresh – Indian Institute of Science, India
Chamakiotis, Petros – University of Bath, United Kingdom
Chen, Wei – Northwestern University, United States

ICED11 Scientific Committee cont.

Childs, Peter R.N. – Imperial College London, United Kingdom
Chirone, Emilio – Universita di Brescia, Italy
Christensen, Bo – Copenhagen Business School, Denmark
Clarkson, Peter John – University of Cambridge, United Kingdom
Claudio, Dell'Era – Politecnico di Milano, Italy
Clausen, Christian – Technical University of Denmark, Denmark
Coatanea, Eric – Aalto University, Finland
Cormican, Kathryn – National University of Ireland, Galway, Ireland
Coutellier, Daniel – Universite de Valenciennes, France
Culley, Steve – University of Bath, United Kingdom
Cutting-Decelle, Anne-Françoise – Ecole Centrale Paris, France
Dankwort, C. Werner – University of Kaiserslautern, Germany
Darlington, Mansur – University of Bath, United Kingdom
Darses, Francoise – CNRS-LIMSI, France
de Vere, Ian James – Swinburne University of Technology, Australia
de Weck, Olivier Ladislas – MIT, United States
Deans, Joe – The University of Auckland, New Zealand
Dekoninck, Elies Ann – University of Bath, United Kingdom
Dong, Andy – The University of Sydney, Australia
Donndelinger, Joseph A. – General Motors LLC, United States
Dorst, Kees – UTS, Australia
Drejer, Anders – Aalborg University, Denmark
Duffy, Alex – University of Strathclyde, United Kingdom
Duhovnik, Joe – University of Ljubljana, Slovenia
Eckert, Claudia – The Open University, United Kingdom
Eder, Ernst – Royal Military College of Canada (retired), Canada
Eigner, Martin – University of Kaiserslautern, Germany
Ekman, Kalevi – Aalto University, Finland
Ellman, Asko – Tampere University of Technology, Finland
Elspass, Wilfried J. – Zhaw, Switzerland
Eppinger, Steven – Massachusetts Institute of Technology, United States
Eri , Özgür – Franklin W. Olin College of Engineering, United States
Evans, Steve – Cranfield University, United Kingdom
Fadel, Georges M. – Clemson University, United States
Fan, Ip-Shing – Cranfield University, United Kingdom
Fan, Zhun – Technical University of Denmark, Denmark
Fargnoli, Mario – University of Rome "La Sapienza", Italy
Finger, Susan – Carnegie Mellon University, United States
Fischer, Xavier – ESTIA, France
Fortin, Clement – CRIAQ, Canada
Frankenberger, Eckart – Airbus, Germany
Frise, Peter R. – University of Windsor, Canada
Fujita, Kikuo – Osaka University, Japan
Fukuda, Shuichi – Stanford University, United States
Galle, Per – The Danish Design School, Denmark
Gardoni, Mickael – ÉTS / INSA de Strasbourg, Canada
Gausemeier, Jürgen – Heinz Nixdorf Institute, Germany
Gerhard, Detlef – Vienna University of Technology, Austria
Gericke, Kilian – University of Luxembourg, Luxembourg
Gero, John – Krasnow Institute for Advanced Study, United States
Gerson, Philips M. – Hanze University Groningen, Netherlands
Gertsen, Frank – Aalborg University, Denmark
Giess, Matt – Université de Technologie de Compiègne, United Kingdom
Girard, Philippe – University Bordeaux, France
Gogu, Grigore – Institut Francais de Mecanique Avancee, France
Goh, Yee Mey – Loughborough University, United Kingdom
Goker, Mehmet H. – Salesforce.com, United States
Goldschmidt, Gabriela – Technion - Israel Institute of Technology, Israel

Gomes, Samuel – Belfort-Montbeliard University of Technology, France
Graessler, Iris – Robert Bosch GmbH, Germany
Green, Graham – University of Glasgow, United Kingdom
Gries, Bruno – Capgemini Consulting, Germany
Grimheden, Martin – KTH Machine Design, Sweden
Grote, Karl-Heinrich – Otto-von-Guericke-University Magdeburg, Germany
Grubiši, Izvor – University of Zagreb, Croatia
Gunn, Wendy – SPIRE, Denmark
Gurumoorthy, Balan – Indian Institute of Science, India
Gzara, Lilia – Grenoble Institute of Technology, France
Hadj-Hamou, Khaled – Institut National Polytechnique de Grenoble, France
Hagelskjaer Lauridsen, Erik – Technical University of Denmark, Denmark
Hales, Crispin – Hales & Gooch Ltd., United States
Hansen, Claus Thorp – Technical University of Denmark, Denmark
Hansen, Poul Kyvsgaard – Aalborg Universitet, Denmark
Hansen, Zaza Nadja Lee – The Technical University of Denmark, Denmark
Hansen, Christian Lindschou – DTU, Denmark
Hatchuel, Armand – Mines Paris Tech, France
Hein, Lars – IPU, Denmark
Helten, Katharina – TU München, Germany
Hemphälä, Jens – KTH Royal Institute of Technology, Sweden
Henderson, Mark Richard – Arizona State University, United States
Hicks, Ben – University of Bath, United Kingdom
Hohne, Gunter – Technische Universität Ilmenau, Germany
Holliger, Christoph – University of Applied Sciences Northwestern Switzerland, Switzerland
Horvath, Imre – Delft University of Technology, Netherlands
Hosnedl, Stanislav – University of West Bohemia, Czech Republic
Howard, Thomas James – Technical University of Denmark, Denmark
Hsuan, Juliana – Copenhagen Business School, Denmark
Huet, Greg – Ecole Polytechnique Montreal, Canada
Ijomah, Winifred – University of Strathclyde, United Kingdom
Ilies, Horea – University of Connecticut, United States
Ion, William – University of Strathclyde, United Kingdom
Isaksson, Karl Ola – Luleå tekniska universitet, Sweden
Jackson, Mats – Malardalen University, Sweden
Jelaska, Damir – University of Split, Croatia
Jensen, Torben Elgaard – Technical University of Denmark, Denmark
Jensen, Ole Kjeldal – Technical University of Denmark, Denmark
Johannesson, Hans – Chalmers, Sweden
Johansson, Glenn – School of Engineering, Jönköping University, Sweden
Jorgensen, Ulrik – Technical University of Denmark, Denmark
Jun, Thomas – Loughborough University, United Kingdom
Kannengiesser, Udo – NICTA, Australia
Karlsson, Lennart – Luleå University of Technology, Sweden
Keates, Simeon – IT University of Copenhagen, Denmark
Keldmann, Troels – Keldmann Healthcare A/S, Denmark
Kim, Harrison – University of Illinois at Urbana-Champaign, United States
Kim, Yong Se – Creative Design Institute, Sungkyunkwan University, SK
Kiriyama, Takashi – Tokyo University of the Arts, Japan
Kokkolaras, Michael – University of Michigan, United States
Kotzab, Herbert – CBS, Denmark
Kovacevic, Ahmed – City University London, United Kingdom
Kreimeyer, Matthias – MAN Truck & Bus AG, Germany
Kreye, Melanie E – University of Bath, United Kingdom
Krishnamurty, Sundar – University of Massachusetts-Amherst, USA
Kristensen, Tore – Copenhagen Business School, Denmark
Krus, Petter – Linköping University, Sweden
Kuosmanen, Petri – Aalto University/School of Engineering, Finland
Ladeby, Klaes – Technical University of Denmark, Denmark

ICED11 Scientific Committee cont.

Larsson, Andreas – Lund University, Sweden
Larsson, Tobias C. – Blekinge Institute of Technology, Sweden
Le Masson, Pascal – Mines Paris Tech, France
Leary, Martin John – RMIT university, Australia
Lee, SeungHee – University of Tsukuba, Japan
Legardeur, Jeremy – ESTIA, France
Lenau, Torben – Technical University of Denmark, Denmark
Levy, Pierre Denis – Eindhoven University of Technology, Netherlands
Lilly, Blaine – Ohio State University, United States
Lindahl, Mattias – Linköping University, Sweden
Lindemann, Udo – TUM, Germany
Lloveras, Joaquim – Technical University of Catalonia, Spain
Mabogunje, Ade – Stanford University, United States
MacGregor, Steven – IESE Business School, Spain
Maher, Mary – University of Maryland, United States
Maier, Anja Martina – Technical University of Denmark, Denmark
Maier, Jonathan – Clemson University, United States
Malmqvist, Johan Lars – Chalmers University of Technology, Sweden
Manfredi, Enrico – University of Pisa, Italy
Marini, Vinicius Kaster – Technical University of Denmark, Denmark
Marjanovic, Dorian – University of Zagreb Faculty of Mechanical Engineering and Naval Architecture, Croatia
Marle, Franck – Ecole Centrale Paris, France
Matta, Nada – Universite de Technologie de Troyes, France
Matthews, Jason – University of Glamorgan, United Kingdom
Matzen, Detlef – University of Southern Denmark, Denmark
Maurer, Maik – Technische Universität München, Germany
McAloone, Tim C. – Technical University of Denmark, Denmark
McAlpine, Hamish Charles – University of Bath, United Kingdom
McKay, Alison – University of Leeds, United Kingdom
McMahon, Christopher Alan – University of Bath, United Kingdom
Medland, Anthony – University of Bath, United Kingdom
Meerkamm, Harald – University Erlangen-Nuremberg, Germany
Mekhilef, Mounib – University of Orléans, France
Merlo, Christophe – ESTIA, France
Millet, Dominique – SUPMECA Toulon, France
Mocko, Gregory Michael – Clemson University, United States
Moehringer, Stefan – Simon Moehringer Anlagenbau GmbH, Germany
Moes, Niels – Delft University of Technology, Netherlands
Montagna, Francesca – Politecnico di Torino, Italy
Mortensen, Niels Henrik – Technical University of Denmark, Denmark
Mulet, Elena – Universitat Jaume, Spain
Mullineux, Glen – University of Bath, United Kingdom
Munk, Lone – Novo Nordisk A/S, Denmark
Murakami, Tamotsu – University of Tokyo, Japan
Nadeau, Jean-Pierre – Ecole Nationale Superieure des Arts et Metiers de Bordeaux, France
Nagai, Yukari – JAIST, Japan
Neugebauer, Line – DTU, Denmark
Newnes, Linda – University of Bath, United Kingdom
Nielsen, Teit Anton – DTU, Denmark
Nielsen, Ole Fiil – Worm Development, Denmark
Norell Bergendahl, Margareta E B – KTH, Sweden
Öhrwall Rönnbäck, Anna B – Linköping University, Sweden
Olesen, Jesper – Bang & Olufsen, Denmark
Olsson, Annika – Lund University, Sweden
Ölundh Sandström, Gunilla – KTH, The Royal Institute of Technology, Stockholm, Sweden
Otto, Kevin Norbert – RSS, United States
Ottosson, Stig – Tervix, Sweden
Ouertani, Mohamed Zied – Cambridge University, United Kingdom
Ovtcharova, Jivka – Karlsruhe Institute of Technology (KIT), Germany

Ozkil, Ali Gurcan – dtu, Denmark
Paetzold, Kristin – University Bundeswehr Munich, Germany
Papalambros, Panos Y. – University of Michigan, United States
Paredis, Chris – Georgia Tech, United States
Pavkovic, Neven – University of Zagreb, Croatia
Pedersen, Rasmus – Worm Development, Denmark
Petiot, Jean-François – Ecole Centrale Nantes, France
Ploug, Ole – Hydro Aluminium, Denmark
Prasad, Brian – Parker Aerospace, United States
Pulm, Udo – BMW, Germany
Raine, John – AUT University, New Zealand
Ramani, Karthik – Purdue University, United States
Ramirez, Mariano – University of New South Wales, Australia
Randmaa, Merili – Tallinn University of Technology, Estonia
Ray, Pascal – Institut Frangais de Mecanique Avancee, France
Reich, Yoram – Tel Aviv University, Israel
Reidsema, Carl – University of Queensland, Australia
Remmen, Arne Remmen – Aalborg University, Denmark
Restrepo, John – Technical University of Denmark, Denmark
Riel, Andreas – EMIRAcle and Grenoble Institute of Technology, France
Riitahuhta, Asko Olavi – Tampere University of Technology, Finland
Rinderle, James – University of Massachusetts, United States
Ritzén, Sofia – KTH, Sweden
Roth, Bernard – Stanford University, United States
Roucoules, Lionel – Arts et Métiers ParisTech, France
Rovida, Edoardo – Politecnico di Milano, Italy
Štorga, Mario – University of Zagreb, Faculty of Mechanical Engineering and Naval Architecture, Croatia
Sachse, Pierre – University of Innsbruck, Austria
Sagot, Jean-Claude – Universite Technologique de Belfort Montbeliard, France
Sakao, Tomohiko – Linkoping University, Sweden
Salehi, Vahid – University of Bath, United Kingdom

Salustri, Filippo Arnaldo – Ryerson University, Canada
Schabacker, Michael – Otto-von-Guericke University Magdeburg, Germany
Schaub, Harald – IABG, Germany
Seering, Warren – Massachusetts Institute of Technology, United States
Setchi, Rossi – Cardiff University, United Kingdom
Shah, Jami – Arizona State U, Tempe, United States
Shea, Kristina – Technische Universitat Munchen, Germany
Shimomura, Yoshiki – Tokyo Metropolitan University, Japan
Shu, L.H. – University of Toronto, Canada
Siadat, Ali – Arts et Metiers ParisTech, France
Sigurjónsson, Jóhannes B. – NTNU, Norwegian University of Science And Technology, Norway
Simmons, John – Heriot-Watt University, United Kingdom
Simpson, Timothy W. – Penn State University, United States
Snider, Chris – University of Bath, United Kingdom
Stal-Le Cardinal, Julie – Ecole Centrale Paris, France
Stankovic, Tino – University of Zagreb, Croatia
Stark, Rainer G. – Berlin Institute of Technology, Germany
Stören, Sigurd – Norwegian University of Science and Technology, Norway
Subrahmanian, Eswaran – Carnegie Mellon and Center for Science Technology & Policy, India
Summers, Joshua – Clemson University, United States
Sundin, Erik – Linköping University, Sweden
Sunnersjo, Staffan – Jonkoping University, Sweden
Söderberg, Rikard – Chalmers, Sweden
Takai, Shun – Missouri University of Science and Technology, United States
Tan, Ah Kat – Ngee Ann Polytechnic, Singapore, Singapore
Tan, Adrian – BIO Intelligence Service, France
Taura, Toshiharu – Kobe University, Japan
Tegel, Oliver – Dr.Ing. h.c. F. Porsche AG, Germany
Thallemer, Axel – Universitaet fuer industrielle und kuenstlerische Gestaltung, Austria

ICED11 Scientific Committee cont.

Thurston, Deborah Lee – University of Illinois, United States
Tichkiewitch, Serge – Grenoble INP, France
Tollenaere, Michel – Institut Polytechnique de Grenoble, France
Tomiyama, Tetsuo – TU Delft, Netherlands
Torry-Smith, Jonas Mørkeberg – Technical University of Denmark, Denmark
Troussier, Nadege – Université de Technologie de Compiegne, France
Törlind, Peter – Luleå University of Technology, Sweden
Udiljak, Toma – University of Zagreb, Croatia
Uflacker, Matthias – Hasso Plattner Institute, Germany
Ullman, David – Robust Decisions Inc, United States
Vajna, Sandor – Otto-von-Guericke Magdeburg, Germany
Valderrama Pineda, Andres Felipe – Technical University of Denmark, Denmark
Vance, Judy M. – Iowa State University, United States
Vaneker, Tom Henricus Jozef – University of Twente, Netherlands
Vermaas, Pieter – Delft University of Technology, Netherlands
Visser, Willemien – CNRS-INRIA, France
Vukic, Fedja – University of Zagreb, Croatia
Wallace, Ken – University of Cambridge, United Kingdom
Wartzack, Sandro – Friedrich-Alexander-Universität Erlangen-Nürnberg, Germany
Weber, Christian – Ilmenau University of Technology, Germany
Weil, Benoit – MinesParisTech, France
Weiss, Menachem – TECHNION, Israel
Whitfield, Robert Ian – University of Strathclyde, United Kingdom
Whitney, Daniel E – MIT, United States
Wikander, Jan – KTH Royal Institute of Technology, Sweden
Wood, Kristin Lee – The University of Texas, United States
Wynn, David C – University of Cambridge, United Kingdom
Wörösch, Michael – DTU Management, Denmark
Yan, Xiu – University of Strathclyde, United Kingdom
Yang, Maria – MIT, United States
Yannou, Bernard – Ecole Centrale Paris, France
Youn, Byeng D. – Seoul National University, South Korea
Young, Robert – Loughborough University, United Kingdom
Yuen, Matthew – Hong Kong University of Science and Technology, China
Žavbi, Roman – University of Ljubljana, Slovenia
Zeiler, Wim – Technical University Eindhoven, Netherlands
Zissimos, Mourelatos – Oakland University, United States

Table of Contents

Preface by the Program Chair	i
Preface by the Conference Chair	ii
Preface by the Design Society President	iii
ICED11 Design Society Programme Committee	v
ICED11 Organising Team	v
ICED11 Theme Chairs	vi
ICED11 Scientific Committee	vii

VOLUME 1: DESIGN PROCESSES

Monitoring a Property Based Product Development – from Requirements to a Mature Product *Hartmut Krehmer, Harald Meerkamm, Sandro Wartzack*	1-1
Item Life Cycles in Product Data Management: A Case Study on How to Implement a Design Data Validation Process *Bertrand Nicquevert, Jean-François Boujut*	1-12
Process Optimization by DSM-Based Modelling of Inputs and Outputs *Maik Maurer*	1-24
Morphological Analysis of a Sustainable School Design *Wim Zeiler*	1-35
Enabling Set-Based Concurrent Engineering in Traditional Product Development *Dag Raudberget*	1-45
Generic Model of The Early Phase of An Innovation Process Regarding Different Degrees Of Product Novelty *Robert Orawski, Jan Krollmann, Markus Mörtl und Udo Lindemann*	1-57
Bayesian Project Monitoring *Peter C Matthews And Alex D M Philip*	1-69
User Centered Design in The Wild *Stompff G., Henze L.A.R., Jong, F. De , Vliembergen E. Van, Slappers P.J., Smulders F.E.II.M., Buijs J.A.*	1-79
Integrated Process and Product Model for The Evlauation of Product Properties *Christoph Westphal, Sandro Wartzack*	1-91
Product Development Processes in Small and Middle-Sized Enterprises – Identification and Elimination of Inefficiency caused by Product Variety *Katharina G. M. Eben, Katharina Helten, Udo Lindemann*	1-101
Neutral Description and Exchange of Design Computational Workflows *Gondhalekar A. C., Guenov M. D.1, Wenzel H. , Nunez M.1, Balachandran L. K.1*	1-113

Decision Based Variable Mechatronic Design Processes
Peter Hehenberger, Babak Farrokhzad, Florian Poltschak — 1-122

Using Simulation to Support Process Integration and Automation
of the Early Stages of Aerospace Design
Warren Kerley, Gareth Armstrong, Carla Pepe, Michael Moss, P John Clarkson — 1-134

Study on The Introduction of Design Management
in The Product Development Process of Brazilian Clothing Companies
Graziela Kauling, Maurício Bernardes — 1-147

Lean Product Development: Hype Or Sustainable New Paradigm?
Amer ´Catió, Michael Vielhaber — 1-157

Structured Concept Development with Parameter Analysis
Ehud Kroll — 1-169

Creating Value through Lean Product Development – Applying Lean Principles
Jörgen Furuhjelm, Håkan Swan, Johan Tingström — 1-180

What are the Characteristics of Engineering Design Processes?
Anja M Maier, Harald Störrle — 1-188

Empirical Verifications of Some Radical Innovation Design Principles
onto the Quality of Innovative Designs
Bernard Yannou, Marija Jankovic, Yann Leroy — 1-199

Decision-Making in Disruptive Innovation Projects: A Value Approach
F. Petetin, G. Bertoluci, J. C. Bocquet — 1-211

A Comparison of Evolutionary and Revolutionary Approaches in Mechatronic Design
Ralf Stetter, Stefan Möhringer, Udo Pulm — 1-221

Consideration of Goal Interrelations in Lifecycle-Oriented Product Planning
Clemens Hepperle, Armin Förg, Markus Mörtl And Udo Lindemann — 1-233

Rules for Implementating Dynamic Changes in DSM-Based Plans
Arie Karniel, Yoram Reich — 1-243

Analysis Of Created Representations Of The Design Object During
The Problem Solving Process
Albert Albers And Aaron Wiedner — 1-256

Influence of Design Evaluations on Decision-Making
and Feedback During Concept Development
Vinicius Kaster Marini, Saeema Ahmed-Kristensen And John Restrepo — 1-266

A Mechatronic Case Study Highlighting the Need for Re-Thinking The Design Approach
Jonas Mørkeberg Torry-Smith, Niels Henrik Mortensen — 1-276

Design Support Tools for Product-Service Systems
Yong Se Kim, Sang Won Lee, Jee-Hyong Lee, Dae Man Han And Hye Kyung Lee — 1-288

Design Process Automation – A Structured Product Description by Properties
and Development of Optimization Algorithms
Sebastian Gramlich, Herbert Birkhofer And Andrea Bohn — 1-299

Solving Global Problems using Collaborative Design Processes
Torben Lenau, Christina Okai Mejborn — 1-310

Lean Approach to Integrate Collaborative Product Development
Processes And Digital Engineering Systems
Thomas Vosgien, Marija Jankovic, Benoit Eynard, Thomas Nguyen Van, Jean-Claude Bocquet   1-321

An Empirical Evaluation of a Framework for Design for Variety and Novelty
Srinivasan V, Amaresh Chakrabarti   1-334

Indicating the Criticality of Changes During the Product Life Cycle
Florian G. H. Behncke, Udo Lindemann   1-344

Characterizing the Dynamics of Design Change
Afreen Siddiqi, Olivier L. De Weck, Bob Robinson, Rene Keller   1-355

The Influence of A Company's Strategy on Creativity and Project Results
in An NPD – Case Study
Nikola Vukšinovi, Nuša Fain, Jože Duhovnik   1-366

When Sensemaking Meets Resource Allocation: An Exploratory Study
of Ambiguous Ideas in Project Portfolio Management
Ernesto Gutiérrez   1-373

The Continous "Fuzzy Front End" as a Part of The Innovation Process
Milan Stevanović, Dorian Marjanović   1-383

Comparisons of Design Methodologies and Process Models
across Disciplines: A Literature Review
Kilian Gericke, Luciënne Blessing   1-393

A Framework for Developing Viable Design Methodologies for Industry
Timo Lehtonen, Tero Juuti, Hannu Oja, Seppo Suistoranta, Antti Pulkkinen, Asko Riitahuhta   1-405

Development of A Framework for Improving Engineering Processes
Carla Pepe, Daniel Whitney, Elsa Henriques, Rob Farndon, Michael Moss   1-417

Coping With Deviation And Decision-Making
Joakim Eriksson, Anette Brannemo   1-429

Ensuring the Integration of Performance and Quality Standards
in Design Process Management: Codesteer Methodology
Aurélien Poulet, Bertrand Rose, Emmanuel Caillaud   1-441

Engaging Actors in Co-Designing Heterogeneous Innovations
Ulrik Jørgensen, Hanne Lindegaard And Tanja Rosenqvist   1-453

Capturing Interactions in Design Preferences: A Colorful Study
Hannah Turner, Seth Orsborn   1-465

Re-Conceptualising Value in The Engineering Design Process: The Value Cycle Map
Ghadir I. Siyam, David C. Wynn, P. John Clarkson   1-475

Facing the Open Innovation Dilemma – Structuring Input at the Company's Border
Andreas Kain, Rafael Kirschner, Alexander Lang, Udo Lindemann   1-487

Integrated System and Context Modeling of Iterations and Changes
in Development Processes
Stefan Langer, Arne Herberg, Klaus Körber, Udo Lindemann   1-499

Embodiment Design through the Integration of OTSM-TRIZ Situation Analysis
with Topological Hybridization of Partial Solutions
Alessandro Cardillo, Gaetano Cascini, Francesco Saverio Frillici, Federico Rotini   1-509

A Framework for Integrated Process Modeling and Planning of Mechatronic Products *David Hellenbrand And Udo Lindemann*	1-521
KPI Measurement In Engineering Design – A Case Study *Bruno Gries, John Restrepo*	1-531
Information Models Used to Manage Engineering Change: A Review of The Literature 2005-2010 *Naveed Ahmad, David C. Wynn, P. John Clarkson*	1-538

VOLUME 2: DESIGN THEORY AND RESEARCH METHODOLOGY

On the Link Between Features and Functions *D.Gabelloni, R. Apreda , G. Fantoni*	2-1
Understanding the Worlds of Design and Engineering - an Appraisal of Models *Martin Gudem, Casper Boks*	2-13
Courses of Product Development Identification – Effects and Visions *Milosav Ognjanovi*	2-23
Integral Design: To Combine Architecture and Engineering for a Sustainable Built Environment *Wim Zeiler*	2-31
How to Validate Research in Engineering Design? *Alex Barth, Emmanuel Caillaud, Bertrand Rose*	2-41
Towards a Scientific Model of Function-Behavior Transformation *Yong Chen, Zhinan Zhang, Zelin Liu, Youbai Xie*	2-51
Design of Functions by Function Blending *Yu Park, Shota Ohashi, Eiko Yamamoto, Toshiharu Taura*	2-61
A Method for Selecting Base Functions for Function Blending in Order to Design Functions *Syo Sakaguchi, Akira Tsumaya, Eiko Yamamoto, Toshiharu Taura*	2-73
A Systematic Approach of Design Theories Using Generativeness and Robustness *Armand Hatchuel, Pascal Le Masson, Yoram Reich, Benoit Weil*	2-87
Accepting Ambiguity of Engineering Functional Descriptions *Pieter E. Vermaas*	2-98
Theoretical Framework for Comprehensive Abstract Prototyping Methodology *Imre Horváth*	2-108
Design Inspired Innovation for Rural India *Sten Ekman, Annalill Ekman, Uday Salunkhe, Anuja Agarwal*	2-120
The Semantic Debate in Design Theories Applied to Product Identity Creation *Grégoire Bonnemaire, Andre Liem*	2-130
Modeling Paradoxes in Novice and Expert Design *Kees Dorst, Claus Thorp Hansen*	2-142
Designer Behaviour and Activity: An Industrial Observation Method *Philip Cash, Ben Hicks, Steve Culley, Filippo Salustri*	2-151

Design Research Reflections – 30 Years On *Crispin Hales, Ken Wallace*	2-163
Development of an Evaluation Framework for Implementation of Parametic Associative Methods in an Industrial Context *Vahid Salehi, Chris Mcmahon*	2-173
Measuring History: Does Historical Car Performance Follow the TRIZ Performance S Curve? *Chris Dowlen*	2-183
Comparing Designing Across Different Domains: an Exploratory Case Study *Jeff Wt Kan, John S Gero*	2-194
Product Failure: A Life Cycle Approach *Luca Del Frate*	2-204
Designing Patent Portfolio for Disruptive Innovation – A New Methodology Based on C-K Theory *Yacine Felk, Pascal Le Masson, Benoit Weil, Patrick Cogez, Armand Hatchuel*	2-214
Virtualisation of Product Development/ Design – Seen from Design Theory and Methodology *Weber, C., Husung, S.*	2-226
Investigating Elementary Design Methods – Using and Extending the Genome-Approach *Sebastian Zier, Hermann Kloberdanz, Herbert Birkhofer, Andrea Bohn*	2-236
Environment Based Design (EBD) Vs. X Development: A Dialog Between Theory and Retrospection *Yong Zeng, Jean Vareille*	2-246
Dimensions of Objectives in Interdisciplinary Product Development Projects *Albert Albers, Quentin Lohmeyer, Bjoern Ebel*	2-256
The Impact of Examples on Creative Design: Explaining Fixation and Stimulation Effects *Marine Agogué, Akin Kazakçi, Benoit Weil, Mathieu Cassotti*	2-266
A Method for Design Reasoning Using Logic: from Semantic Tableaux to Design Tableaux *Hendriks, Lex; Kazakci, Akin Osman*	2-275
Description, Prescription and "Bad" Design *Paul Winkelman*	2-287
LINKographer: An Analysis Tool to Study Design Protocols Based on FBS Coding Scheme *Morteza Pourmohamadi, John S Gero*	2-294
Do Functions Exist? *G.Fantoni, R.Apreda, D.Gabelloni, A.Bonaccorsi*	2-304
E3 Value Concept for a New Design Paradigm *Yong Se Kim, Chang Kyu Cho, Young Dae Ko, Haeseong Jee*	2-314
CK, an Engineering Design Theory? - Contributions, Limits and Proposals *Denis Choulier, Eric Coatanea, Joelle Forest*	2-323
A Theory of Decomposition in System Architecting *Hitoshi Komoto, Tetsuo Tomiyama*	2-334

A Framework for Comparing Design Modelling Approaches Across Disciplines
Boris Eisenbart, Kilian Gericke, Luciënne Blessing — 2-344

A Note on the Debate on Scientific Process Vs. Design Process
Damien Motte, Robert Bjärnemo — 2-356

Conducting Preliminary Design Around an Interactive Tabletop
Thierry Gidel, Atman Kendira, Alistair Jones, Dominique Lenne, Jean-Paul Barthès, Claude Moulin — 2-366

Toward an Adaption-Innovation Strategy for Engineering Design
Philip Samuel, Kathryn Jablokow — 2-377

VOLUME 3: DESIGN ORGANISATION AND MANAGEMENT

How Design Researchers can lead Higher Education to a Greater Impact on Society
Howard T.J., McMahon C.A., Giess M.D. — 3-1

Addressing the Risks of Global Product Development
Hansen, Z.N.L. & Ahmed-Kristensen, S. — 3-11

Design Driven Portfolio Management
Søren Ingomar Petersen, Martin Steinert, Sara Beckman — 3-21

Business Plans Informed by Design
Søren Ingomar Petersen, John Heebøll — 3-31

Influence of the Time Perspective on New Product Development Success Indicators
Afrooz Moatari Kazerouni-, Sofiane Achiche, Onur Hisarciklilar, Vincent Thomson — 3-40

Considerations on Design Management of Furniture Manufacturing Companies in Southern Brazil
Ana Galafassi, Maurício Bernardes — 3-52

Technology Development Practices in Industry
Ulf Högman, Hans Johannesson — 3-62

Challenges in Networked Innovation
Christiane Maurer, Rianne Valkenburg — 3-74

Experiences with Idea-Promoting Initiatives – Why they don't always Work
Liv Gish — 3-83

Improving the Management of Design Project Risks using the Concept of Vulnerability : A Systems Approach
Vidal La., Marle F., Bocquet Jc. — 3-93

A Frequency Analysis Approach to Ensure the Robustness of Interactions-Based Clustering of Project Risks
Marle F., Vidal La. — 3-104

Effective Scheduling of User Input During the Design Process
Young Mi Choi, Ph.D. — 3-116

Engineering Environment for Product Innovation
Michael Bitzer, Michael Vielhaber — 3-123

Requirements for Product Development Self-Assessment Tools
*Christoph Knoblinger1, Josef Oehmen, Eric Rebentisch,
Warren Seering Katharina Helten* — 3-133

Expanding The Social Dimension: Towards A Knowledge Base
for Product-Service Innovation
Åsa Ericson, Andreas C. Larsson, Tobias C. Larsson — 3-143

Designing a Process for a Monopoly to transform to a Free Market Competitor
– The Swedish Pharmacy System
Annalill Ekman, Stefan Carlsson, Sten Ekman — 3-153

Managing Resource Scarcity in Small Enterprises' Design Processes
Lars Löfqvist — 3-164

Collaborative Glitches in Design Chain: Case Study of an Unsuccessful
Product Development with a Supplier
Hélène Personnier, Marie-Anne Le Dain, Richard Calvi — 3-176

Recommendations for Risk Identification Method Selection according
to Product Design and Project Management Maturity, Product Innovation
Degree and Project Team
Viviane Vasconcellos Ferreira Grubisic, Thierry Gidel, André Ogliari — 3-187

Identification and Design of Pilot Projects to Implement Lean Development
Katharina Helten, Katharina G.M. Eben, Udo Lindemann — 3-199

Exploring Collaboration in a Networked Innovation Project in Industry
Katinka Bergema, Maaike Kleinsmann, Cees De Bont, Rianne Valkenburg — 3-211

Business Model Design Methodology for Innovative Product-Service Systems:
A Strategic and Structured Approach
Ji Hwan Lee, Dong Ik Shin, Yoo S. Hong, Yong Se Kim — 3-221

Prototyping in Organizational Process Engineering
Matteo Vignoli, Diego Maria Macrì, Fabiola Bertolotti — 3-233

On the Stability of Coordination Patterns in Multidisciplinary
Design Projects
João Castro, Martin Steinert, Warren Seering — 3-245

Management of Product Development Projects through Integrated
Modeling of Product and Process Information
Kazuya Oizumi, Kei Kitajima, Naoto Yoshie, Tsuyoshi Koga, Kazuhiro Aoyama — 3-253

Packaging Design in Organic Food Supply Chains – A Case Study in Sweden
Annika Olsson, Helena Lindh, Gwenola Bertoluci — 3-264

A Holistic Procedure for Process Integration In Design Cooperation
Christiane Beyer, Karl-Heinrich Grote, Oliver Tegel, Christian Kubisch — 3-274

Exploit and Explore: Two Ways of Categorizing Innovation Projects
Åsa Ericson, Åsa Kastensson — 3-284

Steering the Value Creation in an Airplane Design Project from the Business
Strategies to the Architectural Concepts
Ndrianarilala Rianantsoa, Bernard Yannou, Romaric Redon — 3-294

A Comparison of the Integration of Risk Management Principles in Product
Development Approaches
Denis Bassler, Josef Oehmen, Warren Seering, Mohamed Ben-Daya — 3-306

Supporting Cycle Management by Structural Analysis of the Organisational Domain in Multi-Project Environment *Fatos Elezi, Alvaro Pechuan, Alexander Mirson, Sebastian Kortler, Wieland Biedermann, Udo Lindemann*	3-317
The Central Role of Exploration in Designing Business Concepts and Strategy *Senni Kirjavainen, Tua A. Björklund*	3-325
Integration of Suppliers into the Product Development Process using the Example of the Commercial Vehicle Industry *Nicole Katharina Stephan, Christian Schindler*	3-335
Overcoming the Keep the Market Out Premise (KMOP) in Product Development *A. Lang, R. Kirschner, A. Kain, U. Lindemann*	3-346
Stakeholders' Analysis Tools to Support the Open Innovation Process Management – Case Study *Istefani Carísio De Paula, Samanta Yang, André Korzenowski, Marcelo Nogueira Cortimiglia*	3-354
Inside a PSS Design Process: Insights through Protocol Analysis *Tomohiko Sakao, Svante Paulsson, Hajime Mizuyama*	3-365
Knowledge Management Challenges In New Business Development - Transition Of The Energy System *Ole Kjeldal Jensen, Saeema Ahmed-Kristensen, Nevena Jensen*	3-377
Rethinking Value: A Value-Centric Model of Product, Service and Business Development *Merili Randmaa, Krestine Mougaard, Thomas Howard, Tim C. McAloone*	3-387

VOLUME 4: PRODUCT AND SYSTEM DESIGN

Product Platform Automation for Optimal Configuration of Industrial Robot Families *Mehdi Tarkian, Johan Ölvander, Xiaolong Feng, Marcus Pettersson*	4-1
On the applicability of structural criteria in Complexity Management *Wieland Biedermann, Udo Lindemann*	4-11
Eco Tracing - A Systems Engineering Method for Efficient Tracelink Modelling *Rainer Stark, Asmus Figge*	4-21
Equilibrium Design Problems in Complex Systems Realization *Jitesh H. Panchal*	4-33
Balancing Internal and External Product Variety in Product Development *Iris Graessler*	4-45
Integrated Product and Production Model – Issues on Completeness, Consistency and Compatibility *Stellan Gedell, Anders Claesson, Hans Johannesson*	4-55
A Framework for Designing Product-Service Systems *Gokula Vijaykumar Annamalai Vasantha, Romana Hussain, Rajkumar Roy, Ashutosh Tiwari, Stephen Evans*	4-67

Orthogonal Views on Product/Service-System Design in an Entire Industry Branch *Tim C. Mcaloone, Krestine Mougaard, Line M. Neugebauer, Teit A. Nielsen, Niki Bey*	4-77
Representation and Analysis of Business Ecosystems Co-Specializing Products and Services *Changmuk Kang, Yoo S. Hong, Kwang Jae Kim, Kwang Tae Park*	4-88
Interdisciplinary System Model for Agent-Supported Mechatronic Design *Ralf Stetter, Holger Seemüller, Mohammad Chami, Holger Voos*	4-100
Strategic Planning for Modular Product Families *Henry Jonas, Dieter Krause*	4-112
Modularity Within a Matrix of Function and Functionality (MFF) *Žiga Zadnik, Vanja Čok, Mirko Karakaši, Milan Kljajin, Jože Duhovnik*	4-122
Proactive Modeling of Market, Product and Production Architectures *Niels Henrik Mortensen, Christian L. Hansen, Lars Hvam, Mogens Myrup Andreasen*	4-133
Designing Mechatronic Systems: A Model- Integration Approach *Ahsan Qamar, Jan Wikander, Carl During*	4-145
An Approach for More Efficient Variant Design Processes *Sebastian Schubert, Arun Nagarajah, Jörg Feldhusen*	4-157
The Process of Optimizing Mechanical Sound Quality in Product Design *Nielsen, Thomas Holst; Eriksen, Kaare Riise*	4-167
Development of Modular Products under Consideration of Lightweight Design *Thomas Gumpinger, Dieter Krause*	4-175
The Investigation and Computer Modelling of Humans with Disabilities *A. J. Medland, S.D. Gooch*	4-185
Evaluation of an Automated Design and Optimization Framework for Modular Robots using a Physical Prototype *Vaheed Nezhadali, Omer Khaleeq Kayani, Hannan Razzaq, Mehdi Tarkian*	4-195
Designing Consistent Structural Analysis Scenarios *Wieland Biedermann, Udo Lindemann*	4-205
Design of an Upper Limb Independence-Supporting Device using a Pneumatic Cylinder *Norihiko Saga, Koichi Kirihara, Naoki Sugahara*	4-215
Actuation Principle Selection – An Example of Trade-off Assessment by CPM-Approach *Torsten Erbe, Kristin Paetzold, Christian Weber*	4-222
Automated User Behavior Monitoring System for Dynamic Work Environments *Yeeun Choi, Minsun Jang, Yong Se Kim, Seongil Lee*	4-230
On the Design of Devices for People with Tetraplegia *S.D. Gooch, A. J. Medland, A. R. Rothwell, J.A. Dunn, M.J.Falconer*	4-238
Brownfield Process for Developing of Product Families *Timo Lehtonen, Jarkko Pakkanen, Jukka Järvenpää, Minna Lanz, Reijo Tuokko*	4-248
Approach for the Creation of Mechatronic System Models *Martin Follmer, Peter Hehenberger, Stefan Punz, Roland Rosen, Klaus Zeman*	4-258
Modeling and Design of Contacts in Electrical Connectors *Albert Albers, Paul Martin, Benoit Lorentz*	4-268
Empirical Consideration of Predicting Chain Failure Modes in Product Structures During Design Review Process *Yuichi Otsuka, Shotakiguchi, Hirokazu Shimizu, Yoshiharu Mutoh*	4-278

A Design Methodology for Haptic Devices
Suleman Khan, Kjell Andersson   4-288

A Methodical Approach for Developing Modular Product Families
Dieter Krause, Sandra Eilmus   4-299

Product Model of the Autogenetic Design Theory
Konstantin Kittel, Peter Hehenberger, Sándor Vajna, Klaus Zeman   4-309

Product Development Support For Complex Mechatronic System Engineering– Case Fusion Reactor Maintenance
Simo-Pekka Leino, Harri Mäkinen, Olli Uuttu, Jorma Järvenpää   4-319

Analyzing the Dynamic Behavior of Mechatronic Systems within the Conceptual Design
Frank Bauer, Harald Anacker, Tobias Gaukstern, Jürgen Gausemeier, Viktor Just   4-329

Linear Flow-Split Linear Guides: Inflating Chambers to Generate Breaking Force
Nils Lommatzsch, Sebastian Gramlich, Herbert Birkhofer, Andrea Bohn   4-337

A Knowledge-Based Master Modeling Approach to System Analysis and Design
Marcus Sandberg, Ilya Tyapin, Michael Kokkolaras, Ola Isaksson   4-347

Social Systems Engineering – An Approach for Efficient Systems Development
Thomas Naumann, Ingo Tuttass, Oliver Kallenborn, Simon Frederick Königs   4-357

Improving Data Quality in DSM Modelling: A Structural Comparison Approach
Steffen F- Schmitz, David C. Wynn, Wieland Biedermann, P. John Clarkson, Udo Lindemann   4-369

Enhancing Intermodal Freight Transport by Means of an Innovative Loading Unit
Dipl.-Ing. Max Klingender, Dipl.-Ing. Sebastian Jursch   4-381

Representing Product-Service Systems with Product and Service Elements
Yong Se Kim, Sang Won Lee And Dong Chan Koh   4-390

A Classification Framework for Product Modularization Methods
Charalampos Daniilidis, Vincent Enßlin, Katharina Eben, Udo Lindemann   4-400

A Meta Model of the Innovation Process to Support the Decision Making Process using Structural Complexity Management
Sebastian Kortler, Udo Lindemann   4-410

Pareto Bi-Criterion Optimization for System Sizing : A Deterministic and Constraint Based Approach
Pierre-Alain Yvars   4-420

Sick Systems: Towards a Generic Conceptual Representation of Healthcare Systems
Alexander Komashie, Thomas Jun, Simon Dodds, Hugh Rayner, Simon Thane, Alastair Mitchell-Baker, John Clarkson   4-430

Property Rights Theory as a Key Aspect in Product Service Engineering
Anna Katharina Dill, Herbert Birkhofer, Andrea Bohn   4-441

Product with Service, Technology with Business Model: Expanding Engineering Design
Tomohiko Sakao, Tim Mcaloone   4-449

Analysing Modifications in the Synthesis of Multiple State Mechanical Devices using Configuration Space and Topology Graphs
Somasekhara Rao Todeti, Chakrabarti Amaresh   4-461

VOLUME 5: DESIGN FOR X, DESIGN TO X

Ecodesign in Industrial Design Consultancies – Comparing Australia, China, Germany and the USA *Johannes Behirsch , Dr. Mariano Ramirez, Dr. Damien Giurco*	5-1
Design for Additive Manufacturing Technologies: New Applications of 3D-Printing for Rapid Prototyping and Rapid Tooling *Stefan Junk, Marco Tränkle*	5-12
Reflections on Design for Sustainability- A View from a Distinct Point and the Role of Interior Designer *K Ioannou-Kazamia, J Gwilliam*	5-19
Life Cycle Approach to Support Tooling Design Decisions *Inês Ribeiro , Paulo Peças, Elsa Henriques*	5-28
Designing with a Social Conscience: An Emerging Area in Industrial Design Education and Practice *Mariano Ramirez Jr*	5-39
Global Optimization of Environmental Impact by a Constraint Satisfaction Approach – Application to Ship-Ecodesign *Vincent Larroudé, Pierre-Alain Yvars, Dominique Millet*	5-49
Designing for Resilience: using a Delphi Study yo Identify Resilience Issues for Hospital Designs in a Changing Climate *Mary Lou Masko, Claudia M. Eckert, Nicholas H.M. Caldwell, P. John Clarkson*	5-60
Modelling Time-Varying Value of an End-Of-Life Product for Design for Recovery *Minjung Kwak, Harrison M. Kim*	5-70
The Impact of Safety Standards and Policies on Optimal Automobile Design *Steven Hoffenson, Panos Papalambros*	5-81
Design for Diagnosis *Ralf Stetter, Ulrike Phleps*	5-91
An Engineering-Based Environmental Management System Design *Colin Burvill, John Weir, Martin Leary*	5-103
Getting to Sustain (-Able Systems) via using Survivable and Impose-Able Ones *Richard Tabor Greene*	5-113
Toward Proactive Eco-Design Based on Engineer and Eco-Designer's Software Interface Modeling *Maud Rio, Tatiana Reyes, Lionel Roucoules*	5-124
Designing Sustainable Society Scenarios using Forecasting *Yuji Mizuno, Yusuke Kishita, Haruna Wada, Maki Hirosaki, Shinichi Fukushige, Yasushi Umeda*	5-135
Success Criteria for Implementing Sustainability Information in Product Development *Silje Helene Aschehoug, Casper Boks*	5-145
Universal Design and Visual Impairment: Tactile Products for Heritage Access *Jaume Gual, Marina Puyuelo, Joaquim Lloveras*	5-155
Developing an Ecology of Mind in Design *Emma L Dewberry*	5-165

Management of Energy Related Knowledge in Integrated Product Development – Concept and Selected Instruments *Uwe Götze, Erhard Leidich, Annett Bierer, Susann Köhler*	5-176
Sustainability Innovation in Early Phases *Massimo Panarotto, Peter Törlind*	5-187
Manufacturing Cost Estimation During Early Phases of Machine Design *Michele Germani, Marco Mandolini, Paolo Cicconi*	5-198
Design for Dependability – Identifying Potential Weaknesses in Product Concepts *Tim Sadek, Michael Wendland*	5-210
Impact of Modularised Production on Product Design in Automotive Industry *Waldemar Walla, Thomas Baer, Jivka Ovtcharova*	5-220
Design for Reliability: An Event- and Function-Based Framework for Failure Behavior Analysis in the Conceptual Design of Cognitive Products *Thierry Sop Njindam, Kristin Paetzold*	5-228
A New Approach to Modularity in Product Development – Utilising Assembly Sequence Knowledge *Authors: A. Robert, X.T. Yan, S. Roth, K. Deschinkel, S. Gomes*	5-238
Approach to Visualize the Supply Chain Complexity Induced by Product Variety *Max Brosch, Gregor Beckmann, Dieter Krause*	5-249
Integration of Remanufacturing Issues into the Design Process *Gillian Hatcher, Winifred Ijomah And James Windmill*	5-259
Investigating the Requirements Needed to Make Appropriate End of Life Decisions *Kirsty Doyle, Winifred L. Ijomah, Jiju Antony*	5-265
Product and Process Evaluation in the Context of Modularization for Assembly *Niklas Halfmann, Steffen Elstner, Dieter Krause*	5-271
A Case Study of Design for Affordance: Affordance Features of a Simple Medical Device *Yong Se Kim, Young Chan Cho, Sun Ran Kim*	5-282
Which Guideline is Most Relevant? Introduction of a Pragmatic Design for Energy Efficiency Tool *Karola Rath, Herbert Birkhofer, Andrea Bohn*	5-293
The Concept of Ecological Levers – A Pragmatic Approach for the Elicitation of Ecological Requirements *Shulin Zhao, Herbert Birkhofer, Andrea Bohn*	5-302
Methodology for Choosing Life Cycle Impact Assessment Sector-Specific Indicators *Bruno Chevalier, Tatiana Reyes-Carrillo, Bertrand Laratte*	5-312
Selection of Physical Effects Based on Disturbances and Robustness Ratios in the Early Phases of Robust Design *Johannes Mathias, Tobias Eifler, Roland Engelhardt, Hermann Kloberdanz, Herbert Birkhofer, Andrea Bohn*	5-324
Requirements of a Carbon Footprinting Tool for Designers *Rhoda Trimingham, Sofia Garcia-Noriega*	5-336
Extraction and Analysis Methodology for Supporting Complex Sustainable Design *Helen Liang, David Birch*	5-346

VOLUME 6: DESIGN INFORMATION AND KNOWLEDGE

Scenario-Based Design in Design Pattern Mining
Claudia Iacob — 6-1

Integral Designed Database Morphology for Active Roofs
Wim Zeiler, Emile Quanjel — 6-11

Dual Perspective on Information Exchange Between Design and Manufacturing
Jessica Bruch, Glenn Johansson — 6-21

The Digitial Divide: Investigating the Personal Information Management Practices of Engineers
Hamish Mcalpine, Ben Hicks, Can Tiryakioglu — 6-31

Predicting Emerging Product Design Trend by Mining Publicly Available Customer Review Data
Conrad Tucker, Harrison M. Kim — 6-43

How Product Representation Types are Perceived at the Client's End to Facilitate Communication and Decision Making
André Liem — 6-53

Data Management Planning in Engineering Design and Manufacturing Research
Mansur Darlington, Tom Howard, Alex Ball, Steve Culley, Chris Mcmahon — 6-65

Capturing the Conceptual Design Process with Concept-Configuration-Evaluation Triplets
Ehud Kroll, Alexander Shihmanter — 6-76

Modeling and Management of Product Knowledge in an Engineer-to-Order Business Model
Fredrik Elgh — 6-86

Manifestation of Uncertainty - A Classification
Melanie E. Kreye, Yee Mey Goh, Linda B. Newnes — 6-96

Acquisition of Design-Relevant Knowledge within the Development of Sheet-Bulk Metal Forming
Sebastian Röhner, Thilo Breitsprecher, Sandro Wartzack — 6-108

Knowledge Representation for Supplier Discovery in Distributed Design and Manufacturing
Christian Mcarthus, Farhad Ameri — 6-121

Case Studies to Explore Indexing Issues in Product Design Traceability Framework
Neven Pavković, Nenad Bojčetić, Leonard Franić, Dorian Marjanović — 6-131

Learning from the Lifecycle: The Capabilities and Limitations of Current Product Lifecycle Practice and Systems
James A. Gopsill, Hamish C. Mcalpine, Ben J. Hicks — 6-141

Visualising Ergonomics Data for Design
Hua Dong, Eujin Pei, Hongyan Chen, Robert Macredie — 6-153

Knowledge Configuration Management for Product Design and Numerical Simulation
J. Badin, D. Monticolo, D. Chamoret, S. Gomes — 6-161

Reference Model for Traceability Records Implementation in Engineering Design Environment
Mario Štorga, Dorian Marjanović, Tomaž Savšek — 6-173

The Management of Manufacturing Processes using
Complementary Information Structures
Greg Huet, Clément Fortin, Grant Mcsorley And Boris Toche — 6-183

IT-Based Configuration and Dimensioning of Customer Specific Products – Towards
a Framework for Implementing Knowledge Based Design Assistant Systems
Detlef Gerhard, Christoph Lutz — 6-192

Assessing the Conditions for Dissemination of End-User and
Purchaser Knowledge in a Medtech Context
Carl Wadell, Margareta Norell Bergendahl — 6-200

A Scalable Approach for the Integration of Large Knowledge
Repositories in the Biologically-Inspired Design Process
D. Vandevenne, P.-A. Verhaegen, S. Dewulf, J.R. Duflou — 6-210

Understanding Engineering Systems Through the Engineering
Knowledge Genome: Structural Genes of Systems Topologies
Offer Shai, Yoram Reich — 6-220

Interface Qualification Between the Research Central Team
and Design Offices in Order to Evaluate the Knowledge Sharing
Marie Fraslin, Eric Blanco, Valerie Chanal

Challenges in Semantic Knowledge Management for Aerospace Design Engineering
Isaac Sanya, Essam Shehab, Dave Lowe — 6-241

Means for Internal Knowledge Reuse in Pre-Development
– The Technology Platform Approach
Daniel Corin Stig, Dag Bergsjö — 6-249

Exploiting Neighborhood and Multi-Dimension Granular
Information for Supporting Design Rationale Retrieval
Yan Liang, Ying Liu, Wing Bun Lee, Chun Kit Kwong — 6-262

Providing Design Solution Repostories in the Field of Mechanism Theory
Torsten Brix, Ulf Döring, Michael Reessing, Christian Weber — 6-272

A Parametric Design Framework to Support Structural and
Functional Modeling of Complex Consumer Electronics Products
Kenichi Seki, Hidekazu Nishimura, Shaopeng Zhu, Laurent Balmelli — 6-282

The Challenge of Handling Material Information From Different Sources
A. Janus, D. Tartler, H. Krehmer, S. Wartzack — 6-292

The Retrieval of Structured Design Knowledge
Hongwei Wang, Aylmer L. Johnson, Rob H. Bracewell — 6-303

Identification, Translation and Realisation of Requirements for a
Knowledge Management System in an Engineering Design Consultancy
Thomson, Avril Isabel — 6-313

Development of Engineering Knowledge Models to Achieve Product Innovation
Anna Karlsson, Peter Törlind — 6-322

Perceptions of and Challenges with Knowledge Sharing
– Enterprise Collaboration in a Virtual Aeronautical Enterprise
Pär Johansson, Christian Johansson — 6-332

Software Supported Knowledge Transfer for Product Development *Sönke Krebber, Hermann Kloberdanz, Herbert Birkhofer, Andrea Bohn*	6-342
Search for Similar Technical Solutions by Object Abstraction using an Ontology *Andreas Kohn, Udo Lindemann*	6-350
A Methodology for Discovering Structure in Design Databases *Katherine Fu, Jonathan Cagan, Kenneth Kotovsky*	6-360
Comparative Study of Theoretical and Real uses of Eco-Designed Laundry Detergents *Chapotot, Emilie; Abi Akle, Audrey; Minel, Stéphanie; Yannou, Bernard*	6-370
Designers' Thinking and Acting in Meetings with Clients *Sónia Da Silva Vieira, Petra Badke-Schaub, António Fernandes, Teresa Fonseca*	6-382
A Structure for Representing Problem Formulation in Design *Mahmoud Dinar, Jami Shah, Pat Langley, Glen Hunt And Ellen Campana*	6-392
A New Metamodel to Represent Topologic Knowledge in Artifactual Design *J. M. Jauregui-Becker, K.J.W. Gebauer, F. J. A. M. Van Houten*	6-402
N-Gram Analysis in the Engineering Domain *Martin Leary, Geoff Pearson, Colin Burvill, Maciej Mazur, Aleksandar Subic*	6-414
Applying Context to Organize Unstructured Information in Aerospace Industry *Yifan Xie, Steve J Culley, Frithjof Weber*	6-424
Improving Design Rationale Capture During Embodiment Design *Jeroen Van Schaik, Jim Scanlan, Andy Keane, Kenji Takeda, Dirk Gorissen*	6-436
Exploring the Synthesis of Information in Design Processes – Opening the Black-Box *Raja Gumienny, Tilmann Lindberg, Christoph Meinel*	6-446
Representation of Cross-Domain Design Knowledge through Ontology Based Functional Models *Dipl.-Ing. Milan Marinov, Dipl.-Wi.-Ing. Dan Gutu, B.Sc. Janet Todorova, Dr. Miklós Szotz, András Simonyi, Prof. Dr. Dr.-Ing. Dr. H.C. Jivka Ovtcharova*	6-456
Adapting Aerospace Design Rationale Mapping To Civil Engineering: A Preliminary Study *Nathan Eng, Emanuele Marfisi, Marco Aurisicchio*	6-468
Knowledge Sharing Approaches in Method Development *Peter Thor, Johan Wenngren, Åsa Ericson*	6-480
Application of MOKA Methodology to Capture Knowledge in Design for Poka-Yoke Assembly *Gabriela Estrada, Joaquim Lloveras*	6-490
The Evolution of Information While Building Cross Domain Models of a Design: A Video Experiment *Naveed Ahmad, David C. Wynn, P. John Clarkson*	6-500

VOLUME 7: HUMAN BEHAVIOUR IN DESIGN

Improving Communication in Design: Recommendations From The Literature *Anja M Maier, Denniz Dönmez, Clemens Hepperle, Matthias Kreimeyer, Udo Lindemann, P John Clarkson*	7-1

Culture and Concept Design: A Study of International Teams *Andrew Wodehouse, Ross Maclachlan, Hilary Grierson and David Strong*	7-12
Associative Thinking as a Design Strategy and its Relation to Creativity *Hernan Casakin*	7-22
Social Media Enabled Design Communication Structure in a Buyer-Supplier Relationship *V. Hölttä, T. Eisto*	7-32
Emotion-Driven Elicitation of Elderly People user needs Illustrated by a Walking Frame Case Study *Hjalte P. Gudmundsson, Casper L. Andersen, Sofiane Achiche, Per Boelskifte*	7-44
Choosing Innovation: How Reasoning Affects Decision Errors *Ronny Mounarath, Dan Lovallo, Andy Dong*	7-54
Enabling Objects for Participatory Design of Socio-Technical Systems *Ole Broberg*	7-64
Challenges and Limitations of Applying an Emotion-Driven Design Approach on Elderly Users *Casper L. Andersen, Hjalte P. Gudmundsson, Sofiane Achiche, Per Boelskifte*	7-74
On the Effective use of Design-By-Analogy: The Influences of Analogical Distance and Commonness of Analogous Designs on Ideation Performance *Joel Chan, Katherine Fu, Christian Schunn, Jonathan Cagan, Kristin Wood, Kenneth Kotovsky*	7-85
Ingredients of the Design Process: Going Through Emotional Passage *Cliff Shin*	7-97
Meaning-Based Assessment in Creative Design *Hernan Casakin, Shulamith Kreitler*	7-107
How Important is Team Structure to Team Performance? *Vishal Singh, Andy Dong, John S Gero*	7-117
Applied Test of Engineering Design Skill: Visual Thinking, Characterization, Test Development and Validation *Jami J. Shah, Jay Woodward, S. M. Smith*	7-127
Analysing the use of four Creativity Tools in a Constrained Design Situation *Snider, C.M., Dekoninck, E.A., Yue, H., Howard, T.J.*	7-140
Collaborative Trust Networks in Engineering Design Adaptation *Simon Reay Atkinson, Anja M Maier, Nicholas Caldwell, P John Clarkson*	7-152
The importance of Empathy in IT Projects: A Case Study on the Development of the German Electronic Identity Card *Eva Köppen, Ingo Rauth, Maxim Schnjakin, Christoph Meinel*	7-162
A Sound-Based Protocol to Study the Emotions Elicited by Product Appearance *Weihua Lu, Jean-François Petiot*	7-170
A Scenario of user Experience *Juan Carlos Ortíz Nicolás, Marco Aurisicchio*	7-182
The Relationship Between a Model and a Full-Size Object or Building: The Perception and Interpretation of Models *Yvonne Eriksson, Ulrika Florin*	7-194

Review of Collaborative Engineering Environments: Software, Hardware, Peopleware *Jonathan Osborn, Joshua D. Summers, Gregory M. Mocko*	7-204
Exploring Consumer Needs with Lewin's Life Space Perspective *Kee-Ok Kim, Hye Sun Hwang*	7-214
Proposal of "Expectology " as Design Methodology *Tamotsu Murakami, Satoshi Nakagawa, Hideyoshi Yanagisawa*	7-224
An Approach to Analyzing user Impressions and Meanings of Product Materials in Design *Georgi V. Georgie v, Yukari Nagai*	7-234
A Method to Study Affective Dynamics and Performance in Engineering Design Teams *Malte F. Jung, Larry J. Leifer*	7-244
Cultural "Value Creation" in the Design of Cellular Phones *André Liem, Bijan Aryana*	7-254
Idea Screening in Engineering Design using Employee-Driven Wisdom of the Crowds *Balder Onarheim, Bo T. Christensen*	7-265
Facilitating Creative Problem Solving Workshops: Empirical Observations at a Swedish Automotive Company *Katarina Lund, Johan Tingström*	7-275
Creative Teamwork in Quick Projects Development QPD, 24 Hours of Innovation *Luz-Maria Jiménez, Denis Choulier, Jeremy Legardeur, Mickaël Gardoni*	7-285
Emotional Orientation and Context Analysis for Design Creativity Exercise Test *Jongho Shin and Yong Se Kim*	7-297
Understanding Fixation: A Study on the Role of Expertise *Vimal Viswanathan and Julie Linsey*	7-309
Adoption of a Systematic Design Process: A Study of Cognitive and Social Influences on Design *Thea Morgan, Theo Tryfonas*	7-320
Taxonomy of Cognitive Functions *Torsten Metzler, Kristina Shea*	7-330
Supporting Annotation-Based Argumentation Linking Discursive and Graphical Aspects of Design for Asynchronous Communication *Jean-François Boujut*	7-342
Initial Conditions: The Structure and Composition of Effective Design Teams *Greg Kress, Mark Schar*	7-353
Creativity Techniques for a Computer Aided Inventing System *Davide Russo, Tiziano Montecchi*	7-362
Characterizing Reflective Practice in Design – What about those Ideas you get in the Shower? *Rebecca M. Currano, Martin Steinert, Larry J. Leifer*	7-374
Monitoring Design Thinking Through In-Situ Interventions *Micah Lande, Neeraj Sonalkar, Malte Jung, Christopher Han, Banny Banerjee, Larry Leifer*	7-384

Design-by-Analogy using the Wordtree Method and an Automated Wordtree Generating Tool *E. V. Oriakhi, J. S. Linsey, X. Peng*	7-394
Around You: How Designers Get Inspired *Milene Gonçalves, Carlos Cardoso, Petra Badke-Schaub*	7-404
Information Behavior in Multidisciplinary Design Teams *Ensici, Ayhan, Badke-Schaub, Petra*	7-414
Designing: Insights from Weaving Theories of Cognition and Design Theories *Eswaran Subrahmanian, Yoram Reich, Frido Smulders, Sebastiaan Meijer*	7-424
Understanding the Front End of Design *T. Harrison, M. Aurisicchio*	7-437
Context-Specific Experience Sampling for Experience Design Research *Yong Se Kim, Yeon Koo Hong, Jin Hui Kim, Young Mi Kim*	7-448
Product Profile to Reduce Consumer Dissatisfaction in Terms of Soft Usability Problem and Demographical Factors: an Exploratory Study *Chajoong Kim, Henri Christiaans*	7-458
The Psychological Experience of user Observation *Elizabeth Gerber*	7-468
Problems and Potentials in the Creation of New Objects *Joakim Juhl, Martin Gylling*	7-480
A Protocol for Connective Complexity Tracking in the Engineering Design Process *James Mathieson, Michael Miller, Joshua Summer*	7-492
A New Framework of Studying the Cognitive Model of Creative Design *Ganyun Sun, Shengji Yao*	7-501

VOLUME 8: DESIGN EDUCATION

Identifying and Quantifying Industry Perceptions of Engineering Drawing Skills in Novice Malaysian Engineers *Zulkeflee Abdullah, Colin Reginald Burvill, Bruce William Field*	8-1
Measuring Malaysian Undergraduate Skills in Reading and Interpreting Engineering Drawing *Zulkeflee Abdullah, Bruce William Field, Colin Reginald Burvill*	8-13
Assessing Quality of Ideas in Conceptual Mechanical Design *William Lewis, Bruce Field, John Weir*	8-23
Fluency and Flexibility of Concepts Arising from Personalised Ideation Techniques *Bruce Field*	8-35
Exploiting Hand Sketching in Educating 'Mechanically Oriented' Engineering Students *Farrugia P.J., Borg J.C., Camilleri K.P.*	8-45
Implementing Design Critique for Teaching Sustainable Concept Generation *William Z Bernstein, Devarajan Ramnujan, Monica F Cox, Fu Zhao, John W Sutherland, Karthik Ramani*	8-55

Integrated Systems Design Education
Alastair Conway, Graham Wren, Bill Ion — 8-66

A Process of Conceptual Engineering Design for New Patentable Products
Joaquim Lloveras — 8-78

Strengthening Asian Advanced Design and Manufacture Education Through a Framework Approach
Fayyaz Rehman, Xiu-Tian Yan, Youhua Li, Xincai Tan, Eric Miller, Nick Woodfine — 8-88

Planning Industrial Phd Projects in Practice: Speaking Both 'academia' and 'practitionese'
Ingrid Kihlander, Susanne Nilsson, Katarina Lund, Sofia Ritzén, Margareta Norell Bergendahl — 8-100

Characterization of Leadership within Undergraduate Engineering Design Teams Through Case Study Analysis
Gary Palmer, Joshua D. Summers — 8-110

Do Basic Schemata Facilitate Embodiment Design?
Roman Žavbi, Nuša Fain, Janez Rihtaršič — 8-120

Machine Part Exhibition and Functional Mock-Ups to Enrich Design Education
Gregor Beckmann, Dieter Krause — 8-130

Teaching Design for Environment in Product Design Classes
Michael C. Baeriswyl, Steven D. Eppinger — 8-140

Integration of DFMA Throughout an Academic Product Design and Development Process Supported by a Plm Strategy
Gilberto Osorio-Gomez, Santiago Ruiz-Arenas — 8-151

Learning Levels in Technical Drawing Education: Proposal for an Assessment Grid Based on the European Qualifications Framework (EQF)
Riccardo Metraglia, Gabriele Baronio, Valerio Villa — 8-161

A Coherent and Discriminating Skills Standard for Innovative Design
Denis Choulier, Pierre Alain Weite — 8-173

A Proposal for an Assessment Form for Engineering Design Theses
Robert Watty, Matthias Kreimeyer — 8-184

Adapting Industrial Design Education to Future Challenges of Higher Education
André Liem, Johannes B. Sigurjonsson — 8-194

New Job Roles in Global Engineering – from Education to Industrial Deployment
Kai Lindow, Patrick Müller, Rainer Stark — 8-205

An Ethical Stance: Engineering Curricula Designed for Social Responsibility
Ian De Vere, Ajay Kapoor, Gavin Melles — 8-216

Developing A Drawing Culture: New Directions in Engineering Education
Ian De Vere, Gavin Melles, Ajay Kapoor — 8-226

Shaping The Individual Designer: Participatory Design in Emergency Context.
Briede Westermeyer, Juan Carlos; Cartes, Jorge; Bustamante, Alejandro; Perez, Marcela — 8-236

A 'Theatric' Approach to the Teaching of Design
Jason Matthews, Tony Medland — 8-245

The DesignExchange: Supporting the Design Community of Practice
Celeste Roschuni, Alice M. Agogino, Sara L. Beckman — 8-255

Improving Engineering Education in India Using Information and Communication Technology: A New Framework *Prerak Mehta*	8-265
Foundations for a New Type of Design-Engineers – Experiences from DTU *Ulrik Jørgensen, Søsser Brodersen, Hanne Lindegaard, Per Boelskifte*	8-275
Sharing Experience in Design Education Based on Research and Industrial Practice *Stanislav Hosnedl, Zbynek Srp*	8-287

VOLUME 9: DESIGN METHODS AND TOOLS PART 1

Decision Support for Improving the Design of Hydraulic Systems by Leading Feedback into Product Development *Michael Abramovici, Andreas Lindner, Florian Walde, Madjid Fathi, Susanne Dienst*	9-1
Clustering Customer Dreams – An Approach for a More Efficient Requirement Acquisition *Benjamin Röder, Herbert Birkhofer, Andrea Bohn*	9-11
Proposal About the use of Data Base in Engineering Design *Francesco Rosa, Edoardo Rovida, Roberto Vigano*	9-21
Application of Basic Design Principles for Solution Search in Biomimetics *Manuela Iulia Parvan, Andreas Schwalmberger, Udo Lindemann*	9-30
Use of Constraints in the Early Stages of Design *Glen Mullineux*	9-40
A Knowledge-Based Superposing Sketch Tool for Design Concept Generation Through Reflection of Verbal and Drawing Expression *Yutaka Nomaguchi, Yuko Kotera, Kikuo Fujita*	9-50
Economic Impact Estimation of New Design Methods *Roland Koppe, Stefan Häusler, Frank Poppen, Stephan Große Austing, Axel Hahn*	9-61
Failure Mode and Effects Analysis in Combination with the Problem-Solving A3 *Eirin Lodgaard, Øystein Pellegård, Geir Ringen, Jon Andreas Klokkehaug*	9-71
Supporting Inclusive Product Design with Virtual User Models at the Early Stages of Product Development *Pierre T. Kirisci, Klaus-Dieter-Thoben, Patrick Klein, Markus Modzelewski*	9-80
Using Virtual Reality in Designing the Assembly Process of a Car *Ilse Becker, Ville Toivonen, Simo-Pekka Leino*	9-91
On the Types and Roles of Demonstrators for Designing Medical Devices *Benoît Herman, Julien Sapin, Khanh Tran Duy, Benoît Raucent*	9-101
Computational Representations for Multi State Design Tasks and Enumeration of Mechanical Device Behaviour *Somasekhara Rao Todeti, Amaresh Chakrabarti*	9-111
IFMEA – Integration Failure Mode and Effects Analysis *Stefan Punz, Martin Follmer, Peter Hehenberger, Klaus Zeman*	9-122
ELISE 3d - A Database-Driven Engineering and Design Tool *Authors: M. Maier, Dr. C. Hamm*	9-132

Evaluation of Data Quality in the Context of Continuous Product Validation Throughout the Development Process *Jochen Reitmeier, Kristin Paetzold*	9-143
Geometric Manipulation Method for Evaluation of Aesthetic Quality in Early Design Phases *T. Stoll, A. Stockinger, S. Wartzack*	9-153
A Morphological Approach to Business Model Creation using Case-Based Reasoning *Ji Hwan Lee, Yoo S. Hong,*	9-165
A Metric to Represent the Evolution of Cad/Analysis Models in Collaborative Design. *Nicolas Drémont, Pascal Graignic, Nadège Troussier, Robert Ian Whitfield, Alex Duffy*	9-176
Significance of Requirements for the Implementation of New Technologies using Shape Memory Technology *Sven Langbein, Konstantin Lygin, Tim Sadek*	9-186
Interdisciplinary Systems Modeling using the Contact and Channel-Model for SYSML *Albers, Albert, Zingel, Christian*	9-196
A Novel Hybrid 2D and 3D Augmented Reality Based Method for Geometric Product Development *Pablo Prieto*	9-208
A Methodology to Evaluate the Structural Robustness of Product Concepts *Maximilian P. Kissel, David Hellenbrand, Udo Lindemann*	9-215
Towards Assessing the Value of Aerospace Components: A Conceptual Scenario *Marco Bertoni, Christian Johansson, Alessandro Bertoni*	9-226
Seven Years of Product Development in Industry – Experiences and Requirements for Supporting Engineering Design with 'Thinking Tools' *Sven Matthiesen*	9-236
Design of Innovative Product Profiles: Anticipatory Estimation of Success Potential *Yuri Borgianni, Alessandro Cardillo, Gaetano Cascini, Federico Rotini*	9-246
UMEA - A Follow-Up to Analyse Uncertainties in Technical Systems *Roland Engelhardt, Marion Wiebel, Tobias Eifler, Hermann Kloberdanz, Herbert Birkhofer, Andrea Bohn*	9-257
Selection of Design Concepts using Virtual Prototyping in the Early Design Phases *Mikko Seppälä, Andrea Buda, Eric Coatanéa*	9-267
Early Robustness Optimization of Automotive Modules – Regarding the Key Impact of the Human Factor *Fabian Wuttke, Florian Feustel, Martin Bohn, Steffen Csernak, Andrea Bohn*	9-275
Decision Processes in Engineering Design: A Network Perspective of Stakeholder and Task Interaction *Julie R. Jupp*	9-285
Immersive Product Improvement IPI – First Empirical Results of a New Method *Rafael Kirschner, Andreas Kain, Alexander Lang, Udo Lindemann*	9-295
A Methodology for Designing a Recommender System Based on Customer Preferences *Inès Jomaa, Emilie Poirson, Catherine Da Cunha, Jean-François Petiot*	9-305
Evaluation of Solution Variants in Conceptual Design by Means of Adequate Sensitivity Indices *Tobias Eifler, Johannes Mathias, Roland Engelhardt, Hermann Kloberdanz, Herbert Birkhofer, Andrea Bohn*	9-314

Modular Optimization Strategy for Layout Problems *Julien Bénabès, Emilie Poirson, Fouad Bennis, Yannick Ravaut*	9-324
Maintenance Engineering: Case Study of Fitness for Service Assessments *Francesco Giacobbe, Domenico Geraci, Emanuele Biancuzzo, Mirko Albino*	9-335
An Approach for the Automated Synthesis of Technical Processes *Tino Stankovi, Kristina Shea, Mario Štorga, Dorian Marjanovi*	9-345
Understanding Styling Activity of Automotive Designers: A Study of Manual Interpolative Morphing Through Freehand Sketching *Shahriman Zainal Abidin, Anders Warell, Andre Liem*	9-357
Advanced Applications of a Computational Design Synthesis Method *F. Bolognini, K. Shea, A.A.Seshia*	9-367
Exploring Potentials for Conservational Reasoning using Topologic Rules of Function Structure Graphs *Chiradeep Sen, Joshua D. Summers, Gregory M. Mocko*	9-377
A Decomposition Algorithm for Parametric Design *J. M. Jauregui-Becker, W. O. Schotborgh*	9-389
Innovation Through Design for Emotion *Vincent Berdillon, Anne Guenand*	9-399

VOLUME 10: DESIGN METHOD AND TOOLS PART 2

Can Existing Usability Techniques Prevent Tomorrow's Usability Problems? *Christelle Harkema, Ilse Luyk-De Visser, Kees Dorst, Aarnout Brombacher*	10-1
Application to a Car Body Frame Based on Parameter Guidelines for Deriving Diverse Solutions using Emergent Design System *Koichiro Sato, Yoshiyuki Matsuoka*	10-11
Hierarchical System Concept Generation *Rosenstein David, Yoram Reich*	10-24
Usage Context-Based Choice Modeling for Hybrid Electric Vehicles *Lin He, Wei Chen*	10-35
An Examination of the Application of Plan-Do-Check-Act Cycle in Product Development *Eirin Lodgaard, Knut Einar Aasland*	10-47
Realizing a Truly 3D Product Visualization Environment – A Case for using Holographic Displays *Eliab Z. Opiyo*	10-56
Design Exploration with Useless Rules and Eye Tracking *Iestyn Jowers, Miquel Prats, Alison Mckay, Steve Garner*	10-66
Simulation Based Generation of an Initial Design Taking into Account Geometric Deviations and Deformations *Michael Walter, Thilo Breitsprecher, Georg Gruber, Sandro Wartzack*	10-78
Collaborative Idea Generation using Design Heuristics *Seda Yilmaz, James L. Christian, Shanna R. Daly, Colleen M. Seifert, Richard Gonzalez*	10-91

Understanding Managers Decision Making Process for Tools Selection in the Core Front End of Innovation *Francesco P. Appio, Sofiane Achiche, Tim Mcaloone, Alberto Di Minin*	10-102
Evaluating the Risk of Change Propagation *Arman Oduncuoglu, Thomson, Vince*	10-114
Multilayer Network Model for Analysis and Management Of Change Propagation *Michael C. Pasqual, Olivier L. De Weck*	10-126
Can Designers be Proactively Supported as from Product Specifications? *Amanda Galea, Jonathan Borg, Alexia Grech, Philip Farrugia*	10-139
Design Preference Elicitation: Exploration and Learning *Yi Ren, Panos Papalambros*	10-149
Designing to Maximize Value for Multiple Stakeholders: A Challenge to Med-Tech Innovation *Lauren Aquino Shluzas, Martin Steinert, Larry J. Leifer*	10-159
Customer Value is Not a Number - Investigating the Value Concept in Lean Product Development *Martin Gudem, Martin Steinert, Torgeir Welo, Larry Leifer*	10-167
An Agent-Based System for Supporting Design Engineers in the Embodiment Design Phase *Martin Kratzer, Michael Rauscher, Hansgeorg Binz, Peter Goehner*	10-178
Combining Narrative and Numerical Simulation: A Supply Chain Case *Mette Sanne Hansen, Klaes Rohde Ladeby, Lauge Baungaard Rasmussen, Peter Jacobsen*	10-190
Change in Requirements During the Design Process *Mohd Nizam Sudin, Saeema Ahmed-Kristensen*	10-200
Understanding Adaptability Through Layer Dependencies *Robert Schmidt Iii, Jason Deamer, Simon Austin*	10-209
Interactivity in Early-Stage Design by Real-Time Update of Stress Information for Evolving Geometries *J. Trevelyan, D.J. Scales*	10-221
A Value-Centric QFD for Establishing Requirements Specification *Xinwei Zhang, Guillaume Auriol, Anne Monceaux, Claude Baron*	10-228
IT Support for the Creation and Validation of Requirements Specifications – with a Case Study for Energy Efficiency *Thomas Reichel, Gudula Rünger, Daniel Steger, Haibin Xu*	10-238
A Decision Support System for the Concept Development of RIM Parts *Ricardo Torcato, Ricardo Santos, Madalena Dias, Richard Roth, Elsa Olivetti, José Carlos Lopes*	10-248
Abstract Prototyping in Software Engineering: A Review of Approaches *Els Du Bois, Imre Horváth*	10-258
Early Reliability Estimation in Automotive Industry *Michael Kopp, Daniel Hofmann, Bernd Bertsche, Christian Heß, Oliver Fritz*	10-270
System Dynamics Modeling of New Vehicle Architecture Adoption *Carlos Gorbea, Udo Lindemann, Olivier De Weck*	10-278

Evaluating Methods for Product Vision With Customers' Involvement
to Support Agile Project Management
João Luis G. Benassi, Lucelindo D. Ferreira Junior, Daniel C. Amaral — 10-290

Wants Chain Analysis: Human-Centered Method for Analyzing and Designing Social Systems
Takashi Maeno, Yurie Makino, Seiko Shirasaka, Yasutoshi Makino, Sun K. Kim — 10-302

Redundancy Eliminations and Plausible Assumptions of Design Parameters for
Evaluating Design Alternatives
Alexandros Zapaniotis, Argyris Dentsoras — 10-311

Use of Design Methodology to Accelerate the Development and Market Introduction
of New Lightweight Steel Profiles
Frank Nehuis, Jan Robert Ziebart, Carsten Stechert, Thomas Vietor — 10-321

The use of Storyboard to Capture Experiences
Anders Wikström, Jennie Andersson, Åsa Öberg, Yvonne Eriksson — 10-331

Usability Compliant Supportive Technologies in Simulation-Driven Engineering
Jochen Zapf, Bettina Alber-Laukant, Frank Rieg — 10-341

A Visualization Concept for Supporting Module Lightweight Design
Thomas Gumpinger, Henry Jonas, Benedikt Plaumann, Dieter Krause — 10-349

Exploring a Decision-Making Forum in Early Product Development
ngrid Kihlander — 10-360

The Benfits and Pitfalls of Digital Design Tools
Tucker J. Marion, Sebastian K. Fixson — 10-370

A Haptic Based Hybrid Mock-Up for Mechanical Products Supporting
Human-Centered Design
Daniel Krüger, Andreas Stockinger, Sandro Wartzack — 10-380

Models and Software for Corrugated Board and Box Design
Vahid Sohrabpour, Daniel Hellström — 10-392

Evolutionary Design Synthesis Comparison: Growth and Development Vs. Fixed-Mesh Cells
Or Yogev, Andrew A. Shapiro, Erik K. Antonsson — 10-402

Autonomous Visualization Agents to Enhance the Analysis of Virtual Prototypes
Rafael Radkowski, Jürgen Gausemeier — 10-413

Identifying a Dynamic Interaction Model: A View from the Designer-User Interactions
Jaehyun Park, Richard Boland, Jr. — 10-426

Linkage of Methods Within the UMEA Methodology - An Approach to Analyse
Uncertainties in the Product Development Process
Roland Engelhardt, Tobias Eifler, Johannes Mathias,
Hermann Kloberdanz, Herbert Birkhofer, Andrea Bohn — 10-433

On the Functions of Products
M. Aurisicchio, N.L. Eng, J.C. Ortíz Nicolás, P.R.N. Childs, R.H. Bracewell — 10-443

Benchmarking Study of Automotive Seat Track Sensitivity to Manufacturing Variation
Maciej Mazur, Martin Leary, Sunan Huang, Tony Baxter, Aleksandar Subic — 10-456

Bioinspired Conceptual Design (BICD): Conceptual Design of a Grasshopper-Like
Jumping Mechanism as a Case Study
Aylin Konez Eroglu, Zuhal Erden, Abdulkadir Erden — 10-466

INTERNATIONAL CONFERENCE ON ENGINEERING DESIGN, ICED11
15 - 18 AUGUST 2011, TECHNICAL UNIVERSITY OF DENMARK

PRODUCT PLATFORM AUTOMATION FOR OPTIMAL CONFIGURATION OF INDUSTRIAL ROBOT FAMILIES

Mehdi Tarkian[1], Johan Ölvander[1], Xiaolong Feng[2] and Marcus Pettersson[2]
(1) Linköping University, Sweden
(2) ABB Corporate Research, Sweden

ABSTRACT

Product platform design is a well recognized methodology to effectively increase range and variety of products and simultaneously decrease internal variety of components by utilizing modularization. The tradeoff between product performance and product family commonality has to be carefully balanced in order to for the company to meet market requirements and simultaneously obtain economy of scale. This paper presents a framework based on high fidelity analyses tools that concurrently optimize an industrial robot family as well as the common platform. The product family design problem is formally stated as a multi-objective optimization problem, which is solved using a multi-objective Genetic Algorithm.

Keywords: Automated design, multidisciplinary design optimization, parametric CAD

1. INTRODUCTION

Product family design based on modularization has for a long time been a well recognized method to address the demands of mass customization [1]. Based on the concept of product platforms, it is possible to deliver products within a short time frame and have a broad product range to meet specific customer requirements while maintaining low development and manufacturing costs [1]. A potential drawback of product families is that the performance of individual members are reduced due to the constraints added by the common platform, i.e. parts and components need to be shared by different family members.

This paper focuses on quantitative approaches where the product family design problem is formally stated as an optimization problem where high fidelity analyses tools are used to find a tradeoff between degree of commonality and product performance. The optimization problem involves balancing performance of the members of the product family against cost savings during design and possible re-use of modules in the family.

The outline of the paper is as follows: Following the introduction, section 2 explains the scope of the paper by describing the concept of product family and platform design and the research conducted in the global research arena. A brief outline of how the identified obstacles ought to be tackled is given. In section 3, an overview of the field of Multidisciplinary Design Optimization (MDO) and Knowledge Based Engineering (KBE) is presented. These techniques are two important enablers to pursue practical product family design and optimization. The design procedures and hurdles of modular industrial robot design are presented in section 4. The automated framework along with the optimization procedures adopted is described in section 5. Finally, in section 6 the paper is concluded.

2. PLATFORM DESIGN

The definitions of product platform are plenty and since it is a fundamental term in this paper the following definition has been chosen; *"the use of a standard module set between different products is known as a platform"* [2]. Thereby, a platform is the set of standard components, manufacturing processes, and/or assembly steps that are common in a set of products. The overall aim with product family design is to reduce cost due to the commonality between the variants. However, there is always a trade-off between commonality and performance of individual family members [3] & [4].

There are many benefits or reasons for modularity, e.g. the twelve modular drivers described in [5]. However, most of them have economic implications, either in the design and development stage, in purchasing, during manufacturing or in the aftermarket. In the literature there are many indices defined to measure the degree of modularity within a product family, see [6]. A commonality index is typically based on different parameters such as the number of common components, the component manufacturing volume, the component costs, the manufacturing processes, and so on. The leading principle of the indices is to provide an estimate of the cost savings within a family.

2.1. Modularization

Jose and Tollenaere [7] describe modularization as *"an approach to organize complex designs and process operations more efficiently by decomposing complex systems into simpler portions"*. Modularization as a way to save cost is by no means a new phenomenon. For instance, the truck manufacturer Scania has been working successfully with modularization and the concept of product families since the forties [1]. Much research in product family design has been qualitative to its nature. Hence according to Jose and Tollenaere [7], *"today the methods for platform product development are not practical and future results can be obtained with an integral methodology using a practical design representation linked to an optimization methodology"*. Thus, this paper makes an effort to presents a quantitative approach where product family design is formulated as a formal multi-objective optimization problem.

Before initiating the modular design process, the product has to be evaluated for whether being appropriate for modular design. The modularity level of the product is then determined and strategies for modular design are carried out. To assure practical use of the outlined framework in industry, tight interaction with commercial CAE and simulation tools are of greatest importance. In order to accomplish system optimization incorporating various simulation tools, the concept of Multidisciplinary Design Optimization has been adopted, which will be discussed in the following section.

3. MULTIDISCIPLINARY DESIGN OPTIMIZATION

Li and Huang [8] recognize that by using different platform strategies such as commonality, modularity and scalability, product platforms can be developed and customized with different flexibility for realizing *mass customized products* [8]. This enabler is termed *adaptive platform*. Li and Huang also coined the terms *scalable modules* and *instance module*, established to achieve adaptive platforms. Furthermore, in the design of complex and tightly integrated engineering products it is essential to be able to handle cross-couplings and synergies between different subsystems [9]. A typical example of such products is mechatronic machines like industrial robots. To effectively design and develop such products, efficient tools and methods for integrated and automated design are needed throughout the development process. Multidisciplinary Design Optimization (MDO) is one promising technique that has the potential to drastically improve such a concurrent design. Giesing et al. [10] have defined MDO as *"a methodology for the design of complex engineering systems and subsystems that coherently exploits the synergism of mutually interacting phenomena"*.

3.1. Knowledge Based Engineering for Design of Engineering Products

Knowledge-Based Engineering defines a wide range of methods and processes and could be described in several ways depending on the application focus. In the literature there are various definitions that strive to highlight the multiple sides of KBE. Chapman et al. [11] define KBE as *"an engineering method that represents a merging of object oriented programming (OOP), artificial intelligence (AI) techniques and computer-aided design technologies, giving benefit to customized or variant design automation solutions"*. It therefore presents great potential for improving the product development process as well as reducing the time-to-market, thanks to an enhanced effectiveness of computer aided engineering systems.

It is the authors' opinion that KBE is a means to achieve design reuse and automation, and thereby create prospects for a holistic perspective throughout the design process.

A holistic product perspective by means of design reuse and automation is needed in order to effectively manage product complexity and to introduce MDO. In this field, KBE is believed to be a powerful tool [12]. In the coming sections methods for design automation are proposed.

3.2. Dynamic Top-Down Modeling

By introducing KBE techniques a new mean of CAD modeling is introduced, referred to here as dynamic top-down modeling. When applying a dynamic top-down development process, the actual CAD models can be generated from pre-described High Level CAD templates (HLCt). The critical information on how the HLCt should be instantiated is stored in the inference engine [14]. The geometry model is divided into sub-models that are linked to each other in a hierarchic relational structure [15]. Various components can be attached dynamically to the model and their shape altered by the inherited design variables, supporting the concept of *scalable modules* by Li and Huang [8]. This process continuous until the geometry is completely defined.

4. DESIGN APPLICATION: INDUSTRIAL ROBOTS

The mechanical structure of a modular industrial robot consists of a base followed by a series of modular structure links. Each module consists of drive-train components (servo actuator, combining precision Harmonic Drive gearing with highly dynamic servo motors). Major components of the robot controller are power units, rectifier, transformer, axis computers and a high level computer for motion planning and control.

Designing industrial robots is a complex process involving tremendous modeling and simulation effort. For all the various domains of robot design, the geometry plays a significant role as input provider.

To more effectively understand and manage the complexity of this technology and find the optimal solution for a family of robots faster, a joint novel design framework is being developed at ABB and Linköping University, see Johansson et al. [16], Petterson et al [17], and Tarkian et al. [13].

4.1. Modular Geometry approach for Modular Industrial Robots

One outcome of modularity within a product family is increase of variety and decrease of components. The same principle is adopted here for the modeling of the product family. Since the geometries are saved as HLCts and instantiated with unique internal design variables, the number of model variants is effectively increased by sharing few geometric templates between the model variants.

By importing the HLCt geometries, the robot is defined in three steps, see Figure 1. Firstly the number of axes is determined in a user interface, defining the skeleton model of the robot, stored in the Datum HLCt and placed according to the logic of the inference engine. The type of Component HLCt for each axis is then decided and an appropriate structure, from Structure HLCt, is chosen in the final step. The model of the robot is thereby transformed from an empty initial model into a complete model in three steps, as shown in Figure 1.

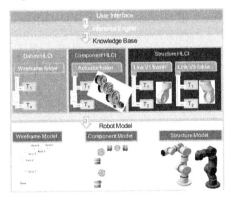

Figure 1. Relations between the robot models and the HLCt libraries.

4.2. Dynamic Model

To simulate the dynamic properties of a robot, a dynamic model has to be utilized. The dynamic model in the outlined framework is made using an in-house simulation tool developed at ABB. The motion of the rigid manipulator can be described by

$$Q = M(q) \cdot \ddot{q} + V(q, \dot{q}) + G(q) + B(\dot{q}) \qquad (1)$$

where M is the inertia matrix, V is the vector of Coriolis and centrifugal forces, G is a vector of gravity forces and B is a vector of viscous friction forces. q is a vector of generalized coordinates e.g. angular position of each joint in the manipulator. For more information about dynamic models and trajectory planning for robots see [18] & [19].

In the Newton-Euler formulation [18], link velocities and acceleration are iteratively computed, forward recursively.

$$a_{e,i} = \left(R_{i-1}^{i}\right)^{T} a_{e,i-1} + \dot{\omega}_i \times r_{i,i+1} + \omega_i \times \left(\omega_i \times r_{i,i+1}\right)$$
$$a_{c,i} = \left(R_{i-1}^{i}\right)^{T} a_{e,i-1} + \dot{\omega}_i \times r_{i,ci} + \omega_i \times \left(\omega_i \times r_{i,ci}\right) - \left(R_0^i\right)^T g_0 \qquad (2)$$

When the kinematic properties are computed, the force and torque interactions between the links are computed backward recursively from the last to the first link.

$$f_i = R_i^{i+1} f_{i+1} + m_i a_{c,i}$$
$$\tau_i = R_i^{i+1} \tau_{i+1} - f_i \times r_{i,ci} + \left(R_i^{i+1} f_{i+1}\right) \times r_{i+1,ci} + I_i \alpha_i + \omega_i \times \left(I_i \omega_i\right) \qquad (3)$$

Where is the angular velocity and $\dot{\omega}$ angular acceleration, a_e and a_c describe the acceleration at the end and at the center of each link respectively, f and describe the force and torque between each link respectively. R is the rotational matrix, I the mass inertia, g_0 the gravity acceleration and r the positional vector.

4.3. Automated and Holistic Design approach

The geometric and dynamic models are seamlessly integrated through a user interface, where various engineering aspects of the robot are analyzed concurrently. Furthermore the geometrical and dynamical aspects of the robot components are stored in a component library, see Figure 2.

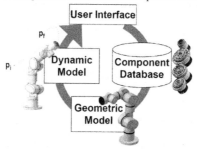

Figure 2. Weight and dynamic properties are concurrently computed following parametric input in the user interface.

5. OPTIMIZATION

In this section the problem formulation is presented following the selection of optimization algorithm for the specified problem, as well as the actual optimization framework.

5.1. Problem Formulation

The problem formulation consists of concurrently optimizing the performance and commonality level of a product family consisting of four robots. The optimization variables are choice of servo actuators for axes 1, 2 and 3 as well as a coefficient defining the relation between length of link 2 and link 4 (see figure 3), amounting to overall 16 optimization variables for the entire product family. The optimization problem is limited to three axes in order to restrict the design space for the optimization algorithm. Ideally all 7 joint could be optimized using the same approach depicted. The four robots' reach and payload requirements are visualized in Table 1.

	Robot 1	Robot 2	Robot 3	Robot 4
Reach [mm]	760	860	960	1060
Payload [kg]	3	5	10	14

Table 1. Payload and reach requirements of the robot family

The problem is multi objective with the performance and commonality being the objective functions. The performance objective, f_1, is the sum of cycle time (CT) and the robot weight (Weight) for all four robots. The performance objective is to be minimized, hence low weight and low cycle time is preferred. The commonality objective is to maximize number of common components in the robot family, for both the links and actuators. f_2 is the percent commonality, ranging from 0 to 100.

$$f_1 = \sum (\lambda_1 CT_i + \lambda_2 Weight_i)$$
$$f_2 = 100 \cdot (k_1 \frac{\sum Link_{shared}}{\sum Link} + k_2 \frac{\sum Actuator_{shared}}{\sum Acutator}) \quad (4)$$

$i = 1,2,3,4$

λ_j & k_u are weighting factors where $\sum k_u = 1$. The weighting factors k_u have been chosen to prioritize link share prior to actuator share. The weight and CT are normalized and λ_j chosen to balance the weighting.

5.2. Multi-Objective Genetic Algorithm

The presented problem consists of discrete variables, and the objectives and constraints are represented by non-linear functions where no analytical derivatives are available. Therefore a Genetic Algorithm has been chosen since generally speaking non-gradient methods are applicable to a broader range of problems as they do not rely on assumptions on the properties of the objective function such as differentiability and continuity, etc. The basic idea of Genetic Algorithms is the mechanics of natural selection [20]. Each optimization variable is coded into a gene as for example a real number or a string of bits. The corresponding genes for all parameters form a chromosome, which describes each individual. Each individual represents a possible solution, and a set of individuals form a population. In a population, the fittest individuals have the highest probability of being selected for mating. Mating is performed by combining genes from different parents to produce a child, called a crossover. Then there is also the possibility that a mutation might occur. Finally the children are inserted into the population to form a new generation.

Moreover since the tradeoff between performance and commonality is difficult to quantify beforehand, preferably the algorithm utilized should generate a Pareto frontier of the design solutions. Optimization methods that can handle this type of problems in general are Genetic Algorithms and specifically Multi-Objective Genetic Algorithms [21]. There are also many examples in the literature where GA:s and MOGA:s are applied to platform design problems [22]& [23]. In this paper NSGA-II are used as the optimizer [24].

5.3. Optimization framework

In previous work [13], a robot design framework has been utilized to design a single optimal modular robot for a specific task and a set of requirements. A product family optimization requires further evaluations to converge since the family members are simulated in sequence. Also, to reach convergence, the number of evaluations increases due to increased number of design variables. Consequently, to shorten the optimization time, the earlier framework [13] had to be completely reworked, which will be further depicted in following sections.

5.3.1. Geometry Database

Although commercial CAD tools are well suited to generate high fidelity geometry for various

Although commercial CAD tools are well suited to generate high fidelity geometry for various analyses tools, they often require extensive update time. Therefore, a geometry database has been created to eliminate the lengthy simulation times required. The database is created by evaluating and storing an array of various geometric configurations. Meaning that the shape and number of the robot structure are varied leading to a new robot configurations of which the geometric properties are stored as illustrated in Figure 3. The geometric properties include mass, center of gravity and inertia.

In Figure 3 all links subjected to parametric modification are colored white. For link 1 and link 3 the shape is altered by modifying the type of actuator. These are modified by altering discrete values ranging from 1 to 15 which will automatically insert the actual detailed actuator geometry which is stored as an HLCt. The logic stated in the inference engine will then update the internal design variables of the links housing the actuator. For link 2 and link 4 the shape of the structure is modified by varying the lengths between 200-450 [mm] and 200-400 [mm] respectively.

The geometry database for each link is computed independently. The separate mass properties of the links are then assembled together during the optimization framework to represent the complete robot.

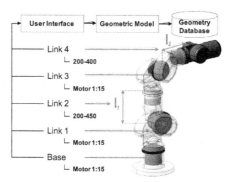

Figure 3. A geometry database is created by altering the design variables of the geometric model through a user interface.

5.3.2. Dynamic Database

Another bottleneck in the optimization framework is the dynamic simulation. A method to shorten the simulation time is thereby of importance. Storing the design configurations in a corresponding dynamic database as done for the geometric database is however not a promising approach. This is due to the kinematic and dynamic couplings between the links, both forward and backward recursively as stated in formula (2) and (3). Thereby the total number of design alternatives stored in the database would amount to tens of thousands. Consequently, in the following section another proposal is made to minimize the number of calls to the dynamic model.

5.3.3. Distributed Optimization

To further speed up optimization process, distributed optimization is utilized. The members of the robot family are thereby each distributed to a *slave PC*, as illustrated on a simplified flow chart in Figure 4. The family optimization presented in this paper consists of one *master PC* and four slaves. If the number of family members is increased, so will the number of slaves, presenting an effective means to keep the optimization time low irrespective of the number of individuals in the product family.

The optimization process starts by the master PC generating an initial population, declared as initial *Design Variables* in Figure 4. The Design Variables are utilized to calculate the commonality objective as stated in (4) and also sent to the slaves. The analyses of the robots take place in parallel, which upon completion will return the performance objective to the master, see Figure 4.

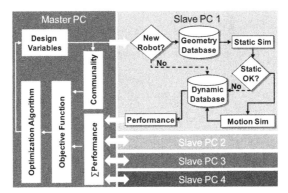

Figure 4. To speed up the evaluation process, the robot family is concurrently computed on 4 slave PCs.

For each slave PC the motion and static simulation results are stored in a dynamic database. As a result when a previously evaluated design is suggested by the optimization algorithm, the results will be retrieved from the dynamic database, thereby skipping both the static and motion simulations. If the design variables represent a new robot then the mass properties are retrieved from the geometry database and sent for static simulation. The static simulation includes a range of robot work space positions. The static simulation evaluates if the chosen motors are strong enough to withstand the gravitational forces. If the configuration does not meet the gravitational forces, then the performance objective is given a penalty value and the dynamic simulation will not be initiated, and hence the computational burden reduced. The results are stored in the dynamic database and the performance objective generated.

If the static simulation is successful then the geometrical data from the geometry database model is sent for motion simulation. The geometry database is used to parameterize the matrix and vectors in equation (1). The equation of motion for the robot is implemented in a dynamic simulation program which also includes path and trajectory planner and calculates properties such as torques, accelerations, speed and cycle-times. A set of motion cycles are simulated for each robot and the results are stored in the dynamic database and a performance objective calculated.

Although distributed computing presents faster evaluation, it is a complex procedure which needs to be properly setup otherwise it will lead to an ineffective and failed framework. The communication process involving the master and the slave is illustrated in more detail in *Figure 5*. The process starts by an initial population generated and the master sending the design variables to the slave as shown in Figure 4. The Master will then wait for a predefined time to receive a signal from the slave indicating that the design variables have been retrieved. If not, the slave PC will be terminated and another slave PC will be initiated to compute the performance objective.

The slave has a defined time at his disposal for each robot evaluation. The computation is terminated by the master as soon the time runs out. The reason for this constraint is due to some motion simulations taking several minutes to perform, whilst a preferred simulation only should take seconds. The lengthy simulation time is due to real time properties of the motion simulator, where simulation time is equal to the cycle time of each motion. Consequently, robot individuals consisting of barely strong enough motors to pass the static simulation requirement are still too weak to generate fast cycle times during the motion simulation. These individuals are weak and thus the simulation is terminated when the simulation time surpasses the time limit. A penalty is then given to the individual, which is stored in the dynamic database and the performance objective is calculated, see *Figure 5*.

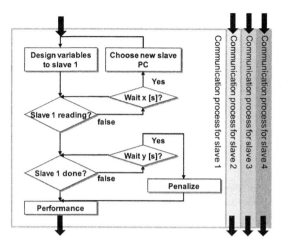

Figure 5. Communication process between the master and the slave.

5.4. Results

The outlined optimization framework is utilized to search for the Pareto frontier of the presented problem. The performance objective is to be minimized with the aim of decreasing the robot weight and cycle time, while the commonality objective is to be maximized to increase module sharing amongst the robots. In order to search for the global optimum, the following number of individuals and generations has been evaluated:

	Opt. 1	Opt. 2	Opt. 3	Opt. 4
Individuals	40	60	100	300
Generations	200	200	200	200

Table 2. Four sets of individuals evaluated.

Final results of the Pareto frontiers, up to the 5th rank, are visualized in Figure 6, where not surprisingly, as the number of individuals increase, the Pareto frontiers move to more optimal locations. However this movement is progressively minimized, suggesting that about 100 to 300 individuals is sufficient for finding the optimal Pareto frontier.

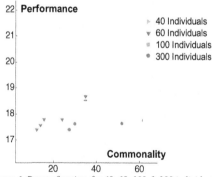

Figure 6. Pareto frontiers for 40, 60, 100 & 300 individuals.

Judging from the 1st order Pareto frontier in Figure 7, the algorithm is well suited to find solutions for both high commonality and best performance. In the robot family with best performance (1), the highest reach robot has more powerful actuators, while the smaller robots are capable in performing the pre-set trajectories with smaller actuators, hence weighing less. However the commonality level is low. For the robot family with highest commonality (2), the actuators and arm lengths are selected in order to maximize commonality, however the overall performance is worse. The arrows on the right side of figure 7 indicate modules that are shared within the product family.

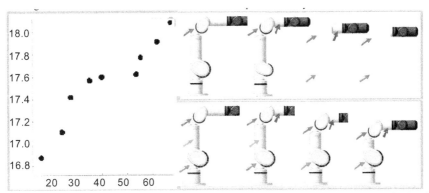

Figure 7. Pareto front of the product family, with best performance (1) and highest commonality (2). The shared modules are marked with an arrow.

6. DISCUSSION & CONCLUSION

In this paper a quantitative approach is presented where robot product family design is formulated as a formal multi-objective optimization problem. The product family design is based on tightly integrated set of high fidelity physics based models, supporting design reuse and automation. By utilizing the automated and reusable models, a MDO framework is established, facilitating automatic search for optimal robot families. An optimization case has been set up where the combination of discrete component selections invokes changes in the geometric structure, together with constraints in the dynamic simulation. The links and power train for a robot family has been optimized, and a Pareto frontier generated by applying the multi-objective genetic algorithm NSGA-II. Based on the objective functions a Pareto front is generated, presenting a range of robot families where the performance and commonality objectives have different importance. Hence, one major advantage of the presented method is that the balance between performance and commonality can be determined after generating the Pareto frontier. Therefore critical decisions can be made later in the design process, allowing engineers to gather more knowledge about the product under evaluation.

For future work, continuous variables of the actuators, e.g. maximum torque and angular velocity can be taken into consideration during the optimization. By taking the continuous variables into account the life time estimation of the drive train components can be computed and added to the performance objective. However by increasing the number of design variables, the optimization framework will have to undergo further modifications to reduce simulation time. To introduce several hierarchical layers of optimization, as well as meta-models for high fidelity models will be some of the options which should be examined further.

Another future investigation is the development of cost measurements, by estimating cost by taking commonality, component prices and life cycle in consideration.

REFERENCES

[1] Andersson, S. and Sellgren, U., "Modular product development with a focus on modelling and simulation of interfaces," Design Society – Workshop on Product Structuring, Copenhagen, January 2003.
[2] Jose A, Tollenaere M (2005) Modular and platform methods for product family design: literature analysis. J Intell Manuf 16:371–390
[3] Fellini R., Kokoloras M., Papalambros P., Perez-Duarte A., Platform Selection Under Performance Bounds in Optimal Design of Product Families, Journal of Mechanical Design, vol. 127, pp. 524-535, July, 2005.

[4] Nelson, S., Parkinson M., Papalambros P., Multicriteria Optimization in Product Platform Design, Journal of Mechanical Design, vol. 123, pp 199-204, June 2001.
[5] Erixon G., Erlandsson A., von Yxkull A., Östgren B. M., Modulindela produkten, (in Swedish), Industrilitteratur, 1994.
[6] Thevenot H., Simpson T., Commonality indices for product family design: a detailed comparison, Journal of Engineering Design, Vol. 17, No. 2, pp 99–119, 2006.
[7] Jose A, Tollenaere M (2005) Modular and platform methods for product family design: literature analysis. J Intell Manuf 16:371–390
[8] Li L., Huang G.Q., Newman S.T. A cooperative coevolutionary algorithm for design of platform-based mass customized products (2008) Journal of Intelligent Manufacturing, 19 (5), pp. 507-519.
[9] Bowcutt, K.G., "A Perspective on the Future of Aerospace Vehicle Design", AIAA 2003-6957, 12th AIAA International Space Planes and Hypersonic Systems and Technologies, Dec. 2003, Norfolk, VA, USA
[10] Giesing, J.P., Barthelemy, J-F.M., "A summary of industry MDO applications and needs", AIAA 1998-4737, AIAA/USAF/NASA/ISSMO Symposium on Multidisciplinary Analysis and Optimization, Sept. 1998, St. Louis, MO, USA
[11] Chapman, C.B., Pinfold, M. "The application of knowledge based engineering approach to the rapid design and analysis of an automotive structure", Journal of Advances in Engineering Software, Vol. 32, Issue 12, December 2001, pp. 903-912, Elsevier
[12] Liening, A., Blount, G.N., "Influences of KBE on the aircraft brake industry", Aircraft Engineering and Aerospace Technology, Vol. 70, No. 6, 1998, pp. 439-444, MCB University Press
[13] Tarkian, M. Ölvander, J., Feng X., Pettersson M., "Design Automation of Modular Industrial Robots", ASME CIE09, San Diego, USA, Sept. 2009
[14] Hopgood, A. A., Intelligent Systems for Engineers and Scientists, Second Edition, Florida, CRC Press LLC., 2001
[15] Ledermann, C., Hanske, C., Wenzel, J., Ermanni, P., Kelm, R., "Associative parametric CAE methods in the aircraft pre-design", Journal of Aerospace Science and Technology, Vol. 9, Issue 7, October 2005, pp. 641-651, Elsevier
[16] Johanson B., Ölvander J., Pettersson M., "Component Based Modeling and Optimization for Modular Robot Design", ASME DAC'07, Las Vegas, USA, September 4-7, 2007.
[17] Petterson M., Andersson J., Krus P., "Methods for Discrete Design Optimization", proceedings of ASME DETC'05, Design Automation Conference, , Long Beach, California, USA, Sept. 2005.
[18] Sicilano B., Modeling and Control of Robot Manipulators, Springer Verlag, 2001.
[19] Spong W. Mark and Vidyasagar M Robot Dynamics and Control, John Willey & Sons Inc, pp 65-71, 1989.
[20] Goldberg, D., 1989. Genetic Algorithms in Search, Optimization and Machine Learning. Addison Wesley
[21] Deb K., Multi-objective Objective Optimization using Evolutionary algorithms, Wiley and Sons Ltd, 2001.
[22] Fujita K., Yoshida H., Product Variety Optimization Simultaneously Designing Module Combination and Module Attributes, Concurrent Engineering: Research and Applications, 12(2), pp105-118, 2004.
[23] Jiao J., Zhang Y., Wang Y., A Generic Genetic Algorithm for product family design, Journal of Intelligent Manufacturing, DOI:10.1007/s10845/-007-0019-7, 2007.
[24] Deb, K., Pratap. A, Agarwal, S., and Meyarivan, T. A fast and elitist multi-objective genetic algorithm: NSGA-II. IEEE Transaction on Evolutionary Computation, 6(2), 181-197, 2002.

INTERNATIONAL CONFERENCE ON ENGINEERING DESIGN, ICED11
15 - 18 AUGUST 2011, TECHNICAL UNIVERSITY OF DENMARK

ON THE APPLICABILITY OF STRUCTURAL CRITERIA IN COMPLEXITY MANAGEMENT

Wieland Biedermann[1] and Udo Lindemann[1]
[1]Technische Universität München

ABSTRACT
Companies face challenges due to increasing complexity through shorter product life cycles, manifold costumer requirements, more solution options and discipline-spanning collaboration. During the development of complex systems efficient tools for analysis and for assessment of solutions are necessary. A common approach is structural analysis, which can be applied in early development phases. System structures are analyzed with structural criteria such as cycles and clusters. Manifold criteria have been introduced in graph theory and applied in complexity management. In industrial applications suitable criteria have to be chosen. In research the significance of the criteria has to be shown. Based on an extensive literature review we show applications of structural criteria in complexity management. We derive requirements onto structural criteria from the applications. We show methods to prove the applicability of the criteria. Researchers get tools for proving and assessing the significance of structural analyses. More effective analyses can be developed. The quality of technical solutions increases and manifold solutions can be developed.

Keywords: structural complexity management, graph theory, structural analysis, design structure matrix, multiple-domain matrix

1 INTRODUCTION
Companies face challenges due to rising external complexity in engineering design. Reasons are shorter product life cycles, manifold costumer requirements, more solution options due to technological advances and combinations of products and services. Companies react by offering more products and introducing discipline-spanning collaboration. This increases their internal complexity. If complexity is not managed successfully it leads to longer development times, cost overruns and wrong decisions with highly detrimental and long-term consequences [1-3].
Structural considerations are an established approach to manage complexity. One of the most used methods in engineering design is the design structure matrix (DSM) [4]. It has been applied to products, organizations, processes and parameters [5]. Its analytical capabilities have been supplemented by graph theory [1] and network analysis [6]. Its modeling capabilities have been supplemented by the domain mapping matrix (DMM) [7] and the multiple-domain matrix (MDM) [1]. Maurer has proposed a structural approach to deal with complexity in technical systems [1,2].
Manifold structural analysis criteria have been proposed in complex systems research. They are from graph theory [8], network analysis [9], matrix theory [2] and motif analysis [10]. The criteria comprise properties of entire structures like planarity or connectedness, subsets of structures like cycles or clusters, metrics like degree or relational density and visualizations like matrices, graphs or portfolios. Maurer [1] and Kreimeyer [3] have proposed collections of structural criteria. Especially, the introduction of motif analysis has led to an almost infinite variety of structural criteria. The need for careful selection of analysis criteria arises.
Following research questions are addressed in this paper:
- What are applications of structural analysis criteria in complexity management?
- Which requirements of structural analysis criteria arise from the applications?
- How can structural analysis criteria be tested for compliance with the requirements?

The scope of this paper is application of structural analysis criteria in engineering design. We include applications in concept design, process management, project management and organization management but are not limited to them. We exclude applications in production, manufacturing and logistics. We focus on criteria, which can be applied to structural models, which do not comprise node or edge parameters like weights, costs or probabilities. We also exclude criteria based on labeled graphs.

The paper is structured as follows. First, we describe our research approach for the literature survey (section 2). We present an overview of applications for structural criteria in complexity management (section 3). We show requirements arising from the applications and present methods for testing criteria for compliance (section 4). We discuss the results and derive questions for future research (section 5). Finally, we conclude this paper by proposing future research and supporting activities (section 6).

2 METHODOLOGY FOR LITERATURE REVIEW

We follow, like Krishnan and Ulrich [11] and Browning and Ramasesh [12], a loosely structured approach to survey the literature relating to complexity management within the defined scope. We focus on works that identify themselves with the term complexity management or complex system. However, because many papers address similar issues without this term, we also surveyed some non-complexity-specific literature in areas such as project management, and systems engineering, where these fit our scope. First, we created a superset of papers related to structural complexity management through following steps:

1. We searched the tables of contents of the DSM knowledge area of DSMweb [13] from 2001 to 2010, which contains journal articles, books, book chapters, reports, theses and conference proceedings. DSMweb provides access to proceedings of International DSM Conference from 2004 to 2010.
2. We searched the tables of contents of ten major journals from 2006 to 2010: ASME Journal of Mechanical Design, Design Studies, European Journal of Operational Research, IEEE Transactions on Engineering Management, Journal of Engineering Design, Journal of Operations Management, Management Science, Operations Research, Research in Engineering Design, and Systems Engineering. These journals span the engineering design, management science, and operations management areas.
3. We conducted a general search of the literature based on key words, looking also at the broader literature on software engineering, engineering processes, and engineering management.
4. We used the reference lists from highly cited papers.

These steps resulted in a master list of about 300 papers, from which we derived a working list of about 80 papers by filtering out ones that were:

1. outside our scope
2. not in archival publications
3. devoted to software tools, vendors or algorithms
4. presenting case studies without developing new methods

In this paper we present the qualitative results of the survey. Therefore, we omit all quantitative results in the remaining paper.

3 APPLICATIONS OF STRUCTURAL CRITERIA

In this section we present the applications. Table 1 shows the nine applications, which we derived from literature. Each application is assigned to a phase of structural complexity management process as proposed by Maurer [1]. We found applications in four of five phases: data acquisition/modeling, deduction indirect dependencies, structural analysis and discussion of practices. We describe each application by giving an estimation of its commonality, naming its aim and describing the purpose of the structural criteria. For each application we present selected references describing the application or contributing by proofing the significance of a structural criterion.

Steering and controlling of the data acquisition/modeling – This is a rather uncommon application. The aim is to plan the modeling process and to put most of the effort to critical parts of the system model. Improved planning increases the efficiency of the modeling process. The structural criteria are used to identify parts of the model, which are likely to be erroneous, or to estimate the impact of potential errors. Biedermann et al. have proposed a measurement system to improve data acquisition workshops [14].

Model checking for consistency and plausibility – This is a rather uncommon application. The aim is to test the model for errors and to correct them. This increases the model quality. The structural criteria are used to test the model for characteristic properties. Braha and Bar-Yam contribute by identifying characteristic degree distributions of design processes [15]. Shaja and Sudhakar contribute by showing that structural characteristics of components are specific for the type of product [16].

Determining of formulas for deducing indirect dependencies – This is a rather uncommon application. The aim is to derive structural models from models, which are already existing or easier to create. This increases the efficiency of the modeling process. The structural criteria are applied to the meta-model of complex system. Biedermann and Lindemann propose a method to identify computations of DSMs using cycles [17]. Mocko et al. use paths to identify computations of DMMs and DSMs [18].

Table 1: Applications of structural criteria in the structural complexity management approach (partially based on [1])

Phase in the structural complexity management process	Applications
Data acquisition/modeling	Steering and controlling of the data acquisition/modeling
	Model checking for consistency and plausibility
Deduction indirect dependencies	Determining of formulas for deducing indirect dependencies
Structural analysis	Identification of prominent elements, which determine the system behavior and properties
	Identification of system partition, which allows for efficient handling
	Comparison of system architectures
	Deduction of an optimal substructure
	Deduction of consistent system specifications
Discussion of practices	Steering of searches for error causes/Estimation the impact of changes and planning of changes

Identification of prominent elements, which determine the system behavior and properties – This is one of the most common applications. The aim is to identify system elements, which are important to the system behavior and its properties. This improves the handling of the system and leads to optimized systems. The structural criteria are used to rate the elements. Sosa et al. use network metrics to estimate the component modularity in product networks [19]. Kreimeyer uses structural metrics to evaluate engineering design processes [3]. Batallas and Yassine use social network metrics to identify key players in product development networks [21]. Gokpinar et al. estimate the likelihood of quality problems with structural metrics of the product architecture network [22]. Kurtoglu and Tumer use failure paths to remove potential failures and to design capabilities to detect and mitigate failures [23]. Lee et al. use structural metrics to measure the importance of parts and modules for change impacts and propagation [24]. Zakarian et al. use matrix metrics to identify elements, which are relevant for product robustness [25]. Sosa et al. use metrics to estimate quality in software architectures [26].

Identification of system partition, which allows for efficient handling – This is the most common application. The aim is to find a partition of the system. This improves the handling of the system and leads to optimized systems. The two major methods are clustering and partitioning. The structural criteria are used to identify clusters or system partitions or to evaluate the resulting partition. Chen et al. use system partitioning for rapid redesign [27]. Pektas and Pultar use structural properties of parameter networks to determine the optimal decision sequence [28]. Bustnay and Ben-Asher use graph theoretical properties to identify independent subsets of a system [29]. Seol et al. combine clustering and partitioning to derive process modules [30]. Gershenson et al. give an overview of modularity measures [31]. Kusiak and Wang use cycles and strong components for process planning [32]. Yassine gives an overview of structure-based objective functions for partitioning and clustering [33]. Browning surveys partitioning methods and their applications [5].

Comparison of system architectures – This is a common application. The aim is to determine the best system architecture. This improves the quality of the final solution. The structural criteria are used to estimate properties of the system or to compare the system structures. Hofstetter et al. compare structural system models to identify opportunities for commonality [34]. Ameri et al. use network metrics to quantify the design complexity [35]. Gokpinar et al. compare product architecture networks and communication networks to quantify the coordination deficits [22]. MacCormack et al. compare software architectures based structural modularity metrics to evaluate the effect of the mode of organization [36]. Summers and Shah use graph metrics to measure the complexity of design problems and to estimate the effort necessary to solve design problems [37]. Hölttä-Otto and de Weck develop two modularity metrics to characterize systems [38]. Shaja and Sudhakar use structural component characteristics to classify complex products [16]. Browning and Yassine use relational density of project activity networks to choose the appropriate priority rule for resource allocation [39].

Deduction of an optimal substructure – This is an uncommon application. It is a standard application in operations research and management. The structural model describes a network including all potential solutions. The model is often supplemented by parameters like costs. We omit examples, which require weighted network model as they are out of the scope of this paper. The aim is to identify the best solution. This increases the effectiveness and quality of the final solution. The structural criteria are used to identify the substructure or to evaluate the substructure. Cappelli et al. use trees to identify the optimal disassembly sequence [40].

Deduction of consistent system specifications – This is a rather uncommon application. The structural model describes a network including all potential solutions. The aim is to identify all consistent solutions. This increases the effectiveness of the result and the efficiency of the development process. The structural criteria are used to identify system specifications. Braun and Deubzer use cliques in variant management [41]. Hellenbrand and Lindemann use cliques to identify consistent concepts of aircrafts [42]. Gorbea et al. use cliques to identify consistent requirement sets and consistent concepts of hybrid electrical vehicles [43].

Steering of searches for error causes/Estimation the impact of changes and planning of changes – This is an uncommon application. The aim is to support engineers in decision making by highlighting the decision's consequences and in finding root problems by guiding the search. This improves the system handling. In operations research search strategies are commonly applied. Here, we focus on semi-automated search. The structural criteria are used to guide the search by focusing on important elements. Maurer describes the use of active and passive degree in feed-forward analysis and mine seeking [1].

4 APPLICABILITY REQUIREMENTS

In this section we describe the requirement arising from the applications. We show how they are handled in literature and how they are tested.

4.1 Computability

The basic requirement is computability of the criteria based on the structural model. It comprises two sub-requirements. First, the criterion must be defined for the type of structure and an algorithm to compute them must be known and implemented. Criteria can be defined for undirected structures only (e.g. blocks) or for directed structures only (e.g. active degree). Directed structures can be transformed into undirected ones to allow for applying all structural criteria. Transforming undirected to directed structures is not feasible as the results do not reflect the directedness. Second, the criteria must be computable in a given time. The computation time depends on the complexity of the structure, the complexity of the algorithm, the implementation of the algorithm and the computer hardware. The available computation time depends on the project and the analyzing engineer.

There are many tools available for structural analysis [44,45]. The issue of implementation hardly arises. The computation time is not limiting the application of structural criteria in engineering design due to advances in computer hardware and algorithmic graph theory. In the papers describing applications in engineering design the computability of structural criteria is not addressed unless many models are involved [46]. In network theory much larger structures are analyzed and computation time is still an issue [9]. There is no specific testing method for this requirement. It is usually tested by self-assessment.

4.2 Distribution and variety requirements

The distribution and variety requirements refer to the forms, which structural criteria may have. In contrast to the computability they depend on the application and the type of structure. Distribution and variety depend on each other. Low variety correlates with uniform distribution and vice versa. Table 2 shows the applications and the derived requirements. The fulfillment of the requirement depends on the system, on the type of structure and the structural model. Three applications pose no requirement onto the distribution and variety of structural criteria: Determining of formulas for deducing indirect dependencies, deduction of an optimal substructure and deduction of consistent system specifications.

Table 2: Distribution requirements depending on the application of structural criteria

Application	Distribution requirement
Steering and controlling of the data acquisition/modeling	Unequal distribution within the system.
Model checking for consistency and plausibility	Uniform distribution across systems.
Determining of formulas for deducing indirect dependencies	None
Identification of prominent elements, which determine the system behavior and properties	Unequal distribution within the system.
Identification of system partition, which allows for efficient handling	Unequal distribution within the system.
Comparison of system architectures	Unequal distribution across systems.
Deduction of an optimal substructure	None.
Deduction of consistent system specifications	None.
Steering of searches for error causes/ Estimation the impact of changes and planning of changes	Unequal distribution within the system.

Unequal distribution within the system – This is the most common requirement as it applies to four applications – the two most common among them. The criteria must occur in many varieties with unequal frequencies. The rarest and the most extreme forms of the criteria (e.g. the highest criticality) characterize elements, which have outstanding importance for the system. The requirement can be tested within one system model. To prove the general applicability several models have to be tested. In literature this requirement is mostly not explicitly addressed. It is implicitly expected to be fulfilled. Criteria, which do not fulfill the requirement, are generally omitted.

Uniform distribution across systems – This requirement applies to criteria for the uncommon application of model checking. The criteria must occur at low variety in all systems of the same type. There is some literature on characteristic properties of complex systems in general. There is hardly any literature on the characteristics of structures, which occur in engineering design. The fulfillment of the requirement can be tested by analyzing a significant proportion of all systems of the same type. The test also depends on the meta-model and the modeling process.

Unequal distribution across systems – This requirement applies to structural criteria for the common application of architecture comparison. The criteria must occur in high variety across systems of the same type. The distribution of the criteria must significantly differ among the systems. The requirement can be tested by comparing a few system models. In the literature this requirement is hardly addressed. It is expected to be fulfilled. Criteria not fulfilling the requirement are generally not presented.

4.3 Significance and relevance requirements

These requirements are the most important. The fulfillment of the two other groups of requirements is necessary but not sufficient for a criterion to be applicable. The requirement is fulfilled if the criterion allows for describing or estimating a system property, which is relevant for the application. If the purpose is reduction of development time the criterion must e.g. correlate with the process duration. If the purpose is increasing product quality the criterion must e.g. correlate with error frequencies. The requirement is addressed in about half of the papers. We found four methods in literature, which have been applied to test and proof the fulfillment: analogy, comparison, simulation and statistical analysis. We describe each method by presenting its rationale, an example and its major challenges and limitations.

Analogy of the criterion and a known phenomenon – This approach builds an analogy between the criterion and a known phenomenon. The implications, properties and effects of the phenomenon are transferred to the criterion. The significance and relevance of the criterion is correlated with the phenomenon's properties. Kusiak and Wang [32] use this approach to develop a structure-based sequencing method. They show an analogy between iterations and cycles in the activity network. Iterations are repetitions of activities and tasks. Cycles are close sequences of information flows among activities. Iterations tend to increase the process duration and the planning uncertainty. Cycles inherit these properties. Efficient dealing with cycles allows for better handling of iterations. Removal of cycles lowers the risk of iterations. The analogy approach does not allow for quantified structural analyses as only tendencies but not quantified parameters are inherited.

System structure comparison – In this approach exemplary structures are created, which possess extreme structural properties. They are expected to represent ideal systems with pure characteristics without trade-offs as they occur in real engineering systems. The structures are compared to real systems. The differences can be quantified to measure the real system's properties in relation to the ideal systems. Hölttä-Otto and de Weck [38] use this approach to measure the degree of modularity of engineering systems and products. They define three exemplary (or canonical in their terms) structures: integral, bus-modular and modular. They compare these with real system structures. They also compare pairs of systems with the same functionality but different technological constraints. Highly-constraint systems tend to be more integrally modularized. Hölttä-Otto and de Weck also include random structures to show that real engineering have significantly different structural characteristics. Comparing real structures to randomly created ones is a common research approach in network theory [9]. The approach allows for semi-quantified result. They are limited to measuring the differences to the exemplary structures. They cannot measure system properties directly as the properties of the exemplary structures are not quantified. The main challenges are to find appropriate reference structures, and to reliably determine the real structures.

Simulation – In this approach simulation models are derived from structural models. The simulation results are compared to structural criteria. The significance and relevance of criteria are shown by correlating them with significant and relevant simulation results. Browning and Yassine [39] use this approach to evaluate priority rules for resource allocation in multi-project environments. They show that relational density of activity networks is one of three criteria to choose appropriate priority rules. They achieved this result by synthesizing and simulating 12,320 project set-ups. The variations and means of the simulation results were analyzed. The analyses showed a significant correlation between relational density and the appropriate choice of priority rules. The simulation approach allows for quantified structural analyses. The main challenge is to create simulation models, which cover the complete parameter space. Both, the space of the potential structures and the space of the simulation models have to be explored.

Statistical analysis – In this approach the statistical relation between structural criteria and system properties is determined. If the results are statistically significant the structural criteria are significant as well. The relevance of the criteria depends on the relevance of the system properties. Sosa et al. [26] use this approach to show the connection between coupling in the component structure of software systems and the quality of the software system. They analyze the structures of 20 software systems (in 108 versions in total). For each version they compare the number of bugs and the number of resolved bugs with the coupling (e.g. in form of cycles) within the structure of the previous version. They show that high actual coupling (originating from the architecture) increases the number of bugs and that high intrinsic coupling (originating from the organization of the engineering project) decreases the capability to fix bugs. The statistical analysis approach allows for quantified results. The main challenges are to determine enough structures to be statistically significant, to reliably create the structural models, to determine the system properties independently of the structure and to avoid hidden parameter biases.

5 DISCUSSION

The results of the literature review comprise nine applications of structural criteria in complexity management, the requirements onto them and an overview of methods to test them.

We identified nine applications, which occur in four out of five phases in the structural complexity management approach. Five applications occur during structural analysis – the three most common among them. The applications in data acquisition (or modeling), deduction of indirect dependencies and discussion of practices are rather uncommon. This unequal distribution across the phases is a consequence of the primary focus of complexity management. Structural models are primarily a tool for system analysis. This is their original purpose [4]. Most of the subsequent research focused on it. Applications during modeling are uncommon; this is a result from the tendency in the literature to omit the model creation from the description and possibly consideration. A lack of support during modeling is mentioned by industrial appliers of structural analysis [47]. Researchers have recognized this lack as well. Participants of the 2010 International DSM Conference voted data acquisition the prime research topic in structural complexity management [48]. The rare application in discussion of practices can be explained by the focus of complexity management. Most applications support tasks during the planning and concept phases of product development. Module definition and initial project or process planning are typical task associated with complexity management. Applications in engineering day-to-day business have hardly been addressed.

We define three categories of requirements for the applicability of structural criteria: computability, distribution and variety, and relevance and significance. Computability and variety are mostly not discussed in the literature. They are expected to be fulfilled. Most papers only present criteria, which fulfill the requirements. In some cases (e.g. [16]) criteria are applied, which are not sensibly computable in the use case. As discussed in section 4.1 criteria for directed networks should not be applied to undirected networks. Most applications do not pose critical variety requirements onto the criteria. The notable exception is model checking, which requires low variety across systems of the same type. This application is uncommon. One reason is that structural models of many systems are needed to test the variety requirement. The third group of requirements comprises relevance and significance of the criteria. We present four methods to test these requirements and four exemplary applications (one for each method) in detail. The methods analogy and comparison test the criteria qualitatively. The methods simulation and statistical analyses test the criteria quantitatively. Most papers use qualitative methods. They can be applied at little expense as only one model is required. The quantitative methods require models of many systems to gain significant results. The methods allow for results, which can be used for optimization of systems. The rarity of quantitative results indicates a lack of models to test structural criteria.

6 CONCLUSION

The survey results show that there is a gap in applications in modeling and a gap in quantitative results based on structural criteria. According to the survey in [48] supporting the data acquisition is one of most pressing issues in complexity management research. The existing proposals (e.g. [14,17]) show promising results but need to be extended and validated. Optimizing systems based on their structural properties is a pressing issue as well. Optimization requires objective functions and quantitative measurements of the system. The few quantitative results (e.g. [26,39]) show that applying structural criteria allows for measuring relevant system properties. Yet, there are hardly any results available in literature. We derive two research questions from these results:
- How can structural modeling of complex systems be supported to become more efficient?
- How can researchers be supported in creating, testing and proofing quantitative structural analysis approaches?

We propose three measures to close the existing gaps and to answer the research questions:
- A collection of structural models of engineering systems, which serves as a reference set for testing structural criteria
- A collection of characteristic structural properties of engineering systems, which supports model checking and serves as a basis for example structure synthesis
- A tool for creating exemplary, characteristic and random structures, which serves as a base for simulation analyses of complex systems

Our results show that structural criteria are widely applied in complexity management. We show the requirements arising from the application and how they are dealt with in research. We support researchers in finding the right method to test their hypotheses. We support appliers to find the right criterion to analyze their systems. We show the limitations of the application and the research results. We propose measures to overcome the limitations. Thereby, the quality of the research results will rise. New and better tools to analyze complex systems will be developed and improve complex products and their development.

ACKNOWLEDGEMENTS
This research was made possible through the generous funding by the German Research Foundation (DFG) in the context of the project A2 ("Modelling and analysis of discipline-spanning structural criteria and their impacts on product development processes") within the Collaborative Research Centre SFB 768 ("Managing cycles in innovation processes").

REFERENCES
[1] Maurer M. Structural Awareness in Complex Product Design, 2007 (Dr.-Hut, Munich).
[2] Lindemann U., Maurer M. and Braun T. Structural Complexity Management - An Approach for the Field of Product Design, 2009 (Springer, Berlin).
[3] Kreimeyer K. A Structural Measurement System for Engineering Design Processes, 2010 (Dr.-Hut, Munich).
[4] Steward D.V. Design Structure System: A Method for Managing the Design of Complex Systems. IEEE Transactions on Engineering Management, 1981, 28(3), 71-74.
[5] Browning T.R. Applying the design structure matrix to system decomposition and integration problems: a review and new directions. IEEE Transactions on Engineering Management, 2001, 48(3), 292-306.
[6] Collins S.T., Yassine A.A. and Borgatti S.P. Development Systems Using Network Analysis. Systems Engineering, 2008, 12(1), 55-68.
[7] Danilovic M. and Browning T.R. Managing Complex Product Development Projects with Design Structure Matrices and Domain Mapping Matrices. International Journal of Project Management, 2007, 25(3), 300-314.
[8] Gross J.L. and Yellen J. Graph Theory and Its Applications, 2005 (CRC Press, Boca Raton).
[9] Cami A. and Deo N. Techniques for Analyzing Dynamic Random Graph Models of Web-Like Networks: An Overview. Networks, 2008, 51(4), 211-255.
[10] Milo R., Shen-Orr S., Itzkovitz S., Kashtan N., Chklovskii D. and Alon U. Network motifs: simple building blocks of complex networks. Science, 2002, 298(5594), 824-827.
[11] Krishnan V. and Ulrich K.T., Product development decisions: A review of the literature. Management Science, 2001, 47(1), 1-21.
[12] Browning T.R. and Ramasesh R.V. A Survey of Activity Network-Based Process Models for Managing Product Development Projects. Production and Operations Management, 2007, 16(2), 217-240.
[13] Kreimeyer M., The New Community Portal DSMweb.org. In Proceedings of the 11th International DSM Conference, Greeneville, October 2009, pp. 9-12 (Hanser, Munich).
[14] Biedermann W., Kreimeyer M. and Lindemann U. Measurement System to Improve Data Acquisition Workshops. In Proceedings of the 11th International DSM Conference, Greeneville, October 2009, pp. 119-130 (Hanser, Munich).
[15] Braha D. and Bar-Yam Y. The Statistical Mechanics of Complex Product Development: Empirical and Analytical Results. Management Science, 2007, 53 (7), 1127-1145.
[16] Shaja A.S. and Sudhakar K. Classifications of Systems from Component Characteristics. In 20th Anniversary INCOSE International Symposium, Chicago, July 2010. (International Council On Systems Engineering).
[17] Biedermann W. and Lindemann U. Cycles in the Multiple-Domain Matrix - Interpretation and Applications. In Proceedings of the 10th International DSM Conference, Stockholm, November 2008, pp. 25-34 (Hanser, Munich).
[18] Mocko G.M., Fadel G.M., Summers J.D., Maier J.R.A. and Ezhilan T. A Systematic Method for Modelling and Analysing Conceptual Design Information. In Proceedings of 9th International DSM Conference, Munich, October 2007, pp. 297-309 (Shaker, Aachen)
[19] Sosa M.E., Eppinger S.D. and Rowles C.M. A Network Approach to Define Modularity of Components in Complex Products. Journal of Mechanical Design, 2007, 129(11), p. 1118-1129.

[20] Sosa M.E., Agrawal A., Eppinger S.D. and Rowles C.M. Network Approach to Component Modularity. In 7th International Design Structure Matrix Conference, Seattle, 2005 (The Boeing Company, Seattle).
[21] Batallas D.A. and Yassine A.A. Information Leaders in Product Development Organizational Networks: Social Network Analysis of the Design Structure Matrix. IEEE Transactions on Engineering Management, 2006, 53(4), 570-582.
[22] Gokpinar B., Hopp W.J. and Iravani S.M.R. The Impact of Misalignment of Organizational Structure and Product Architecture on Quality in Complex Product Development. Management Science, 2010, 56(3), 468-484.
[23] Kurtoglu T. and Tumer I.Y. A Graph-Based Fault Identification and Propagation Framework for Functional Design of Complex Systems. Journal of Mechanical Design, 2008, 130(5), 051401.
[24] Lee H., Seol H., Sung N., Hong Y. and Park Y. An analytic network process approach to measuring design change impacts in modular products. Journal of Engineering Design, 2010, 21(1), 75-91.
[25] Zakarian A., Knight J. and Baghdasaryan L. Modelling and analysis of system robustness. Journal of Engineering Design, 2007, 18(3), 243-263.
[26] Sosa M.E., Browning T.R. and Mihm J. A Dynamic, DSM-based View of Software Architectures and Their Impact on Quality and Innovation. In Proceedings of the 10th International DSM Conference, Stockholm, November 2008, pp. 313-325 (Hanser, Munich).
[27] Chen L., Macwan A. and Li S. Model-based Rapid Redesign Using Decomposition Patterns. Journal of Mechanical Design, 2007, 129(3), 283-294.
[28] Pektas S.T. and Pultar M. Modelling detailed information flows in building design with the parameter-based design structure matrix. Design Studies, 2006, 27(1), 99-122.
[29] Bustnay T. and Ben-Asher J.Z. How many systems are there? Using the N2 method for systems partitioning. Systems Engineering, 2005, 8(2), 109-118.
[30] Seol H., Kim C., Lee C. and Park Y. Design Process Modularization: Concept and Algorithm. Concurrent Engineering, 2007, 15(2), 175-186.
[31] Gershenson J.K., Prasad G.J. and Zhang Y. Product modularity: measures and design methods. Journal of Engineering Design, 2004, 15(1), 33-51.
[32] Kusiak A. and Wang J. Decomposition of the Design Process. Journal of Mechanical Design, 1993, 115(4), 687-695.
[33] Yassine A.A. Multi-Domain DSM: Simultaneous Optimization of Product, Process & People DSMs. In Proceedings of the 12th International DSM Conference, Cambridge, July 2010, pp. 319-332 (Hanser, Munich).
[34] Hofstetter W.K., Wooster P.D., de Weck O.L. and Crawley E.F. The System Overlap Matrix – A Method and Tool for the Systematic Identification of Commonality Opportunities in Complex Technical Systems. In Proceedings of 9th International DSM Conference, Munich, October 2007, pp. 215-224 (Shaker, Aachen).
[35] Ameri F., Summers J.D., Mocko G.M. and Porter M. Engineering design complexity: an investigation of methods and measures. Research in Engineering Design, 2008, 19(2-3), 161-179.
[36] MacCormack A., Rusnak J. and Baldwin C.Y. Exploring the Structure of Complex Software Designs: An Empirical Study of Open Source and Proprietary Code. Management Science, 2006, 52 (7), 1015-1030.
[37] Summers J.D. and Shah J.J. Mechanical Engineering Design Complexity Metrics: Size, Coupling, and Solvability. Journal of Mechanical Design, 2010, 132 (2), 021004.
[38] Hölttä-Otto K. and de Weck O. Degree of Modularity in Engineering Systems and Products with Technical and Business Constraints. Concurrent Engineering, 2007, 15(2), 113-126.
[39] Browning T.R. and Yassine A.A. Resource-constrained multi-project scheduling: Priority rule performance revisited. International Journal of Production Economics, 2010, 126(2), 212-228.
[40] Cappelli F., Delogu M., Pierini M. and Schiavone F. Design for disassembly: a methodology for identifying the optimal disassembly sequence. Journal of Engineering Design, 2007, 18(6), 563-575.
[41] Braun T. and Deubzer F. New Variant Management Using Multiple-Domain Mapping. In Proceedings of 9th International DSM Conference, Munich, October 2007, pp. 363-372 (Shaker, Aachen).

[42] Hellenbrand D. and Lindemann U. Using the DSM to support the selection of product concepts. In Proceedings of the 10th International DSM Conference, Stockholm, November 2008, pp. 363-374 (Hanser, Munich).
[43] Gorbea C., Hellenbrand D., Srivastava T., Biedermann W. and Lindemann U. Compatibility Matrix Methodology Applied to the Identification of Vehicle Architectures and Design Requirements. In DESIGN 2010 - Proceedings, Dubrovnik, May 2010, pp. 733-742 (Design Society).
[44] Wynn D.C., Nair S.M.T. and Clarkson P.J. The P3 Platform: An approach and software system for developing diagrammatic model-based methods in design research. In Proceedings of ICED'09, Volume 2, Design Theory and Research Methodology, Stanford, August 2009 (Design Society).
[45] Lau H.T. A Java Library of Graph Algorithms and Optimization, 2006 (CRC Press, Boca Raton).
[46] Browning T.R. and Yassine A.A. A random generator of resource-constrained multi-project network problems. Journal of Scheduling, 2009, 13(2), 143-161.
[47] Herfeld U. From the Real Product to Abstract Architecture and Back Again. In Proceedings of the 10th International DSM Conference, Stockholm, November 2008, p. xv (Hanser, Munich).
[48] Wynn D.C., Kreimeyer M., Eben K., Maurer M., Lindemann U. and Clarkson J., eds. Proceedings of the 10th International DSM Conference, 2010 (Hanser, Munich).

Contact:
Wieland Biedermann
Technische Universität München
Institute of Product Development
Boltzmannstr. 15
D-85748 Garching, Germany
Phone +49 89 289-15129
Fax +49 89 289-15129
biedermann@pe.mw.tum.de
http://www.pe.mw.tum.de

Wieland Biedermann is a scientific assistant at the Technische Universität München, Germany, and has been working at the Institute of Product Development since 2007. He has published several papers in the area of structural complexity management.
Udo Lindemann is a full professor at the Technische Universität München, Germany, and has been the head of the Institute of Product Development since 1995, having published several books and papers on engineering design. He is committed in multiple institutions, among others as Vice President of the Design Society and as an active member of the German Academy of Science and Engineering.

ECO TRACING - A SYSTEMS ENGINEERING METHOD FOR EFFICIENT TRACELINK MODELLING

Rainer Stark[1,2] and Asmus Figge[1]
(1) Technische Universität Berlin, Germany (2) Fraunhofer IPK, Germany

ABSTRACT
Using expertise and combining functionalities from different domains has led to a significant increase of information engineers have to deal with. It is hardly possible to identify influenced components in activities like change requests. A model containing tracelinks between the elements of involved partial models as an essential part of PLM based Systems Engineering helps to overcome this deficit. The main obstacle for a broad introduction of traceability is the significant workload involved in creating tracelink models as every element combination has to be examined for dependencies. This calls for an approach to support developers in creating tracelink models more efficiently.
The presented approach *Eco Tracing* allows developers to significantly *Eco*nomize modelling effort. In order to do so, the method uses the hierarchical structure of many models and a top-down analyzing approach to exclude element combinations prior user examination. Furthermore *Eco Tracing* allows choosing the desired level of detail flexibly while modelling. *Eco Tracing* is a promising approach helping to establish traceability in product development by reducing modelling effort significantly.

Keywords: traceability, dependency modelling, product development, mechatronics

1 INTRODUCTION AND MOTIVATION

A challenge many companies face today is the increasing complexity of products. Complexity is defined as a property of a system depending on the amount of system elements, the number of relations between them and the multitude of its possible states [1]. Regarding mechatronic product development there are numerous sources for complexity, such as the product's augmented, cross-domain functionalities.
Embedding and integrating functionalities from different domains is a major source of innovation [2], but on the other hand it has led to a significant increase in information that developers have to deal with. It is impossible to oversee all implications of activities like change requests since influenced components from other domains cannot be identified. Thus, changes in complex systems are very difficult to handle [3]. This is especially challenging since changes occur very often: usually more than 50% of system's requirements get modified before it is eventually put into service [4]. Nonetheless there is little tool and methodological support for that challenge.
During product development several partial models are created (e.g. function structure) all describing the final product. Between some elements of these partial models there is a dependency with regards to their content – e.g. function "cool passenger compartment" satisfies requirement "provide means for the regulation of passenger compartment temperature". An approach to deal with the mentioned challenges is the continuous linkage between all partial models. Changes can be propagated efficiently and implications can be detected easily based on dependencies between partial models. These dependencies can be represented and documented through tracelinks, which mainly contain information about the linked source and target model elements. The degree to which a relationship can be established between those models is called traceability [3]. Especially in software engineering traceability is viewed as a measure of system quality and process maturity and is mandated by many standards such as MIL-STD-498, IEEE/EIA 12207, and ISO/IEC 12207. But also in Systems Engineering tracelink models can be of significant assistance. For example by automatically filling in information into QFD or FMEA spreadsheets helping to ease the use of established but infrequently used quality and reliability methods.
The main obstacle for a broad introduction of traceability in system development is the significant workload involved when creating and maintaining tracelinks as well as the discrepancy between the persons modelling and using tracelinks [3]. In 2000 the US Department of Defence spent about 4% of

their system development costs on traceability [5], emphasising the necessity to have a good and efficient traceability strategy [6].

The situation described calls for an approach to support product developers with methods and tools helping to reduce the necessary effort for primarily building up and controlling complex systems and dependencies between their comprising models. For this reason, the approach *Eco Tracing* presented in this paper focuses on how to use hierarchical characteristics to model tracelinks in an efficient way. It has been integrated in the software prototype Model Tracer, which is a tool for tracelink modelling [7].

In this paper, the state of the art and research is described in section 2 and the shortcomings in the area of tracelink identification are discussed in order to illustrate the motivation for *Eco Tracing*. Section 3 introduces the Model Tracer, which is a tool for tracelink modelling. During the project ISYPROM the *Eco Tracing* method was integrated into the Model Tracer and tested in an industrial environment. In section 4, *Eco Tracing*, its basic principles, its effectiveness as well as its implementation are detailed. Section 5 provides an overview over the application of the *Eco Tracing* method in context of a systems engineering example. Section 6 concludes the findings presented in this paper and hints at further research in the area of tracelink identification and maintenance.

2 STATE OF THE ART AND RESEARCH

Procedure models are supposed to help developers coping with complexity in product development. Cooperation between different domains or departments is promoted through their application [8]. In most design steps models are created (e.g. requirements model, functional model, structural model), which represent different development perspectives on the same system. These models are developed based on each other. In this way, e.g. functions of a system are derived from the requirements, which have been defined before. Since this happens implicitly, dependencies between partial models are usually not documented [9].

A way to cope with this problem is to establish traceability between different partial models by explicitly linking the information contained in the models. Maurer proposes the use of matrices for the connection of information. He differentiates between Design Structure Matrices (DSM), Domain Mapping Matrices (DMM) and Multi Domain Matrices (MDM), which are a combinatorial advancement of the first mentioned. Goal of his studies is to analyse, control and improve the dependencies of complex systems. For this purpose he suggests a number of analysing techniques that allow for the identification of connected structures by rearranging matrices or the tracing of impact chains. For visualisation purposes he uses matrices or strength based graphs, if the number of connected elements becomes too large. Both visualisations can be derived from each other and are included in the commercial software Loomeo [10], [11].

The Change Prediction Method (CPM) tool developed at the Engineering Design Centre in Cambridge combines different visualisation techniques that help understanding a system. The CPM tool aims at decision assistance by providing effective and intuitive information visualisation regarding the prediction of change propagation [12].

METUS by ID-Systems provides functionalities to specifically model functions, subfunctions, components and modules of a product as well as dependencies between their elements. These tracelinks are used by several analysis functions in order to e.g. optimize costs or the weight of the system in focus. To reduce the work to model and analyse the system, a PLM integration has been implemented, which enables METUS to acquire product information [13], [14].

ToolNet, which was developed in a research project by DaimlerChrysler and EADS, completely follows this approach of acquiring information from existing sources. Partial models are created in the established authoring software while only tracelinks between the models are created and stored in ToolNet [15].

A complementary approach is followed by Reqtify. In this approach, tracelinks are not managed in an additional application but stored in the original documents with the help of references to elements of other documents. Reqtify is then used to visualise these tracelinks, perform several analyses and to export reports [16].

All described traceability approaches aim at an improvement of system's understanding and model consistency during the development. This means that there are a lot of ways to use the already modelled tracelinks in a beneficial way. But none of the presented approaches propose any effective solutions on how to identify the dependencies between the partial models' elements in order to model the tracelinks in the first place. Even though especially the manual creation and the effort necessary to achieve high connection quality (avoidance of wrong and missing dependencies) still pose high challenges for their application [3], [10]. Present suggested approaches for this problem, like the use of interdisciplinary workshops or the collection of existing data (e.g. QFD, TRIZ) still imply high efforts. Only in software development there are some approaches to generate tracelinks automatically, e.g. as the result of model transformation or the statistical interpretation of change histories in order to identify dependencies between items [3]. But these approaches either have not yet been adapted to challenges posed in mechatronic system development or can't be easily applied to it because of different procedures (e.g. no model transformations). For this reason it is necessary to offer more methods and tools that help limiting the additional work effort in order to further integrate the creation of tracelinks in today's system development reality.

3 MODEL TRACER – A TOOL FOR DEPENDENCY MODELLING IN SYSTEMS ENGINEERING

In order to cope with the mentioned challenges, a prototypical software tool was developed – called Model Tracer. The general approach of Model Tracer provides means to define tracelinks between different partial models (Figure 1). Each qualitative tracelink represents a general dependency between a pair of elements from two different partial models [7].

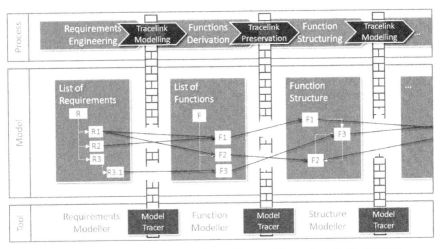

Figure 1: Process steps and models enhanced by tracelinks created with Model Tracer.

Since different partial models are not necessarily created with the same tool it is important to allow tracelink modelling while avoiding to introduce "just another tool" that aims at replacing well established authoring tools.

Therefore Model Tracer acquires all models from their specific authoring tools and visualises them in its own graphical user interface (GUI) as shown in Figure 2. During the project ISYPROM interfaces for common standards like ReqIF (Requirements Interchange Format) and PLMXML have been implemented to exemplarily prove feasibility. Basically the approach allows for importing and using any hierarchical model in Model Tracer if an interface is implemented. Model Tracer only saves information regarding the tracelinks and references to the connected models.

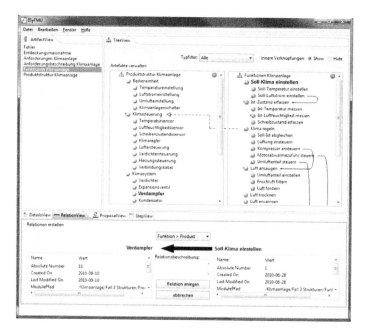

Figure 2: GUI of Model Tracer

With every start of Model Tracer, partial models are loaded from their proprietary databases. Modified and new elements are highlighted which helps guaranteeing actuality of data. As mentioned before, especially the laborious identification of dependencies is one of the biggest challenges and the main hindrance for an industrial application of methods to connect information.

4 ECO TRACING - A SYSTEMS ENGINEERING METHOD TO MODEL TRACELINKS EFFICIENTLY

To increase efficiency in tracelink modelling in the course of Systems Engineering two basic categories have to be considered. There are technological approaches using diverse algorithms to identify dependencies between elements; this is used for example in model transformation or automatic tracelink recovery [3]. On the other hand there are methodical approaches which help users to create tracelinks efficiently. The latter can be especially useful when initially modelling dependencies. For that activity a structured procedure is needed to make sure all dependencies are identified. A methodical approach to *Eco*nomize tracelink modelling is introduced in this paper: *Eco Tracing*. In Section 5 details on how to apply *Eco Tracing* in Systems Engineering will be discussed.

4.1 Basic Principles for Eco Tracing

The analysis of other approaches in the field of system development reveals that traceability is rarely used in industry yet. A major reason for its lack of application is the high effort needed for the manual creation of tracelinks [12], [10]. A widely used approach makes use of interdisciplinary workshops where all possible combinations of elements of both partial models have to be considered individually. For each pair of elements a decision regarding their dependency has to be made [10]. Since many combinations of elements do not feature a dependency, their examination is unnecessary and thus causing additional costs. This means the biggest challenge is to identify the majority of elements, which do not have to be analysed, in advance. Furthermore, a high flexibility in choosing the necessary level of detail is desirable when modelling tracelinks. This flexibility gives users the opportunity to decide in which areas to model in detail and where to stop at a high hierarchy level to avoid high efforts.

With this in mind, characteristics of nested hierarchical structures, which many models feature (e.g. requirements, functions or assemblies), in combination with a consequent top-down-approach can be used to exclude independent element combinations: In nested hierarchical structures, parent elements consist of or contain child elements (e.g. assemblies consist of parts) [17]. This leads to the conclusion that no child element depends on a certain element, if the dependency was dismissed for its parent element (which consists of its child elements).

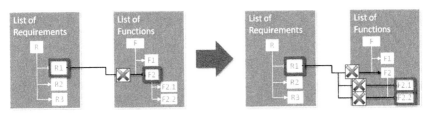

Figure 3: If there is no dependency between R1 and F2, no dependency exists between R1 and all child elements of F2.

Figure 3 is illustrating this conclusion: if there is no dependency between requirement R1 in the list of requirements and function F2 in the list of functions, there will be no dependency between R1 and sub-functions F2.1 and F2.2 either.

This conclusion has been applied to requirements models, functions structures and product structures successfully. Since it is the main principle *Eco Tracing* is based on, it is recommendable to verify its applicability for any new model prior application.

In the following section *Eco Tracing* is described in detail.

4.2 The Method Eco Tracing

As mentioned before *Eco Tracing* is a top-down-approach, meaning that users are starting every analysis at a high hierarchical level and work their way down to lower levels. In the example illustrated in Figure 4 the examination of dependencies following the *Eco Tracing* approach would therefore start with the combination of A1 and all parent elements in model B (B1, B2, and B3). If no dependency is identified between A1 and B1, no tracelink is set (illustrated by a red cross in Figure 5). Since parent elements contain or consist of their child elements, no dependencies from A1 to any of B1's child elements are possible. That is why these combinations do not have to be analysed at all. To maintain this knowledge for later reference all combinations from A1 to every child element of B1 are marked with a flag meaning they have been examined for dependencies and dismissed (illustrated by a red cross in Figure 5).

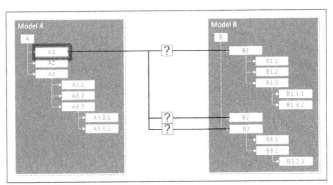

Figure 4: Examination for dependencies.

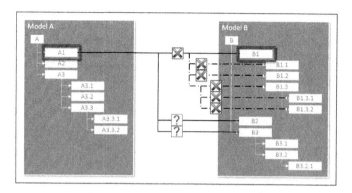

Figure 5: Flagging dismissed dependencies.

If a dependency is approved, a tracelink is set for this combination illustrated by a green check mark in Figure 6. The conclusion that "no child element depends on a certain element, if the dependency was dismissed for its parent element" is now inverted: it is very likely that at least one child element of B3 features a property that has led to the affirmation of the dependency to A1. Therefore all possible combinations of A1 and all direct child elements of B3 (B3.1 and B3.2) are flagged (illustrated by green flags) meaning at least one dependency exists within those combinations (Figure 6).
With the help of the flags it is possible to differentiate between:
- Combinations of elements which have been examined and were rejected,
- Combinations of elements which have not been examined yet,
- Combinations of elements where dependencies are likely but no detailed examination has been performed.

This differentiation of states is especially important as it enables users to discontinue the examination for dependencies at any time, allowing them to continue at a later time or to finish at any level of detail.

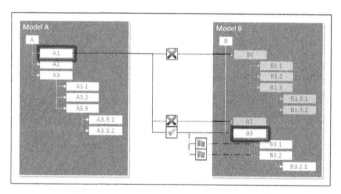

Figure 6: Setting a tracelink and automatically flagging its child elements.

After investigating all combinations of A1 and Bx the approach proceeds with the investigation of all green flagged combinations of A1 and Bx.x (in this case A1 – B3.1 and A1 – B3.2). When traversed all the way to the lowest user-chosen hierarchy, the investigation restarts with the next parent element in model A (in this case A2) from the beginning.

4.3 Eco Tracing vs. conventional Dependency Examination

In this section *Eco Tracing* is compared to the conventional way for dependency examination (e.g. workshops [10]) with the help of a simple example. This example is illustrated in Figure 7 and comprises two models with a total of 19 elements (white blocks) and 17 dependencies (red lines). For the purpose of calculation it is assumed that during the application of the conventional method all element combinations have to be examined, even though in praxis the analysis of some combinations could probably be dismissed by the experts.

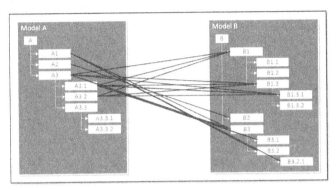

Figure7: Dependencies between model A and model B.

Table 1 shows the same example as Figure 7 in a different visualisation type and with some additional information. All rows contain elements of model A whereas the columns comprise model B. It is assumed all elements are of interest and are examined. If a dependency between two elements exists a '1' is entered in their common cell otherwise a '0'. All combinations examined with *Eco Tracing* are in white cells, automatically dismissed ones in grey cells.

The conventional way means to examine every combination of elements from model A with those from model B. Since there are eight elements in model A and eleven elements in model B, *88 combinations* have to be checked for dependencies.

If *Eco Tracing* is applied not all combinations have to be checked since several dependencies are already dismissed on a higher hierarchical level. That is why only *32 combinations* have to be checked (number of white cells in Table 1) which means the effort for examining *56 combinations* is saved.

Table 1: Comparison between Eco Tracing and a conventional method for dependency examination.

	B1	B1.1	B1.2	B1.3	B1.3.1	B1.3.2	B2	B3	B3.1	B3.2	B3.2.1
A1	0						0	1	0	1	1
A2	0						1	0			
A3	1	0	0	1	1	0	0	1	1	0	
A3.1	1			1	1	0		1	1		
A3.2	1			1	1	0		0			
A3.3	0							0			
A3.3.1											
A3.3.2											

Under the assumptions made, the saving is about 60 %. But the effort which can be saved by *Eco Tracing* depends on the hierarchical structure of the models and in which hierarchy levels the dependencies exist. For example, if a parent element has few child elements, the effort possible to be saved is less than if it would have many child elements. Therefore no general conclusion about the effectiveness of the method can be quantified.

4.4 Eco Tracing Prototype

Eco Tracing as described in Section 4.2 is implemented as a plug-in of Model Tracer introduced in Section 3. The aim of *Eco Tracing* is to guide users in a wizard-like environment through the process of examining dependencies while reducing the necessary effort.

When starting *Eco Tracing* a selection dialog is displayed. In a first step parts of model A need to be selected. Based on this selection and under consideration of the flags that have been set before, a reduced view of model B is displayed. User support is given by excluding all elements of B that have already been examined or cannot have a dependency to one of the selected elements of model A, reducing the overall amount of information. Additionally users can further limit their selection within this reduced view of model B.

This selection dialog is especially helpful, when examining combinations of very complex models with a large quantity of elements. When for example dependencies between requirements and functions are examined it is possible to choose only the functional requirements, skipping the non-functional requirements and thus increase clarity.

Once a selection is made dependency examination begins. Figure 8 shows a screenshot of *Eco Tracing* in use. In the left and right column the selected models are displayed. The elements comprising the combination being examined at the respective moment are highlighted in green. For every combination users have three possibilities:

- Skip: Skipping the examination regarding a dependency between the two elements. All combinations of their child elements will also not be examined.
- Dismiss: If there is no dependency between the elements, the tracelink is dismissed. All combinations of their child elements will also be dismissed.
- Approve: If there is a dependency between the elements, a tracelink is set.

For motivational purposes a progress bar has been added visualising the rapidly growing ratio of already examined and overall combinations.

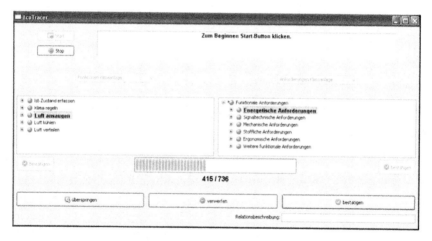

Figure 8: Examination of dependencies.

Eco Tracing provides users with the opportunity to economize their modelling efforts significantly. Numerous element pairs can be excluded from examination before users even get to see them. Users are also given the possibility to choose the level of detail for tracelink modelling on the fly - skipping entire parts of a model if desired. By allowing different levels of detail within one and the same model combination, dependency modelling can be realised as accurate as necessary with as little effort as possible. Through the guided procedure implemented in the *Eco Tracing* wizard users cannot get lost. They are systematically guided through all possible combinations of elements selected and always get provided with relevant context information. Once a decision has been made *Eco Tracing* saves this information to preclude redundant modelling work and to allow cancelling the process at any time.

5 APPLICATION IN SYSTEMS ENGINEERING

In Figure 9 an interdisciplinary system development process is illustrated. It is an adaption of the V-model according to VDI 2206 [2]. In the early phases of system development (represented by the left side of the V-model) a large amount of different models, determining the system's properties, is developed. Furthermore, designing is an iterative process and prone to numerous changes. That is why the potential for cost savings with the help of traceability, once the tracelinks are modelled, is especially high. *Eco Tracing* can efficiently provide tracelink information enabling continuity in early design phases, which are shortly described in the following paragraphs.

In the phase of *Product Planning* an alignment of the planned product with the business strategy is carried out. Business and system requirements are developed. During *System Architecture Reflection* the predecessor's system architecture is used as an initial point for discussions about the new product. Innovations and new systems are roughly integrated. These changes and plans for the new product are consolidated in the *Requirements* phase into new and changed high-level requirements. These are further detailed during the *Requirements Cascade* until they can be allocated to product functions. In order to satisfy all requirements it is necessary to add, remove, change and restructure functions during the *Development of Function Structure*. In the *Development or Adjustment of System Architecture* new solution elements for executing functions are added, out-of-date ones are removed or changed and system boundaries are defined. The last step of the *Continuous System Design*, is the *Partitioning of Models*. All models developed during preceding phases (requirements, functions, solution elements and behaviour models) are partitioned to the disciplines: mechanics, electronics / electrics (including software development), services and process & resources.

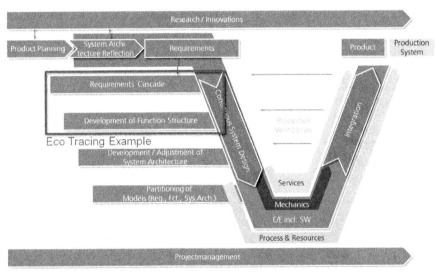

Figure 9: Allocating the described Eco Tracing example in a system development process.

During ISYPROM a hypothetical industry-like example was developed according to the interdisciplinary system development process in order to verify elaborated methods. In the following, this example is used to explain the application of *Eco Tracing*. The described region of the process is marked with a red rectangle in Figure 9. The chosen example deals with an engineering adaptation of an existing air conditioning system for a car. Cause for the adaptation is the ratification of EU directive 2006/40/EC. A high level requirement is added requesting the compliance of the air conditioning with 2006/40/EC. As all high level requirements it is further detailed causing several new and revised system level requirements. One of which is the requirement to exclusively utilize refrigerant fluids with a maximum global warming potential (GWP) of 150. All requirements are added to the existing requirements model. Following, the new requirements have to be connected to the function structure to maintain traceability.

Starting *Eco Tracing* a user needs to select both partial models which need to be examined. In this case: requirements and function structure. The selection is narrowed down to the new and changed requirements, while the unchanged function structure with its functions and sub-functions is selected as a whole. The *Eco Tracing* wizard then leads the user through the examination process. The user has to decide for each combination if a dependency between the elements is existent. For example, a tracelink has to be set between the described system level requirement "GWP of 150 or less" and the main-function "cool passenger compartment". With the help of *Eco Tracing* only 12 of 30 main- and sub-functions contained in the function structure have to be examined in combination with this specific requirement. For five of these combinations tracelinks are established and seven are dismissed. Thus 18 combinations can be ignored, saving 60 % examination effort and accelerating the process. As mentioned before, this number strongly depends on the amount of levels in the models' hierarchy. That's why no general conclusion about the effectiveness can be made, but in several tests during ISYPROM the possible savings were between 60 % and 75 %.

6 CONCLUSION AND OUTLOOK

Recent projects have shown that traceability is of particular interest to industry when mechatronic systems are developed. The usage of Model Tracer aims to help developers to achieve effective cross-model traceability and complexity management.

In this paper a number of existing approaches to manage tracelinks have been presented, which provide sophisticated functions for an improved model consistency as well as understanding and analysis of systems. While they concentrate on the beneficial use of tracelinks, there is little support for their identification and modelling.

In order to cope with this deficit, the approach *Eco Tracing* has been detailed which utilises characteristics of nested hierarchies, often found in models in mechatronic system development. In doing so, many combinations of elements which usually have to be analysed during tracelink modelling can be excluded prior examination. The possible savings in effort highly depend on the amount of levels in the hierarchy. But several tests could show that the possible savings compared to a conventional method (where all element combinations have to be examined) were between 60 % and 75 %. An overview of the features of *Eco Tracing* and its benefits is provided in Table 2.

Table 2: Features and Benefits of Eco Tracing

Feature	Benefit
Economize Modelling Efforts	
Flag all child elements of rejected parent elements	Skip unnecessary examination of a huge amount of element combinations
Choose Level of Detail flexibly	
Top-down processing offers the opportunity to skip branches of the model if desired	User can choose the level of detail for tracelink modelling on the fly
Follow a guided Procedure	
Eco Tracing wizard	User is systematically guided through all possible combinations of selected elements
Conserve Knowledge	
Examined element combinations are flagged and those decisions are saved	Preclude redundant modelling work

"Each method for automatic relationship discovery provides one piece of the puzzle, but to obtain a complete picture, we still need to fit the pieces together and develop methods to integrate them to provide a complete solution." [3] Therefore the presented approach *Eco Tracing* is yet another piece of the puzzle towards a complete solution raising significance of traceability in industrial system development.

Future research will focus on aspects of assistance for dependency identification and maintenance e.g. by use of methods from semantic web applications in order to identify false and missing tracelinks.

The aim is to reduce the necessary effort to create and maintain dependency models and to gain a maximum benefit of them with respect to the overall quality of and a reduced development time for the developed system.

ACKNOWLEDGEMENTS

The research and development project ISYPROM is funded by the German Federal Ministry of Education and Research (BMBF) within the Framework Concept "Research for Tomorrow's Production" (funding number 02PC105x) and managed by the Project Management Agency Karlsruhe (PTKA). The author is responsible for the contents of this publication.

REFERENCES

[1] Ulrich, H. and Probst, G. J. B. *Anleitung zum ganzheitlichen Denken und Handeln: Ein Brevier für Führungskräfte*, 1995 (Paul Haupt, Bern).
[2] Verein Deutscher Ingenieure. VDI 2206: Design methodology for mechatronic systems, 2004 (Beuth Verlag, Berlin).
[3] Aizenbud-Reshef, N., Nolan, B. T., Rubin, J., and Shaham-Gafni, Y. Model traceability. *IBM SYSTEMS JOURNAL*, 2006, 45(3), 515–526.
[4] Sutinen, K., Gustafsson, G., and Malmqvist, J. Computer Support for Requirements Management in an international Product Development Project. In *Design Engineering and Technical Conferences and Computers and Information in Engineering Conference, DETC'04*, Salt Lake City, September 2004, pp. 1-12.
[5] Ramesh, B. Implementing Requirements Traceability. *Cutter IT Journal*, 2000, 13 (5).
[6] Sutinen, K., Almefelt, L., and Malmqvist, J. Implementation of requirements traceability in systems engineering tools. In *Produktmodeller 2000*, Linköping, November 2000, pp. 313-330.
[7] Stark, R., Beier, G., Figge, A., Wöhler, T. Cross-Domain Dependency Modelling - How to achieve consistent System Models with Tool Support. In *European System Engineering Conference, EUSEC 2010*, Stockholm, May 2010, pp. 1-14.
[8] Lindemann, U. *Methodische Entwicklung technischer Produkte: Methoden flexibel und situationsgerecht anwenden*, 2007 (Springer-Verlag, Berlin, Heidelberg).
[9] Eigner, M. and Stelzer, R. *Product Lifecycle Management. Ein Leitfaden für Product Development und Life Cycle Management*, 2009 (Springer-Verlag, Berlin, Heidelberg).
[10] Maurer, M. S. *Structural Awareness in Complex Product Design*, 2007 (Dissertation, Technische Universität München).
[11] TESEON GmbH. *Loomeo*. http://www.teseon.com. Accessed December 20th 2010.
[12] Keller, R., Eger, T., Eckert, C. M., and Clarkson, P. J. Visualising Change Propagation. In *International Conference on Engineering Design, ICED'05*, Melbourne, August 2005, pp. 1-12.
[13] ID-Systems GmbH. *METUS - The System Design Method*. http://www.id-consult.com/metus-software/metus-methodik/. Accessed January 10th 2011.
[14] Tretow, G., Göpfert, J. and Heese, C. In sieben Schritten systematisch entwickeln. *CAD-CAM Report*, 2008, 8, 36–39.
[15] van Gorp, P., Altheide, F., and Janssens, D. Traceability and Fine-Grained Constraints in Interactive Inconsistency Management. In *Second ECMDA Traceability Workshop, ECMDA-TR 2006*, Bilbao, 2006, 1-13.
[16] Geensoft. *Reqtify. Effective Solution for Managing Requirements Traceability and Impact Analysis across Hardware and Software Projects Lifecycle*. http://www.geensoft.com/en/article/reqtify. Accessed December 17th 2010.
[17] Allen, T.F.: *A Summary of the Principles of Hierarchy Theory*. http://www.isss.org/hierarchy.htm. Accessed April 19th 2011.

Contact: Asmus Figge
Technische Universität Berlin
Chair of Industrial Information Technology
Pascalstr. 8-9, 10587 Berlin
Germany
Tel.: Int +49 30 314 26290
Fax: Int +49 30 393 0246
Email: asmus.figge@tu-berlin.de
URL: http://www.iit.tu-berlin.de

Prof. Dr.-Ing. Rainer Stark is head of the Division Virtual Product Creation and director of the Chair for Industrial Information Technology at the Technische Universität Berlin since 02/2008. After his studies in mechanical engineering (Ruhr University Bochum and Texas A&M University) he received the Dr.-Ing. degree from the Universität des Saarlandes Saarbrücken. During his industrial activities of many years he worked in different leading positions in the automotive industry. His research topics are the intuitive and context-related information modelling, intuitively usable and functionally experienceable virtual prototypes, the function-oriented Virtual Product Creation as well as development processes and methodologies for the product modelling.

Asmus Figge studied mechanical engineering with a specialisation in engineering design at Technische Universität Dresden. Since 2008 he works as a research fellow at the Chair for Industrial Information Technology at the Technische Universität Berlin. His research interests include traceability in systems engineering.

INTERNATIONAL CONFERENCE ON ENGINEERING DESIGN, ICED11
15 - 18 AUGUST 2011, TECHNICAL UNIVERSITY OF DENMARK

EQUILIBRIUM DESIGN PROBLEMS IN COMPLEX SYSTEMS REALIZATION

Jitesh H. Panchal
Washington State University, Pullman, WA, USA 99163

ABSTRACT

Equilibrium design is a class of problems where the design of complex systems is not directly controlled by designers but emerges from the self-interested decisions of stakeholders. While such problems have been common in economics and social sciences, they have not yet been addressed in engineering design. This is because the focus in engineering design is on technical performance with the assumption that designers directly control the design space. However, with the increasingly interconnected nature of the technical, social, economic and environmental aspects, equilibrium design problems become more important for designers. Instead of solving a specific equilibrium design problem, the goals in this paper are to highlight the importance and uniqueness of this class of problems and to present a general formulation within engineering design context. Specifically, we present a general formulation using concepts from non-cooperative game theory, mathematical tools for solving them, and various example problems relevant to engineering design that can be modeled as equilibrium design problems.

Keywords: Systems design, non-cooperative games, Nash equilibrium, mathematical programming

1 INTRODUCTION

How can we design large-scale complex systems whose structures and behaviors are not directly controlled by designers, but emerge dynamically from the local decisions and self-organization of individual entities? This question has been central to many parts of economics and social sciences. The design of markets, mechanisms, auctions, and organizations, all deal with essentially the same question. Increasingly, this question is also becoming relevant for engineering designers dealing with large-scale complex systems that involve technical, social, economic, and environmental aspects.
Traditionally, engineering design research has primarily been focused on systems whose design space is directly in control of the designers. However, there is an increasing importance of complex systems that are not designed, but emerge out of the individual decisions of different stakeholders. A prime example of such systems is the Internet which has evolved as a result of independent decisions of multiple stakeholders. Other examples include traffic systems, peer-to-peer networks, and communication networks. The key characteristic of such systems is that the overall performance is dependent on the design, which in turn is dependent on decisions made by individual decision-makers. The design (and hence, performance) of such systems can be directed by affecting the decisions of the individual stakeholders through different mechanisms such as the provision of incentives.
The natural framework for analyzing systems that involve multiple decision-makers is non-cooperative game theory [1]. Non-cooperative games have been used in engineering design, primarily as a way to represent decentralized design scenarios [2, 3] where designers are modeled as decision-makers. Decentralized design is characterized by four conditions [3]: a) designers have knowledge of only their own local objectives, b) designers act unilaterally to minimize their objective function, c) designers have complete control over specific local design variables, and d) designers communicate by sharing the current value of their local design variables. The decisions are in equilibrium if none of the designers can unilaterally improve their payoff by changing their own decisions. This equilibrium is referred to as the Nash Equilibrium. Current research on non-cooperative game theory in engineering design is focused on identifying the Nash equilibrium and its stability properties. However, the goal from a systems design standpoint is to achieve desired system performance by influencing stakeholder decisions. We refer to the corresponding problem as the "equilibrium design problem" in this paper.

The goals of this paper are three fold: 1) to define the equilibrium design problem, 2) to discuss mathematical tools that can be used for solving equilibrium design problems, and 3) to discuss various problems in engineering design that can be modeled as equilibrium design problems. The organization of the paper is as follows. In Section 2 we provide a background on non-cooperative game theory and the concepts of equilibrium in games. The general equilibrium design problem is formulated in Section 3. Two mathematical tools for solving the equilibrium design problem are discussed in Section 4. Examples of problems in engineering systems design that can be formulated as equilibrium design problems are discussed in Section 5. Closing thoughts are presented in Section 6.

2 BACKGROUND: NON-COOPERATIVE GAMES AND EQUILIBRIA

2.1 A Brief Overview of Non-Cooperative Games

In game theory, non-cooperative games [1] are models of situations where individuals make independent decisions without collaboration or communication. The individuals are referred to as players whose decisions may affect each other. A non-cooperative game consists of n players; each player has a finite set S_i of pure strategies. A combination of all the *strategies* of players in the product space $S = S_1 \times S_2 \times ... \times S_n$ is called the *strategy profile* of the game. Corresponding to each player i, there is a payoff function, p_i, which maps the player's strategies to real numbers. The payoffs capture the preferences of decision makers, with higher payoffs being more preferable to lower payoffs. In the literature on decision-making, payoffs are commonly represented as utility functions [4]. A *mixed-strategy* of player i is a probability distribution over the player's pure strategies.

At the core of non-cooperative games is the concept of equilibrium. A set of strategies is in equilibrium if no player has an incentive to unilaterally change the strategy. Mathematically, a strategy profile $s^* \in S$ is a *Nash equilibrium* if

$$\forall i, s_i \in S_i, s_i \neq s_i^* : p_i(s_i^*, s_{-i}^*) \geq p_i(s_i, s_{-i}^*) \qquad (1)$$

where $p(s_i, s_{-i}^*)$ represents a change in player i's strategy from s_i^* to s_i, while keeping all other players' strategies the same. The equilibrium is called strict Nash equilibrium if the symbol $\geq$ is replaced with $>$ in equation (1). Nash equilibrium can be defined either for pure strategies or for mixed strategies. Nash proved that every finite game has at least one mixed-strategy Nash equilibrium [1]. A game can have multiple Nash equilibria.

2.2 Challenges Associated with Nash Equilibria

Nash equilibrium is only one type of equilibrium for non-cooperative games. A generalization, referred to as *correlated equilibrium* was first suggested by Aumann [5]. In this case, the players choose their strategies based on a public signal from a trusted party. The trusted party chooses a strategy profile according to a probability distribution and informs it to the corresponding players. Individual players choose their strategies based on this information. If no player has an incentive to unilaterally deviate from his/her strategy, then the strategy set is called a correlated equilibrium. The advantages are that correlated equilibria always exist for finite games [6], they may be more efficient than Nash equilibria [7], and unlike Nash equilibria they can be efficiently computed and learnt [8, 9].

Despite the generality of the concept of correlated equilibrium, the concept of Nash equilibrium is used as the standard notion of equilibrium. Papadimitriou and Roughgarden [10] suggest that this is because "everybody uses it," it is used as a baseline for refinements and generalizations (such as the correlated equilibrium), and it is an open computational problem in computational game theory. Hence, in this paper, we focus on the Nash equilibrium to illustrate the problem of designing equilibria in decentralized systems design.

One of the key challenges in non-cooperative game theory is the *complexity of finding Nash equilibria*. Nash [1] commented that "The complexity of the mathematical work needed for a complete investigation increases rather rapidly..." Even with the developments in computers during the past 60 years, calculating Nash equilibria is still challenging. Various algorithms have been proposed for finding Nash equilibria but none of them is known to run in polynomial time [11]. Recently, the computational complexity of the problem of calculating Nash equilibria, even for a two player game, is classified as PPAD (Polynomial Parity Arguments on Directed graphs) complete [11-13].

The second challenge is related to the *dynamics of the processes* leading to the equilibria [14]. The dynamics refers to the sequence of decisions made by individual players in response to the decisions made by preceding players. Arguably the simplest and the most popular game dynamics is the "best response" (BR) dynamics [15] where at a given time, a small portion of players adjust their strategy to a strategy that is the best response to the current strategy of the other players. Other types of dynamics are also studied in the literature where the players anticipate future strategies of other players and respond accordingly. Convergence to an equilibrium depends on a) the game (i.e., the payoff functions of the individuals), and b) the dynamic process. Not all dynamic processes for a game converge to Nash equilibria. Similarly, a given dynamic process may or may not converge depending on the game. General results on the convergence of dynamics to Nash equilibria are available only for a small class of games. It has been shown that for a class of games called *potential games* [16] simple dynamics such as best response are guaranteed to converge to a Nash equilibrium. The rate of convergence and the dynamic stability of the equilibria are other related issues.

The third major challenge is related to the *inefficiency of equilibria* [17]. The prisoner's dilemma is a well-known example illustrating that the equilibrium achieved by the decentralized decisions of players may be less than the socially optimal solution that can be achieved by a central authority. The most commonly used notion of optimality is Pareto optimality. A set of strategies is Pareto optimal if it is impossible to strictly increase the payoff of a player without strictly decreasing the payoff of another player. The extent of inefficiency of the equilibria can be measured using measures such as a) the price of anarchy, and b) the price of stability. The price of anarchy is the ratio between the worst objective function of the equilibrium and that of an optimal outcome. On the other hand, the price of stability is the ratio of the best objective function value to that of an optimal outcome. Hence, price of anarchy is based on a pessimistic view whereas the price of stability is based on an optimistic view.

As a summary, the key challenges associated with equilibria in non-cooperative games are: a) efficiently determining the Nash equilibrium points, b) ensuring that the dynamics converges to the Nash equilibria, and c) ensuring that the Nash equilibria are closer to the efficient solutions.

2.3 Illustrative Example

In this section, a simple example with two designers (players), each with an objective function, is presented to illustrate the concept of Nash equilibrium. Consider two designers with payoff functions F_1 and F_2 as shown in Table 1. The payoff functions are polynomials in two variables x_1 and x_2. The strategy set is determined by the values of the design variables that each designer can control. The two designers are responsible for different variables; the first designer can choose values for x_1 and the second designer can choose the values for x_2. The strategy set for the first designer is $x_1 = [0\ 1]$ and the strategy set for the second designer is $x_2 = [0\ 1]$.

Table 1- Payoff functions and strategy sets of the two designers in the illustrative example

Designer 1	Designer 2
Payoff Function: Maximize: $F_1 = x_1 x_2 - x_1^3$	Payoff Function: Maximize: $F_2 = x_1 x_2 - x_2^3$
Strategy set: $x_1 = [0\ 1]$	Strategy set: $x_2 = [0\ 1]$

The strategy of each designer is to choose the values of corresponding design variables such that their payoffs are maximized for a given value of the design variable chosen by the other designer. This is referred to as the "best response" to the other designer's strategy. Hence, based on the first order optimality condition, Designer 1 chooses x_1 for a given value of x_2 such that $\frac{\partial F_1}{\partial x_1} = 0$ and Designer 2 chooses x_2 such that $\frac{\partial F_2}{\partial x_2} = 0$. The strategy profiles corresponding to these optimality conditions are referred to as the Best Response Correspondences (BRCs) of the designers. The BRCs for the example problem are shown in Figure 1. The points of intersection of the BRCs for the two designers are the Nash

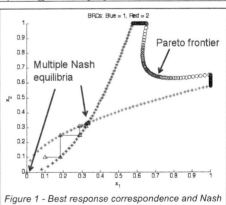

Figure 1 - Best response correspondence and Nash equilibria for the illustrative example

equilibria for the game. In the illustrative example, there are two Nash equilibria: $(x_1, x_2) = (0, 0)$ and (1/3, 1/3).
If the designers sequentially choose the best response to the other designer, the dynamics is termed as a best response dynamics. The best response dynamic can be represented using the following iterated map: $x_{1,n+1} = \sqrt{\frac{x_{2,n}}{3}}$; $x_{2,n+1} = \sqrt{\frac{x_{1,n+1}}{3}}$. The iterated map may or may not converge to the Nash equilibrium. In Figure 1, it is shown that starting from an initial point $(x_1 = x_2 = 0.1)$ the iterated map converges to one of the Nash equilibria: $(x_1, x_2) = (1/3, 1/3)$. The Pareto optimal solutions for the game are shown in the figure. At these points none of the designers can improve their payoff without adversely affecting the other player's payoff. Comparing the Nash equilibria with the Pareto solutions, it is observed that the Nash equilibrium (1/3, 1/3) is closer to the Pareto frontier.

3 EQUILIBRIUM DESIGN PROBLEM IN NON-COOPERATIVE GAMES

The discussion in the previous section is focused on a set of fixed equilibrium points. Now consider a scenario where it is possible to modify the individual designers' payoffs through some incentives. In such cases, the individuals' strategies vary based on the payoffs. Hence, the corresponding Nash equilibria also change. The new Nash equilibria may have different efficiency (price of anarchy and price of stability), convergence, and stability characteristics compared to the original equilibria. By appropriately choosing the incentives to modify the designers' payoffs, the *resulting Nash equilibria can be designed* to possess the desired characteristics. This design problem is referred to as the equilibrium design problem. The higher level authority that has the power to provide incentives to modify individual payoffs is referred to as a "game designer."

For illustrative purposes, we extend the example from Section 2.3 to an equilibrium design problem. Assume that the payoff functions of the two designers contain parameters c_1 and c_2 that can be selected by the game designer (see Table 2). The ranges of these parameters are $c_1 = [0\ 3]$, $c_2 = [0\ 3]$. For $c_1 = c_2 = 1$, the payoffs are similar to the previously discussed scenario. The best response strategies of the two designers for different values of the parameters are shown in Figure 2. The intersection of the best response correspondence of the two designers for different combinations of c_1 and c_2 are shown in Figure 3. Each intersection point corresponds to a combination of the parameters. The region highlighted in the figure is the sub-space of the joint strategy space where each point can be achieved as Nash equilibrium by choosing appropriate values of the parameters. We refer to this region as the *Nash feasible space* of the equilibrium problem. This notion is similar to the notion of feasible space in optimization problems because no point outside this space can be achieved as Nash equilibrium.

Table 2- Modified objective functions of the two designers

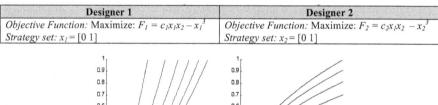

Designer 1	Designer 2
Objective Function: Maximize: $F_1 = c_1 x_1 x_2 - x_1^3$	*Objective Function:* Maximize: $F_2 = c_2 x_1 x_2 - x_2^3$
Strategy set: $x_1 = [0\ 1]$	*Strategy set:* $x_2 = [0\ 1]$

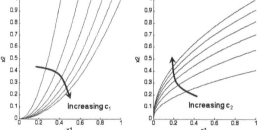

Figure 2 – The set of BRCs of the two designers (left: Designer 1 and right: Designer 2)

Based on the discussion in the previous section, the desired properties of the designed Nash equilibrium, and hence the **goals for the equilibrium design problems** are:

1. *Closeness to the best solution:* The Nash equilibrium should be close to the best possible (efficient) solution. The goodness of the solution can be defined in various different ways. Pareto efficiency, discussed in Section 2.2, is one of the most widely used concepts for quantifying the goodness of a solution in a decentralized system. If Pareto efficiency is used to define the ideal solution, the price of stability of the Nash equilibrium should be as close to 1 as possible. However, it is important to recognize that there are other ways of defining efficiency. Researchers in economics have proposed concepts such as Kaldor-Hicks efficiency, X-efficiency, allocative efficiency, distributive efficiency, dynamic efficiency, and productive efficiency. All of these notions of efficiency

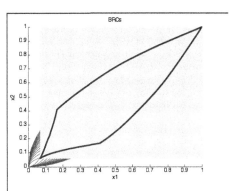

Figure 3 – The Nash feasible space resulting from the intersection of a range of BRCs

relate the individual preferences to the overall performance of the social system. Within systems design, the system level goal may or may not correspond to the efficiency of the solution (further discussed in Section 5.1). Hence, we do not limit ourselves to the notion of Pareto efficiency because within systems design, the quality of the solution may be dictated by system-level goals.

2. *Convergence of the equilibrium:* The dynamics of information exchange and decision making plays a significant role in the equilibrium design problems. There are various dynamic processes, the best response correspondence being the simplest one. Ideally, the dynamics of the process should result in the convergence of the solution to Nash equilibrium. However, it has been shown that depending on the problem, the dynamic processes may or may not converge. Additionally, for problems with multiple equilibria (as in Section 2.3), the process should converge to the one closer to the desired solution. Hence, achieving the convergence properties is an essential goal for the equilibrium design problems.

3. *Stability of the equilibrium:* The Nash equilibrium should be stable, i.e., small perturbations should not result in divergence from the equilibrium. Further details on stability are provided in Section 4.2.

As a summary, an equilibrium design problem can be defined by decision makers, their strategy space, individual payoffs, system-level goals, dynamic processes, and ways in which individual payoffs can be modified to affect the equilibrium and its characteristics (closeness to the best solution, convergence, and stability).

4 MATHEMATICAL TOOLS FOR EQUILIBRIUM DESIGN PROBLEMS

Having identified the key characteristics of an equilibrium design problem, the key question is: How can the equilibrium design problem be systematically formulated and solved? In this section, we discuss two tools, one from optimization theory and another from non-linear control theory that can be used as foundations for formulating and solving equilibrium design problems. In Section 4.1, we discuss mathematical programming with equilibrium constraints (MPEC) for finding the location of the best equilibria, and in Section 4.2 we discuss Lyapunov stability theory for assessing the stability and convergence characteristics of the equilibria.

4.1 Mathematical Programming with Equilibrium Constraints (MPEC)

The MPEC is a type of constrained nonlinear programming problem where some of the constraints are defined as parametric variational inequality or complementarity system [18]. These constraints arise from some equilibrium condition within the system, and hence, are called equilibrium constraints. MPEC is a special type of bi-level programming problems [19] consisting of a higher level optimization problem, whose constraints are defined in terms of solutions to lower-level optimization problems. MPEC is applicable to a variety of problems in engineering such as optimal design of mechanical structures, network design, motion planning of robots, and facility location and production. MPEC is also used to study equilibrium problems in economics. MPEC is closely related to the Stackelberg game [20] where a leader makes a decision first and then the followers make their decisions based on the leader's decision. The leader corresponds to the upper-level optimization problem in MPEC and the followers correspond to the lower-level problems. Examples of problems of economic equilibrium where MPEC has been used include maximizing revenue from tolls on a traffic system, optimal taxation, and demand adjustment problems.

Mathematically, a MPEC problem can be represented using two sets of variables, $x \in \mathbb{R}^n$ and $y \in \mathbb{R}^m$. Here, x belongs to the upper-level problem and y solves the lower-level equilibrium problem. The solution of y depends on the value of x chosen for the upper-level problem. The overall objective function $f(x,y)$ is minimized.

$$\min_{(x,y)} f(x,y) \qquad (2)$$

Subject to: $(x, y) \in \Omega$, and $y \in S(x)$ (3)

where Ω is the joint feasible region of x and y; and $S(x)$ is a set of variational inequalities that represent the equilibrium problem. In the case of the equilibrium design problem discussed in Section 3, the set x is the set of parameters c_1 and c_2 affecting the individual payoffs and controlled by the game designer. Based on the values of these parameters, the players determine their best responses to the other players. Hence, the variables x_1 and x_2 in Section 3 correspond to the variable y in the MPEC formulation above. The intersection of the best responses is the Nash equilibrium point. Here, the function $f(x,y)$ represents a system-level function that quantifies the goodness of the solution (as defined in Section 3).

The set $S(x)$ corresponds to the feasible Nash space. As discussed earlier, the Nash equilibrium corresponds to the best response of each designer to the decisions made by other designers. The Nash equilibrium point can be formulated as a variational inequality using the first order necessary conditions for optimality such as Karush–Kuhn–Tucker (KKT) conditions [21]. Having formulated the equilibrium design problem as a MPEC, the next step is to solve it. Solving the MPEC problems is challenging because of the non-linearities in the problem, non-convex feasible space, combinatorial nature of constraints, disjointed feasible space, and multi-valued nature of the lower equilibrium problem [18]. There has been some progress in developing efficient algorithms for solving MPEC problems. Examples include variations of NLP algorithms, and interior point algorithms [22].

As a summary, MPEC can be used as a mathematical tool to model equilibrium design problems. We discussed how the formulation can be used to account for the goodness of the Nash equilibrium. MPEC results in the best Nash Equilibrium. However, this only addresses the first requirement listed in Section 3. It does not account for the dynamics of the problem. The solution does not provide any insight into the convergence and stability of the equilibrium. To address this limitation, we utilize some of the tools from non-linear control theory. We specifically focus on the Lyapunov stability theory in the next section.

4.2 Lyapunov Stability Theory

In Section 2.3, we illustrate how the best response dynamics can be modeled as an iterative map where the values of the variables during some iteration are given in terms of the values in the previous iteration. This represents a dynamic system for which the stability characteristics of the equilibrium points can be evaluated using the Lyapunov Stability Theory [23]. For nonlinear systems, various notions of stability have been developed. These include Lyapunov stability, asymptotic stability, exponential stability, and global asymptotic stability. Lyapunov stability means that trajectories in the phase space starting at two points close to equilibrium will stay sufficiently close to it. Asymptotic stability is a stronger notion of stability where in addition to Lyapunov stability the trajectories starting close to the equilibrium also converge to the equilibrium as time goes to infinity. An equilibrium point is exponentially stable if the trajectories converge to the origin faster than an exponential function.

The Lyapunov theory consists of a direct method and an indirect method for evaluating the stability of nonlinear dynamic systems. In the indirect method, the nonlinear system is approximated as a linear system near the equilibrium points and the stability of the linear system is determined. For a nonlinear system in the state space representation $\dot{x}_i = f_i(x_1, x_2, ..., x_n)$ where x_i is a state variable, then the equilibrium point is given by $f_i(x_1, x_2, ..., x_n) = 0 \; \forall \; i$. If the equilibrium is at the origin, the stability can be determined from the eigenvalues of the Jacobian matrix $J = \left(\frac{\partial F}{\partial X}\right)$ where F is the set of functions f_i and X is the set of state variables x_i. If all the eigenvalues are negative, then the system is stable. If at least one of the eigenvalues is positive, then the system is unstable. In contrast to the indirect method, the direct method accounts for the nonlinearities in the system using the notion of a Lyapunov function. For a system, if there is a positive-definite function of state variables which decreases along all state trajectories, the system is stable. That function is called a Lyapunov function of the system. A system can have more than one Lyapunov function. The key challenge in using the direct method is that there is no general method to find the Lyapunov function for a system. It requires intuition and trial-and-error. In the case of iterated maps, which are discrete-time systems, the concept of Lyapunov exponents is used to determine stability of a system. Assuming that two trajectories start at nearby points x_0 and $(x_0 + \delta_0)$. Then, the separation between the two points after n iterates is $|\delta_n| = |\delta_0|e^{n\lambda}$. The equilibrium point is stable if $\lambda < 0$. As elaborated by Strogatz [24], the Lyapunov exponent for a system trajectory starting at x_0 is:

$$\lambda = \lim_{n \to \infty} \left\{ \frac{1}{n} \sum_{i=0}^{n-1} ln|f'(x_i)| \right\} \quad (4)$$

In order to incorporate stability as an integral part of the solution of the equilibrium design problem the Lyapunov stability criteria need to be integrated within the MPEC framework. One possibility is to develop constraints on the eigenvalues and to integrate them as constraints within the MPEC. Another potential approach is to determine a set of good solutions from MPEC and then to perform stability analysis to determine the best solution from the stability standpoint. The third approach is to determine basins of attraction within the design space and then use them as a feasible design space for the MPEC. Currently, there is a lack of methods that account for both stability and efficiency in an integrated manner for the equilibrium design problems. Further investigation is necessary.

5 EQUILIBRIUM DESIGN PROBLEMS WITHIN ENGINEERING DESIGN

The equilibrium design problem can be found in various problems related to engineering design. In this section, we discuss some classes of problems that have equilibrium design at their core. All of these classes of problems can be viewed from a collective systems perspective and it is possible to modify individual preferences through the provision of incentives. An overview of the characteristics of the problems is provided in Table 3.

Table 3 - Overview of the classes of equilibrium design problems in engineering design

	Decision Makers	Individual Preferences	System-level goals	Dynamic Processes	Mechanisms for equilibrium design
Requirements allocation	Designers working on different aspects of the system	Subsystem goals	System-level design goals	Updating individual decisions based on others' decisions	Requirements decomposition and targets
Evolutionary networks	Entities making decisions about linking with different nodes	Payoffs for individuals are network dependent	Performance of the network	Addition and removal of nodes and links	Incentives to individuals to affect linking behaviors
Collective Innovation	Individuals participating in collective innovation projects	Satisfying intrinsic and extrinsic needs	Development of the entire project	Self-organization of communities and growth of products	Incentives to individuals to participate in different activities
Decentralized Energy	Individual consumers	Minimize cost of owning and operating energy resources	Technical, Economic, Environmental, Social	Market processes of purchasing and selling energy	Taxes, incentives, laws, market rules

5.1 Requirements allocation in decentralized design

Requirements allocation, as used in systems engineering, is the process of decomposing system-level requirements into requirements for lower-level subsystems and components. Requirements allocation is a part of the requirements-engineering process. There are two types of requirements at the system level – a) requirements that can be directly assigned to individual subsystems and components, and b) requirements that need to be divided among multiple components. An example of the former type of requirements for an automotive system is "provide energy" which can be fulfilled by a lower level subsystem such as an "engine". An example of the latter type is "weight should be lower than 10,000 kg" which can be divided into upper bounds of weights for individual systems. Such requirements are also referred to as allocable requirements. Here, weight is a system attribute which is a function of attributes of the components. Collopy [25] refers to these attributes as extrinsic attributes. Other examples of extrinsic attributes are cost, efficiency, and reliability.

The performance of the overall system is dependent on the allocation of requirements. Hence, the requirements allocation should be such that it maximizes the overall system performance. Traditionally, requirements allocation has been carried out by system designers based on their insights and the knowledge from prior systems. The effect of alternate requirements allocations on the system performance is rarely considered. The requirements allocation problem can be modeled from two different perspectives: optimization and non-cooperative game theory, as discussed next.

Optimization Perspective: Consider a single organization representing a completely collaborative scenario where the goals of all designers are to achieve the system-level objectives, and information can be freely shared among them at any stage during the design process. In such a scenario, the requirements allocation problem involves finding the best decomposition of requirements for extrinsic attributes. From an optimization perspective, the widely utilized approach for requirements allocation is to determine lower and upper bounds for the extrinsic attributes for subsystems. These bounds are such that if all the lower-level designers designed their subsystems to satisfy their corresponding bounds, the system level requirements are automatically satisfied. The bounds are used as constraints in sub-system level design problems. Collopy argues that using requirements for extrinsic attributes as constraints for subsystems and components results in inferior systems as compared to using them as parts of objective functions [25]. Whether the requirements are modeled as constraints or parts of objective functions, existing multi-disciplinary optimization approaches such as collaborative optimization [26], analytical target cascading [27], etc. can be used to model the scenario.

Non-cooperative Game Perspective: Consider another scenario of systems design carried out by multiple distributed entities (e.g., organizations or teams) where each entity has its own underlying goals and there are barriers (both organizational and technical) to complete information exchange throughout the design process. This scenario is representative of many complex automotive and aerospace systems designed by multiple organizations. Due to the limited information flow between designers, the resulting solution is equilibrium. Here, requirements allocation modifies individual payoffs and hence, acts as a way of modifying the equilibrium. In such a scenario, the requirements allocation problem can be modeled as an equilibrium design problem. The systems designer's decision is to determine the best allocation of requirements such that when the individual designers make their decisions, the equilibrium is close to the desired solution. The approaches discussed in Sections 4.1 and 4.2 can be used to model the equilibrium design problem. Each equilibrium design problem has unique challenges. One of the challenges is the lack of detailed models of individual subsystems before the systems are designed. Without the availability of these subsystem models, it is challenging to model the impact of alternate ways of requirements allocation on the equilibrium and its properties.

5.2 Design of complex evolutionary networks

Recently a number of large-scale complex networks have been identified whose structures are not directly controlled by designers, but emerge dynamically from the local decisions and self-organization of individual entities. A prime example of such a network is the Internet, whose structure is a result of individual connection decisions made by individual entities. The overall topology of the Internet affects its reliability, the effectiveness of search, etc. The Internet is just one example of such networks. Other examples include social networks, ad hoc networks, trade networks etc. All of these networks are similar in the sense that: a) the topology is a result of local behaviors, and b) the topology has a significant effect on their performance. Such networks are also referred to as *endogenous networks* [28]. The system-level objective is to guide the evolution of such networks towards desired structures with desired behaviors and performance.

Existing approaches for network design are focused on centralized network design applied to networks such as transportation [29]. In these problems, the design variables are nodes and links, and the objectives are minimization of the cost of transportation, minimization of the distance travelled, etc. On the other hand, the design of complex evolutionary networks is governed by individual decisions which can be modified by providing incentives. The individual decisions are also based on the decisions made by decision makers. Hence, the decisions can be modeled as equilibrium problems. The individual decisions affect the formation of nodes and links, thereby affecting the network structure, and hence, the network performance. The goal is to maximize the performance of the overall network. For example, one of the performance goals for the Internet is to maximize the effectiveness of search. Hence, the design problem in such evolutionary networks is fundamentally different from traditional network design, and is more appropriately represented as an equilibrium design problem. The challenge associated with such problems is the presence of large and discrete design spaces.

5.3 Collective Innovation

Collective innovation is an emerging paradigm in product realization where complex systems are developed in a bottom-up manner by communities of independent individuals, as opposed to hierarchical organizations. The paradigm is epitomized by successful examples from open-source software development (e.g., Linux, Apache), crowdsourcing (e.g., Innocentive), and open encyclopedias (e.g., Wikipedia). The fundamental difference between traditional product realization processes and collective innovation processes is that the former are based on top-down decomposition and structured task assignment while the latter are based on self-organization of individually motivated participants into communities. The participants' contributions are not based on the pre-specified tasks, and the product evolves over time based on the contributions of the participants [30].

Collective innovation can be viewed from a complex systems perspective where individuals are decision makers with their own preferences, needs, and capabilities. The individual goals can range from enjoyment-based or community-based intrinsic motivation to extrinsic motivations such as career advancement and skill improvement [31]. Based on their interests and competencies, individuals make decisions such as whether to participate or not, which project to participate on, and whom to collaborate with. They self-select (or define) the activities they would like to participate in. Individuals interact with each other in two ways – directly and indirectly. The direct interactions are through one-on-one discussions, online forums, and other web-based mechanisms. The indirect interactions are mediated through the product structure, which is an essential aspect of the environment. Based on these decisions and interactions, the product evolves. At the same time, as the individuals decide to collaborate with each other on different tasks, the community structure also grows with time.

Panchal et al. [32] highlight that due to the interdependencies between different product modules, the product sequence in which they are developed has an impact on the growth rate of the product. The modules on which other modules depend should be developed first. In addition to the dependencies between modules, both the product structure and the community structure are also interdependent [32]. The growth of communities affects the way in which products evolve and the growth of products affect the evolution of communities. Some product structures and community structures are better than others in terms of collective innovation [33]. If the system-level goal is rapid growth of the product being designed, collective innovation is also associated with an equilibrium design problem. The individual behaviors can be modified by providing different types of incentives (such as awards and recognition). As the individuals make decisions to participate on different activities, the equilibrium design problem is to determine the incentives that can be provided to them at different points in time to participate on appropriate modules at appropriate time. Through a systematic design of the incentive structure, the dynamics of collective innovation processes can be directed towards faster growth and the achievement of targeted structures of products and communities.

5.4 Decentralized energy

The current energy infrastructure in the United States (and most other countries around the world) is primarily based on a centralized model of energy generation and distribution. The centralized energy model is characterized by large-scale power plants from which energy is distributed to the consumers through a centrally controlled network of cables. While the centralized energy model has been used for over two centuries, it has a number of limitations – a) waste of energy, b) large transmission losses, c) expensive distribution infrastructure, d) low resilience to failure, and e) impacts on the environment [34]. With the increasing use of small-scale energy generation from renewable sources and increasing deregulation of the energy sector, an alternative paradigm of energy generation and distribution is emerging. It is called *decentralized energy* and consists of distributed generation resources [34]. It is believed that decentralized energy can address the limitations of centralized energy. Since energy is generated closer to the consumers, the waste heat generated in the process can be used for space heating purposes. Since the energy does not need to be transmitted over long distances, the transmission losses are lower. Further, the cost of distribution infrastructure is lower. However, decentralized energy is faced with challenges such as poor control on the power quality due to the intermittency of renewable energy sources.

Energy infrastructure is associated with multiple levels of decisions such as network reconfiguration, service restoration, operation planning, and expansion planning [35]. These decision problems vary in their objectives and time horizons. The optimal network reconfiguration is an operation-related problem of finding the branches of the network to be opened to supply the loads with minimum energy losses. Optimal service restoration involves identifying the best strategy to meet the demands after a fault to minimize the effect of fault propagation. Operation planning involves choosing the optimum structural changes considering a constant load in the network. Expansion planning involves deciding how to grow the network considering future changes in demand. The timescale considered for expansion planning is about 20 years.

Within a centralized energy paradigm, all these decisions are made by central authorities who are in-charge of the infrastructure. However, in a decentralized infrastructure, the fundamental difference is that different stakeholders make their own decisions and the overall system-level performance is dependent on the individual decisions. For example, the consumers play an active role by acting as producers. They make their own decisions on a) which technologies to invest in, b) how much energy to generate, c) how much energy to buy and from whom, and d) how much energy to sell [36]. Other stakeholders include power producers (e.g., utility companies), grid operators, and regulators (e.g., government and other regulating authorities). The decisions made by different stakeholders are often conflicting. For example, consumers' decisions are often driven by economic aspects such as minimizing their energy costs. In addition to economic objectives, other objectives such as technical, environmental and social objectives are also important from a systems level. Technical objectives include minimum system losses, voltage stability, unbalance conditions, power quality, and energy needs. Environmental objectives include minimization of emissions and hazardous materials. Social objectives include fairness and quality of life.

Based on the decisions made by the individual stakeholders, the system reaches an equilibrium point which defines its overall behavior. The individual decisions can be directed through a number of mechanisms such as policy tools, incentives (e.g., tax breaks), penalties (e.g., tariffs), markets rules, and laws. The corresponding equilibria can be changed through these mechanisms. Hence, the decisions within decentralized energy can be modeled as an equilibrium design problem. The decisions of the policy makers can be represented as the higher-level optimization problem and the decisions of the individual consumers can be represented as lower-level equilibrium problem.

5.5 Other problems

There are various other examples of systems with similar structure. The sustainability standards development process such as LEED is a natural example of equilibrium design problem. The standards influence the design decisions made by architects and material decisions by builders. By appropriately choosing the standards, the decisions can be directed towards better environmental performance. Similarly, most of the problems that relate to policy design, determination of the right amount of taxes (e.g., carbon tax) are also equilibrium design problems.

6 CLOSING COMMENTS

As the scope of problems considered by engineering design researchers is extended beyond just technical design to include broader aspects such as organizational design, policy, and economics, the equilibrium design problem becomes pervasive. The first step towards addressing these challenges is to recognize the common structure of all these problems and the existence of tools in different fields that can be used to address some of the associated challenges. That is the primary goal of this paper.

Traditionally, problems related to equilibrium design are studied in economics, social sciences, and computer science. Specifically, the field of mechanism design [37] within economics deals with the "design of games" with desired outcomes. Design of multi-agent systems [38] involves designing the behaviors of individual decision-making agents to achieve the desired system-level behaviors. In this paper, we show that equilibrium design problems are also central to engineering systems design and hence, needs attention from the engineering design community.

Equilibrium design problems are complex in nature. One of the key challenges is the multi-objective nature of desired system-level outcome. Different problems are associated with unique challenges that require different ways of addressing them. As discussed in Section 5, equilibrium design problems in requirements allocation are challenging due to the lack of subsystem models before they are designed. The problems in evolutionary networks are challenging due to discrete and vast design spaces. There are various opportunities for research in addressing such challenges in specific classes of problems. Additionally, the classes of problems discussed in the paper are only representative examples of equilibrium design problems based on the author's own research. Many more such problems can be identified.

REFERENCES

[1] Nash, J. Non-Cooperative Games. *The Annals of Mathematics*, 1951, 54(2), 286-295.
[2] Marston, M. and Mistree, F. Game-Based Design: A Game Theoretic Extension to Decision-Based Design. In *ASME DETC, Design Theory and Methodology Conference*, Baltimore, MD, 2000, DETC2000/DTM-14578.
[3] Chanron, V. and Lewis, K. A Study of Convergence in Decentralized Design Processes. *Research in Engineering Design*, 2006, 16, 133-145.
[4] Fishburn, P.C. *Utility Theory for Decision Making*, 1970 (Wiley, New York).
[5] Aumann, R. Subjectivity and Correlation in Randomized Strategies. *Journal of Mathematical Economics*, 1974, 1(1), 67-96.
[6] Hart, S. and Schmeidler, D. Existence of Correlated Equilibria. *Mathematics of Operations Research*, 1989, 14(1), 18-25.
[7] Nisan, N. Correlated Equilibria — Some CS Perspectives [online]. 2009. Available from: http://agtb.wordpress.com/2009/05/10/correlated-equilibria-some-cs-perspectives/ [November 02, 2010]
[8] Foster, D.P. and Vohra, R.V. Calibrated Learning and Correlated Equilibrium. *Games and Economic Behavior*, 1997, 21(1-2), 40-55.
[9] Hart, S. and Mas-Colell, A. A Simple Adaptive Procedure Leading to Correlated Equilibrium. *Econometrica*, 2000, 68(5), 1127-1150.
[10] Papadimitriou, C.H. and Roughgarden, T. Computing Correlated Equilibria in Multi-player Games. *Journal of the ACM*, 2008, 55(3), 14(11-29).
[11] Daskalakis, C., Goldberg, P.W. and Papadimitriou, C.H. The Complexity of Computing a Nash Equilibrium. *SIAM Journal on Computing*, 2009, 39(1), 195-259.
[12] Papadimitriou, C.H. On the Complexity of the Parity Argument and Other Inefficient Proofs of Existence. *Journal of Computer and System Sciences*, 1994, 48(3), 498-532.
[13] Conitzer, V. and Sandholm, T. New Complexity Results about Nash Equilibria. *Games and Economic Behavior*, 2008, 63(2), 621-641.
[14] Skyrms, B. Chaos in Game Dynamics. *Journal of Logic, Language, and Information*, 1991, 1, 111-130.
[15] Hopkins, E. A Note on Best Response Dynamics. *Games and Economic Behavior*, 1999, 29(1-2), 138-150.
[16] Monderer, D. and Shapley, L.S. Potential Games. *Games and Economic Behavior*, 1996, 14(1), 124-143.
[17] Roughgarden, T. and Tardos, E. Introduction to the Inefficiency of Equilibria. In Nisan, N., Roughgarden, T., Tardos, E. and Vazirani, V.V., eds. *Algorithmic Game Theory*, 2007, pp. 441-457 (Cambridge University Press, New York, NY).
[18] Luo, Z.-Q., Pang, J.-S. and Ralph, D. *Mathematical Programs with Equilibrium Constraints*, 1996 (Cambridge University Press, Cambridge).
[19] Dempe, S. *Foundations of Bilevel Programming*, 2002 (Kluwer Academic Publishers, Dordrecht, The Netherlands).
[20] Fudenberg, D. and Tirole, J. *Game Theory*, 1993 (MIT Press, Cambridge, MA).
[21] Bertsekas, D.P. *Nonlinear Programming: Second Edition*, 1999 (Athena Scientific, UK).
[22] Dirkse, S.P. and Ferris, M.C. Modeling and solution environments for MPEC: GAMS and MATLAB. In Fukushima, M. and Qi, L., eds. *Reformulation: Nonsmooth, Piecewise Smooth, Semismooth and Smoothing Methods*, 1999, pp. 127-148 (Kluwer Academic Publishers, Dordrecht, The Netherlands).

[23] Slotine, J.-J.E. and Li, W. *Applied Nonlinear Control*, 1991 (Prentice Hall, Upper Saddle River, NJ).
[24] Strogatz, S. *Nonlinear Dynamics and Chaos: With Applications To Physics, Biology, Chemistry And Engineering* 1994 (Westview Press, Cambridge, MA).
[25] Collopy, P. Adverse Impact of Extensive Attribute Requirements on the Design of Complex Systems. In *7th AIAA Aviation Technology, Integration, and Operations Conference (ATIO)*, Belfast, Northern Ireland, 2007, pp. 1326-1332, No: 7820.
[26] Kroo, I. and Manning, V. Collaborative Optimization: Status and Directions. In *8th AIAA/NASA/ISSMO Symposium on Multidisciplinary Analysis and Optimization*, Long Beach CA, 2000, AIAA-2000-4721.
[27] Liu, H., Chen, W., Kokkolaras, M., Papalambros, P.Y. and Kim, H.M. Probabilistic Analytical Target Cascading -- A Moment Matching Formulation for Multilevel Optimization Under Uncertainty. *ASME Journal of Mechanical Design*, 2006, 128(4), 991-1000.
[28] Hojmana, D.A. and Szeidlb, A. Endogenous networks, social games, and evolution. *Games and Economic Behavior*, 2006, 55(1), 112-130.
[29] Ahuja, R.K., Magnanti, T.L. and Orlin, J.B. *Network Flows: Theory, Algorithms, and Applications*, 1993 (Prentice Hall Inc., Upper Saddle River, New Jersey).
[30] Panchal, J.H. Agent-based Modeling of Mass Collaborative Product Development Processes. *Journal of Computing and Information Science in Engineering*, 2009, 9(3), 031007.
[31] Lakhani, K.R. and Wolf, R.G. Why Hackers Do What They Do: Understanding Motivation and Effort in Free/Open Source Software Projects. In Feller, J., Fitzgerald, B., Hissam, S. and Lakhani, K., eds. *Perspectives on Free and Open Source Software*, 2005, pp. 3-21 (MIT Press, Cambridge, MA).
[32] Panchal, J.H. Co-Evolution of Products and Communities in Mass-Collaborative Product Development - A Computational Exploration. In *International Conference on Engineering Design (ICED'09)*, Stanford, CA, 2009, ICED'09/147.
[33] Le, Q. and Panchal, J.H. Modeling the Effect of Product Architecture on Mass-Collaborative Processes. *Journal of Computing and Information Science in Engineering*, 2011, 11(1), 011003.
[34] Pepermans, G., Driesen, J., Haeseldonckx, D., Belmans, R. and D'haeseleer, W. Distributed Generation: Definition, Benefits and Issues. *Energy Policy*, 2005, 33(6), 787-798.
[35] Chicco, G. Challenges for Smart Distribution Systems: Data Representation and Optimization Objectives. In *12th International Conference on Optimization of Electrical and Electronic Equipment (OPTIM)*, Brasov, Romania, 2010, pp. 1236 - 1244.
[36] Chicco, G. and Mancarella, P. Distributed Multi-Generation: A Comprehensive View. *Renewable and Sustainable Energy Reviews*, 2009, 13(3), 535-551.
[37] Hurwicz, L. and Stanley, R. *Designing Economic Mechanisms*, 2006 (Cambridge University Press, New York, NY).
[38] Wooldridge, M. *An Introduction to MultiAgent Systems*, 2002 (John Wiley & Sons Ltd., Glasgow, UK).

Contact: Jitesh H. Panchal
Washington State University
School of Mechanical and Materials Engineering
100 Dairy Road, Pullman, WA 99164 USA
Phone: +1-509-715-9241; Fax: +1-509-335-4662; E-mail: panchal@wsu.edu
URL: http://www.mme.wsu.edu/people/faculty/faculty.html?panchal

Jitesh H. Panchal is an Assistant Professor in the School of Mechanical and Materials Engineering at Washington State University. He received his B.Tech. from IIT Guwahati (India), and MS and PhD in Mechanical Engineering from Georgia Institute of Technology, Atlanta. His research interests are in the field of collective systems innovation and multilevel design. He is a member of ASME and ASEE.

INTERNATIONAL CONFERENCE ON ENGINEERING DESIGN, ICED11
15 - 18 AUGUST 2011, TECHNICAL UNIVERSITY OF DENMARK

BALANCING INTERNAL AND EXTERNAL PRODUCT VARIETY IN PRODUCT DEVELOPMENT

Iris Graessler
Robert Bosch GmbH, Stuttgart, Germany

ABSTRACT

Applicability of "Mass Customization" to mechatronic systems is proven by various product examples in automobile industry. For example, chassis performance is adjusted to the driver's specific wishes or to present driving circumstances, such as road condition. As a basic principle, hardware forms functional framework while software defines specific functional contents and characteristics.

Balancing internal with external product variety emerges as critical success factor in this context. From external point of view, as much variety shall be provided as end customers are willing to pay for. From internal point of view, each product variant induces consequential costs and thus lessens profitability. In this contribution, a methodology of designing a construction kit for customer specific solutions based on classic German design theories is proposed. A modular product architecture forms the logical context of the construction kit for customer specific solutions. Deduced products are individualized by selection and connection of standardized, discretely and continuously varying components. Thus economic variation becomes feasible also on a high technical level.

Keywords: Product Variety, Complexity, Mass Customization, Design Methodology, Construction Kit for Customer Specific Solutions

1 INTRODUCTION

As mechanics, electrics, electronics and software follow a synthesis trend to mechatronics, customer specific variation can increasingly be offered at competitive prices. The basis of this approach is formed by the competition strategy of Mass Customization. Mechatronic systems make up one of the most promising application area within the branch of industrial goods.

1.1 Mass Customization

The term "Mass Customization" combines the two contrasting approaches of Mass Production and Customization. Mass Production implies cost reduction due to scale effects and gained production experience. Customization focuses on exact fulfillment of customer's requirements and results in an unique competitive position. Mass Customization therefore aims at producing products to meet individual customer's needs with mass production efficiency [1]. Thus customized products are offered at prices comparable to standard products and continuous individual relationships are established between each customer and the manufacturer [2-5]. The combination of cost leadership and differentiation results in a simultaneous, hybrid competition strategy (figure 1).

For producing companies, the focus of Mass Customization lies on individualizing material core products. Often, tailored services related to the core products are offered in addition. Prerequisites of economic success of Mass Customization are mature markets and flexible technologies. Mature markets are characterized by heterogeneous, rapidly changing customer requirements, which can hardly be predicted. Flexible product technologies, such as adaptable materials or mechatronic systems, allow easy adaptation to the individual customer's preferences. Furthermore, generative or Laser driven production technologies make economic production possible in spite of varying characteristics and low lot sizes.

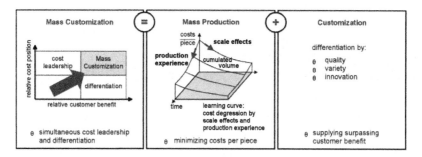

Figure 1. Definition of Mass Customization (based upon [6, 7])

1.2. Mechatronic Systems

Mechatronic systems emerge from functional shift and extension of mechanics to electrics, electronics and software. As a result of closely interacting disciplines, adaptive and intelligent systems are formed (figure 2). Due to functional integration of mechanics, electrics, electronics and software, the borderline between standard and variable system functions can be moved into areas of low efforts. Software thus advances to a variety driver within mechatronics and increasingly depends on application specific knowledge. Therefore, mechatronic systems are one of the most promising application field of Mass Customization for producing companies.

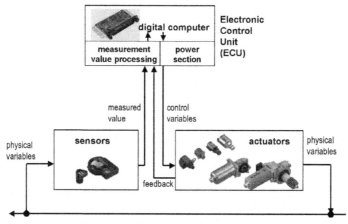

Figure 2. Principle of a mechatronic system

2 CUSTOMER SPECIFIC CONSTRUCTION KIT

In order to economically realize a wide range of variation, development must focus on order neutral creation of construction kits for customer specific solutions. With the term "customer specific construction kit" a construction kit is described, from which a defined range of customized products can be deduced. Deducing a customer specific variant implies reusing requirements, specifications, functions, principles, components up to product documentation and operation plans following a modular product architecture (figure 3). A detailed overview of modularization practices can be found in [8]. The extent of reuse ranges from taking standards to selecting discretely varying options and adapting continuously varying options.

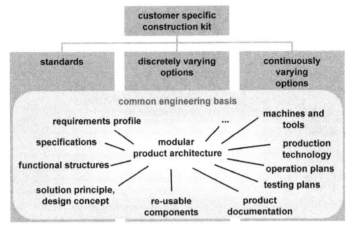

Figure 3. Costumer specific construction kit

In contrast to conventional construction kits, the customer specific construction kit is based on a prospective and revolutionary development approach. In industrial series production, construction kits are typically derived from already processed orders. The affiliated development approach can be subsumed as "retrospective" because sales volumes and specification ranges are already well known and serve as valuable basis of analysis. "Prospective" however means that the construction kit is developed in order to meet future requirements of new products. Synergies between different variants shall be opened up from the first deduced product on. Potential sales volumes are not yet known and have to be pre-estimated using market surveys. "Revolutionary" adds the challenge of developing such a foresighted design frame all at once instead of a step by step implementation.

A common product architecture of all products to be deduced serves as logical backbone of the customer specific construction kit. Deduced variants and applications are individualized by selection of standardized, discretely and continuously varying components and cross-disciplinary variation mechanisms.

Defining validity limits of the customer specific construction kit forms the basis of effecting a compromise between cost degression and individualization. Besides aspired lot size, the following four dimensions of validity have to be fixed (figure 4). The range of individualization (1st dimension) characterizes built-in variety. It reflects the spectrum of selection alternatives as well as limiting combination rules and exclusions. The defined range of individualization decisively influences how many application development projects can be served by the same construction kit. As counterpole to individualization, the 2nd dimension "level of product hierarchy" stands for standardization. As pointed out by [9] modularization in new product development can take place at many different levels. Therefore the product hierarchy level of standardization indicates, whether standardized components can be found on level of parts, subassemblies or entire platforms. From production point of view, the number of preferred production technologies and the flexibility of production method are represented by the 3rd dimension "range of production". With the 4th dimension "temporal stability" intended economic life-time, questions of generation planning, upward- and downward compatibility are addressed.

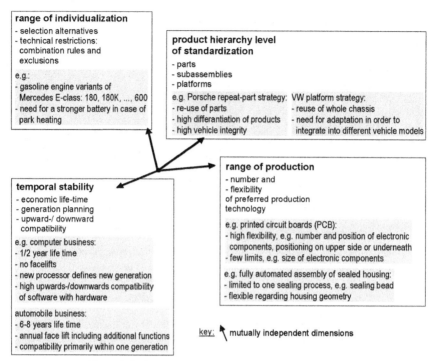

Figure 4. Defining validity of customer specific construction kit

3. DESIGN METHODOLOGY

In order to put such a customer specific construction kit into industrial practice, a corresponding design methodology has been developed [10]. The subsequently presented design methodology supports a systematic, methodical procedure of order neutral creation of construction kits for customer specific solutions with regard to specifics of mechatronic systems. The methodology meets the following three demands. First, classic German design theories are integrated. Thus, the methodology is based on a systematic, methodical procedure. Second, known methods and tools of creating standards and discretely varying components are used and supplemented by new approaches of designing continuously varying components. For each design phase, a selection of appropriate solution approaches is provided in a clear and well structured manner. Third, cross-disciplinary variation mechanisms are created by integrating and coordinating involved disciplines. According to the respective design phase, needs of coordinating partial solutions between involved disciplines are changing. Therefore, focus of design methodology lies on defining appropriate interfaces between partial solutions in order to form a balanced overall solution. In the following, these characteristics are described using examples.

3.1. Phases

Due to underlying prospective and revolutionary development approach, the design task is characterized by a high degree of innovation which results in the need of early design phases, e.g. "establishing function structures" or "finding working principles". As reference, VDI guideline 2210 E is taken, in which a multitude of German design approaches were unified [11], compare also VDI 2206 [12]. Based on this phase model, the design methodology for customer specific construction kits is structured into six phases (figure 5). During the initial phase "planning customer specific requirements" clear limits between standards and variation are drawn within specification. This separation of product characteristics into "standards", "discretely varying" and "continuously varying" is kept up during entire subsequent design process. Based on specifications, a functional product architecture is derived (phase 2). Partial functions and functional structures are partitioned according to provided variability and involved disciplines. According to [13], possible product modularity depends on similarity between the physical and functional architecture of the design. In phase 3, appropriate working principles are selected and cross-disciplinarily connected. Special attention is paid on mutual interactions between chosen principles, effects and algorithms.

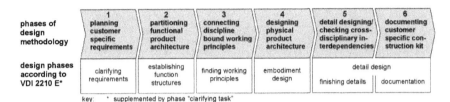

Figure 5. Phases of design methodology

Physical product architecture derived from the principle solution makes up the logical backbone of the customer specific construction kit (phase 4). On condition of this common structure, standardization and individualization are balanced. Standard, discretely varying and continuously varying components are designed to be mainly independent from each other and to be recombined. Due to standardized interfaces, the construction kit can be expanded order neutrally as well as order specifically. In phase 5, modular structures and components are detailed. Functionality and compatibility of connected variation mechanisms are checked. Finally, results worked out in phases 1-5 are comprehensively documented in phase 6. Besides product documentation, procedures and rules of handling the construction kit are defined.

3.2. Methods Use and Results

Adequate methods and tools are assigned to each phase of the design methodology for customer specific construction kits. On the one hand, they are structured applying the view-points "entire mechatronic system", "mechanics/ electrics/ electronics" and "software". On the other hand, they are separated according to their application into "standards", "discretely varying" and "continuously varying". Examples of methods and tools are Quality Function Deployment, generalized elements of Product Line Approach known from software development, morphological boxes, modularization, up-/downscaling, architecture evaluation and compatibility checks. In figure 6, methods of phase 2 "partitioning functional product architecture" are shown. Functional product architecture is an architecture of system functions, whose partial functions are adapted to individual preferences by variation and adaptation.

In figure 7, the methods "enhanced functional subdividing" and "enhanced functional structuring" are applied to the case study "power window actuator". The methods are based on functional subdivision and functional structuring introduced by [14, 15]. Using enhanced functional subdividing, the entire function is subdivided into partial functions, until these can be separated into "standard (S)", "discretely varying (V)" and "continuously varying (I)" partial functions. As a guideline, system variety shall be isolated in distinguished partial solutions. Thus, an embodiment structure is prepared from early on, in which standard components are kept distinct from individual components. In case of the power window actuator, individualization is realized in user interface (switch) and squeeze protection. For example, the switch can be configured in a manner that the window moves downwards

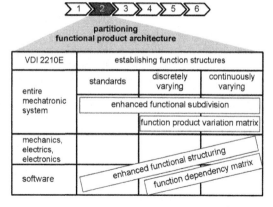

Figure 6. Methods use in phase 2

if the switch is pressed shortly or if pressing is kept up continuously. The configuration of the switch is chosen according to the customer's or end customer's wishes. Set-actual comparison for squeeze protection however has to be adapted to the window lifter's engine power. Also, the roadster's function of inside pressure regulation when closing the doors has to be taken into account. While a squeeze situation requires that the window immediately moves backwards and stays open, inside pressure regulation takes the windows to open shortly and close again instantly when doors are closed.

Partitioning partial functions into standard, discretely varying and continuously varying depends on the disciplines in which the functions shall be realized. Therefore, enhanced functional subdividing and enhanced functional structuring are mutually interacting with each other. Following enhanced functional structuring, partial functions are structured into the corresponding disciplines. In the process the general guideline is applied, that high degrees of individualization are to be implemented in software. Mechanical, electric, electronic subsystems shall primarily be used to realize standards or discretely varying partial functions. Supplementary, individualization options shall be separated in one discipline only. Thus coordination efforts within development and testing are minimized.

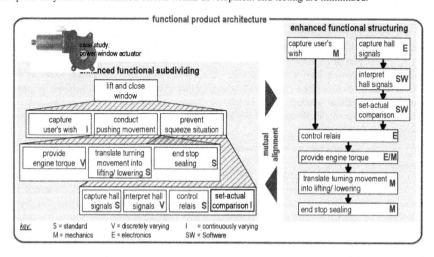

Figure 7. Case Study Power Window Actuator

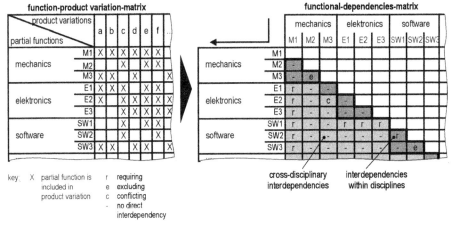

Figure 8. Example of methods use in phase 2

Besides enhanced functional subdividing and structuring, mutual logical dependencies between optional and necessary partial functions have to be determined and checked. In order to fulfill this purpose, matrices are applied. The idea of "function-product variation-matrix" (figure 8) is based on the "product and feature matrix" of software development [16]. In function-product variation matrix alternative choices of functions of the customer specific construction kit are mapped. Due to the pursued prospective design approach predicted product variations are taken instead of selected pilot customers. In addition to the overview of alternative functional ranges, transparency of mutual dependencies is given in a functional-dependencies-matrix (based on feature graphs [16]).

Core result of design methodology's phase 2 is a functional product architecture to be used by all customized variants. An overview of resulting partial results of phase 2 is given in figure 9.

3.3. Cross Disciplinary Variation Mechanisms

Besides general tolerance of cross disciplinary variation mechanisms, in particular compatibility of interconnected variety ranges has to be ensured. Basis of coordinating partial solutions between involved disciplines is established by breaking down the entire function into partial functions during phase 2 as described above.

Mutual compatibility of discipline bound working principles, effects and algorithms is checked and ensured in phase 3. Only working principles which are compatible among each other, are selected and connected. Due to cross linked disciplines, working principles must not only be compatible within each discipline, but also cross disciplinary. Principle solutions are only valid, if variety ranges of partial working principles complement one another to the functionally required overall variation span. Interactions between alternative states of interconnected working principles are checked, whether they weaken, exclude or intensify each other. Critical constellations must be replaced by alternative working principles.

In the context of physical product architecture (phase 4), logical variation possibilities are embodied in terms of components and coordination is optimized as a whole. In this step, cost effects of assigning variety to disciplines become evident. The strategy of cross disciplinary variation mechanisms is put into action by the following approaches. As already known from conventional construction kits, customers select components out of a collection of alternatives as well as additional components can be mounted. As characteristics of construction kits for customer specific solutions these mechanisms

VDI 2210E	establishing function structures

entire mechatronic system	- enhanced functional subdividing
	- enhanced functional structuring
	- functional range of expected product variations
	- specification of
	   - cross-disciplinary functions
	   - functional interfaces between the disciplines
	   - discipline bound standardization and variation share

mechanics, electrics, electronics	- specified partial functions
	- partial functions assigned to disciplines to become:
	   - standards
	   - discretely varying
	   - continuously varying
	- structures of partial functions
	- specified interfaces between partial functions
	- specified interfaces between disciplines
	- dependencies between partial functions

software	- defined software features
	- software partial functions to become
	   - standards
	   - discretely varying
	   - continuously varying
	- structures of partial functions
	- feature graph
	- specified interfaces within software partial functions
	- specified interfaces with other disciplines

Figure 9. Partial results of phase 2

are completed by adaptation of discretely varying components and configuration of continuously varying components. Prerequisites are standardized interfaces, independent components and limited interactions between all kinds of components.

In phase 5, connected variation mechanisms are investigated on level of the entire mechatronic system and optimized if necessary. In detail, combination of discrete and continuous variation ranges must correspond to the overall variation span defined in the specification.

Variation mechanisms determined along development are documented using parameter tables (for discretely varying components) and technical restrictions or constraints (for continuously varying components) in phase 6. Besides direct and desired interdependencies, also unwanted interactions of multiple parameter variation are documented.

4. CONTROLLING THE BALANCE OF INTERNAL AND EXTERNAL VARIETY

In order to control consequent realization of product variation, a measurement system consisting of key figures must be put up and integrated into business processes. Within these processes, not only responsibilities, but also reporting and decision structures have to be defined. In figure 10, the resulting process is illustrated. During conceptual design of customer neutral platform development the appropriate degree of re-use is planned. As all platform standards, preferred production processes and preferred components, are stored in a universal techniques catalogue, new proposals for standards are identified. Proposals must be released by a technical committee before they become a part of the techniques catalogue. If open questions are left, proposals are reworked. Key figures of planned re-use are used as milestone criterion of platform concept release.

Once the platform concept is released, derived variant or application projects can use the planned re-use as development guide. First of all, specific product requirements are compared with planned re-use. If deviations from techniques catalogue are detected, they have to be released by the technical committee. As next step, the planned conformity factor of the customer project resulting from comparison has to be released at the milestone "conceptual design". The conformity factor describes in how far a derived variant or application follows the planned logic of re-use. From this point on, realized conformity factor is continuously reviewed against planned conformity factor. Thus a self controlling system is implemented.

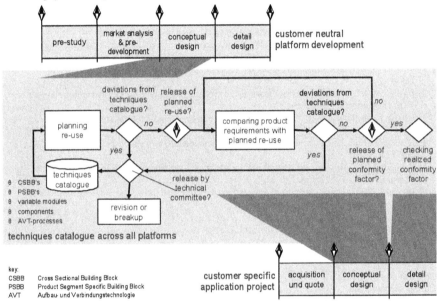

Figure 10. Business process of controlling internal variety

5. CONCLUSIONS AND FUTURE PERSPECTIVES

In this contribution, a design methodology for customer specific construction kits is presented. Applying this design methodology, producing companies are enabled to open up potentials of mass customizing mechatronic systems.

However, successfully initializing, implementing and keeping up a customer specific construction kit, relies on a profound change of values and behaviour in all classical key functions of an enterprise. Therefore, much thought and effort must be spend on change management in order to implement the new design methodology. Special emphasis must be put on sales and marketing, engineering, production and logistics.

REFERENCES

[1] Tseng, M. M., Lei, M., Su, C., 1997, A collaborative control system for Mass Customization manufacturing. Annals of the Cirp, 46/1997/1, pp. 373-376
[2] Toffler, A., 1971, Future shock, Pan Books, London, Basingstoke, Oxford
[3] Davis, S., 1987, Future perfect, Addison-Wesley, Reading, Massachusetts
[4] Pine, J. B. II, 1991, Paradigm shift: from mass production to mass customization, Master Thesis MIT Cambridge
[5] Kotha, S., 1995, Mass customization, implementing the emerging paradigm for competitive advantage. Strategic Management Journal 16 (1995), pp. 21-42
[6] Corsten, H., 1998, Grundlagen der Wettbewerbsstrategie, Teubner, Stuttgart, Leipzig, 1998
[7] Fleck, A., 1995, Hybride Wettbewerbsstrategien, Zur Synthese von Kosten- und Differenzierungsvorteilen, Gabler, Deutscher Universitäts Verlag, 1995
[8] Brun, A., Zorzini, M., 2009, Evaluation of product customization strategies through modularization and postponement, International Journal of Production Economics, Vol. 120, pp. 205-220, 2009
[9] Hsuan, J., 1999, Impacts of supplier-buyer relationships on modularization in new product development, European Journal of Purchasing & Supply Management 5, pp. 197-209
[10] Graessler, I., 2004, Kundenindividuelle Massenproduktion, Entwicklung, Vorbereitung der Herstellung, Veränderungsmanagement, Springer, Berlin, Heidelberg, New York et al.
[11] VDI 2210, VDI-Richtlinie 2210 Entwurf, Datenverarbeitung in der Konstruktion, VDI Düsseldorf 1975
[12] VDI 2206, VDI-Richtlinie 2206 Entwurf, Entwicklungsmethodik für mechatronische Systeme, VDI, Düsseldorf, März 2003
[13] Ulrich, K., 1995, The role of product architecture in the manufacturing firm, Research Policy 24, pp. 419-440
[14] Beitz, W., 1972, Übersicht über Konstruktionsmethoden, Konstruktion 24 (1972), pp. 68-72, 109-114
[15] Pahl, G., 1972, Analyse und Abstraktion des Problems, Aufstellen von Funktionsstrukturen, Konstruktion 24 (1972), pp. 235-24
[16] Bosch, J., 2000, Design and use of software architectures, adopting and evolving a product-line approach, Addison-Wesley, Harlow, London, New York, 2000

Contact:
Dr.-Ing. Iris Graessler
Robert BOSCH GmbH
CP/PUQ
Postfach 106050
70049 Stuttgart
Germany
Tel.: +49 711/811-38339
Email: iris.graessler@bosch.com

Dr.-Ing. Iris Graessler studied mechanical engineering and graduated as a PhD (Dr.-Ing.) at Aachen University of Technology (RWTH Aachen). In 2003 she qualified as a university lecturer (Privatdozentin) at RWTH Aachen. Since 11 years she has been working for BOSCH in several managerial functions in the fields of Product Development, Lean Production and Continuous Improvement Process. Since August 2011 she teaches and researches as full professor for design methodology and product development at the Cologne University of Applied Sciences.

INTERNATIONAL CONFERENCE ON ENGINEERING DESIGN, ICED11
15 - 18 AUGUST 2011, TECHNICAL UNIVERSITY OF DENMARK

INTEGRATED PRODUCT AND PRODUCTION MODEL – ISSUES ON COMPLETENESS, CONSISTENCY AND COMPATIBILITY

Stellan Gedell[1], Anders Claesson[2] and Hans Johannesson[1]
(1) Chalmers University of Technology, Sweden (2) Saab Automobile AB, Sweden

ABSTRACT
Product development of complex products and their corresponding production systems continue to provide challenges in industry as well as interesting and challenging research questions. Recent research in the area has aimed at increased understanding and development of an integrated product and production system-modeling framework supporting cross-functional collaboration and concurrency. In this context, a well-known challenge in industry is the problem of how to ensure correct and complete sets of parts for manufacturing of different product variants. In striving towards integrated modeling capabilities this is one of several fundamental problems to be addressed. Thus, this problem has been in focus for the research work reported on in this paper. The work includes a framing of the concepts of completeness, consistency, and compatibility. Based on this framing a case study is conducted exploring the possibilities and implications involved in using the modeling framework to include supporting functionality. The case study is ongoing and preliminary findings are included in this paper.

Keywords: product development, product model, systems theory

1 INTRODUCTION
Systems engineering is an important field of research. It aims at a more systematic development process, characterized by a method- and model-based cross-functional collaboration and concurrency, supported by information management tools. Information management tools are an important pre-requisite to enable the required information sharing as well as provide for necessary traceability. Furthermore, explicit information carried in formal information management tools are a fundamental pre-requisite and starting point for knowledge capture and reuse. The dependencies identified between product and production implies a need to strive for cross-functional collaboration, which highlights the need for information models and tools capable of describing the product and the production systems using one integrated model.
Based on a systems theory approach, recent research has resulted in an integrated modeling framework supporting collaborative design of product and production systems [1]. Included in that work dependencies and interactions within product and production systems has been elaborated and an integrated product and production systems model is presented. However, a deceitfully simple question to request the model to produce a complete list of parts required to manufacture a specific product variant reveals the need to provide additional thoughts on the concept of *completeness*.
The work presented here is a first step towards increased understanding of the issues involved and their implications on some additionally needed capabilities in the modeling framework. In other words, the scope of the paper is mainly to problematize on *completeness*, *consistency* and *compatibility*. The ongoing case study serves to contribute empirical data to the discussion of the problem and as a source of increased understanding of the validity of proposed solution approaches as well as potential hidden challenges and pitfalls. First, a short description of the modeling framework in [1] is presented. Then, the issues of completeness, consistency, and compatibility are presented and elaborated. An ongoing exploratory case study aiming to enhance the modeling framework is presented. Finally, some conclusions and reflections are provided.

2   INTEGRATED PRODUCT AND PRODUCTION MODEL

Some of the important aspects of the modeling framework [1] used as a starting point for the work presented in this paper are outlined below for convenience and in order to highlight some important features and aspects of the framework. This is done through describing three important cornerstones. The first cornerstone illustrates some important fundamental aspects of the framework. The framework was originally proposed by Claesson [2], and further enhanced by Gedell [3], aiming to support *structured development of complex, variant rich and platform-based design by means of re-use and information sharing*. It has some important advantages to point out:

- It can represent any system of interest. This capability can be used to represent any abstract system while not being limited to, for example, physical parts.
- Multiple similar designs can be represented by one parameterized model, a model with a design bandwidth defined by its parameters. This can provide an overview and definition of the product range the design is capable of supporting, which is an important aspect in platform design. With reduced duplication of work the workload is minimized. Quality is improved since the possibility for mistakes is reduced as a consequence of a reduced number of models to maintain.
- A system is described by means of *design solutions*. The amount of details, in other words, the granularity of the description depends of the purpose of the model. Consequently, there is no *right* amount of details, it depends on what the model is intended to support. The level of detail is to some extent guided by the need to provide sufficient description of the performances to be expected from a particular design solution. Furthermore, the design solution – in its context – will collaborate with other design solutions resulting in emergent properties. Comprehensive system models include these emergent properties as well as how these arise from the collaborating subsystems.
- A design rationale model is used to explain why a design solution is chosen, in terms of what set of requirements that the design has to meet. The design rationale model consists of design solutions, functional requirements, constraints and relation objects [3]. The relation objects carries the information why a design solution is considered a good choice to meet the requirements. When design rationale [4] is included within the model its usability in terms of modification and re-use in highly improved. Those that are interested in the design can easier understand why a design solution is chosen when the reasoning behind that choice is presented together with the functional requirements and constraints that it fulfils.
- Extensive designs can inadvisable be model as monolithic units. Though a design is not the sum of its parts, as will be discussed in the next chapter, it is practical to break complex phenomena into parts. For example, the parts may easier be identified as usable in multiple designs with the advantage of economy by scale as one driver. Another example is when multiple stakeholders, organizations and companies want to have a clear division of responsibility. Extensive designs can be described as systems composed of sub-systems, which – in their turn – are composed of sub-systems in a recursive fashion. Composition includes how a system presents its configurability to potential super-systems as well as how a system selects and configures sub-system.

The second cornerstone in the integrated product and production model [1] are the interactions between the product model and the production model. The framework is based on systems theory and Hitchins [5] gives some valuable input to the importance and effects of interacting in the statement:

"A system is an open set of complementary, interacting parts with properties, capabilities, and behaviors emerging both from the parts and from their interactions".

There are several important aspects that can be extracted from this sentence.

- *A systems behavior* is a consequence of the system itself *and* its interactions with other systems as well as its own *internal* structure and *internal* interactions. In other words, it is not meaningful, nor possible to describe and understand the behavior of a system without considering its *context* as well as its *internals*.
- Thus, the behavior of a system is not simply the sum of the behavior of its parts, as opposed to reductionism. Decomposing, without a mechanism to model emerging behavioral characteristics of the system is a simplification and will consequently have shortcoming.
- Finally, even though not explicitly mentioned in the citation above, systems behave differently in different stages in their lifecycles (Figure 1). This can be illustrated if we consider a system

model of a car. When a car is being produced in a plant it can be seen as two systems (the car and the plant) that interact with each other. The car is in its manufacturing lifecycle, whereas the plant is in its use lifecycle phase. Focusing on the system model of the car, this model of the car has previously been in its definition (or development) stage of its lifecycle. Then, after being in its manufacturing lifecycle stage it will be entering its supply and use cycles of its life.

The temporal duration of super-systems formed by interacting systems varies. Super-systems can be formed with the intention to have a relatively long duration like the use lifecycle phase of consumer products. A super-system, which is formed to describe the production lifecycle phase of a product, will have a short duration. For example, when parts are placed in a fixture, to be positioned before welding, they together can be seen as a temporary system. Similarly, every interaction that takes place during a products production phase and the production systems are possible to view as short-lived super-systems.

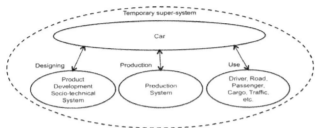

Figure 1, Super-systems formed of interacting systems during certain lifecycle phases.

The third cornerstone is the ability to allow multiple overlapping (partial) models. The interacting design solutions (Figure 2) can for modeling purposes be encapsulated in order to represent interacting systems for different purposes as indicated by the shaded areas with different colors to the right in the figure. Nothing restricts a design solution to participate in several different encapsulations.

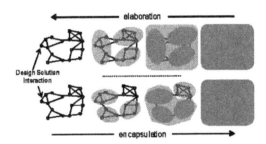

Figure 2, Two alternative elaborations and encapsulations [5].

Together these three cornerstones provide a foundation for the integrated product and production model [1]. How products and production processes relates to each other is facilitated by viewing the production processes as something going on within a production system, and similarly viewing the product (system) and production system as a temporarily formed super system, i.e. a system in its own right.

Figure 3, which is borrowed from [1], is used to illustrate an integrated model. The *body-in-white* is composed of the *roof panel production system* (160) and the *roof panel*. That exemplifies how systems originating from different organizational parts of the company (product and production) seamlessly form an integrated model. The interaction *align with pin & hole*, describes how the production system's (the *fixture's*) positioning pin interacts with the product's (the *roof panel's*) positioning hole. Finally, the body-in-white and the *body shop* can exemplify a temporary system, as they together forms a system during the body-in-white's production phase.

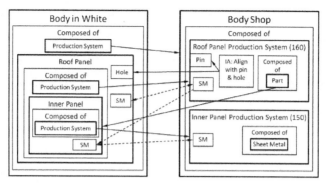

Figure 3. An integrated model representing a subset of a car and a subset of a plant [1].

3 COMPLETE, CONSISTENT AND COMPATIBLE

One of the key drivers behind the originally proposed framework [2] was the understanding of the product development process as a *journey* from an *incomplete* and *inconsistent* state of affairs towards a final gate "start of production" (or SOP) when the state of affairs should be characterized by a *complete* and *consistent* model capable of providing the information required to run large volume series production. Enabling a model to support an incomplete and inconsistent situation is rather easily achieved, for example, by simply avoiding putting any formal requirements on the model. The challenge, however, was not to just simply allow for an incomplete and inconsistent model. The challenge is rather to provide modeling mechanisms that allow the modeler to understand and identify when the model is incomplete and/or inconsistent in order to provide him opportunities to recover and correct such a state if it is of importance to do so. From an overall perspective, the speed of a product development process is equal to the speed of convergence from an incomplete and inconsistent model to a complete and consistent state.

The term *complete* can, deceitfully, be perceived as an absolute term, "When everything is there it is complete". However, an absolute definition vanishes at a closer look upon the issues involved. For example, using *Google define* we find two representative results from a search on complete: (1) *having every necessary (...) part*, and (2) *with all the necessary parts*. Instead of an absolute meaning, *complete* depends on a relative or subjective opinion of what is *necessary*.

Requirements and their solutions generally co-develop during the design activities. We initially focus on the solution part of a product description in order to provide a starting point for reasoning about completeness. The manufacturing of a product can be used to illustrate a concrete situation where completeness may be a problem. *When a product is manufactured as an assembly of parts, how do we know and ensure that exactly the required set of parts are selected and assembled?* The required set of parts required to assemble the product can be said to be *complete* in the sense that these parts are both required and sufficient to form the product – they are *necessary*. Adding or subtracting a part from this set would make the set either incomplete (one or more parts lacking) or redundant (one or more parts to many). This may seem as a trivial thing to achieve. However, when producing complex and variant rich products (like cars) there are a couple of thousand parts required per product and thousands of product variants possible to produce – each variant requiring a specific set of parts in order to be *complete*.

For the reasons mentioned above, the question – *What parts constitute a complete set of parts for the assembly of a product variant?* – is by far not trivial and most relevant. The situation described clearly shows that the answer to the question is that – *it depends*. It depends on which variant of a product that is to be produced. Clearly, different product variants will have different sets of parts depending on which features the product shall have as well as on the set of requirements it must fulfill. For example, a set of parts, that is complete for a low content product, will most certainly be incomplete for a high content product. The conclusion drawn from this is that *complete* is a relative concept that describes if some fundamental need is met. Our challenge is to explore if, and how, this issue can be handled using our modeling framework.

Thus, for the modeling framework to be able to make a statement about the *completeness* of a design, it is required to include a request, or expectation, of the modeled design. This request or expectation will establish a statement on what is *necessary*. The *criterion for completeness* is met when the modeled design solutions leaves none of the *explicitly stated requests* or *expectations* unanswered. The modeling framework, therefore, must include modeling mechanisms to formulate and establish these requests and expectations as well as modeling mechanisms allowing the modeled design solutions to provide *responses* to such requests and expectations.

Having provided a base for reasoning on completeness we now turn our interest to the concept of *consistency*. In logic, a consistent theory is one that does not contain a *contradiction*. The lack of contradiction can be defined in either *semantic* or *syntactic* terms. Consistency, in general, is also used in a slightly different way meaning a harmonious uniformity or agreement among things or parts or something of a regularly occurring, dependable nature. However, in this context we refer to consistency in the former meaning that consistency among a set of statements implies that no contradiction *logically* follows from these statements.

To include reasoning about consistency in our modeling framework we must examine it to identify where we are making statements that may lead to a contradiction. First we observe that unless we provide more than one statement about "the same thing" there is no possibility for a logical contradiction to arise. An example may be that we define a parameter, let's say a *length* of some design solution. As long as this parameter has no associated value, the model in a way could be said to be inconsistent, since a value is more or less required. However, it is probably more straightforward to view this situation as incomplete – we lack a value for a parameter. Then, someone assign this parameter a value. Now, unless the model contains other statements on this parameter and its value, the model is consistent and the value assigned to the parameter is simply a statement of fact – a kind of axiom. This clearly shows that in order to fruitfully discuss whether or not a model is consistent, the model must include mechanisms to make several different statements about the same entity. When all of these statements agree, we conclude that the model is consistent. Otherwise, we conclude that we have an inconsistency in our model. Looking at the model as a tool used to support product development during early phases on the journey from incomplete and inconsistent towards a more complete and consistent situation it is of no interest to just eliminate inconsistency, but rather provide support to identify inconsistencies in order to support understanding of the causes and thereby moving the sequence of design decision further.

As stated above, the modeling framework must allow that multiple statements can be made on the same entity (or fact). Another example of this might be allowing different stakeholders to use different evaluation methods to obtain a performance value. Even though the individually returned performance values may differ this is not a sufficient ground in itself to conclude that they are inconsistent. Provided that all these statements (performance values) according to *some criterion* agree with each other we still may conclude that the model is consistent. If they to some extent, again according to some criterion, disagree we may conclude that our model is inconsistent. Consistency as defined in our context can only be evaluated when there are more the one statement available about an entity, and a method as well as a criterion to determine whether there is a contradiction among these statements or not.

Moving the focus to the third issue – *compatibility* – we first recall that our integrated product and production-modeling framework is based upon a system oriented modeling approach. The framework provides several opportunities to represent complex and configurable systems that are defined and described as collections of collaborating *sub*-systems. The modeling mechanisms provided to define and describe these collaborations are primarily through *interfaces* and *interactions*. The modeling mechanism referred to as *composition* is used to identify which systems to include in such a collaborative collection. A consequence of collecting a set of systems with an expectation that they will collaborate is that the corresponding set of interfaces and interactions thereby obtained will – in a sense – *connect* and fulfill their expected behavior. Interfaces that in this way have been able to connect and fulfill their expected behavior can be viewed as *compatible*, i.e., they are capable of providing the requested and expected levels of collaboration. However, this is a conclusion that might not be possible to draw looking on a single interaction only. The reason for this is that we need to allow observation of emerging properties that we may have modeled on a "higher" system level. Even though an interface may seem locally ok, it may be the case that the results on important emergent properties are not as requested or expected. The implication here is that conclusions on compatibility when viewing a system collection can be successively made starting at an individual interaction among a couple of interfaces and then successively propagating towards higher system levels ensuring that compatibility among the collaborating systems are maintained on all system levels. Another interesting issue regarding compatibility is that if offered capability meets or exceeds requested capability the request and response are compatible, otherwise offered capability is incompatible with request.

An important aspect to consider is that the quest for a complete, consistent and compatible model must not limit the models capability to handle incompleteness, inconsistency and incapability during the model's design lifecycle. In the design phase incompleteness and inconsistency must be allowed, for example due to conflicting design alternatives or stakeholders' prioritizations. To rigid processes or tools will hinder design activities and likely create frustration, for example due to reduced organization efficiency or deviation from prescribed processes and rules. To summarize, it is necessary for a design model to support incomplete and complete, inconsistent and consistent, and incompatibility and compatibility.

4 IMPLICATIONS FOR THE INTEGRATED MODEL

The modelling framework as it has evolved to the state described in [1] does not really include any mechanisms to support the kind of reasoning described above regarding completeness, consistency, or compatibility. In order to enable our modeling framework to include mechanisms to support issues on completeness, consistency, and compatibility we must provide some form of automated, or semi-automated, reasoning support. Formally, *automated reasoning* is a research area in its own right (e.g., see [7]). The objective of automated reasoning is to write computer programs that assist in solving problems and in answering questions requiring reasoning [8]. In a semi-automated reasoning such a program is used in an iterative fashion; that is, you can instruct it to draw some conclusions and present them to you, and then, based on your analysis of the conclusions, it can in the next run execute your new set of instructions. Alternatively, you can use such a program in a batch mode, that is, you can assign it an entire reasoning task and await the final result. The intention in our case is to enhance the modeling framework with some *basic capabilities* to provide for a first step towards a *semi-automated reasoning* with focus on our issues concerning completeness, consistency, and compatibility. An interesting overview of different forms of automated reasoning is provided in [9].

Reasoning is a process of drawing conclusions from facts. For the reasoning to be sound, these conclusions must inevitably follow from the facts from which they are drawn. In other words, reasoning is *not* concerned with some conclusion that has a good chance of being true when the facts are true. Indeed, reasoning as used here refers to *logical reasoning*, not of *common-sense reasoning* or *probabilistic reasoning*. The only conclusions that are acceptable are those that follow *logically* from the supplied facts.

This rather strict definition of reasoning is not really what we are aiming for in our ambition to provide a modeling framework capable of supporting concurrent product and production development. An engineering solution does not seek to claim that it is *logically right* – an engineering solution is one solution – among many potential solutions – that valued in the context of a set of expectations and requirements is *good enough*. What makes it so difficult is the vast amount of design parameters possible to decide upon and the many performances upon which expectations and requirements are placed. A further complication is that many of the most important performances upon which we place expectations and requirements are emergent properties on higher system levels and thus very difficult to attribute to any particular set of design parameters where the design decisions actually are taken. The consequences of the design decisions emerge from a whole range of design decisions rather than from any one decision in particular.

The mechanisms available to us within our modeling framework to start our journey towards providing some form of semi-automated reasoning along the thinking outlined above are primarily our parameters. In [2] three semantically different kinds of parameters were distinguished: *design parameters*, *performance parameters*, and *variant parameters*. The understanding of these are that *design parameters* are those parameters that a design engineer or a decision maker can influence and decide upon their values in order to form design solutions in accordance with their intentions. *Performance parameters* are additional parameters that provide information about the consequences or outcomes from the design solutions in terms of observable properties of interest. The understanding of how parameters depend on other parameters is captured introducing a new modeling element referred to as a *parameter map*. *Variant parameters* are a kind of convenience mechanism that, for example, enables us to refer to huge sets of parameters with one simple statement of a value of a variant parameter. Conversely, the value of a variant parameter may be derived from an observation of the values of a set of other parameters, thereby providing a sort of automated categorization of which variant we currently are dealing with. With these basic modeling elements in place the first step towards an ability to provide a simple form of semi-automated reasoning is in place.

Another mechanism we will need to introduce is a possibility to define *expressions* and/or *constraints*. For example, we will need to establish a constraint expressing that a certain performance value must exceed a required value. Another example might be the performance parameter *weight* that we would like to minimize while also requiring it to be below a threshold value. Since we also want to support multiple opinions and allow for more than one statement on an entity, we furthermore need to provide explicit mechanisms supporting which statements we are taking into account during an evaluation of a constraint or expression as well as how we arrive at a certain conclusion. Yet, another issue to provide for in the modeling framework is how to initiate and trigger evaluations of constraints and expressions as well as how conclusions and results from parameter mappings are allowed to propagate forming a chain of successive mappings and conclusions. The steps taken regarding these concerns in the work reported here only touch upon these subjects in a basic and simple manner. More elaboration on this particular topic is beyond the scope of this work.

5 AN EXPLORATORY CASE STUDY

The purpose of the exploratory case study is to apply the thoughts on completeness, consistency, and compatibility outlined above. Further, the aim is to enhance the modeling framework in practice and examine the capabilities achieved through the enhanced framework. The approach taken in pursue of this exploration was to apply the framework in an attempt to model a car program (products) and the production system required to manufacture these products. The intent is to apply the framework and use it to describe and define current and next generation products and production system(s) as well as the platform(s) upon which these are based and derived. The study is conducted in collaboration between academic and industrial partners and based upon accumulated industrial experiences as well as research results obtained from many years of research in the area. The models created shall include solution bandwidths, architecture definitions, and definition of the platform(s). The approach is to define and maintain a complete and consistent holistic model while continually refining, detailing, and extending the model through elaboration and encapsulation.

As mentioned in the introduction, information management tools are required prerequisites for dealing with these models. Since the authors are unaware of any existing tool that can be used to capture, maintain and manage the information model defined in the framework, the study also include the creation of a prototype tool with enough functionality to work with the modeling aspects in focus of the study. Creating and using this prototype tool will provide valuable insights in itself and be a learning platform both in terms of modeling methodology and in terms of usability requirements on a future and more efficient tool.

From a scientific point of view the expectations on the case study are that it will provide both empirical validation of the proposed integrated product and production modeling framework and new insights in new questions for future research. From an industrial point of view the case study will enable an update on the modeling frameworks state-of-art and subject it to some relevant industrial issues in order to gain understanding about current modeling capabilities as well as experiences and knowledge about important issues to develop further in the future.

6 CONDUCTING THE CASE STUDY

The aim of the case study is to examine the capability of the enhanced modeling framework regarding the defined issues on completeness, consistency and compatibility. In order to achieve this, the model will have to include both system level aspects such as product variants and performance expectations and detailed design decision on design parameters and the consequences of these in resulting performances. Furthermore, it is of interest to include possibility to model both physical and functional interactions on physical part level as well as emerging performances on higher system levels. A choice to model the chassis system in a car was made while it provides all the above opportunities and also includes well-known system level performance expectations (*braking distance* of a car). A chassis system of a car also provides many opportunities to model product variation.

Besides the modeling of an example system using the framework, the case study includes creating a prototype information management tool. There are two main reasons. First, the authors are not aware of any existing tool with the functionality to host the modeling framework and its required functionality. Second, the expected extensions to the modeling framework required in order to address the described issues on completeness, consistency and compatibility are not known in detail as to what functionalities the modeling framework and the information tool must be capable of providing. The case study is expected to shed some more light and understanding on these aspects.

Creating an information management tool for the proposed framework is by no means a simple and straightforward task. The first issue to deal with is that the definition and documentation of the framework is provided through the description and references provided above. As a consequence, any missing or ambiguous elements must be given complementary and assumed definitions. The second issue to deal with is that this work in itself is of an exploratory character having the implication that it is not entirely known beforehand exactly what has to be included in the information management tool, nor what functionalities it is expected to be able to provide. Both issues combine to a very ambiguous, unclear, and incomplete situation and foundation for creating an information tool. Thus, if this tool were to be created by a third party, the amount of work required to bring clarity to these issues would be almost overwhelming and require a lot of time and resources to be spent on creating more formal requirements for this tool. The approach taken in this exploratory study is to define and create the tool in parallel with the ongoing modeling and conceptual work.

In order for this to be feasible the information tool is conceptually divided into two major areas of functionality: information capture and information visualization. The work conducted so far has been to enable information capture of all (or most) modeling entities defined in the modeling framework as well as extending the framework with some modeling entities discovered to be of vital importance in order to address the research questions on completeness, consistency, and compatibility.

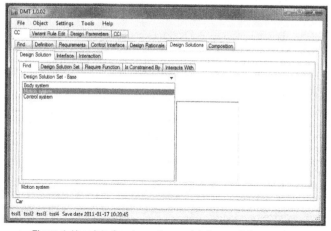

Figure 4. User interface to capture and define modeling entities.

The prototype information tool is developed using a C#-environment and is initially a single user and standalone application using simple files for data storage.

The user interface of the tool follows a more or less one-to-one mapping of the information model defined in the framework. Several sets of tabbed pages (Figure 4) have been used to provide an easy contextualization for each of the model entities to be defined.

The approach to create a product model including necessary elements to explore completeness, consistency, and compatibility takes is illustrated by Figure 5. The approach utilizes as a starting point, those physical parts of the chassis system that are required in order to manufacture a vehicle, illustrated by *rotor* and *brake pad* in the figure. Since the chassis exists in several different variants several sets of physical parts will be included in the model. The variability is represented in the figure by parameters, e.g. two rotor diameters. Depending on the product variant to build, this starting point provides requirements on the model to define how product variants will utilize different sets of parts. Thus, the model must include several additional model elements representing higher system levels, exemplified by *chassis*, until the vehicle system level, i.e. *car*, is modeled and described. On vehicle level both performance expectations (e.g. *braking distance* in fr:braking) and vehicle variants (e.g. *sportiness* derived from *driving experience*) are added to the model, thereby providing a starting point for examination of completeness as well as consistency. Starting with adding model elements on the vehicle level requesting a certain level of performance (in this case exemplified by *braking distance*) a request for a performance response has been defined. Until such response is provided the model is incomplete. In order to resolve this incomplete state the model must include additional elements capable of providing a connection between part level performances and delivered performance on vehicle level.

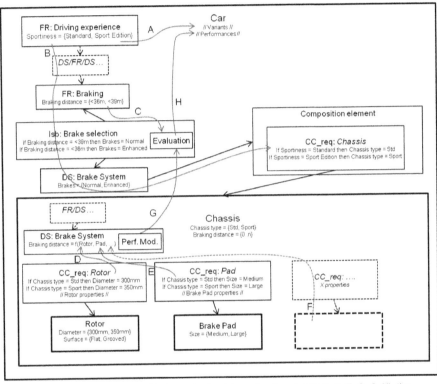

Figure 5. A subset of a car with the evaluation information flow (arrow D, E, F, G, C, H), the source for a variant parameter (arrow A), and a top-down design approach (arrow B).

Two elements in Figure 5 contribute in evaluating the designs performance relative the requirements, *evaluation* in *brake selection* and *performance model* (perf.mod.) in *brake system*. The information flows from the design solutions to the performance model (arrow D, E and F), to the *evaluation* (arrow C and G), and the result from the evaluation (arrow H) to *car*. Together these arrows form a bottom-up evaluation.

Attempting to establish this connection from design solutions to the car clearly showed that it is virtually impossible to form such a performance model by following a physically oriented product breakdown structure. As a consequence, the model must be capable of managing several overlapping modeling elements in order to provide for both a response on which parts to use for manufacturing and for calculation to evaluate achieved performance.

The *brake system's performance model* in Figure 5 is elaborated further in Figure 6, as a mean to model emergent properties. In order to model emergent properties, parameter mappings are utilized in design solution elements of abstract sub-systems, in this case *chassis*. This system representation provides the ability to host performance models that for example map the different angles of the chassis corners' (toe in, camber, caster etc.) contributions to performance measures on ride and handling. These and other similar design parameters contribution to the chassis behavior and performance requires an abstract dynamic chassis model (Figure 6) to be included.

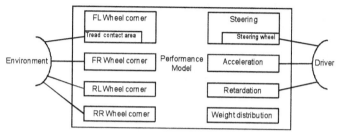

Figure 6, Illustration of some elements in a dynamic chassis model for Brake System.

Attempting to establish this connection from design solutions to the car clearly showed that it is virtually impossible to form such a performance model by following a physically oriented product breakdown structure. As a consequence, the model must be capable of managing several overlapping modeling elements in order to provide for both a response on which parts to use for manufacturing and for calculation to evaluate achieved performance.

The modeling elements outlined above constitute the prerequisites for a software agent to evaluate if necessary sub-systems are included. In other words, the modeling framework includes modeling mechanisms to formulate and establish requests and expectations as well as modeling mechanisms allowing the modeled design solutions to provide responses to those requests and expectations.

7 CASE STUDY RESULTS AND CONCLUSIONS

In performing the case study it is evident that deep insights of the design intent as well as the design itself are required in order to create appropriate descriptions and models. The focus for the case study was to model the vehicle chassis system and how it contributes to vehicle behavior through the different parts used. It was, however, an interesting experience for the researcher creating the model to realize that the task required knowledge way beyond his own, even though the researcher is an experienced senior engineer with long automotive experience. Even to briefly describe expected performances on car level and list a number of sub-systems requires deep and extensive knowledge. A respect for expert knowledge is a lesson to remember. In other words, our conclusion is that the product model preferably should be created and maintained as close to the source of knowledge as possible, i.e. by the designers themselves.

Further, concerning expert knowledge, modeling of performance will range from the complete vehicle down to individual parts. That range is seldom covered by a single person, but of a number of specialists that together covers the range from details to complete design. This puts even more emphasis on capabilities for supporting highly dynamic collaboration when using the described framework.

A strong benefit of creating a tool in parallel with the modeling research lies in the clarity that is required by the software tool in order to ensure proper capture and functionality for the modeling elements identified in the research. Furthermore, the requirement to actually capture the modeling entities using a software tool provides a higher level of clarity also in the approaches taken to the system modeling as such. In doing so, it becomes almost brutally clear where the modeling framework is supportive and where it has some weaknesses or missing elements or concepts.

The case study is still ongoing and results presented here are preliminary and based on the work done so far.

8 REFLECTIONS ON RESULTS AND CONCLUSIONS

In the early framing of the research scope presented a brief literature search was made in order to find a starting point and baseline. The outcome was rather disappointing and it was difficult to find a set of appropriate references upon which this research could be founded.

The completeness of a design is not absolute, but depends on the expectations of the design. The designs completeness can to a limited extent be evaluated against a list of expectations. There is, however, no known way to ensure whether this list in itself is complete or not. Obviously this is a recursive problem where it is only possible to state that completeness *depends*. The consequence is the need, presented above, to define a *criterion of completeness* in each and every case.

To illustrate and support the statement that the completeness of a design is not absolute, we can refer to Roozenburg and Eekels [6]. A design does not have functions (and thereby behavior and performance) on its own. Rather, a designs behavior (function) depends of the design itself (the four boxes in the upper left corner in Figure 7) in combination with its *mode and condition of use*. How a design is to be used is outside the control of the design itself. Actually the number of possible combinations of mode and condition of use is an infinite number. Based on this reasoning, the findings presented must be seen as initial steps in understanding and addressing issues on completeness, consistency, and compatibility.

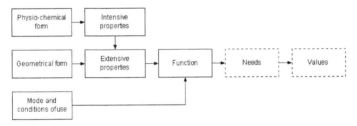

Figure 7: Product functioning [6].

A simplistic view and illustration of the problems dealt with above is presented in Figure 8. This generic feedback-loop shows the causality between expectations and performance responses from the design. It is our intention to apply the same approach of reasoning to the problems regarding consistency and compatibility.

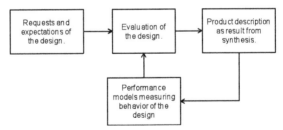

Figure 8. Conceptual illustration of the involved information blocks when evaluating completeness, consistency and compatibility.

The approach and findings presented has been found to provide a valuable starting point for further research on these topics.

ACKNOWLEDGEMENTS

This work was carried out at the Wingquist Laboratory VINN Excellence Centre within the Area of Advance – Production at Chalmers, supported by the Swedish Governmental Agency for Innovation Systems (VINNOVA). The support is gratefully acknowledged.

REFERENCES

[1] Gedell, S., Michaelis, M., Johannesson, H., *Integrated Model for Co-Development of Products and Production Systems – A Systems Theory Approach*, 2010, (Accepted by Journal of Concurrent Engineering)
[2] Claesson, A., *A Configurable Component Framework Supporting Platform-Based Product Development*, 2006 (Doctoral Thesis, Division of Product and Production Development, Chalmers University of Technology, Göteborg, Sweden)
[3] Gedell, S., *Platform-Based Design - Design Rational Aspects within the Configurable Component Concept*, 2009 (Licentiate Thesis, Division of Product and Production Development, Chalmers University of Technology, Göteborg, Sweden)
[4] Andersson, F., *The Dynamics of Requirements and Product Concept Management*, 2003 (Doctoral Thesis, Division of Product and Production Development, Chalmers University of Technology, Göteborg, Sweden)
[5] Hitchins, D. K., *Advanced Systems – Thinking, Engineering, and Management*, 2003 (Norwood, MA: Artech House)
[6] Roozenburg, N. F. and Eekels, M. J., *Product Design: Fundamentals and Methods*, 1995 (Chichester: Wiley)
[7] Portoraro, F., *Automated Reasoning*, 2010 Winter Edition (The Stanford Encyclopedia of Philosophy, Zalta, E. N. (ed.)), http://plato.stanford.edu/archives/win2010/entries/reasoning-automated/
[8] Wos, L., Overbeek, R. Lusk, E., Boyle, J., *Automated reasoning: Introduction and Applications*, 1992 (McGraw Hill)
[9] Bonacina, M. P. and Martelli, A., *Automated reasoning*, 2006 (Intelligenza Artificiale, III(1-2):14-20, Marzo/Giugno)

Contact: Stellan Gedell
Chalmers University of Technology
Product and Production Development
412 96 Göteborg
Sweden
Tel: Int +46 (0)736 278525
Email: stellan.gedell@chalmers.se
URL:http://www.chalmers.se/ppd/SV/organisation/avdelningar/produktutveckling/personal/doktorander/gedell-stellan

Stellan Gedell is currently a PhD student at the Department of Product and Production Development at Chalmers University of Technology in Gothenburg, Sweden. He started his PhD studies at Chalmers University of Technology in 2008 focusing on integrated platform-based product development.

INTERNATIONAL CONFERENCE ON ENGINEERING DESIGN, ICED11
15 - 18 AUGUST 2011, TECHNICAL UNIVERSITY OF DENMARK

A FRAMEWORK FOR DESIGNING PRODUCT-SERVICE SYSTEMS

Gokula Vijaykumar Annamalai Vasantha[1], Romana Hussain[1], Rajkumar Roy[1], Ashutosh Tiwari[1], and Stephen Evans[1]
(1) Cranfield University, UK

ABSTRACT

In this competitive globalizing scenario, manufacturers are adopting a strategy of bundling products and services into an integrated solution to create sustainable competitive advantage. Servitizing manufacturers are increasingly transforming their processes and practices to build product-service systems (PSS). During this transformation they require substantial support to face stringent challenges. Research in the PSS domain is heading towards the development of a design theory and methodology that facilitates the systematic creation of viable PSS conceptual designs. In this paper, various proposed design methods are reviewed and research gaps are summarized. Primarily, it has been observed that the importance of the capabilities of the stakeholders involved in designing PSS has not been noted in the proposed methods. Regarding this capability view point, a framework for designing PSS has been proposed. This framework highlights the important features required in designing PSS such as co-creation, responsibilities and competences. Every step in the framework has been explained with a case study involving laser systems used for manufacturing cutting operation.

Keywords: Product-Service System, design, capability, co-creation

1 INTRODUCTION

Manufacturers are feeling the strain of the recent recession and need alternate strategies to cope with globalization, reducing profit margins and for retaining and attracting customers. Servitization is a promising approach to help manufacturers to achieve these objectives. Servitization emphasizes the importance of service and aids in integrating products and services to satisfy customer needs better. It is a strategy of bundling products and services into an integrated solution to create sustainable competitive advantage. It aims to provide required customer value through reduce cost, optimized resources which can be sustained for both consumption and production. The term servitization is also referred as the service economy or Product-Service Systems (PSS). These concepts intend to emphasise a use or outcome to the customer. Many definitions for PSS are proposed in literature. Commonly PSS is defined as a "system of products, services, networks of "players" and supporting infrastructure that continuously strives to be competitive, satisfy customer needs and have a lower environmental impact than traditional business models" [1]. The major merits for the manufacturer of this approach are increased revenue, prolonged and strategic relationships with the customer and product/service improvements based on improved understanding of customer requirements.

Based on the spectrum of product and service mixtures in the offerings, PSS has been commonly classified into three types: Product Oriented, Use Oriented and Result Oriented [2]. This classification is based on product ownership and functionality, business models and product and service substitution. The emphasis in all types is on the '*sale of use*' rather than the '*sale of product*'. A major perspective of this concept is to consider the system as a whole, rather than just physical products [3]. The partial substitution of product and service shares over the lifecycle and the dynamic adaptation to changing customer demands and provider abilities are with the details that defining PSS [4].

These points illustrate that the manufacturer's core competences are moving away from manufacturing to systems design and integration. The primary element required to widen their core competences lies in the process of co-creation. This is because the design of a PSS is a co-creation process between manufacturers, suppliers and customers. PSS could be a win-win-win solution for all the stakeholders involved. The ability of the manufacturer to deliver a PSS very much depends on the capability of the available service network. It has been highlighted that the ability of a manufacturer to action a strategy of servitization is dependent on the capability of the available service network as over 75% of a product is designed and sourced from the supply chain and they contribute to through life support through the design and delivery of services [5]. Design of a system along with products and services within a network context makes this process complex. Many factors are to be considered in designing PSS such as stakeholders (culture, relationships, role and communication), environment (B2B, B2G, B2C), business model, life cycle stages, support system, infrastructure, technology and risks. The focus of this work is on the business to business (B2B) environment.

PSS involves complex B2B relationships within a service network and has to consider the capability to deliver the service over a long timeframe across geographies. Currently the conceptual design in practice is ad-hoc and lacks a systematic approach to consider the service network and customer capabilities and issues in the PSS design process. The lack of systematic approach is also valid for take-back service operations such as product remanufacturing and recycling [6]. The research has also observed that service is often added after the product is designed, there is a lack of communication between after-sales and design teams, and the designers' mindset is still very product centric. The research has also identified current lack of knowledge to trade-off between physical (product) and non physical (service) functionalities to create required customer value or reduce cost and opportunities for resource optimisation during the PSS design process.

Within our research group, PSS design is defined as a process to synthesise and create sustained functional behaviour through tangible products and intangible services. Sustained functional behaviour should represent how the system achieves its purpose continuously. PSS design involves design of business models, design of products and services, design of processes and the interactions between elements involved in the system. It has been emphasized that the requirement is to innovate the system, not just the business model. The aim of our research is to develop a formal approach to conceptual design of PSS considering existing and potential service network capability and past knowledge from the use of similar provisions. This PSS conceptual design framework for the manufacturer should address the capabilities and requirements of the service network and customer using a co-creation process.

In this paper, an initial framework for designing conceptual PSS has been proposed to emphasize the capabilities of the stakeholders. The rest of the paper is structured as follow: Section 2 summarizes various PSS design methodologies proposed in literature, Section 3 details the proposed framework explaining with a case study of laser system which is used for cutting operations and Section 4 concludes with a discussion and future work to be carried out.

2 LITERATURE SURVEY

In this section, PSS design methodologies proposed in literature are discussed and some of the research challenges are highlighted with respect to the focus of this paper. Komoto and Tomiyama [7] proposed Service CAD which supports designers to generate conceptual design of PSSs. They argue that for PSS design processes, designers define *activity* to meet specified *goal* and *quality*, and define *environment* as being the circumstance within which the *activity* is realized. The elements used in Service CAD are *service environment, provider, receiver, channel, contents, activity, aim of the service receiver's activity, target, promised goal, realised service, quality* and *value added*. They also developed ISCL (Integrating Service CAD with a life cycle simulator) which has functions to support quantitative and probabilistic PSS design using life cycle simulation.

Maussang et al. [8] consider the whole system and detail the physical objects and service units necessary to develop a successful PSS. They argue that this methodology can support the design of PSSs to start from the design of the architecture to go to the detail of physical objects (products) specifications. They used operational *scenarios* to go deeper into the system description once main elements of the system (physical objects and service units) have been identified. External *functional analysis* is used to list the external functions that the customer and actors involved in the product lifecycle expect from the 'product' without considering elements available to provide them. They argue that a specific external analysis must be carried out for each step of the product life cycle (use, manufacture, maintenance, recycling, etc.). They characterize each function or constraint by *criteria, level* and *allowance*. They argue that this characterisation leads to the detail of specifications and product performance expected by the customer.

Shimomura et al. [9] aim to propose a method for designing service activity and product concurrently and collaboratively during the early phase of product design. To enable this, a unified representation scheme of human process and physical process in service activity is proposed. They expressed a state change of a customer by parameters called *Receiver State Parameters* (RSPs), which represent customer value. They propose a *view model* which handles functions and attributes to represent RSPs. They include three phases in service design process: *identifying customer value, design of service contents* and *design of service activity*. They also developed a method to evaluate these processes with Quality Function Deployment. Sakao et al. [10] developed a service model consisting of four sub-models: *flow model, scope model, scenario model, and view model*. They emphasize that the critical concept is not the function of a product, but rather the state change of the receiver. The state change can be fulfilled either by products or by service activities. They have implemented these models in their prototype software tool which is named Service Explorer.

Aurich et al. [11] introduce a process for the systematic design of product related technical services based upon its *modularization* to link with corresponding product design processes. They propose an Object oriented technical service model to support the specification of technical services during their actual designing. The service components mentioned in the model are: the component *description* which provides a general overview of a technical service both verbally and graphically; the component *reference* covers the description of the products, product components or users' profiles addressed by the technical service along with the intended effects on them; the component *function* describes the measures for realizing the service functions; and the component *resources* covers both physical and nonphysical resources necessary for realizing a service. They developed a systematic service design process to specify technical services according to the presented service model. They suggest that adapting already existing product design processes to account for the special characteristics of technical services would lead to maximum acceptance for application within the enterprise.

Welp et al. [12] argue that an Industrial PSS (IPS^2) constitutes any combination of product and service shares and propose that the IPS^2 concept development is responsible for generating principle solutions that meet customer specific requirements. They present a model based approach to support an IPS^2 designer generating heterogeneous IPS^2 concept models in the early phase of IPS^2 development. They frame three planes for systematic conceptual development: IPS^2 function plane, IPS^2 object plane and IPS^2 process plane. Three different types of model elements are defined: *system elements, disturbance elements* and *context elements*. The combination of all types of model elements, planes and their respective relations constitutes a heterogeneous IPS^2 concept model.

Alonso-Rasgado et al. [13] described a design process for Total Care Product (TCP) creation that integrates hardware and service support by providing a robust design methodology. Five stages identified in the design of service support systems for a functional product are: concept creation for the service support system, identification of subsystems required, integration of the subsystems that together will provide the service, modeling of the proposed service system and testing and implementation. The fast-track design process consists of a methodology that breaks down the iterative process between customer and supplier into a number of distinct stages necessary for the creation of the TCP. Fast-track design process is framed as: business ambitions of the client, potential business solutions, enhanced definition of the potential TCP, business case risk analysis of options, business case validation and evaluation of alternatives and contract. They consider two main variables of the system to consider in simulations: time taken to perform the service and the quality and flow of information within the system.

Muller et al. [14] have proposed a method for the development of PSS called PSS Layer method. This method is intended to apply to the early development phases which comprise of the clarification of the design task and the conceptual design phase. It defines a metamodel of nine main element classes for a PSS. The classes are: n*eeds, values, deliverables, actors, lifecycle activities, core products, periphery, contract and finance*. All classes are graphically layered to simplify the representation. They argue that this model provides the user with a structured outline and an overall picture of PSS idea or concept. Tan et al. [15] proposed four dimensions of PSS that had to be considered: *value proposition, product life cycle, activity modelling cycle* and the *actor network*. They argue that these elements cover the essential design elements of a PSS. They suggest that a change in one dimension influences the others and the designer has to ensure that each of the dimensions of a new PSS concept support each other in order to be consistent.

Some of the observations from the various methodologies discussed in literature are as follow:
Integrating products and services seems to be the major objective for most of the proposed methodologies.

- The driving factors (risks and uncertainties) of PSS are not properly modelled.
- Most of the approaches are based upon a systems perspective.
- Only a few methodologies stress the importance of co-creation between stakeholders and feedback loops between the steps involved in the process.
- The roles of the stakeholders involved in designing PSS offerings are not clearly defined in the methodologies. In particular, the capabilities of the stakeholders are not considered during design stage.
- The influences of business models on product and service offers are not studied in detail.

These issues stress the enormous amount of research still required in developing PSS design methodology. To stress the importance of co-creation and the capabilities of the stakeholders in designing PSS, the following framework has been developed. It should be noted that PSS design involves offerings to the customers and also the system development which delivers the offerings for the contractual period. The next section details the framework structure and elaborates the steps through a laser system case study applied to manufacturing cutting operation.

3 FRAMEWORK FOR DESIGNING PSS

From literature it has been identified that a framework is required to emphasise the importance of co-creation process and capabilities of the stakeholders involved in designing PSS. A framework has been developed from our understanding through industrial case studies. This framework intends to facilitate:

- Structuring the purposes of interactions between the customer, manufacturer and suppliers,
- An understanding of the value of PSS offerings as appreciated by the customer,
- An understanding of the competences of the stakeholders and
- Assist in implementing developed PSS offerings.

Figure 1 illustrates this initial framework for the PSS co-creation design process. The subsequent sub-sections detail each step in the framework using a case study example of a laser system which is used for cutting operations for manufacturing purposes.

3.1 Customer Needs

Identifying and understanding the customer needs are the primary steps in the design process. Apart from identifying the value needed by the customers, in PSS design prime importance has to be stressed in the added value to be received by the customer in long term. The value addition needs to be emphasized in all the dimensions of economic, social and environmental sectors. In literature, Shimomura et al. [9] details this stage through a state change of a customer. Alonso-Rasgado et al. [13] specified this stage through understanding the business ambitions of the client. Technical PSS considers how to make the best use of capital-intensive assets so more value can be released and more revenue generated per cost unit of the asset throughout its lifecycle. PSS design should focus on integrating business models, products and services together considering throughout the lifecycle stages which create innovative value addition to the system. Influences of the business models on the products and services requirements specification need to be highlighted. As mentioned in [2], the business focus has to be shifted from the actual goods or services sold to the "need behind the need" that has to be fulfilled. Figure 1 represents that it would be ideal if all the stakeholders involved in every step of the framework. This involvement will provide wider visibility and aids to build a robust network to offer PSS. Aurich et al. [11] stress the importance of information procurement which is defined as "providing the manufacturer with customer information from product usage such as experiences, expectations or suggestions." This helps to develop the complete list of customer requirements.

We argue that every case study report in PSS should specify three parameters to indicate the applicability of their work to different types of PSS. The three factors which would differentiate each case study are: maturity of considered products, customer's intelligence and industrial domains (B2B, B2C and B2G). Such contextual information for the laser system case study in this work is outlined below:

- The laser systems under consideration are mature products as are the laser processes which are structured and mostly in-built to the system.
- The customers are laser job-shop owners who procure laser systems from the original equipment manufacturer and supply semi-finished goods to the end product manufacturer. The laser job-shops have many years experience in this field. They could explicitly specify their requirements precisely.
- The context is business to business environments, as illustrated in the above point.

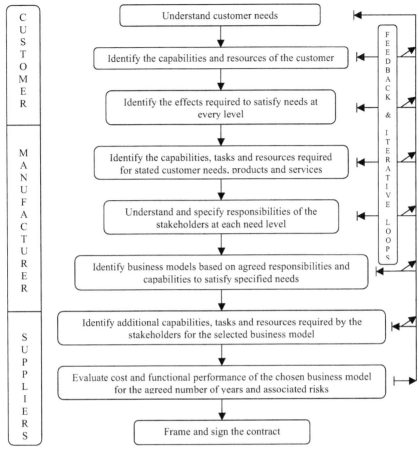

Figure 1. Initial framework for PSS co-creation design process

The specification of laser systems requirements is an important step towards defining PSSs which could be suitable for the customer and provider. Within industry, laser systems are commonly specified by following parameters:

- Power range
- Maximum sheet thickness
- Repeatability
- Working range
- Maximum work piece weight
- Maximum speed
- Precision (depends on work piece, pre-treatment, sheet size and position)
- Maximum axis acceleration
- Laser gas used
- Wavelength and Focal length
- Occupied volume and
- Environmental temperature to be maintained (at specified degrees).

Importantly it has to be emphasised that in addition to these parameters, following parameters are most required in order to develop PSS requirements. These are:

- Reliability
- Flexibility/modularity
- Robustness
- Portability
- Energy/ consumables/ wastages
- Maintainable/repairable
- Amount of usage/cycles of usage
- Interchangeable
- Updatability
- Component age
- Use context details (temperature, humidity etc) and
- Knowledge transfer from the manufacturer.

These parameters need to be explicitly specified by the manufacturer and should be negotiated with the customer to satisfy their operational needs. The operational needs should depict the variation of usage levels across the intended period. For higher PSS offerings, the defects to the end product through laser processes should also be specified. The parameters could be:

- Porosity
- Cracking
- Spatter
- Excess metal
- Sagging
- Undercut
- Humping and
- Distortion (residual stresses).

Even though specifying these parameters could help the manufacturer to develop better laser systems, the real value for customers would lie in meeting the following parameters:

- High productivity
- Less expertise required
- Less set-up and operating time
- Less space required
- Quality outcomes
- Operating versatility
- Less energy consumption
- Less consumables
- Latest technology
- Protection and safety
- Process stability
- Preventive defect faults
- Cost transparency
- Service scheduling certainty

These parameters represent the *functionalities* to be achieved by the PSS that need to be explicitly stated at this stage. Satisfying some of these needs could lead to a conflict of interest between the customer and provider. Therefore, the careful structuring of needs and careful negotiation would be required for PSS development. It should be noted that in PSS, specification characteristics should be for mass customization rather than mass production. Some of the features involved in mass customization have been highlighted in [16].

3.2 Existing capabilities of the customer

After identifying the customer's needs, the next step is to understand the existing customer's capabilities. This understanding will help to develop products and services aligned to their capabilities. Capability is defined as the continuing ability to generate a desired operational outcome. The capabilities could be realised through people, processes, tools, and technology. It should be noted that these parameters are highly coupled and should be visualized together. This integration is possible if the list of tasks to be carried out is identified and the efficiency of each task is measured. This analysis will highlight the gaps within the customer capabilities that need to be filled by the PSS offering. This stage will highlight the customer's life cycle activities along with the product. When analyzing the life cycle activities it is important to improve the PSS on an overall system level and avoid sub-optimizing towards any of the single activities e.g. production. The main difficulty at this stage is the division of competence available to perform each task based on resource availability. A more open environment between the customer and the manufacturer will help to understand this competence better. The factors to be considered for each task could be performance, technicality, human resources, financial and quality.

A laser system is an assemblage of a laser generator unit, beam delivery system, beam manipulation system, motion system, process monitoring system and a control system. Some of the tasks to map the customer's capabilities in the laser system case study are detailed in Table 1. Developing the complete list of tasks and their respective status will help in understand the capability gaps of the customer. This status will inform the next phase to develop better combination of products and services. It should be noted that the steps mentioned in this framework are highly dependent on each other. For simplicity and clarity, these steps are subdivided and illustrated. Therefore feedback loops exist between every step in the proposed framework.

Table 1. Example tasks to understand customer's capabilities

Tasks	Status
The development of a laser process for specific applications	Such processes are standardized.
Preparation of the work piece for laser cutting.	The necessary equipment is available and operators are well-trained to make work pieces ready for laser machining.
Work piece loading	Automated loading tools are available to fix the work piece with intended precision.
Work piece alignment	Manual alignment is performed and this can be a problematic area.
Cleaning and adjusting the optical parts	This is an error prone zone: the risks of damaging the optics are high.

3.3 Identify products and services

From the steps 1 and 2, customer needs and their capabilities will be stated and specified. The next step is to identify the products and services which will satisfy their needs and fill the gaps identified in the required capabilities. The important point in this step is to identify the trade-off between the products and services because the capability shifts as this boundary shifts. Figures 2 and 3 illustrate these capability shifts using a laser system and it's supporting maintenance activities. Although maintenance can increase asset availability, when maintenance is being administered, it can also contribute to the assets unavailability. In scenario 1, the customer finds the amount of maintenance unacceptable as there is too much disturbance to business operations. Scenario 2 shows how the capability for a certain level of availability has shifted from the maintenance service to the asset: here, the asset is redesigned to require less maintenance. The overall outcomes of the laser system have not changed between these scenarios but some of it has been redistributed from service to asset.

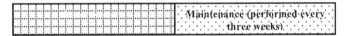

Figure 2. Scenario 1: The capability for asset availability division between product and service

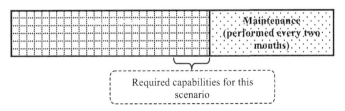

Figure 3. Scenario 2: Laser system and maintenance schedule

The above scenarios represent that the current level of products and services needs to be investigated and future scope should be envisioned. This will help in the design of the right product and service mix which should enable superior designs which increase the availability of laser systems and satisfy the needs of customers. Various services are specified in the laser market such as:

- Software support
- Tools and spare parts supply Installation
- Technical services
- Remote services
- Upgrading
- Customer training
- Consult to fine tune the machine's parameters to optimize speed etc.
- Maintenance types
- Remanufacturing
- Recycling

Asset performance and service activities are highly coupled. Therefore questions such as how maintenance (and calibration and servicing) affect availability and performance should be carefully addressed. There are various issues highlighted in current practice due to invisibility of product knowledge to the customer. For example, currently if a laser system breaks down, the manufacturer will often replace a whole module instead of just an individual component which increases the repair cost to the customer. The manufacturer has argued that this replacement reduces the downtime as otherwise there could be difficulty in diagnosing the problem and the repair cost could be more than the replacement cost because of testing and so forth. Additionally, there is an argument which states that the more reliable a product is, the more costly it is to repair. These arguments need to be negotiated between the customer and the manufacturer to find the right mix between products and services. Apart from identifying the products and services, this stage should make out the key performance indicators for the products and services. These indicators should act as a benchmark for throughout the life cycle stages.

3.4 Identify capabilities of the manufacturer and the suppliers

Identifying products and services that need to be delivered given the understanding of the customer's capabilities will help the manufacturer to develop their own capabilities along with that of the supply network. As over 75% of a product is designed and sourced from the supply chain, this contributes to through life support through the design and delivery of services [5]. The supply network needs to be developed by sharing the required capabilities from the manufacturer. The commonalities and differences between the capabilities of the manufacturer, customer and suppliers need to be explicitly shared and understood between them. This network formulation of stakeholders at this stage plays a vital role in developing sustained PSS offerings. As in Table 1, a detailed list of tasks to deliver products and services needs to be created and the status of each of task should be identified. This would lead to an understanding of the available resources between the stakeholders. Table 2 provides examples of the manufacturer's and suppliers' resources and their respective status using the laser system case study.

Table 2. Examples of the manufacturer's and suppliers' resources

Resources	Status
Labour Experience in developing laser systems Knowledge possession in creating advanced technology	The manufacturer has vast experience in developing laser generator unit. Knowledge regarding the beam guidance unit and motion system is advanced in the supply network.
Infrastructure IT support system	IT support is weak between the manufacturer and supply network.
Laser system Reliability and Consumables	Laser systems are extremely reliable (90% - 99.5%). Consumables are readily available.
Location Mobility	The customer's location may be remote. Mobility is an issue to transfer resources.

Identifying a complete list of tasks and resources as well as the respective status of each will aid the development of a more substantial network between the manufacturer and suppliers to satisfy customer needs. This will also help to assess the *service network capability*. It should be noted that the resource status mapping should consider past, present and future scenarios.

3.5 Specify responsibilities

The capability assessment of all the stakeholders by developing complete lists of the tasks and resources required would subsequently help to align the roles and responsibilities between them. This alignment of responsibilities take place over the life span of the PSS offering and will precisely define the network relationships. Various soft elements play vital roles in relationship development such as trust, confidence, commitment, culture and self-esteem. The development of an open network will be more valuable as the responsibilities map should be visible to all in the network even though, ultimately, all responsibilities are the concern of all in the network.

3.6 Identify business models
The framing of business models should be based on the responsibilities alignment between the stakeholders. Business models play a central role in defining PSS as they describe the rationale of how an organization creates, delivers, and captures value: economic, social, or other forms of value. Commonly used business models within the PSS domain are: Product-, Use- and Result-oriented which emphasise cost, ownership and customization elements. In these business models, the business elements should contain parameters which will influence business processes, issues and solutions. In laser systems, various important parameters which influence buying behaviour are purchase cost, running costs, efficiency, consumables (e.g. gases, flash lamps, diodes, optics) and delivery options. Predictable costs, cost transparency and maximal security are the other important factors considered during business model selection. Thus, the framing of business models should considering all of these parameters as well as the demarcation of capabilities and responsibilities.

3.7 Identify additional capabilities required
To fulfil the requirements of the selected business model, additional capabilities would need to be acquired amongst the stakeholders. Existing and new capabilities and resources from each stakeholder should be carefully aligned and integrated. Shifts in the capabilities between stakeholders which lead to acquire additional resources to match the activities needs should be noted. Difficulties for the stakeholders to quickly expand to meet increasing capabilities demands should be handled with possible resource variations and time constraints.

3.8 Evaluation and contract finalization
Evaluation should be part of every step in the proposed framework. To emphasise this evaluation process, it is dealt with separately in the framework. The evaluation should focus on three dimensions: economic, social and environmental. From a business perspective, the major evaluation criteria will be profit, revenue, customer satisfaction, quality of products and services, value-in-use and risk reduction. Both tangible and intangible merits and demerits should be evaluated. In the laser system case study some of the evaluation questions could be:
- What are the risks in the manufacturer in retaining the asset?
- What is the frequency of maintenance and servicing and the associated costs? What other lifecycle costs should be considered?
- Does the manufacturer's supply base have the capability to support possible PSS solutions?
- What are the issues in achieving the stated availability of the laser system?

The final step would be to frame the contract using terms and conditions that are relevant to all of the stakeholders involved. These terms would also stipulate all of the legal obligations to be met by the stakeholders. All of the terms have to be very carefully noted and defined. The contract should be concise, unambiguous, consistent, simple, complete, easy to interpret and easy to maintain.

4 DISCUSSION AND FUTURE WORK
In this paper, various proposed PSS design methods have been reviewed and the research gaps have been summarized. Primarily, it has been observed that the importance of capabilities of the stakeholders involved in designing PSSs have not been noted in the proposed methods. Emphasising this capability view point, a framework for designing PSSs has been proposed. This framework highlights the important features required in designing PSSs such as co-creation, responsibilities and capabilities. Importance of the feedback is stressed by the iterative loops between every step in the proposed framework. Every step in the framework has been illustrated with a case study involving laser systems used for cutting operations in manufacturing. We believe that this framework would facilitate and structure the interactions between the customer, manufacturer and supplier. It also helps to understand the capabilities of the stakeholders and aids an understanding of the value of PSS offerings as appreciated by the customer. This initial framework will be developed iteratively by applying it to various case studies involving various other companies who are in the process of refining the development of their PSS offerings. A computer assisted design tool will be developed to help stakeholders to use this framework for developing PSSs. The design tool is intended to facilitate a novel representation of PSS modelling.

REFERENCES
[1] Goedkoop M. van Haler C. te Riele H. and Rommers P. Product Service-Systems, ecological and economic basics. *Report for Dutch Ministries of Environment (VROM) and Economic Affairs (EZ)*, 1999.
[---] *International Conference on Manufacturing Research (ICMR 2006)*, Liverpool John Moores University, 2006, pp.17 – 22, ISBN 0-9553215-0-6.
[4] Meier H. Roy R. and Seliger G. Industrial Product-Service Systems - IPS^2, *CIRP Annals Manufacturing Technology*, 2010, 59(2), pp. 607-627.
[5] Cohen M.A. Agrawal N. and Agrawal V. Winning in the Aftermarket, *Harvard Business Review* 84 (5), pp.129-138, 2006.
[6] Sundin E. Lindahl M. and Ijomah W. Product design for product/service systems - design experiences from Swedish industry, *Journal of Manufacturing Technology Management*, 2009, 20 (5), pp. 723-753.
[7] Komoto H. Computer Aided Product Service Systems Design (Service CAD and its integration with Life Cycle Simulation), *PhD Thesis*, Delft University of Technology, Delft, the Netherlands 2009.
[8] Maussang N. Peggy Z. and Brissaud D. Product-service system design methodology: from the PSS architecture design to the products specifications. *Journal of Engineering Design*, 2009, 20(4), pp.349 – 366.
[9] Shimomura Y. Hara T. and Arai T. A unified representation scheme for effective PSS development. *CIRP Annals - Manufacturing Technology*, 2009, 58, pp.379–382.
[10] Sakao T. Shimomura Y. Sundin E. and Comstock M. Modeling design objects in CAD system for Service/Product Engineering, *Computer-Aided Design*, 2009, 41, pp.197-213.
[11] Aurich J. Fuchs C. and Wagenknecht C. Life cycle oriented design of technical Product-Service Systems, *Journal of Cleaner Production*, 2006, 14(17), pp.1480-1494.
[12] Welp E.G. Meier H. Sadek T. and Sadek K. Modelling approach for the integrated development of industrial Product-Service Systems. *Proceedings of 41st CIRP Conference on Manufacturing Systems*, Tokyo, Japan, 2008.
[13] Alonso-Rasgado T. Thompson G. and Elfstrom B-O. The design of functional (total care) products. *Journal of Engineering Design*, 2004, 15 (6), pp.515-540.
[14] Müller P. Kebir N. Stark R. and Blessing L. PSS Layer Method – Application to Microenergy Systems, book title *Introduction to Product/Service-System Design*, 2009, Springer Publisher.
[15] Tan A. McAloone T.C. and Hagelskjær L.E. Reflections on product/service-system (PSS) conceptualisation in a course setting, *International Journal of Design Engineering*, 2009.
[16] Sundin E. Lindahl M. Comstock M. Shimomura Y. and Sakao T. Integrated Product and Service Engineering Enabling Mass Customization, on CD-ROM of *19th International Conference on Production Research (ICPR-19)*, 2009, Valparaiso, Chile.

Contact: Professor Rajkumar Roy
Cranfield University
Head of Manufacturing Department
Cranfield, MK43 0AL
UK
Tel: Int +44 1234 758335
Fax: Int +44 1234 758292
Email: r.roy@cranfield.ac.uk
URL: http://www.cranfield.ac.uk/sas/aboutus/staff/royr.html

Professor Rajkumar Roy is leading the Manufacturing Department and Director of the EPSRC Centre for Innovative Manufacturing in Through-life Engineering Services. He is interested in developing whole life cost models for products and services, real life application of soft computing techniques for modeling and optimization and innovating product-service systems business models.

INTERNATIONAL CONFERENCE ON ENGINEERING DESIGN, ICED11
15 - 18 AUGUST 2011, TECHNICAL UNIVERSITY OF DENMARK

ORTHOGONAL VIEWS ON PRODUCT/SERVICE-SYSTEM DESIGN IN AN ENTIRE INDUSTRY BRANCH

Tim C. McAloone, Krestine Mougaard, Line M. Neugebauer, Teit A. Nielsen and Niki Bey
Technical University of Denmark

ABSTRACT

Product/Service-Systems (PSS) is an emerging research area, with terms such as 'functional sales', 'servicizing' and 'service engineering' all contributing to the foundation and our current understanding of PSS as a phenomenon. The field is still in its formative stages and definitions, understandings and approaches to PSS are still fluid.

Much of the literature in the field of PSS has, until now, focused largely on the actual transition from product to PSS and has typically resided in the field of engineering design. Symptomatic of the current literature is the concept of service as the adding-on of non-physical activities and relationships between supplier and customer. There is evidence in the literature, that multi-stakeholder approaches, customer activity understanding, actor-network charting and value chain collaboration are important factors to include in PSS strategies. However, actual case examples of these factors are sparse and limited to conceptual examples.

This paper describes five orthogonal views on PSS design, fostering integrated product/service thinking across organisational boundaries, via a systematic approach to user-oriented product and service development.

Keywords: product/service-systems, product service engineering, innovation strategy, value networks

1 INTRODUCTION

PSS is an emerging business concept in industry. Industrial companies in "high pay" countries tend towards an intensified focus on core competencies and -operations locally, with a subsequent outsourcing of most other (labour intensive) operations and tasks to external suppliers or network partners [1]. This outsourcing of activities implies that the partnering supplier companies perform an alignment and development of customised solutions to fit the needs of their contractors. But also private consumers are readily subscribing to a rising number of service offerings, requiring companies to change their operations and products accordingly [2].

The notion of PSS design, -development and -operation requires a much broadened view of the design object for product development, resulting in *the product* being augmented with respect to time, infrastructure, value and artefact considerations [3]. A key question when considering PSS design is whom to involve in the design process and which competencies to ensure within the organisation, for both the design and the operation of the PSS. Augmenting the product offering in the *time domain* is particularly challenging for manufacturing companies that traditionally transfer both ownership of and responsibility for the product to the customer/user at point of sale. In terms of *infrastructure*, it is necessary for the PSS provider to ensure a system of support elements and ancillary operations that will allow the provider to support the customer and the customer to interact with the provider whilst using the product and/or experiencing the service. In the *value domain*, it is reported that a PSS offering can create interesting benefits to both the customer and the company, as new opportunity parameters arise, regarding the provision of value to the customer [4]. And in terms of the *artefact*, it is important to understand how augmenting the product into an integrated product/service offering actually affects the physical makeup (and therefore design) of the product; or seen from a different perspective, how to Design for Service?

Compared to straight product design, when designing a PSS, the company necessarily makes decisions that have direct and disposed effects on many more aspects of the value chain. The notion of life cycle thinking is therefore important to embrace, as is the idea of value chain collaboration [1]. Furthermore the insights gained by attaining a deeper understanding of user needs and activities, as currently in focus in the field of user-oriented innovation [5], seem to be central to the value-augmentation that a PSS strategy aims at bringing to the end-user. Current literature on PSS shows evidence, that multi-stakeholder approaches, customer activity understanding, actor-network charting and value chain collaboration are important factors to include in PSS strategies [4, 6, 7]. However, actual industrial cases particularly investigating these factors and their influence on PSS development and performance are sparse and limited to conceptual examples.

Taking the above status and conditions for PSS in industry, it seems that there is a need already now to consolidate some of the experiences and findings into a systematic way of viewing PSS design, which in the authors' view is an augmented view on product development, expanded in the domains of time, artefact, value and infrastructure. We describe in this paper a research project where an entire industry branch is under scrutiny, in order to allow for actual experiments in the area of value chain collaboration; complex user activity understanding; and experience exchange (as described in [1]), regarding PSS design and organisational modelling. The case we describe is the Danish maritime industry branch, where twelve industrial companies (mainly component manufacturers) are collaborating to improve their PSS ability, in recognition of the need to ensure sustained competitiveness, in a time where much ship production is moving East. We present five views on PSS design in an attempt to gain insight into PSS creation for a whole branch. We call these five views *orthogonal views* as we purposely refrain, in the first instance, from attempting to reconcile the five views in relation to each other. Our aim with this approach is to push the boundaries of product design theory towards new considerations when working on integrated product/service design, which is carried out across new organisational boundaries, in relation to traditional design.

We describe the five orthogonal views with the help of the industrial context of the maritime industry, in order to both gain real-world insights and to ensure the relevance of this work for industrial application. At the same time we reflect on our work's relevance for the design research community, by highlighting some initial observations regarding the process of PSS design.

2 CURRENT STATE-OF-THE-ART OF PSS

Within the design research community, the activity of expanding product development in the direction of service has been dubbed *functional products* as e.g. by Alonso-Rasgado et al. [8], *functional sales* as e.g. by Sundin and Bras [9] or *product/service-systems* as e.g. by Goedkoop et al. [10], Manzini and Vezzoli [11] and Mont [12], where each of the separate research groups have taken their own approach towards the phenomenon, focusing on different characteristics and types of PSS solutions. Based on a review of the emerging views on PSS, Matzen & McAloone [2] indicate four dimensions in current PSS research, which we subsequently interpret as follows:

1. A theoretical understanding of the operations related to opportunities inherent in PSS approaches to business, exploring and explaining opportunity parameters pointing towards e.g. the dematerialisation of offerings, optimisation of performance or consumption.
    This dimension can be interpreted as *PSS as a potential of benefit*.

2. A theoretical understanding of the phenomenon of combined product and service offerings, exploring and explaining the inherent virtues and inferiorities of physical products throughout their life cycles and how these can be supported and relieved by service offerings.
    This dimension can be interpreted as *PSS as an augmented product development theory*.

3. A prescription of the structures and management technologies necessary to enable companies and company networks to develop, deliver and operate PSS solutions.
    This dimension can be interpreted as *PSS as a strategy*.

4. A prescription of the processes which will enable development teams to identify and take advantage of the potential benefits referred to above. Furthermore a prescription of working tasks and documentation models aiding the development team in the concretisation, communication and realisation of PSS solutions.

This dimension can be interpreted as *PSS as design tools*.

These four interpretations are taken from a review of somewhat disparate research projects and individual cases, none of which have handled more than one of the four dimensions at a time. Besides the above four interpretations of PSS research, there is little coherence in the prescriptive PSS approaches or empirical insights so far, underlining that the field is still in its formative stages, with definitions, understandings and approaches to PSS still being fluid.

The majority of the current research literature on PSS originates from the design research or manufacturing research communities, whereas earlier work was founded in the operations/marketing [12] or sustainability/environmental fields [11]. In the domain of engineering it is important to trace and attempt to foresee the adjustments that organisations must undergo, in order to remain competitive in the business creation process, when the boundary conditions for business creation are in a state of change, in relation to a traditional production-sales situation. Tan [4] states that it is no longer sufficient to have a systematic approach in place for product development and production alone, when increasing proportions of revenue come from before- and after-sales service. Wise and Baumgartner [13] describe the necessity for organisations to transform what earlier has been viewed as a cost-centre (after-sales activities have been traditionally viewed in this way by many companies) over to a profit centre, where an adjustment of the relationship to the customer is necessary. Thus, as the business foundation for more and more companies is increasingly based on service revenues than on those gained by selling the physical artefact itself, there is reason and motivation to much better understand and control the processes surrounding such an augmented business model [14].

The current literature on PSS includes examples of procedures for the integration of product and service features in product development [8, 11], but these approaches do not consider a number of key areas for business, such as the commercial considerations, the strategic organisational issues, or the possibilities of collaboration across the value chain.

3 CASE AND METHOD INTRODUCTION

The Danish shipbuilding industry has traditionally focused on delivering products to their customers, based on the longevity and high technical qualities of their physical artefacts. But as with most industries, the continuing market globalisation in the shipbuilding industry both opens opportunities, in terms of a rising number of potential customers and represents threats, due to the growing number of competitors, worldwide. Maritime component manufacturers are experiencing a growing demand from customers with respect to after-sales service, and they also see a great business opportunity in creating more systematic and integrated product/service development activities.

In the light of the above, a so-called *innovation consortium* named PROTEUS (PROduct/service-system Tools for Ensuring User-oriented Service) has been established and is working to jointly develop new knowledge about how after-sales service can be effectively integrated into business development and industrial organisations, so as to become a source of revenue, rather than a cost to the company. The innovation consortium is funded by the Danish research council to create research insights and innovation results simultaneously, throughout a prolonged (3-5 year) collaborative project, consisting of representatives from Danish research institutions, a technical service partner (consultancy), international university partners and twelve maritime companies. The twelve participating companies in the project are interested in understanding, through examples, how to effectively and systematically integrate service development into their product development and business creation processes. The idea with establishing PROTEUS as a research project has been to create a unique opportunity, both industry-wise and research-wise, to begin to address some of these issues on an entire industry branch. The project is organised into five work packages (Figure 1), each with its unique focus on PSS.

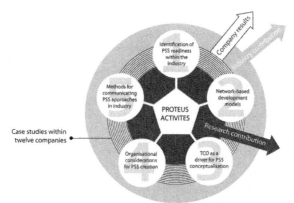

Figure 1 - The organisation of the PROTEUS project

The research method adopted for the project is one of participatory research, where high amounts of company interaction are necessary. Depending on the work package, the research tools that are applied in this project range from surveys and interviews (particularly for collecting insight into state-of-the art practice); workshops and case studies (in order to prompt new ideas and test hypotheses); and ethnographic studies (to gauge the validity and proportions of claims for methodology development and coordination in the individual companies). The overall research design follows Blessing and Chakrabarti's design research methodology framework [15], where a series of iterations of the framework are planned. The research is multiple and trans-disciplinary, carried out in collaboration between a technical university and a business school, in particular by four senior research staff, three PhD students, one industry consultant and several Bachelor and Master students (connected through case studies/student projects). When describing the research views in the following sections, we will therefore elaborate both the innovation activities and the research activities of the project.

At the time of writing this article the project has been running for approximately half a year of its total period of 3½ years. The explorative phase of the initial descriptive study of project, entailing an initial analysis study (21 interviews and 14 site visits), has been completed. Furthermore individual case studies are underway, and a catalogue of case projects are about to be launched. The five orthogonal views on PSS, described in the following sections provide the foundation for our research on this project – and therefore also for later publications on the results achieved.

3.1 PSS readiness and triangulation of companies, industry and market

The first view we take on PSS is to create an understanding of the readiness for a transition towards PSS, of companies, industry and market, with the purpose of establishing the boundary conditions for PSS, and to frame subsequent PSS strategies. By establishing the boundary conditions for PSS in the particular industrial context, it is possible to identify a set of questions to guide the research process within the industry. PSS strategies are of particular interest, both to the participating companies and as an object for observation, from a cross-company viewpoint. With this view we have identified a series of *PSS readiness aspects*, which help assess PSS readiness in the branch and in individual organisations.

The first PSS readiness aspect is to identify prerequisites for companies, which create a foundation for PSS development. Within this case these prerequisites consist of companies' competences; know-how and product embedded knowledge; experiences with after-sales and service development activities; insight into the ship's lifecycle; and ship owner's activities. Many of the employees of the case companies, (e.g. from Sales, After-sales, Operations, etc.) have backgrounds as ship mechanics, superintendents etc. which provides insight into the user activities, such as requirements, needs; customer's considerations; and mindset. The main question here is: What prerequisites are needed in an organisation to introduce PSS development?

The second aspect is to elicit a set of criteria for introducing PSS development and measuring the changes made. The criteria should emphasise the main targets for PSS development, both within the individual company and in collaboration across a number of companies within the same industry. Hence, how should a PSS be shaped in order to fit the current practice of companies, meeting the needs and requirements of the customer and the movements in the markets?

The third aspect identified is to define the market movements that drive the need for PSS introduction. PSS thinking can be used for targeting new market segments, (e.g. offshore, in the maritime case) or to change current market dimensions, e.g. towards integration of service into current activities and customer relationships. Research shows increasing competition on low-cost products as a driver for finding alternative market segments, e.g. offshore [6]. Asian companies now dominate the shipbuilding industry, forcing Western companies to find new entry points to their market, e.g. through European ship owners [16]. To meet market demands, European companies could develop PSS offerings to meet the needs and requirements of the European ship owners' activities, e.g. throughout the life cycle of a ship. Hence, what are the market dimensions for current markets and how are these market dimensions changed towards integrating service as part the offering to the customer?

With the identification of these three PSS readiness aspects as lenses for viewing PSS readiness of companies, markets, and industry, we aim to describe the prerequisites, success criteria and market dimensions necessary for the foundation of a framework for introducing PSS into existing industrial situations. This framework is planned to be developed based on the empirical experiences from workshops and case studies. It is our aim to create a generic framework, targeting the key elements for a successful implementation of PSS design, but customisable dependent on the specific strategy of the individual company.

3.2 Value chain collaboration

The second view we take on PSS focuses our attention towards a new understanding of value creation, where the product and service are seen as a system, contributing with a combined value proposition/offering, by sustaining and enhancing the utility of the offering through the whole product/service life cycle [4]. In this view the value creation process is prioritised to be the main parameter of customer satisfaction, in place, perhaps, of product quality. Porter explains value creation as occurring through a company-centric approach, defined and strengthened through value-chain analysis, where the company is the producer and the customer, the receiver of the value [17]. However, as the business landscape is changing through (among other things) globalisation and dematerialisation, Porter's concept of value creation neither encompasses the creation of intangible products, nor the possibility of the customer as a co-creator [1]. Research is revealing how Porter's definition of the classical sequential value-chain is insufficient when focusing on creating competitive advantage within the service economy [18]. Normann provides an alternative to the sequential value-chain model by describing a so-called *value star* [19], where the value creating system consists of input from multiple actors in a network, rather than a sequential chain, and where the customer is seen as a vital stakeholder, providing resource input to the system. Adopting the value star metaphor provides a mental model of how to exploit and promote the resources and knowledge of multiple stakeholders (including the user) within the early stages of the innovation process and through the whole life cycle [20]. Approaching PSS as a value network/value star is not seen before in literature.

Within the PSS conceptualisation activity a proactive approach towards creating value chain collaboration is one of the elements that will be investigated within this research. One of the hypotheses for this view on PSS is: The competitiveness of a PSS offering will increase when the significant stakeholders within the value chain or potential new value contributors are identified and involved in the co-development of the offering. Within this view the following research questions are of importance: By what techniques can the value chain structure be identified? Thereafter, how can the stakeholder network, the value perception and their relations be visualised and reconfiguration possibilities be explored and exploited?

The innovation content of this view on PSS aims to aid (the maritime) industry in appraising the benefits of focusing on co-creation of value between several stakeholders within a value system, and to ultimately provide normative approaches towards the orchestration of value stars. The creation of new business models, capable of encompassing the opportunities of collaborative inter-organisational networks within the conceptualisation activity of PSS will be in focus, together with the whole life cycle perspective of re-conceptualising the value proposition. The case companies will be facilitated in the transformation of strategy, *from value chain towards value star*, together with other relevant companies from the branch. The insight required to create these normative tools will be created via a series of research activities, such as mapping the companies' technology- and service platforms; charting the companies' competencies and organisational structure profiles; and studying the affinities of the companies in the dimensions mentioned above.

3.3 Total Cost of Ownership & PSS Conceptualisation

Total cost of ownership (TCO) is a concept whereby the economical costs occurred throughout the whole life cycle of a product or system are considered up-front, as opposed to e.g. first purchase cost. TCO has been studied within the PSS field before, where the lack of ability or willingness of customers to think in terms of TCO has been found to be a significant barrier for PSS implementation in a business-to-business setting [6]. By removing the focus from first purchase cost to the total cost of ownership, new business ideas and product/service solutions can arise that not only satisfy the customer but also create a so-called *customer lock-on* [7]; the achievement of continued/prolonged relationships between company and customer.

In the maritime case there is a new tendency for the customer (the ship owner) to move towards TCO as a concept when procuring ships. This gives a great opportunity for the companies to create and present integrated product/service offerings to the ship owners. The main questions to be explored in this third view on PSS are therefore: How to utilise TCO as a driver for PSS concept creation? How to go about PSS conceptualisation, given that the customer is willing and able to operate their products on a TCO basis?

Due to the complexity of the maritime industry, development and implementation of TCO concepts has thus far proven difficult. An example of one of the challenges met is the fact that sub-suppliers' customers are most often not the end users, since the products are first sold to the shipyard and subsequently to the ship owner. Furthermore the definition of "a ship owner" is nowadays much more diffuse than just a decade ago, as it today ranges from one large company, running all aspects of the ship, to a capital investment group owning the ships and outsourcing the operation (e.g. ship management, crew management, spare parts purchase, maintenance) to many different companies, thus resulting in a scattered ship responsibility. The latter type of ship owner makes TCO concepts difficult to sell, since the one paying is seldom the one benefitting.

Nevertheless, there is a great desire to become competent in creating TCO-driven concepts in the maritime industry, both from the sub-suppliers' and the ship owners' points of view. It is therefore also the goal of this view on PSS, to create examples of actual TCO concepts for application in the maritime industry which can: (i) serve as an illustration of how TCO concept development can lead to financial and/or environmental improvements (environmental impact is a very new but rapidly increasing concern in the maritime industry) and thereby serve as an argument to invest in service, both for sub-suppliers and users/customers; and (ii) give the participating companies examples of tools and methods that can be used for TCO-driven PSS concept development.

3.4 Organising for PSS

As yet the PSS research activities charted through the academic literature do not cover the role of the organisation for PSS. Literature indicates that there are different processes involved in PSS development, compared to traditional product development [2, 4], but not how this may affect the requirements on the way in which the PSS development and operation process should be organised.

This fourth PSS view concentrates, therefore, on the tasks of product development organisation for PSS and PSS organisation in the company. The hypothesis is that a successful shift in focus from technology/artefact-based business creation over to service-oriented business creation will demand a redefinition of tasks within the organisation, as well as a re-evaluation of competencies inherent in the company and competencies needed, together with a re-definition of the organisation and management structure in the company. The research will focus on how to enable companies to efficiently integrate service performance into their product and business development processes, including an understanding of the types of changes that this will cause on an organisational/management level.

While research has studied the theory behind and methods towards PSS development and also into the business part of PSS (e.g. revenue models) there has been no research into the cross-functional setting in which the above should be incorporated, i.e. the organisation. The research activity connected to this PSS view is building on the empirical insight of the experiences of companies already attempting or carrying out PSS development, and on literature regarding the organisation of product development. The aim here is to identify and extract organisational considerations and activities that are unique to a PSS situation.

The innovation-oriented content and goal is to develop concrete tools, methods and organisational/ business models for industry use. The sizes of the participating companies vary greatly and solutions should be suitable for them all. The focus is therefore on creating a generic framework that can enable small as well as large companies (no matter what type of business model or organisational structure they have), so they can make the necessary changes to approach PSS development. Results, such as the above framework, are to be tested and further developed, in collaboration with the participating companies during the project, giving them the opportunity to evaluate the usability, e.g. relevance and shortcomings of models and tools. It is expected that the project and the ongoing collaboration with the companies will positively affect their readiness for PSS implementation.

3.5 Implementing change in the Product Development process

In order for intended changes to take effect in real-life contexts, it is necessary to focus on implementation in the daily routines of the involved organisations. Especially this aspect, however, seems not to have been fully successful in known PSS projects so far, since actual up-and-running industry cases are very sparse, let alone PSS cases for an entire industry branch. We therefore take a fifth view on PSS, specifically focused on how to effectively and systematically integrate service development into product development and business creation processes.

From a research perspective, we see two key tasks: firstly, the development of measurable parameters to control and monitor PSS processes; and secondly the development of methods to disseminate such parameters and related experience within the industry. As we deal with a multitude of stakeholders, we understand and treat 'dissemination' both in terms of 'organisation-to-organisation exchange' (e.g. along the value chain) and in terms of 'within one particular organisation' (e.g. among different departments of a company). One challenge here is to cover *all* the involved stakeholders in the Product/Service-System design, as the lack of only one stakeholder's commitment can make the whole PSS fail. Having an entire industry represented in our project, very favourable preconditions are in place to succeed on this task.

From an industrial perspective, we see the actual dissemination of the above methods and experience as the main challenge. Therefore, we have allocated this key task to one dedicated consortium partner, who is used to transferring research-based knowledge into industrial innovation. Dissemination activities will, for instance, comprise freely accessible information on a dedicated homepage, general as well as tailor-made workshops and printed media such as handbooks and executive reports.

With the entire industry represented through the case companies, plus their branch organisation, we have created not-seen-before preconditions to capture and facilitate experience exchange amongst and within the involved industrial companies and organisations regarding PSS design, PSS development and PSS operation.

4 CURRENT STATUS AND INITIAL INSIGHTS

With an initial descriptive study having recently been completed, the project has now moved into a prescriptive phase, where individual projects, e.g. case studies are underway or about to be launched. The initial observations were carried out over a period of six months, conducted via semi-structured interviews of multiple representatives from all twelve participating companies, site visits and workshops. Various models have been generated of the industry as we observe it, from analysing the observation data. These models include *Ship life cycle* (charting the whole life cycle of a chosen ship); *Customer Activity Cycles* (charting the activities of the ship owner, from recognition of need, e.g. for transport vessel, through activities related to the logistics, operation, maintenance, etc.); *Actor Network* (a representation of the relational exchanges between all stakeholders in the shipping arena); and *Strategy map for service design* (initial considerations elicited from the twelve companies, regarding their considerations or plans for service). The models act both as aids for the analysis of the initial observations in the companies and as visualisation aids, regarding the missing data and information, still needed to complete a coherent and consistent picture of all twelve companies. In the following we reveal some of the first insights from the research, followed by the planned next steps for the prescriptive work ahead.

First impressions

General lack of systematic product development processes: Only (the largest) three of the twelve companies have a formalised and systematic product development approach, and this even only to a certain degree. The development in the remaining companies is highly personalised and relies heavily on the know-how of the employees, who commonly have a background within the maritime industry, often with previous employment on a ship. New development tends to be based on the companies' past experiences and ideas of what the customer may want, rather than systematic analyses of current and future user needs. Very little research has been carried out regarding market and user needs, since it is commonly accepted that the employees' comprehensive knowledge of the industry is sufficient. Many minor and major redesigns are developed and made on the factory floor by experienced production workers. This type of development is possible due to the know-how of the workers and the fact that many of the employees are former ship mechanics and due to the products' relatively low complexity. The reasons for redesign are normally a specific set of requirements requested by a specific customer. Otherwise new concepts and redesigns chosen for development are currently based on the potential profit generated through direct sales, not from projected service revenue. For the most part, the product development activities are based on informal conceptualisation, rather than by a systematic (e.g. technology-/user-driven) approach to product development and are mostly redesigns.

Internal/external lack of communication: Internally in the companies there is a common lack of communication between the departments concerning product development and offerings related to service. Better communication between organisational departments around the potential of integration of service offerings in the development process, sales activities, etc. is required. For example, the After-sales departments of many of the case companies handle feedback from customers regarding product-related service claims, such as repairs, maintenance, spare parts etc. These departments have important and obvious potential inputs to the development process, communicating the knowledge from After-sales into the early identification of user needs, as well as the later dispositions in the development process. A significant barrier to improving organisational communication is an inherent organisational deviation in perceptions of service, ranging from "an after-sales add-on" to "a central opportunity for future sources of income/profit". This seems to be due to, for instance, different cultural compositions in the companies' global organisations, e.g. the difference between Asian and Western mindsets towards service. Furthermore, poor internal communication about service causes a neglect of external communications of service offerings towards the customer, e.g. when the sales department neglects to promote and sell the developed product/service offerings.

Traditional product-oriented business structure: Eleven of the twelve companies have a classical product and company-centric business structure and are part of a classical supply chain, where the vast majority of value-adding activities are within the company. Three of the twelve companies have integrated product/service offerings as part of their business portfolio. One example is a large company offering both product and application expertise. This company offers management of a repair during docking of the ship, based on the competencies and geographically dispersed location of the company's application advisors. In this case, amount of man-hours used is reduced, due to optimised, competent application, and the service offering of *management of ship docking and quality assurance of the application* is introduced.

Advantages of locally sourced production facilities: The majority of the companies retain parts of their production facilities in Denmark, despite the competition of Asian suppliers with low-cost production. As a reason, local production is stated to enable the companies to maintain a handful of competitive parameters, such as high quality, time, maintained know-how and flexibility in supply. Also, local production enables high flexibility in supplying spare-parts to e.g. an urgent need in a break-down situation, demanding overnight production and shipment of products/components. Many of the companies also have large parts of their service networks in Denmark. Though the market is flooded with low-cost products, there is still a demand for high-end quality products, e.g. some ship owners demand that the product is produced in Europe. The years of knowledge accrued at the local production facilities ensure the product's reliability and stability in performance that is essential for the company's image and reputation within the industry. Product certification regarding quality and safety is an important component of the maritime industry and it is a way for the high quality production companies to position themselves on the market against low cost/low quality producers.

Multiple shared offerings, affinity indication and collaborative opportunities: One third of the companies have already indicated an interest in collaborative activities through e.g. shared service stations, but are not yet capable of pursuing this. The companies have a large overlap of offerings, both product-wise and service-wise. The companies, however, have not been aware of many of the overlaps identified in our initial study, since currently no complete overview of the offerings exists for the branch (at least, not according to the interviews). Furthermore, most of the companies are involved in several of the same activities within the customer activity cycle and the ship's life cycle, supporting the same customer needs. Their product and technology portfolios overlap in various places, indicating a possibility for horizontal value chain collaboration. Opportunities for vertical collaboration in the value chain also exist through, for example, the market need for training academies for service technicians, which many of the companies offer. These academies are used to train technicians in performing certified installations and maintenance overhauls as part of enhancing the skills and competencies of the technicians in order to ensure correct installation of the product and thereby optimise its performance. However, barriers that complicate alliances between the companies are at the same time emerging due to different expectations; visions for the outcome of the partnership; willingness; and readiness, plus issues connected to the responsibilities of a product's performance and shared distribution.

Reflections of future needs

The next activities within the PROTEUS project build on the common denominators existing between the companies. These will be identified through charting the companies in further detail; by creating a structured overview of the companies' portfolios of offerings; by exploring collaboration opportunities; through the creation of a shared language of services; and by creating knowledge-sharing on new technologies and product improvements. Besides this, a structured overview of the companies' organisational structure, resources and competencies is important, in order to gain a better picture of the company readiness regarding transformation towards a PSS approach. Furthermore, a mapping, structured in relation to the activities of the ship owner, will reveal the areas of potential new business for the companies, where they may already have the necessary competencies inherent to meet this customer need.

A major task for the project is to continue to enhance the understanding of the maritime industry, regarding PSS opportunities. The maritime market is in a particularly high state of flux in these years, due to the migration of maritime business to the East and the effects of the global financial crisis (leading to cancelled orders of new-build ships and increased retrofit activities), and the companies need to be prepared for future changes. Product/service portfolios ought to reflect this, with regards existing products and new market potentials. The companies interviewed did all consider their position in the market and the decision could be to withdraw from one market segment and focus on a new segment, where competencies were put to better use, or to limit competition, e.g. entering offshore markets. This could also be the strategy behind implanting product/service-systems into companies' business models, as the companies are already competing in an area where a holistic perspective and understanding of customers' needs and requirements is a competitive dimension.

5 NEXT STEPS

After the first descriptive phase of the project, we have initiated several steps towards new activities bringing the project into its next phase. First of all, a number of concrete case studies have been identified through a close and intensive dialogue with all participating companies – combined with project suggestions from a research perspective. Three case studies have been initiated and will be commenced in January 2011. For the future of the project, the research methods will increasingly target specific issues identified that call for different approaches, due to the nature of the area of interest. Issues that cover a span of in-depth research into the managerial, communicational and structural composite of the organisation using an action research approach, to the other end of the spectrum, to issues demanding a broader understanding of the interests and requirements for establishing a cross-company collaboration to fulfil the future, presently unacknowledged customer needs. Hence, the orthogonal views on PSS in the industry branch are needed, in order to understand the different aspects through different projects with the shared goal of researching the development of integrated product/service offerings and the characteristics related.

6 REFLECTIONS/DISCUSSION

This paper has presented five orthogonal views on the field of product/service-systems focusing on a specific industry. While one can argue that there exist other views than the ones presented in section 3.1-3.5, these five views were chosen based on earlier and current research, as well as an in-depth dialogue with industry representatives. While other views would contribute with further relevant aspects, research has shown a need for further exploration within the selected areas as well as their potential to unlock new possibilities within the field of PSS. It has been our aim to present these orthogonal views on PSS, together with some initial insights, seen from a whole industry perspective, which has not been seen in PSS literature hitherto. We believe that this approach is of high value for anyone interested in this research field.

The fact that the research seeks to take an entire industry approach has many benefits, but has also proved challenging, with respect to research methods, data handling and company relations. The mere act of coordinating meetings with each of the participating companies; keeping them informed and engaged in the project, while still providing them with industry relevant results; and at the same time creating scientific results, has proven an exercise in itself.

The maritime industry in itself is challenging, in terms of complexity, conservatism and the fact that it operates on an international playing field. But it is also an industry where western suppliers are rising to face the competition from Asia and are ready to look beyond the traditional solutions. This compels the industry (and certain researchers!) to think innovatively for creative solutions that can inspire a new type of innovation into the industry. The fact that representatives from the industry acknowledge the need and the potential for PSS solutions is a reason to focus on this challenging industry.

The empirical studies within the research project have shown that customers' activities are part of the value creating process in multiple ways, both adding and decreasing value. The shipyard, which is a bottleneck for any product/service-system implemented in the maritime industry, is at this time the strongest influencer of choice of supplier, and thereby also of the choice of product and quality and price of it. The decision parameters of the shipyard, which are typically based on *first purchase price* or contract with specific suppliers, are not always beneficial for the ship owner, and can result in a decrease of the overall value system. The ship owner and ship crew are in multiple ways directly part of the value creating system, as they decide the service-level agreement; maintain the ship; choose service supplier; and are therefore the direct influencers of the utility of the ship within its whole life cycle, which indicates a possibility of structured customer value creation.

The Danish suppliers have a great opportunity to create value propositions through their existing knowledge and competencies, aimed directly at the customer's activities. This can support the customer need during maintenance and create possibilities for e.g. preventive maintenance, with the main goal of retaining a unique customer experience. Hence an opening is created for opportunities regarding the creation of new relationships and different networks of stakeholders, through reconfiguration. Barriers within the companies' different motivations for change, their organisational structures, company sizes, and their individual understanding of market need is important to understand, before starting to re-position the firm towards a PSS - individually or collectively in alliances.

ACKNOWLEDGEMENT
The research described in this paper is carried out as part of the PROTEUS Innovation Consortium, financed by the Danish Agency for Science, Technology and Innovation's Council for Technology and Innovation.

REFERENCES

[1] Prahalad, C.K. and Ramaswamy, V. *Co-Creation Experiences: The Next Practice in Value Creation*, 2004. Journal of interactive Marketing 18(3): 5.
[2] Matzen, D. and McAloone, T.C. A Tool for Conceptualising in PSS Development, in *17. Symposium Design for X*, October 2006, Neukirchen, Germany, TU-Erlangen.
[3] McAloone, T.C. and Andreasen, M.M. Defining Product/Service-Systems", in *13. Symposium Design for X*, 2002, Neukirchen, 10-11. October 2002, TU-Erlangen, pp. 51-60.
[4] Tan, A.R., *Service-Oriented Product Development Strategies*, 2010, PhD Thesis, Technical University of Denmark, ISBN: 978-87-90855-32-1.
[5] Chesbrough, H. *Open Innovation: The New Imperative for Creating and Profiting from Technology*, Harvard Business School Press, 2003.
[6] Matzen, D. *A Systematic Approach to Service Oriented Product Development*, 2009, PhD Thesis, Technical University of Denmark, ISBN 978-87-90855-30-7.
[7] Vandermerwe, S. Jumping into the Customer's Activity Cycle: A new role for customer services in the 1990s, *Columbia Journal of World Business*, 1993, Vol: 28 – 2.
[8] Alonso-Rasgado, T., Thompson, G. and Elfström, B.O. The Design of Functional (Total Care) Products, *Journal of Engineering Design*, 2004, Vol. 15 - 6.
[9] Sundin, E. and Bras, B. Making Functional Sales Environmentally and Economically Beneficial through Product Remanufacturing, *Journal of Cleaner Production*,2005, Vol. 13(9).
[10] Goedkoop, M.J., van Halen, C.J.G., Riele, H.R.M. and Rommens, P.J.M. Product Service Systems, *Ecological and Economic Basics*, 1999.
[11] Manzini, E. and Vezzoli, C. A Strategic Design Approach to Develop Sustainable Product Service Systems: Examples Taken from the 'Environmentally Friendly Innovation' Italian prize, *Journal of Cleaner Production*, 2003, Vol. 1(8).
[12] Mont, O. *Product-Service Systems: Panacea or Myth?* 2004, PhD Thesis, Lund University.
[13] Wise, R. and Baumgartner, P. Go Downstream: The New Profit Imperative in Manufacturing, *Harvard Business Review*, 1999, Vol. 77, No. 5, pp. 133-143.
[14] Karmarkar, U. S. and Apte, U.M. Operations Management in the Information Economy: Products, Processes and Chains, *Journal of Operations Management*, 2007, 25(2), pp. 438-453.
[15] Blessing, L. and Chakrabarti, A. *DRM: A Design Research Methodology*, 2009, Springer.
[16] ECORYS SCS Group, *Study on Competitiveness of the European Shipbuilding Industry within the Framework Contract of Sectoral Competitiveness Studies*, EU Report ENTR/06/054, Directorate-General Enterprise & Industry, 2009.
[17] Porter, M.E. *Competitive Advantage: Creating and Sustaining Superior Performance*, 1998.
[18] Stabell C.B. and Fjeldstad, Ø.D. Configuring Value for Competitive Advantages: On Chains, Shops and Networks, 1998, *Strategic Management Journal*, Vol 19, 413-437.
[19] Normann, R., *Reframing Business*, 2001, John Wiley & Sons Ltd., England.
[20] McAloone, T.C., Mougaard, K., Restrepo, J. and Knudsen, S. Eco-Innovation in the Value Chain, in *Proceedings of DESIGN 2010 11th International Conference on Design*, Dubrovnik, 2010.
[21] PROTEUS Innovation Consortium website: www.dtu.dk/proteus

Corresponding author: Tim McAloone
Technical University of Denmark
Department of Management Engineering
Building 424
DK-2800 Kongens Lyngby
Denmark
Tel: +45 4525 6270
Email: tmca@man.dtu.dk
URL: www.staff.dtu.dk/tmca

Tim is Associate Professor of Product Development in the Department of Management Engineering, Section of Engineering Design & Product Development. Tim works closely with Danish and international industry, finding new methods and models for a wide range of product development issues, such as sustainability, product/service-systems and product innovation.

INTERNATIONAL CONFERENCE ON ENGINEERING DESIGN, ICED11
15 - 18 AUGUST 2011, TECHNICAL UNIVERSITY OF DENMARK

REPRESENTATION AND ANALYSIS OF BUSINESS ECOSYSTEMS CO-SPECIALIZING PRODUCTS AND SERVICES

Changmuk Kang[1], Yoo S. Hong[1], Kwang Jae Kim[2], and Kwang Tae Park[3]
(1) Seoul National University, South Korea (2) Pohang University of Science and Technology, South Korea (3) Korea University, South Korea

ABSTRACT

Recent dramatic changes in a mobile industry, initiated by smart phones, are drawing enormous attention to business ecosystems. A perspective to view firms as members of cross-industry ecosystem was first suggested by Moore [1] for describing co-evolution behavior in high-technology business areas. Although an ecosystem has been pervasive in any industry at any time, today's eco-systems are getting more horizontal and complex. This study develops a representation model of such complex ecosystems and a framework for analyzing interrelated productivity of ecosystem members. The proposed model especially emphasizes interdependency between product and service offerings that ecosystem members cooperatively deliver. This interdependency determines the interrelation between members' productivity, and finally sustainability of a whole ecosystem. This study describes a general procedure for representing and analyzing an ecosystem and discusses the difference between the traditional and smart phone mobile ecosystems based on the result of the analysis.

Keywords: Business ecosystem, requirement interdependency, productivity interrelation

1. INTRODUCTION

Recent dramatic changes in a mobile industry, initiated by smart phones, are drawing enormous attention to business ecosystems. Mobile contents business, which had stagnated for a decade in spite of desperate effort of network operators, burst open after the introduction of smart phones. In this business, contents providers, network operators and device vendors form an ecosystem for producing and delivering mobile contents to customers. Whereas competition in a traditional mobile industry had been device-to-device or service-to-service, the smart phones are competing with their own ecosystems in which various and interesting contents are continuously supplied by independent providers and developers. Device vendors and network operators now spend a lot of money on supporting third-party contents developers in order to promote their own ecosystems.

While anyone may intuitively understand a concept of the business ecosystem by the familiarity to its biological origin, a perspective to view firms as members of cross-industry ecosystem was first suggested by Moore [1] for describing co-evolution behavior in high-technology business areas. He stated that ecosystem members co-evolve their capabilities to support new products and satisfy customer needs around a new innovation. They also invest towards a shared future instead of each member's own because it is valued by the rest of the community [2]. Those behaviors are exactly same in smart-phone ecosystems. All the members of iPhone and Android ecosystems try hard to expand their ecosystems in customer and developer communities as well as the leaders like Apple and Google.

Although an ecosystem has been pervasive in any industry at any time, today's ecosystems are getting more horizontal and complex. A common and traditional form of the ecosystem has been a supplier network in which individual firms are vertically linked with a seller-buyer relationship. Recently emerging innovation modes and business models, however, enforce to organize horizontal relationships between ecosystem members. A representative one is open innovation. It is a strategy to find sources of innovation from outside innovators as well as in-house capability [3]. Since outside innovators are prone to be loosely coupled with the firm, a firm needs to manage them like customers. Besides, advertisement sponsored business models [4] and integration of products and service drive horizontal cooperation between independent firms.

It is more complicated to build a healthy ecosystem with such horizontal and complex relationships. As Iansiti and Levien [2] advocated, the first condition of a healthy ecosystem is high productivity.

The productivity includes individual members' profit as well as the system's gross profit. In a horizontal relationship, if a member does not have proper profit from an ecosystem, she easily leaves and finds others like customers. Then, quality of the ecosystem offering quickly drops because it should have been complemented by the left member. Such an ecosystem cannot be healthy. Because profits of ecosystem member are prone to be interrelated each other, one who wants to build a healthy ecosystem needs to understand their relationships.

This study develops a representation model and a framework for analyzing interrelated productivity of ecosystem members based on this model. In literature, an ecosystem often has been represented as a network of value exchange between ecosystem members. However, it is a result of interdependencies between product and service offerings that all ecosystem participants cooperatively deliver. As shown in many studies [2, 5-8], their interdependencies determines positions of their providers in an ecosystem and share of profit they can have. Jacobides *et al.* [7] defined these interdependencies as requirement architecture. Therefore, the proposed representation model augments requirement architecture to value network model. The interrelated productivity of ecosystem members can be systematically analyzed by the proposed framework based on this representation.

2. LITERATURE REVIEW

An economic community of cooperating firms has been common from the beginning of industrialization, even before Moore [1] defined it as an ecosystem. The most well-known form of such a community is a supply chain, which is also referred as an ecosystem in Moore's work [1]. An ecosystem is, however, a more extended organization of firms and individuals beyond a vertical hierarchy of supplier-buyer relationships [9]. It also includes horizontal relationships with stakeholders who are not direct suppliers or customers, but suppliers of complementary products or services, firms outsourcing our business functions, financial supporters, technology providers, and even competitors [10]. As Iansiti and Levien [10] noted, defining boundaries of a firm's ecosystem is impossible, and they should be identified by examining which organizations are most closely intertwined with the firm.

The reason why a firm has to consider its ecosystem health as well as its own competitiveness is that it shares a fate with other ecosystem members [2]. In this sense, Adner [5] stated that a firm needs to concern risks involved in coordinating and integrating complementary innovators when formulating its innovation strategy. The most important one in determining the fate of an ecosystem is a *keystone* player. Following Iansiti and Levien's [2] definition, they provide 'a stable and predictable platform on which other members create niches depending, regulate connections among them, and work to increase diversity and productivity.' Because they provide a platform, they are also called as a *platform leader* in literature. Due to the importance of the role of keystones or platform leaders, most of the ecosystem strategies have been studied and proposed with their perspectives.

As the most comprehensive literature on ecosystem strategies, Iansiti and Levien [2] provided three measures of the ecosystem health, which are productivity, robustness, and niche creation, and described foundations for competing with other ecosystems. Gawer and Cusumano [6] presented what strategic levers platform leaders in industry like Intel, Microsoft, and Cisco use to maintain their leading position, and in their succeeding research [11], they defined two strategies to become a platform leader as coring and tipping through investigation on various practices. The coring strategy is to solve a common problem to create many niches and provide the solution as a platform, and the tipping strategy is to win competition with other platforms and ecosystems. Eisenmann [12] presented challenges in managing platforms when they are proprietary or shared, and Eisenmann *et al.* [13] extended this study by providing strategic elements that should be considered when opening a platform at sponsor, provider, and user levels. In a perspective of managing relationships with other members, Boudreau and Lakhani [14] addressed a decision whether a platform provider should organize outside innovators as a collaborative community or a competitive market according to their motivation and platform's business model, in an open-innovation ecosystem.

Meanwhile, the aim of this study is to analyze architecture of relationships between members and investigate its implications to ecosystem health. Iansiti and Levien [2] also chose architecture as the first foundation for competition. In this approach, several studies represent firm-by-firm relationships in an ecosystem and find implications of the network structure. Iyer *et al.* [15] analyzed network of famous software firms like IBM, Microsoft, SAP and found managerial implications in constructing ecosystem structure. Similarly, Basole [16] visualized inter-firm relationships in a mobile ecosystem and measured its structural properties. Whereas this representation and analysis intuitively shows positions of firms in a network, their roles and characteristics of relationships are ambiguous to analyze ecosystem health.

The more informative representation involves roles of members. Basole [16] also showed consolidation of firms in terms of their roles for abstracting the complex inter-firm network. Basole and Rouse [17] constructed a network consists of relationships between roles in a service delivery process. By analyzing networks of various industries, they asserted low complexity to consumers as a condition for a healthy ecosystem. Allee's [18] value network analysis and Donaldson et al.'s [19] customer value chain analysis more comprehensively represents an ecosystem by noting what values are transferred between players having a certain role. The proposed representation model augments the requirement architecture to such a value network between members playing a certain role.

The analysis of interrelated productivity of ecosystem members is still in infancy. While Allee [20] mentioned that each member's decisions on revenue and cost influence other members' productivity and system dynamics approach is applicable to analyze these relationships, no concrete model has been proposed in her work. On the other hand, Tian et al. [21] proposed a framework for analyzing these relationships based on game theory models and multi-agent simulation. More concrete analysis was conducted for special relationships. Parker and Van Alstyne [22] and Rochet and Tirole [23] analyzed pricing decisions on a platform that is independently sold to both customers and complementary product or service providers, considering their network effects. In a traditional supply chain, Cachon and Lariviere [24] proposed revenue sharing between sellers and suppliers and analyzed its impact on their profits. This study develops a generic analysis framework that can be applied to other types of ecosystems, and this framework investigates interrelation between more than two members' productivity considering their requirement architecture.

3. REQUIREMENT ARCHITECTURE AUGMENTED REPRESENTATION MODEL FOR BUSINESS ECOSYSTEMS

As Jacobides et al. [7] asserted and Tee and Gawer [8] assured by i-mode service example, complementarity and mobility between co-specializing assets plays an important role in benefiting from innovation. It is a fundamental reason for constructing an ecosystem beyond a dyadic relationship between sellers and buyers. Because more than one product and service complement each other and co-specialize a system, their producers cooperate each other as ecosystem members. What and how much profits they achieve may change according to the architecture, and it affects ecosystem health. Therefore, the requirement architecture must be considered in analyzing and improving an ecosystem. Nevertheless, existing ecosystem representation models lack this part and only show value exchange network between players.

This study develops an ecosystem model that comprehensively represents requirement architecture as well as roles of players and their value network altogether, and ecosystem analysis and improvement framework based on this model. As Simon [25] said, "every problem solving effort must begin with creating a representation for the problem," the requirement architecture needs to be first represented in order to analyze its impact on ecosystem health. The procedure of representation starts from defining requirement architecture, specifies roles of players by engaging them to product and service components, and finishes by identifying value exchange between players. This section explains how an ecosystem is represented by the proposed model in this sequence, and illustrates an example of mobile ecosystem.

3.1 Definition of requirement architecture

The requirement architecture is defined by product and service components and their requirement relationships. As denoted in Table 1, entities of a network are product and service components which are cooperatively delivered by an entire ecosystem. For examples, a mobile ecosystem has cell phone devices, platforms, network service, and mobile contents as its product and service components, and a business solution ecosystem has operating systems, databases, server computers, business solutions, and add-on modules.

Their relationships are defined by requirement (*require*) and proprietary requirement (*p-require*) relationships. These two relationships implements Jacobides et al.'s [7] two aspects of asset dependence, which are complementarity and mobility. If a product or service component complements another component, this relationship is represented by a directional requirement relationship from the complementing component to the complemented component. For example, mobile contents complement

Table 1 Definition of requirement architecture

Entity	Description
Product (Service) component	
Product / Service	A product or service component that is delivered by an ecosystem

Relationship	Description
require	
require (produce)	A product/service *requires* a product/service *to produce* itself
require (own)	A product/service *requires* a product/service *to own* itself
require (use)	A product/service *requires* a product/service *to use* itself
p-require	
p-require (produce)	A product/service *requires* a product/service *to produce* itself
p-require (own)	A product/service *requires* a product/service *to own* itself
p-require (use)	A product/service *requires* a product/service *to use* itself

mobile devices, and this relationship is interpreted as 'mobile contents require mobile devices for its usage.' The *require* relationship is notated by a dashed line as illustrated in Table 1. This model also distinguishes reasons of requirement. If a component is required for producing another component, it is notated by a *produce* tag in parentheses.

The proprietary requirement implements low mobility of assets. Low mobility means dedication to a specially offered product or service component that cannot be replaced with other ones. For example, Intel's CPUs (central processing units) are immobile to peripherals because other CPUs are not compatible and no one can substitute Intel's reputation. In an opposite way, peripherals are highly mobile to Intel because there are a number of providers of compatible and standard-quality peripherals. In this case, such an immobile asset like an Intel's CPU is denoted by 'proprietarily required by other components.' Hence, *p-require* relationships are a subset of *require* relationships. The *p-require* relationship is notated by a solid line as illustrated in Table 1.

3.2 Definition of roles of players

Each member of an ecosystem plays one or more roles in producing or consuming interrelated product and service components. This model concentrates their roles instead of their individual distinctiveness as other representation models [17-19]. Therefore, all the members who play the same role are abstracted as a player entity whether it is a single firm or an open community of individuals. Following Moore's [1] definition on the ecosystem, "economic community produces goods and services of value to customers, who are themselves members of the ecosystem," the players are restricted to direct stakeholders who produce, own, or use product and service components. Other stakeholders such as auxiliary enablers in Basole and Rouse's [17] model are excluded from the scope of the proposed representation and analysis. The player entity is denoted by a graphical icon as illustrated in Table 2.

The role of a player is denoted by how it is engaged with product and service components. There are three types of engagement relationships: producing (*produce*), owning (*own*), and using (*use*) a product or service component. If a player has a role of producing a product, a directed arrow marked by *produce* notation is drawn from the player to the product. Likewise, owning and using relationships are denoted by arrows between player entities and product and service components as illustrated in Table 2. One caution is that the owning relationships are applicable to only products. One of the characteristics of service is inseparability between production and consumption [26]. Therefore, no one can have ownership of the service. For products, however, one can own a product and borrow it to actual users like a car leasing company.

The roles are interlinked with the requirement architecture. As defined earlier, product and service components have requirement relationships by which a producer, owner, or user of a component has to

Table 2 Definition of roles of players

Entity	Description
Player	
👤	A single or group of ecosystem members who play a certain role of producing, owning, or using product and service components
Product (Service) component	
Product Service	A product or service component that is delivered by an ecosystem

Relationship	Description
Produce/own/use	
👤 —produce→ Product/Service 👤 —own→ 👤 —use→	A player *produces* a product/service A player *owns* a product A player *uses* a product/service

use another component. These relationships partially determine roles of players. For example, a mobile device user should be a user of mobile network service since the device requires the network service to use itself. As another example, a doctor who is a producer of medical service is also a user of medical devices which are required to produce the service. Therefore, roles are defined after identifying the requirement architecture.

This notation of the roles is flexible and convenient to represent multiple roles of one player and evolution of the roles. In a mobile ecosystem, while a network service operator or a device manufacturer also plays a role of providing content market service like Apple's Appstore, it is hard to figure out who plays this role in other representation models in which roles are defined only by the names. The proposed model does not have this problem since it can be represented by linking the corresponding player and content market service with a *produce* relationship. Moreover, when roles of a player evolve, it can be conveniently represented by rearranging the linkage between players and product and service components.

3.3 Definition of value exchange between players

The players of an ecosystem gain profit from exchange of values between themselves. The value exchange has been most comprehensively represented by the existing models. In existing models like Allee's [18] value network and Donaldson *et al.*'s [19] customer value chain, a relationship between a producer and a user is represented by exchange of a product or service and payment for it. The proposed model, however, represents this relationship by specifying their roles and defining a payment relationship according to the roles. In addition, only values that directly contribute to player's profit are modeled because the goal of the representation is to analyze productivity of an ecosystem. Latent values like user's complaints are out of the scope of this model.

Table 3 Definition of value exchange between players

Entity	Description
Player	
👤	A single or group of ecosystem members who play a certain role of producing, owning, or using product and service components

Relationship	Description
pay	
👤 —pay (value)→ 👤	A player *pays* a certain *value* that directly contributes to profit to another player

The relationship of value exchange is payment (*pay*). As illustrated in Table 3, the payment relationship is denoted by an arrow marked by *pay* notation. It also notes what value is paid in parentheses. Although the most common value is monetary price of a product or service, other intangible values can be paid if it contributes to player's profit. For example, advertisement is a kind of service produced by an advertiser, and a consumer who watches the advertisement pays his attention to the advertiser. From a player's perspective, in-bound and out-bound payment arrows indicate his revenue and cost, respectively. This notation enables to conveniently identify revenue and cost of participating in an ecosystem and analyze its productivity.

A special type of the payment relationship is paying subsidy to a third-party player. Where other payment relationships exist between a producer and an owner or a user, subsidy is paid for owning or using other player's product or service. The subsidy payment is a common way to attract more participants into an ecosystem [11]. It is usual that a network service operator pays subsidy for buying a mobile phone to network subscribers. Because the subsidy payment is not directly linked with roles of the players, it should be defined according to player's ecosystem strategies. As Eisenmann *et al.* [27] explained, who will subsidize and be subsidized changes according to market and other environmental situations.

Summarizing the above steps of ecosystem modeling, Figure 1 illustrates a smart phone ecosystem represented by the proposed model. The ecosystem delivers a smart phone device, network service, and mobile contents to customers. This example assumes a device includes its platform like Apple's iPhone and iOS. The mobile contents for smart phones are dedicated to a device instead of network service unlike traditional mobile contents. Therefore, a specific device is proprietarily required for producing or using mobile contents. Network service is also required for owning and using the contents, but is not proprietary. The device and network service are mutually required for using each other.

4. A FRAMEWORK FOR ANALYSIS OF INTERRELATED PRODUCTIVITY BETWEEN ECOSYSTEM MEMBERS

Success of an ecosystem does not depend only on excellence of its offerings. It largely depends on

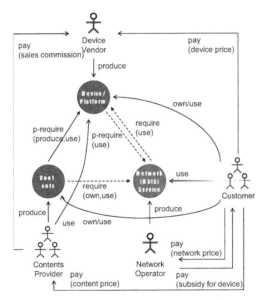

Figure 1. Illustration of modeling procedure: a smart phone ecosystem

how many and competitive members participate in the ecosystem. To be precise, the excellence of the offerings is determined by who are its participants. This phenomena is also called two-sided network effect [22]. The HDTV example is an extreme case in which it could not be diffused in the market over ten years because its content ecosystem was not ready [5]. As Iansiti and Levien [2] pointed out, a healthy ecosystem is nothing but an ecosystem that attracts more members than competing ones.

Iansiti and Levien [2] presented three aspects of the ecosystem health: productivity, robustness, and niche creation. In other words, more firms and individuals want to participate a highly productive, robust, and variety creating ecosystem. This study aims to analyze the ecosystem health in a perspective of productivity among them. It is because the productivity is a prerequisite for other aspects, and the robustness and niche creation can be interpreted as ability to maintain high productivity in an evolving ecosystem.

4.1 Measurement of productivity of players

The first step of the analysis is to identify benefit and cost elements that affect productivity of each player and measure their values. While productivity can be defined in various ways, the most widespread measure is ratio between the benefit and cost, in other words, return on invested capital. More benefit and less cost increases productivity and attracts more members to join an ecosystem. Necessary benefit and cost elements are identified from the proposed ecosystem representation.

First, roles of a player like producing, owning, or using product and service components induce their corresponding benefit and cost. The cost elements are more obvious. Producing a product or service requires material, labor, and other miscellaneous costs, owning it charges maintenance and depreciation costs, and using it also needs time and effort of users. The cost elements of owning and using a product or service beyond its purchasing price have been well studied in total cost of ownership (TCO) literature, which is first introduced by Ellram [28].

The benefit elements are more implicit and unobservable. The benefit of usage is often referred as a concept of utility. As a product or service provides more functions, higher performance, and better quality, its user receives more utility. However, the benefit of producing and owning is usually indirectly observable. Whereas production experience may accumulate learning know-how and ownership may give reputation to owners, they are usually insignificant to take into account in the analysis.

Second, the value exchange between players reveals clearly observable benefit and cost elements. When a player pays a certain value to the other player, the value is a cost element of the payer and a benefit element of the receiver. Accordingly, price of a product or service is benefit of a producer and cost of an owner or user. It should be noted that the benefit and cost elements incurred by the value exchange and the roles of each player are not distinguished in the following analysis. The two step procedure is merely for not overlooking possible elements.

4.2 Influence analysis between benefit and cost elements

In the next step, the identified benefit and cost elements and their possible values are analyzed in a perspective that how they are interrelated. Obviously, benefit and cost elements have positive and negative influence to productivity, respectively. The benefit and cost elements also have interrelation between themselves. A higher price increases producer's benefit while increasing user's cost. In a complex ecosystem, such impact is propagated in a more complicated manner.

In order to represent these relationships, we adopt a causal loop diagram which is commonly used for analyzing system dynamics. A diagram denotes the benefit and cost elements as nodes and their positive and negative influence relationships as links between them. In addition, productivity of each player and some important system variables that affect benefit and cost are also denoted as nodes. It depends on an objective of the analysis which system variables are taken into account. It could be severity of government regulation or social consciousness if an ecosystem delivers environmental product and service components. The causal loop representation for business structure is also found in Casadesus-Masanell and Ricart's work [29]. They modeled a business model of a single firm as a causal loop of business choices and their consequences.

The influence links between nodes can be determined by many different factors. First of all, business models of individual players determine positive and negative influence of benefit and cost elements, respectively. Casadesus-Masanell and Ricart [29] defines those influence links as a business model itself. In ordinary business models, price and cost elements have clear influence links. A product or service price increases seller's productivity while decreasing buyer's productivity. Cost for producing such a product or service usually improves its utility, which increases buyer's productivity. Influence of other elements such as subsidy and commission should be identified according to business models of individual players.

The requirement architecture also determines a structure of influence links. The requirement architecture augmented ecosystem representation model is also required for this analysis. As described in the previous section, the requirement architecture is defined by requirement and proprietary-requirement relationships between complementing product and service components. If a product or service component A requires another component B, A's utility positively influences to B's utility. Requirement is a kind of the means-end relationship. As an end (requiring component) is more valuable, its mean (required component) deserves more value. In an airline industry, airline companies often advertise overseas travel despite no earnings from it, since pleasure of the travel increases utility of the airline service. Also in a smart phone industry, quality and variety of applications determines preference to smart phones which are required for using them.

Finally, influence of internal and external system variables to benefit and cost elements could be found. The most significant system variable is volume of members joining an ecosystem, and the most representative influence is network externality. Network externality is simply defined by effect of the number of users of a product or service on its utility [30]. Direct network externality accelerates growth of volume of users because increased utility attracts more users. It is also defined for two-sided markets as two-sided network externality [22]. If there are two groups of suppliers and users, increase of supplier volume attracts users by expectation for better offerings, and increase of user volume attracts suppliers by expectation for expanded market.

A mobile ecosystem including contents business is a good example to illustrate influence links induced by all these factors. Productivity of all four players, who are a customer, a device vendor, a network operator, and a contents provider, and volume of customers and contents providers are identified as underlined system variables as illustrated in Figure 2. Because productivity and volume can be considered as identical state variables, only volume variables are left in Figure 2. Influence links of price, utility, and cost elements of ecosystem offerings, which are a smart phone device, network service, and other elements identified by business models, which are subsidy for device and contents sales commission, to system variables are also defined. Those can be considered as a set of business models of individual players.

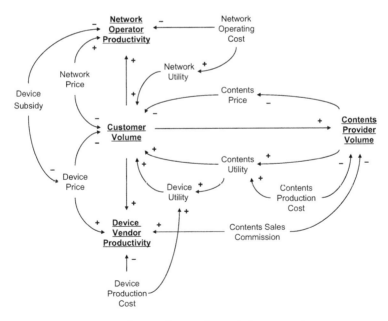

Figure 2. Influence diagram of a smart phone ecosystem

The example of Figure 2 also shows influence links initiated by requirement architecture and network behavior. As illustrated in Figure 1, mobile contents for smart phones require a proprietary device platform. In this situation, utility of contents directly improves utility of a smart phone device. Consequently, a device vendor has pretty high incentive to improve contents utility. Another feature of this ecosystem is that it has two-sided network externality which is illustrated in two closed loops between customer and contents provider volume. First, more customers attract more contents providers to join for bigger revenue of contents sales. As more providers join, they provide various and high quality contents having higher utility, and their prices also drop by severer competition. These consequences establish a virtuous cycle that increases customer volume again. A device vendor and a network operator endeavor to attract more contents providers as well as customers who directly increase their productivity because of this two-sided network externality.

4.3 Assessment of sensitivity of benefit and cost elements to productivity

The final step is analysis of sensitivity of benefit and cost elements and sustainability. Because of interrelation between benefit and cost elements and system variables, each player's decision or performance on their offerings has complicated consequences to his own and other players' productivity. This study defines impact of such consequences as sensitivity to productivity. Because each player wants to maximize its own productivity, the architectural sensitivity finally determines sustainability of an ecosystem.

The underlying basic of the assessment is to track influence links that increase or decrease benefit and cost elements and system variables. By using the influence diagram, one can easily find direct and indirect changes of elements and variables induced by a change of the target element. Let us take a look at 'contents sales commission' element in Figure 2. Tracking positive and negative influence of its increase is illustrated in Figure 3. Its positive influence, increased commission income, is directed to 'device vendor productivity'.

On the other hand, it negatively influences 'content provider volume' since it charges addition cost to contents providers. The decreased volume damages 'contents utility' and raises 'contents price', both

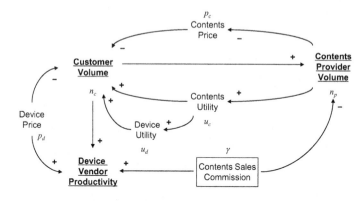

Assume 'Contents sales commission' is increased by γ

Positive influence	Negative influence
	Contents provider volume (N_p) = Provider elasticity to commission rate × γ
	Customer volume (N_c) = Network externality to provider volume × N_p (aggregation of price increase and utility decrease)
Device vendor productivity (P_d) = Average downloads per customer × $p_c \gamma n_c$	Device vendor productivity (N_d) = $p_d N_c$

Figure 3. Sensitivity analysis example of 'Contents sales commission'

of which make less customers to join the ecosystem. Those consequences imply that an ecosystem with high sales commission is hard to grow without other incentives. Additionally, the damaged 'contents utility' lowers 'device utility' that also reduces 'customer volume'. The decreased 'customer volume' has negative influence to productivity of a device vendor who formerly gained additional commission income.

As seen in the above analysis, a decision on a benefit or cost element has both positive and negative influence to productivity of the player leaving the price-demand curves aside. Each player makes decisions comparing the both sides. The assessment on sensitivity enables estimation on each player's decisions. When influence relationships are quantitatively known and decisions are engaged with only two players, the optimal decision can be found using game theory models as Tian *et al.* [21] showed. However, rough assessment on positive and negative sensitivity identified by an influence diagram could provide meaningful implications in usual cases, in which the form of quantitative relationships and specific parameter values are not exactly known, and more than two players are engaged with a decision. The next section shows how failure of contents business in a traditional mobile ecosystem is explained by such rough assessment.

The analysis on player decisions also provides implications on sustainability of an ecosystem. In commercial ecosystems, each player wants to maximize its profit, which is interpreted as productivity in this study. A decision of a player that maximizes his own productivity may harm productivity of other players. If it cannot be compensated by other incentives, they will leave the ecosystem. The architectural difference between ecosystems, i.e., who owns what benefit and cost elements and how they are interrelated, make them have different decisions. For example, contents providers pay less sales commission for smart phones than feature phones, which promotes their participation in smart phone ecosystems (why it happens will be explained in the next section). Such difference determines sustainability of an ecosystem.

5. DISCUSSIONS: WHY DO TRADITIONAL MOBILE ECOSYSTEMS FAIL TO ESTABLISH CONTENTS BUSINESS?

In this discussion, we want to concentrate on a decision on level of 'contents sales commission'. It is a one of the major cost elements of a contents provider. As illustrated in the previous section, it is related with many other elements and system variables through complex propagation paths. The lower commission rate promotes more participation of providers, and it also attracts more customers by the two-sided network externality. Obviously, an ecosystem with lower commission rate is more advantageous to successfully grow. Nonetheless, a player who charges it to the providers cannot make it free because it is also a source of his benefit. In a traditional ecosystem, a network operator charges the commission since he distributes the contents that are proprietarily dedicated to the network. In a smart phone ecosystem, this role is transferred to a device vendor or platform holder who provides a platform on which contents and applications are operated.

In conclusion, a traditional ecosystem is prone to charge a higher commission than a smart phone eco-

Table 4. Comparison of influence of commission rate increase (by $\$\gamma$) in traditional and smart phone ecosystems

Traditional ecosystem	Smart phone ecosystem
Contents provider volume	Contents provider volume
Positive : 0	Positive : 0
Negative : -1% × γ	Negative : -1% × γ
Customer volume	Customer volume
Positive : 0	Positive : 0
Negative : -1% × γ	Negative : -1% × γ
Network operator productivity	Device vendor productivity
Positive : 99%·10M × $\$\gamma$ = 0.99γM	Positive : 99%·10M × $\$\gamma$ = 0.99γM
Negative : -$5 × 1%·10M × γ = -0.5\gamma$M	Negative : -$300 × 1%·1M × γ = -3γM

* base customer volume: 1M customers, base provider volume: 1M providers
** average downloads per customer: 10 dls
‡ average network price per downloads: $5, average device price: $300
‡ provider elasticity to commission rate: 1%, network externality to provider volume: 100%

system. Let us refer Figure 3 that shows positive and negative influence of commission rate increase. A traditional mobile network operator and a smart phone device vendor have different parameters for those influence relationships. Table 4 shows how much quantitative impact is propagated to each system variable under some assumptions. Focusing on productivity of the network operator and the device vendor who decides the commission rate, we can find that increase of γ commission rate has much bigger negative impact on device vendor's productivity ($\$3\gamma$ millions) much more decreases than network operator's ($\$0.5\gamma$ millions), while their positive impact is same. Although raise of the commission rate makes contents providers and customers to leave in both of the ecosystems, its impact is larger for smart phone vendors because its unit price much higher than the price of network usage. It means that the network operator in a traditional ecosystem has higher incentive to raise the commission rate pursuing more profit. As noted earlier, the contents business is hard to prosper in an ecosystem of high contents sales commission. It is one of the reasons that traditional ecosystem was unsuccessful to grow contents business.

6. CONCLUSIONS AND FUTURE DIRECTIONS

In this study, we proposed a representation model of an ecosystem augmenting the requirement architecture that was usually omitted in previous representation models, and a framework for analyzing productivity and decisions of players that determine sustainability of the ecosystem. The representation model supports to find benefit and cost elements of players that constitute their productivity and predefines their influence relationships. One who wants to analyze his ecosystem and predict its sustainability first represents the ecosystem with the proposed model and utilizes the analysis framework based on the representation.

Planning and designing an ecosystem has been considered as an art greatly relying on decision maker's intuition. Whereas some studies have proposed value chain oriented representation models for supporting systematic analysis, they lack investigation on a complicated structure of interrelation between benefits and costs of individual players. This study provides a general framework that describes how to measure and represent such interrelation and assess sensitivity of each player's decision propagated through the relationships. The framework would simulate research on more elaborated ecosystem analysis models.

In discussion, we interpreted unsuccessful establishment of contents business in a traditional mobile ecosystem driven by a network operator with respect to its requirement architecture and interrelation of benefit and cost elements. A monumental change in a mobile industry driven by introduction of smart phones has been widely discussed in many studies and news articles. The presented interpretation reveals network operator's dilemma in determining network usage price by clarifying his benefit and cost elements, and his higher incentive to keep a high contents commission rate by assessing its sensitivity to his productivity. It shows that intuitively and qualitatively conjectured previous arguments can be explicitly revealed and quantitatively derived by the proposed analysis framework.

A further study is directed to identify effective strategies that improve sustainability of an ecosystem beyond analyzing it. The analysis results may reveal what bottlenecks that retard growth of an ecosystem are. Various strategies could be adopted to resolve them, and many of them could be found in existing practices. The proposed ecosystem model is also appropriate for representing product and service strategies embedded in an ecosystem since it clearly denotes the requirement architecture of product and service offerings. Therefore, representation of existing ecosystems with the proposed model will help identify the effective ecosystem strategies. The analysis framework will also be utilized for evaluating their effectiveness.

REFERENCES

[1] Moore, J.F. Predators and prey: a new ecology of competition. Harvard Business Review, 1993, 71(3), pp.75-86.
[2] Iansiti, M. and Levien, R. The keystone advantage, 2004b (Harvard Business School Press, Boston, MA, USA).
[3] Chesbrough, H. Open innovation: the new imperative for creating and profiting from technology, 2003 (Harvard Business Press, Boston, MA, USA).
[4] Cusumano, M. The changing software business: Moving from products to services. Computer, 2008, 41(1), pp.20-27.

[5] Adner, R. Match your innovation strategy to your innovation ecosystem. Harvard Business Review, 2006, 84(4), pp.98-107.
[6] Gawer, A. and Cusumano, M. Platform leadership: How Intel, Microsoft, and Cisco drive industry innovation, 2002 (Harvard Business Press, Boston, MA, USA).
[7] Jacobides, M., Knudsen, T. and Augier, M. Benefiting from innovation: value creation, value appropriation and the role of industry architectures. Research Policy, 2006, 35(8), pp.1200-1221.
[8] Tee, R. and Gawer, A. Industry architecture as a determinant of successful platform strategies: a case study of the i-mode mobile Internet service. European Management Review, 2009, 6(4), pp.217-232.
[9] Moore, J.F. The death of competition: leadership and strategy in the age of business ecosystems, 1996 (Harper Business, New York, NY, USA).
[10] Iansiti, M. and Levien, R. Strategy as ecology. Harvard Business Review, 2004a, 82(3), pp.68-81.
[11] Gawer, A. and Cusumano, M. How companies become platform leaders. MIT Sloan Management Review, 2008, 49(2), pp.28-35.
[12] Eisenmann, T. Managing proprietary and shared platforms. California Management Review, 2008, 50(4), pp.31-53.
[13] Eisenmann, T., Parker, G. and Van Alstyne, M. Opening platforms: how, when and why?, (Harvard Business School Entrepreneurial Management, 2008).
[14] Boudreau, K. and Lakhani, K. How to manage outside innovation. MIT Sloan Management Review, 2009, 50(4), pp.69-75.
[15] Iyer, B., Lee, C. and Venkatraman, N. Managing in a 'smal world ecosystem': lessons from the software sector. California Management Review, 2006, 48(3), pp.28-47.
[16] Basole, R. Visualization of interfirm relations in a converging mobile ecosystem. Journal of Information Technology, 2009, 24(2), pp.144-159.
[17] Basole, R. and Rouse, W. Complexity of service value networks: conceptualization and empirical investigation. IBM Systems Journal, 2008, 47(1), pp.53.
[18] Allee, V. Reconfiguring the value network. Journal of Business Strategy, 2000, 21(4), pp.36.
[19] Donaldson, K., Ishii, K. and Sheppard, S. Customer value chain analysis. Research in Engineering Design, 2006, 16(4), pp.174-183.
[20] Allee, V. Value network analysis and value conversion of tangible and intangible assets. Journal of Intellectual Capital, 2008, 9(1), pp.5-24.
[21] Tian, C., Ray, B., Lee, J., Cao, R. and Ding, W. BEAM: A framework for business ecosystem analysis and modeling. IBM Systems Journal, 2008, 47(1), pp.101-114.
[22] Parker, G. and Van Alstyne, M. Two-sided network effects: a theory of information product design. Management Science, 2005, 51(10), pp.1494-1504.
[23] Rochet, J. and Tirole, J. Platform competition in two-sided markets. Journal of the European Economic Association, 2003, 1(4), pp.990-1029.
[24] Cachon, G. and Lariviere, M. Supply chain coordination with revenue-sharing contracts: strengths and limitations. Management Science, 2005, 51(1), pp.30-44.
[25] Simon, H. The science of the artificial, 1996 (MIT Press, Boston, MA, USA).
[26] Zeithaml, V., Parasuraman, A. and Berry, L. Problems and strategies in services marketing. The Journal of Marketing, 1985, 49(2), pp.33-46.
[27] Eisenmann, T., Parker, G. and Van Alstyne, M. Strategies for two-sided markets. Harvard Business Review, 2006, 84(10), pp.92.
[28] Ellram, L. Total cost of ownership: elements and implementation. Journal of Supply Chain Management, 1993, 29(4), pp.2-11.
[29] Casadesus-Masanell, R. and Ricart, J. From strategy to business models and onto tactics. Long Range Planning, 2010, 43(2-3), pp.195-215.
[30] Katz, M. and Shapiro, C. Network externalities, competition, and compatibility. American Economic Review, 1985, 75(3), pp.424-440.

INTERNATIONAL CONFERENCE ON ENGINEERING DESIGN, ICED11
15 - 18 AUGUST 2011, TECHNICAL UNIVERSITY OF DENMARK

INTERDISCIPLINARY SYSTEM MODEL FOR AGENT-SUPPORTED MECHATRONIC DESIGN

Ralf Stetter[1], Holger Seemüller[1], Mohammad Chami[1] and Holger Voos[2]
(1) Hochschule Ravensburg-Weingarten, D (2) University of Luxembourg

ABSTRACT

This paper presents research results from an ongoing project concerning the development of agent systems which aim to support mechatronic design. Today mechatronic design usually suffers from a lack of domain-spanning IT-support. The abundance of logical connections between the disciplines is only present in the designer's minds or in unstructured documents. One approach to document and use the connections between the disciplines are agent based systems. Such systems use independent software entities representing either components of the product or certain process segments which interact in a system called agent system. Due to their flexibility and the ability to achieve solutions which satisfy multiple objectives such systems are promising approaches to address the complexity challenge of mechatronic design. However, the application of such systems requires the documentation of the interdisciplinary connections in an interdisciplinary system model. In this paper SysML is proposed for this specific task. This proposal is supported by the application to an exemplary product development – the parametric development of a Quadrotor – an unmanned multi-purpose flying object.

Keywords: Mechatronic design, Agent Based Systems, SysML, Interdisciplinary System

1 INTRODUCTION

"Mechatronic Design" describes the synergetic creation and integration of mechanical engineering, electrical engineering and information technology for the specification and description of any kind of physical products and processes [1]. Today mechatronic design usually suffers from a lack of domain-spanning IT-support. Each discipline is using its specific IT-tools (e. g. CAD systems) and data formats and the interdependencies between the disciplines are not documented in a consistent manner (frequently unstructured documents such as MS Word or MS PowerPoint files are used). This approach has significant drawbacks on the consistency of interdisciplinary model data and affects the efficiency of the overall mechatronic design process. Many approaches to unify the tools and data during the last two decades were developed but have not made their way into the product development departments in industrial practice.

From research and industry the idea of an interdisciplinary system model holding cross-domain information about the system and important relationships has been arisen [2], [3]. So the unstructured documents and implicit knowledge of the designers can be summarized and captured in a structured way for later reuse.

However a gap between this interdisciplinary system model and the respective IT-tools of the involved domains still remains. The research presented in this paper proposes an approach to minimize this gap by using a multi-agent system to manage the system model which is based on SysML. In that way the coordination and data consistency between the involved designers and engineers shall be improved and the overall mechatronic design process shall be enhanced.

The remainder of this paper is structured as follows. Section 2 explains the context of the work and presents the needed background information. Section 3 shows the overall concepts of our approach where Section 4 explains them with the help of an example. Section 5 lists possible use cases of the approach. Section 6 concludes the paper and gives an outlook.

2 EXISTING APPROACHES AND BACKGROUND WORK

The following section serves as basis for the further sections through explaining important characteristics of interdisciplinary engineering, summarizing the basics of SysML and giving on overview of agent based systems.

2.1 Interdisciplinary Engineering

Due to the domain-spanning characteristics of mechatronic systems, interdisciplinary approaches and methods are needed to address the needs of all involved engineering domains equally together with their strong interrelationships even in early phases of the mechatronic design process. Although this idea is not new in research and has been frequently discussed, e. g. in [2], [3] and [4], it has been hardly realized due to the lack of tools supporting collaboration across disciplines.

From the engineering process perspective, the VDI 2206 guideline [2] gives a methodological support for the cross-domain development of mechatronic systems. The guideline is mainly based on the V model used commonly in software development which is adapted to the special needs of mechatronic product design. One primary idea is the creation of an interdisciplinary principal solution during the system design phase which describes the system in a domain independent way. For the domain-specific design phase this solution shall be partitioned into the respective domains which then design their parts concurrently. However no method or tool is proposed for the creation of the principal solution.

Chen et al. [3] propose a constraint modeling-based approach by modeling the components of mechatronic systems as objects with attributes, and by identifying and modeling the constraints between these attributes.

From an organizational perspective, Pahl et al. describe in [4] the interdisciplinary cooperation for the organizational and operational structures and focus on the prerequisites for forming interdisciplinary teams by structuring the system in such a way that its properties can be modeled precisely and by defining the interfaces between the process steps more precisely and ambiguously.

Currently, a need for further research is still apparent, because many approaches from academia have not found their way into industry yet. In this paper, we see the handling of the interdisciplinary relationships between the different domain specific models as an important aspect to enhance interdisciplinary product engineering. These relationships have to be captured and documented on one side but also be made usable for engineers on the other side.

2.2 SysML

The System Modeling Language (SysML) is "a general-purpose graphical modeling language for specifying, analyzing, designing and verifying complex systems that may include hardware, software, information, personnel, procedures and facilities" [5]. SysML supports the model-based systems engineering approach by offering modelers the ability to create an overall system model of an interdisciplinary system. This overall system model covers different aspects like requirements, parametric, structure or behavior together with their interrelationships. SysML represents a subset of UML2 with an extension needed to satisfy the requirements of the UML for Systems Engineering Request for Proposal (UML for SE RFP) that was initiated in 2003 by OMG [5].

SysML, as any other modeling language, needs a methodology to collect, analyze, and model the different aspects of a system. Domain experts have already described several methodologies for modeling with SysML. Weilkiens introduced SYSMOD in [6] as a pragmatic approach that uses SysML to model systems from analysis to design. A computational product model in [7] shows a way to use SysML for the conceptual design phase according to VDI2221 [8]. In [9], an approach is presented that uses SysML to create discipline neutral system-level models for complex mechatronic systems. A model integration framework is demonstrated in [10] by using SysML profiles to model different discipline specific views in SysML which can be transformed into respective domain-specific models. Thramboulidis proposed in [11] a synergistic integration of the constituent parts of mechatronic systems by using SysML to specify the central view-model of the mechatronic system.

However, despite the effective role that SysML plays in holding the important interdisciplinary relationships, there is still a lack of usage of these relationships in current industrial practice (according to several discussions with engineers and managers).

2.3 Agent-based Systems

Multi-agent-systems which are based on intelligent software agents have been the object of research since the middle of the nineties of the last century. The core is composed of autonomous software components which interact in a goal directed manner and communicate with each other and other components in the scope of a multi-agent system. Using this communication and appropriate algorithms, these software agents are able to coordinate their behavior and to solve problems together without any central control. The main advantage of this decentral approach (decentral in the sense that no centralized control exists) is its large flexibility. Commonly the result (the solution of the problem) is not an optimum result in the sense of an optimization calculation but a result which fulfills all the complex boundary conditions of such a system. A profound overview of the state of the art can be found in Weiss [12], Luck&d'Inverno [13], Trencansky&Cervenka [14] and Padgham et al. [15]. Different software architectures for multi-agent systems are available [12]. The BDI concept describes an architecture where agents are described and developed in terms of their beliefs, desires and intentions [16].

Intelligent software agents are already applied in pilot plants for the planning and control of production systems [17], [18], [19], [20], [21]. In this application agents are the representatives of production machines as well as of parts to be produced. These agents then start negotiations, in order to achieve an appropriate assignment of production orders to production machines. Further areas of application for intelligent software agents can also be found in defense and robotics applications [16].

The application of intelligent software agents in engineering processes is investigated in first research works by Huang et al. [22], Tseng et al. [23], Mild&Taudes [24] and Zhang et al. [25]; the main emphasis is on web-based collaborative work. No special emphasis is given to supporting the development of mechatronic products; furthermore the inclusion of elaborate engineering tools as well as the process planning and control are only covered in a rudimentary manner.

3 AGENT-BASED ENGINEERING OF MECHATRONIC PRODUCTS WITH SYSML

This approach shall combine the advantages of interdisciplinary system modeling with SysML together with the principles of agent based systems. The basic idea is an agent-based management of a SysML model. Based on the modeled information about the mechatronic system the agents shall negotiate and handle the interdisciplinary dependencies across the system during the development process [26]. Therefore the SysML model has to be transformed into single agents which posses the knowledge previously modeled in SysML (Figure 1 – upper left).

The purpose of using a multi-agent system lies in the concept that each component of the system can be represented by an autonomous agent who takes care about relationships to other agents. In case of a conflict the involved agents can negotiate and forward new information to the respective engineer. In that way data consistency between the involved domains is guaranteed and the interrelationships between discipline specific model data can be considered.

For the realization of this idea, the SysML model has to be created in a form that the multi-agent system can deal with. Therefore the semantics of the SysML model has to be defined clearly. The following sections describe the theoretical foundations we base our approach on, the semantics of the SysML model according to the presented foundation as well as the way the model is transformed into a multi-agent system.

Additionally the multi-agent system can basically be used for engineering design process planning and enactment (e. g. based on SPEM compare section 5). Here an agent represents an engineering activity and uses product information from the system model to guide engineers through the process. Figure 1 presents an overview about the architecture of the multi agent system. However, the focus of this paper is the management of the system model with agents (left side in Figure 1).

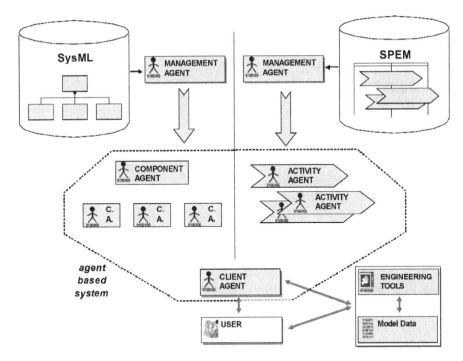

Figure 1. Agents based systems architecture

3.1 Theoretical Model of Mechatronic Systems

In order to deal with a mechatronic system the composition of such systems has to be defined first. According to Stetter&Voos [27] a mechatronic system is assumed to consist of single components connected with interfaces. These components do not necessarily have to belong to a single discipline. So for the design of an electrical motor, a CAD engineer may be responsible for the geometry while a control engineer is modeling the behavior of the device with differential equations. While both engineers are working concurrently on their domain specific model, strong interdependencies such as the dimension of an axis exist within the single interdisciplinary component.

Furthermore it is assumed that an overall system can be hierarchically decomposed into system, subsystem and component level whereas the semantics of these terms are project dependent and cannot be defined in general. These compositional elements can be connected via structural interfaces as well as material, energy and information flow relationships [2].

Additionally interdisciplinary relationships as they can appear within a single component may also exist between distinctive components. So, if the maximum force of the electrical motor changes due to an adoption of the electrical model of this component also the source code of the motor controller has to be adapted.

In order to deal with these aspects, a theoretical model is needed that examines the fundamentals of the system. A simplified model of a mechatronic system is shown in left side of Figure 2 and is described more in detail from a mechanical construction point of view in [4].

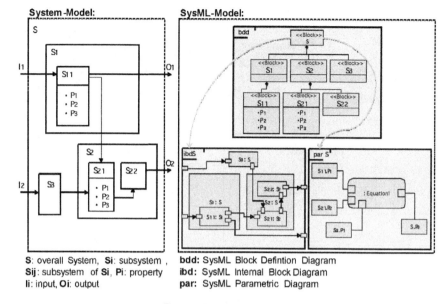

S: overall System, Si: subsystem,   bdd: SysML Block Defintion Diagram
Sij: subsystem of Si, Pi: property   ibd: SysML Internal Block Diagram
Ii: input, Oi: output   par: SysML Parametric Diagram

Figure 2. Theoretical Model

The contribution of this paper deals with the following fundamental aspects of an interdisciplinary system:
- the hierarchy of interdisciplinary components together with their respective parameters and
- the interrelationships between components: structural relationships (interfaces) and flow relationships (material flow, energy flow and/or information flow) as well as
- the interrelationships between disciplines (intra-component relationships or inter-component relationships).

3.2 SysML System Model

As mentioned in section 2.2, a methodology is required for SysML to guide the modeler, with a significant freedom, during the modeling process. In this approach the block definition diagram, the internal block diagram and the parametric diagram from SysML are being used with a given semantics to model a mechatronic system. It is assumed that during a conceptual design phase, a cross-domain system model has been developed according to the assumptions presented in section 3.1. This section describes how this theoretical system model can be realized with SysML for further use with the multi-agent system.

The block definition diagram serves as an appropriate means to model the decomposition of the overall system. Here, a SysML block is used to represent a mechatronic system, subsystem or component as it can be used to represent any type of "thing" that exists in the real world [28]. The composition association expresses the hierarchical relationships between system, subsystem and component layer. The right part of Figure 2 shows an example of a block definition diagram representing the system breakdown structure of the theoretical system of section 3.1.

The interrelationships between components can be modeled with the internal block diagram. Therefore, SysML offers adequate means to express the intended interrelationships between components by using the port concepts which exists of standard ports and flow ports [5] to express different kind of interfaces. The structural relationships will be modeled with a binding connector between two standard ports. Flow relationships including a flow of material, energy or information between blocks can be modeled with an itemflow between flow ports.

For modeling interrelationships between disciplines the parametric diagram serves as an ideal basis as properties of blocks can be related with dependencies as well as constraints using the constraint block from SysML. Constraint blocks offer the possibility to express relationships in e. g. a mathematical equation. In our approach two different kinds of relationships have to be considered: a mathematical relationship as well as a pure dependency relationship of information which can appear between two properties as well as between a property and a block. The latter can be easily expressed by the usage of the dependency relationship. The former relationship will be modeled by using a constraint block and directly entering the mathematical relationship.

Currently, one of the most benefits of SysML is its domain independent syntax. Consequently it can serve the role as a common language between different designer's knowledge. More specifically, in the way of modeling, two aspects of knowledge propagation have to be considered: A top-down way from system modelers and a bottom-up way from domain specific modelers. Different types of knowledge are captured and hold in one modeling language, knowledge sources can be expertise, system level designers, and other various stakeholders. This idea improves directly the issues of knowledge-sharing, traceability, navigation and the transition between the domain independent design phase and the domain specific phase of the V-model [26].

3.3 Transformation

To enable an agent-based management of the interdisciplinary system (SysML) model, it is transferred into an executable multi-agent system, i. e. information about product characteristics and structures are transferred to the agent based system. This transformation is reached via a central management agent who is capable to read an existing model stored in XMI format [29]. The management agent extracts necessary information and creates the respective agents from this information. Additional model data which is not necessary for creating the agents will not be considered. In that way modelers are still free to add extra model data to the system model which may facilitate the expressiveness of the overall system model.

The management agent is running on a central server where the modeler can upload the current model. Triggered by an upload the management agent analyses the model and updates the agents.

- The transformation rules are straightforward:
    1. Each block which is not part of another block is considered to be at the highest level of the system hierarchy and therefore transformed into an agent.
    2. Beginning from point 1 all part associations are transformed into single agents
    3. The parts of parts are also transformed into single agents
    4. Step 3 is repeated recursively until the leafs of the hierarchy tree is reached
- The beliefbase of any single agent contains the following information: the properties of the block, the parts properties of the block, the interfaces of the block, the flow relationships and the modeled relationships between discipline-specific parameters

So a net of SysML agents is created representing the components of the overall system and which is running on a central server and can be accessed by any engineer within the network.

To work with these component agents, client agents are running on the workstations of the respective engineer to serve as an interface between the engineer who is directly working with his domain-specific model data and the multi-agent system. The purpose of the client agent is the negotiation with the component agents when important changes within his field of responsibility occur. Each transformed component agent has the main goal to observe and take care on the modeled properties. If a client agent notifies a component agent about a change of one of its properties, the component agent will check its beliefbase for existing interrelationships between this property and others within the same component or others. In case of a relationship it will inform the respective component agent about the change. This agent now has the responsibility to inform the respective engineer via the associated client agent.

Obviously a danger is present that such systems may explode beyond usage of human clients when the complexity of the product increases. However, the complexity of all kinds of mechatronic products will most probably further increase in the coming years and decades. Approaches which realize a part of the connection between domains and negotiation of complex product characteristics without human interference by means of intelligent systems are therefore promising. In future such approaches have to be combined with approaches which can reduce the complexity and create certain views for human clients.

4 EXAMPLE

The UAV (unmanned aerial vehicle) Quadrotor project, running at the Hochschule Ravensburg-Weingarten in Weingarten, Germany, is currently in the design stage of a rotor-craft equipped with four powered rotors laid up symmetrically around its center. This kind of unmanned multi-purpose flying object is characterized by its good controllability and may find its application in surveillance operations of traffic or sports events. The Quadrotor is considered as a mechatronic system with rather low complexity. It is a typical example of a mechatronic product consisting of distinct mechanical, electrical and IT-based elements. Different students with different backgrounds are involved in the development process and directly supervised by the professor in charge. After having planned the project, a list of concrete requirements and functional analysis was achieved. From these requirements different tasks were assigned to the students in order to be realized in a given period of time. A group of students took the structural body design aspect of the Quadrotor; a detailed CAD-model of the Quadrotor was generated using ProEngineer. This CAD-model was designed in a completely parameterized manner. It is now possible to change one major dimension (e. g. the length of one of the main brackets) and all other components will change so that the general performance parameters of the Quadrotor will still be achieved. This logic is right now realized in ProEngineer by means of Layout Parameters (Figure 3) and external logic in Microsoft Excel.

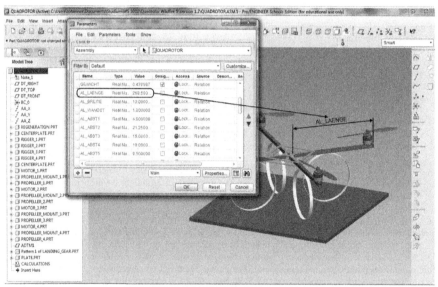

Figure 3. Quadrotor Design in ProEngineer

In this parametric model it is possible to change parameters (e. g. the parameter AL_LAENGE in figure 3) in one of the IT-systems (ProEngineer or Excel) and the other parameters will be automatically adapted and even other motors my be chosen from a certain selection. Through this always a Quadrotor design will result which is able to lift its own weight and a given additional weight.

This logic of the parametric model can in future be used by an agent system in order to allow the usage of this logic also in other domains. (compare also section 4.3 - Figure 6 shows an example of the current parametric logic depiction).

The Quadrotor project is used to evaluate the approach described in this paper. A SysML model for the Quadrotor is achieved to represent the breakdown structure of the Quadrotor and to describe the existing interdisciplinary relationships between its components. The aim of this model is to enhance the collaboration between the different disciplines for a better integration on different levels. In the following, the modeling mechanism of the Quadrotor in SysML is described more in details.

4.1 The Quadrotor Block Definition Diagram:

Figure 2 shows the block definition diagram of the Quadrotor, where the breakdown structure is modeled. This diagram can be read as follows: the Quadrotor project is made up of three interacted systems: the operator, the Quadrotor and the environment where the Quadrotor is suppose to fly. The Quadrotor, as a system, is composed of several mechatronic components. Some components such as the motor are expressed with one block but with several composition associations. Furthermore, other components are again decomposed into their parts, as shown in Figure 4, till the necessary level of decomposition is acknowledged. Each element of the Quadrotor is modeled with a block and each block contains its attributes, ports and interfaces with other blocks. Figure 4 on the right shows another view of the block motor, its specifications are listed in the attributes compartment and two flow ports represent the input/output interfaces with its environment.

The transformation algorithm reads this diagram and creates for every component a single agent. Concretely, beginning from the root of the break down structure each part association represented with a black diamond will be transformed into a respective agent. In that way, e.g. one agent representing the camera and four agents representing each motor will be created. Additionally, the properties and modeled ports will be stored in the beliefs of the respective agent.

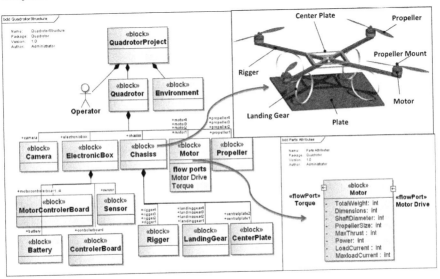

Figure 4. The SysML block definition diagram of the Quadrotor project

4.2 The Quadrotor Internal Block Diagram

An internal block definition diagram can be created to represent the internal structure of this block and the interfaces between its parts. Figure 5 shows an internal block definition diagram of the block Quadrotor where the internal structure is modeled and the interrelationships between its components are represented. One of the block properties of SysML, the part property, is shown in the internal block definition diagram with a solid-outlined box [5] and is used to represent a part of an existing block from the block definition diagram.

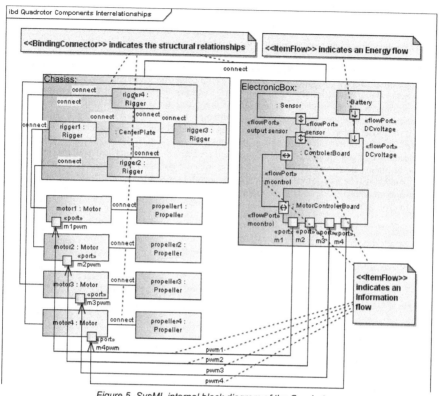

Figure 5. SysML internal block diagram of the Quadrotor

Structural relationships indicating hardware interfaces between parts, such as an assembly connection of a motor with its respective rigger and its propeller is modeled with a binding connector. An information flow of the DC-motor speed control, the PWM signal, between the motor controller board and the motor is modeled with a SysML itemflow. An energy flow can be seen between the battery and the main controller board.

In this way, the more interrelationships information is added to the SysML model with the internal block diagram, the greater the relationships-beliefbase between the agents is achieved.

4.3 The Quadrotor Parametric Diagram

During the domain-specific design of the Quadrotor, engineering students from different disciplines were dealing with their own specialized set of tools to design, simulate or analyze. As described in previous work [26], there is a need to increase the interoperability between existing tools and an all-in-one development suite tool for mechatronic system would not be affordable and even not accepted by industry. We see by using the parametric diagram of SysML an approved approach to represent the interrelationships between the disciplines and later transfer it to the agents' beliefbase. It is important to notice that the parametric diagram is not replacing any of the existing tools, instead it is linking between them by modeling the interdependencies between the properties of the components each tool have to deal with. On the left side of Figure 6 an inter-component relationship diagram is modeled to calculate the total weight of the Quadrotor (as a critical aspect during the development). This diagram holds the weight properties of the components from a wide range of tools. On the right side of Figure 6, an intra-component relationship is modeled to integrate between two disciplines, one from a control engineer student working with Matlab/Simulink to simulate the control algorithms for flying the Quadrotor and other one from an electronic engineer student dealing with the design and software implementation for the motors.

The information coming from the parametric diagram is used during the agent transformation to add the knowledge about the relationships between the disciplines to the agents.

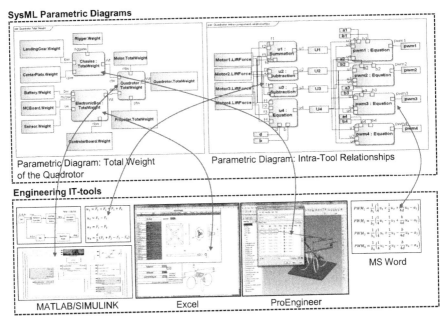

Figure 6. SysML parametric diagrams of the Quadrotor with different it-tools

5 PROCESS INTEGRATION

The multi-agent system helps engineers during the domain-specific design phase of the overall mechatronics design process. However a pure representation of the system model with agents is not sufficient for a successful mechatronics design process as only product information is stored in the model. Also, for a convenient application of the system, process information has to be integrated in a way that domain-specific activities can be allocated and enacted. This information is modeled with SPEM [30], a modeling language for software engineering processes which has the potential to fit to the needs of mechatronics design activities.

The right part of Figure 1 shows the idea of activity agents representing engineering activities. Together with the client agents which serve as an interface to the engineers, relationships between these three types of agents are established to enact the design process. An activity agent can belong to one or more component agent while one component agent can also be related to several activity agents. Additionally client agents are assigned to one or more activity agents. With these relationships, discipline-specific engineers can be instructed to perform activities which are related to modeled system elements within their domain, e.g. CAD for mechanics.

As mechatronic components involve diverse domains, the activities have to be aligned carefully. It is obvious that interdisciplinary relationships have to be considered. The activity agents can negotiate with the component agents about their interrelationships and then align themselves accordingly into the right order. Here three different use cases have been identified in which the multi-agent system can facilitate the cooperation.

1. Parallel and independent activities can be executed concurrently without any interrelation in between.
2. Completely dependent activities have to be aligned in a sequential order as an activity may need the result of another activity.

3. Partial dependent activities are the most interesting case for this approach as the multi-agent system can help managing the dependencies in between these process steps. As the relationships are already known by the system model the respective activities may start concurrently with the knowledge that one activity needs some information from another activity. Triggered by the occurrence of this needed information (e.g. a parameter value) the multi-agent system can automatically inform the requesting activity. Furthermore in the case of a change of this information during the following steps, the former requesting activity can get informed. In that way concurrent engineering is facilitated and the cooperation between the disciplines is enhanced.

6 OUTLOOK

This paper presented an approach to enhance interdisciplinary system modeling for mechatronic with SysML by using the model with a multi-agent system during domain-specific activities. In that way data consistency is warranted. Example mechatronic design processes show the general feasibility of this innovative approach. Obviously, a number of issues still have to be addressed and elaborate software prototypes have to be realized.

Future work will deal intensively with the integration of process information into the existing multi-agent system which currently just represents product information. In detail the transformation of the process description modeled in SPEM notation into distinctive agents and the way to establish the relationships between the component agents, the activity agents and the client agents have to be analyzed and implemented into the existing prototype.

The usage of agents for requirement validation is also an interesting facet of the approach as SysML offers the modeling of requirements and will be validated.

ACKNOWLEDGMENTS

The described project is funded by the German federal ministry of education and research (BMBF) in the scope of the programme "Ingenieur*Nachwuchs*".

REFERENCES

[1] Möhringer S. and Stetter R. A Research Framework for Mechatronic Design. In: *Proceedings of the 11th International Design Conference DESIGN 2010*, Dubrovnik – Croatia, 2010.
[2] VDI 2206. *Design methodology for mechatronic systems*. Beuth, Berlin, 2004.
[3] Chen K., Bankston J., Panchal J. H. and Schaefer D. *A Framework for the Integrated Design of Mechatronic Systems, in Collaborative Design and Planning for Digital Manufacturing*, (L. Wang and A. Nee, Editors), Springer, 2009. pp. 37-70.
[4] Pahl G., Beitz W., Feldhusen J., and Grote K.-H. *Engineering design: A systematic approach, (3rd ed.)*, Springer-Verlag, 2007.
[5] Object Management Group. "OMG Systems Modeling Language (OMG SysML™)," available at http://www.omgsysml.org, November 2008.
[6] Weilkiens T. *Systems Engineering with SysML/UML*. Elsevier, Ed. Morgan Kaufmann Publishers Inc, June 2008.
[7] Wölkl S. and Shea K. A Computational Product Model for Conceptual Design Using SysML. In *Proceedings of the ASME 2009 International Design Engineering Technical Conferences & Computers and Information in Engineering Conference*, San Diego, 2009.
[8] VDI Design Handbook 2221. *Systematic Approach to the Design of Technical Systems and Products*. Düsseldorf: VDI-Verlag 1987.
[9] Follmer M., Hehenberger P., Punz S. and Zeman, K . Using SysML in the Product Development Process of Mechatronic Systems. In *Proceedings of the 11th International Design Conference DESIGN 2010*.
[10] Aditya A. Shah, Schaefer D and Christiaan J.J. Paredis. Enabling Multi-View Modeling With SysML Profiles and Model Transformations. *International Conference on Product Lifecycle Management 2009*.
[11] Thramboulidis K. The 3+1 SysML View-Model in Model Integrated Mechatronics. *Journal of Software Engineering and Applications (JSEA), 2010*.
[12] Weiss G. (Ed.). *Multiagent Systems*. Cambridge, Mass., USA: MIT Press, 2000.
[13] Luck M. and d'Inverno M. *Understanding Agent Systems*. Berlin: Springer Verlag 2001.

[14] Trencansky I. and Cervenka R. Agent Modeling Language (AML): A Comprehensive Approach to Modeling MAS. Informatica, vol. 29, no. 4 (2005) pp. 391-400.
[15] Padgham L., Parker D., Müller J. and Parsons S. (eds.), Proc. of 7th Int. Conf. on Autonomous Agents and Multiagent Systems. AAMAS 2008, Estoril, Portugal, 2008.
[16] Rao A. and Georgeff M. BDI Agents: From Theory to Practise. In V. Lesser, editor, *Proceedings of the First International Conference on Multi-Agent Systems (ICMAS'95)*, pages 312-129, San Francisco, CA, USA, 1995
[17] Mild A. New Product Development Using Adaptive Neutral Agents in an Artificial Factory. A Virtual Experiment. Facultas 2001.
[18] Pechoucek M., Vokrinek J. and Becvar P. ExPlanTech: Multiagent Support for Manufacturing Decision Making. IEEE Intelligent Systems. 2005, vol. 20, p. 67-74.
[19] Pape U. Agentenbasierte Umsetzung eines SCM-Konzeptes zum Liefermanagement in Liefernetzwerken der Serienfertigung. Universität Paderborn: HNI, 2006.
[20] Martinez Lastra J., Lopez Torres E., and Colombo Armando W. A 3D Visualization and Simulation Framework for Intelligent Physical Agents. In: *Holonic and Multi-Agent Systems for Manufacturing*. Springer 2005.
[21] Jennings N.R., Bussmann S. Agent-Based Control Systems. In *IEEE Control Systems Magazine*, June 2003, S. 61-73.
[22] Huang G., Huang J. and Mak K. Agent-based workflow management in collaborative product development on the Internet. *Computer-Aided Design 32 (2000)*, p.133–144.
[23] Tseng K., El-ganzoury W. and Abdalla H. Integration of Non-networked Software Agents for Collaborative Product Development. *Computer-Aided Design and Applications Vol. 2 (2005)*, Nos. 1-4.
[24] Mild A. and Taudes A. An agent-based investigation into the new product development capability. *Computational & Mathematical Organization Theory, Volume 13, Number 3, (2007)*, pp. 315-331.
[25] Zhang X., Luo L. and Duan S. Evaluation of Product Development Process Using Agent Based Simulation. In: *Lecture Notes in Computer Science, Computer Supported Cooperative Work in Design IV: 11th International Conference, CSCWD 2007*, Melbourne, Australia, April 26-28, 2007.
[26] Chami M., Seemüller H. and Voos H. A SysML-based integration framework for the engineering of mechatronic systems. In *IEEE/ASME International Conference on Mechatronics and Embedded Systems and Applications MESA 2010*, p. 245–250.
[27] Stetter R., Voos H. AGENTES - Agent Based Engineering of Mechatronic Products. *In Proceedings of the 8th International Symposium on Tools and Methods of Competitive Engineering (TMCE) 2010*.
[28] Holt J. and Perry, S. *SysML for Systems Engineering*. Professional Applications of Computing Series 7, Institution of Engineering and Technology Press, UK, 2008.
[29] Object Management Group OMG, XMI Specification V.2.1.1, 2007.
[30] Object Management Group OMG, SPEM Specification V.2.0, 2008.

Contact: Ralf Stetter
Hochschule Ravensburg-Weingarten
Department of Mechanical Engineering
88241 Weingarten
Deutschland
Tel: Int +49 751 501 9822
Fax: Int +49 751 501 9832
Email: stetter@hs-weingarten.de
URL: http://www.hs-weingarten.de

Ralf is Professor of Design and Development in Automotive Engineering in the Department of Mechanical Engineering at the Hochschule Ravensburg-Weingarten. He is currently Vice-dean of the department. He is deputy director of the Steinbeis-Transferzentrum "Automotive Systems". His main research interests are mechatronic design and the implementation of product development methods.

INTERNATIONAL CONFERENCE ON ENGINEERING DESIGN, ICED11
15 - 18 AUGUST 2011, TECHNICAL UNIVERSITY OF DENMARK

STRATEGIC PLANNING FOR MODULAR PRODUCT FAMILIES

Henry Jonas and Dieter Krause
Hamburg University of Technology

ABSTRACT
Highly customised products mostly lead to increasing complexity for the production company. Two common design approaches for reducing the internal complexity of a product family are Design for Variety and Modularisation. However, in many application cases it is desirable to consider an optimisation of the variety already in the product planning phase affecting a wide range of products. The approach presented in this paper uses a representation of both the structure of products and economic key figures. Using this method, different strategic scenarios of the product program can be planned and compared to each other. The derivation of strategies for the future and their evaluation is performed using key economic figures and the technical conceptualisation of platform components.

Keywords: Modular Product Families, Product Planning, Variety Management

1 INTRODUCTION
For improving competitiveness, costs and flexibility, most production companies have a more or less applied variety management. An important focus of variety management is the early phases of the product development process as this is where the most influence on an optimised product structure can be achieved. An approach for strategic planning and optimisation of a company's product range is presented here. The methodical procedure defines the products and their variants with a focus on platform design. In subsequent development phases, the product structure will be elaborated further using Design for Variety and Life Phases Modularisation.

2 BACKGROUND
In the product creation phase, variety management basically has two strategies: variety generation and variety avoidance (Figure 1) [5]. Variety avoidance means optimised product design to avoid unnecessary technical variety of parts or assemblies of the products. Variety generation contents steps towards the definition of the product variants. Variety generation is part of the product planning phase and determines the input required for the product development phase.

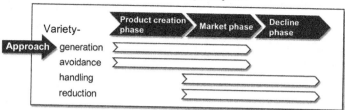

Figure 1. Phases of Variety Management

Often companies tend to plan and develop product families independently from each other, which hinders the use of synergies, e.g. by use of common platform modules. Therefore this paper gives an approach for methodical platform planning considering the future development of the whole product program.
The approach gives systematic support for variety generation (product planning phase) and its interaction with the methods of variety avoidance (product development phase). Therefore, the current methods for variety avoidance next will be investigated. Common methods in this context are modular

design and Design for Variety. Modular design is not necessarily a sub-domain of design for variety, because modularity can support various development aims, such as after-sales or purchase. Modular design is widely used to reduce internal complexity with a modular configuration which generates external variety with only a few basic assemblies. Basic modularisation approaches consider technical-functional [8] & [9] or product strategic perspectives [3]. In technical-functional approaches, matrix systems are often used to cluster the components of the product according to their interdependencies. A modularisation method that considers the life phases of the product is presented in [1]. The core of the method is the Module Interface Graph (MIG), which visualises different modular designs of a product according to the phases of the product life. In a merging step, the life phase perspectives are integrated into a final design. Figure 2 shows a Module Interface Graph in example of an aircraft galley [1] [6].

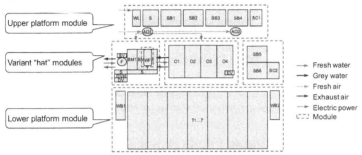

Figure 2. Module-Interface Graph (MIG) of an aircraft galley

A Design for Variety approach is described in [7] [2]. Central to the approach is the Variety Allocation Model (VAM), a four-layer model of differentiating attributes, variant functions, variant working principles and variant components. Using the VAM, the actual product structure is investigated and will be optimised for the ideal configuration of 1:1 mapping of differentiating attributes to variant components. Figure 3 is an example of the Variety Allocation Model [2].

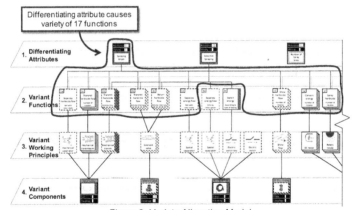

Figure 3. Variety Allocation Model

The Design for Variety approach described above can be used in combination with the Life Phases Modularisation method. This allows the design optimisation of the components (VAM) and the module definition visualised by the MIG. However, this procedure is only applicable to smaller product families. A product family is understood here as being a sum of products that share common parts and functions and operate in similar market segments [4]. For many application cases, it would be useful to enhance the described approach with an investigation of the whole product program. This would allow the development of platform approaches for more than one product family. Figure 4 is an integrated approach for developing modular product families, which consists of a sequence of the three elements Product Program Planning, Design for Variety and Life Phases Modularization. The elements of Design for Variety and Modularization operate at the product family level. The element of Product Program Planning prior to these defines the variety and implements strategic platforms into the program. The methodical approach of Product Program Planning is the focus of this paper and will be described in the next sections.

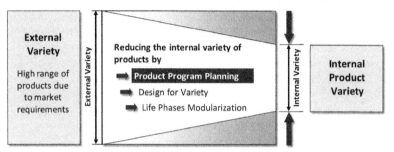

Figure 4. Integrated Approach for development of Modular Product Families

3 APPROACH FOR PLANNING PROGRAM SCENARIOS

This section describes the approach for the variety-optimised planning of the product program. Section 3.1 presents a visualisation tool that uses a representation of both the product structure and economic key figures. Section 3.2 gives heuristic support for generating projections for the future structure of the program. These projections are compared to each other by using the visualisation tool presented. Section 3.3 describes an approach for evaluating the scenarios developed. The basis for the evaluation is the conceptualisation of strategic platforms. Due to the fact that different program scenarios generate different possibilities for using strategic platforms and carry over modules, the use of economic key figures supports the assessment.
Chapter 4 gives an example of practical application of the method.

3.1 Development and Visualisation Tool for Product Programs

For the strategic planning of product program scenarios, a tool is needed which represents the products and their hierarchy and offers economic key figures. In contrast with pure mechanical design, the product planning phase is less constrained and so future scenarios of the product program can be better evaluated if economic information is included. Figure 5 presents the developed tool.
In contrast to a classical tree structure, the two dimensions of angular increment and radial length are used to show information about number of units and sales revenue of a product section. They can either be actual values of the current state or estimated values for the future. Using this visualisation tool and based on the running business, new scenarios for the future structure of the product program can be derived, visualised and evaluated. The product program can be restructured in different scenarios using platform and carry over parts, together with an update of the business strategy.

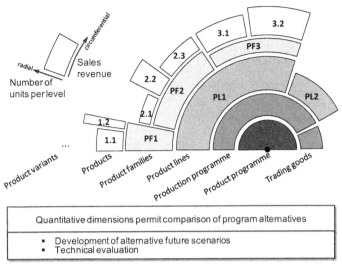

Figure 5. Visualisation tool for the Product Program

The use of quantitative elements in this visualisation creates a quick overview of the basic configuration of scenarios of the product program for the development team, allowing comparison of alternatives for the product planning. Changes can be made on different levels of hierarchy, either changes for the complete program near root level or minor changes near product level. The next section gives heuristic support for the creation and manipulation of strategic product scenarios. Following this, Section 3.3 describes an approach for evaluation.

3.2 Derivation of Scenarios

According to [12], the development of a product program of a company can be categorized into one of four major types. These types can coexist in one company, for example, where products have different technology cycles.

- Type 1: *Custom Engineering:* The product portfolio is based on continuing growth. New products are developed and added to the portfolio in response to customer enquiries. Cuts to the product range are not necessarily made routinely.
- Type 2: *Release Engineering:* The product portfolio is based on constant width and regular updates. The market constantly forces readjustment of the portfolio. This causes routine new product approvals and product eliminations.
- Type 3: *Variety Maintenance:* The product portfolio remains nearly constant. The company uses an active standard of products with only few changes over the time.
- Type 4: *Basic-type Engineering:* The product mix is based on innovation. The market demands new, innovative products. The variety of products is mainly influenced by the dynamic of the market.

To apply the method presented, the product program first needs to be classified with the help of the development types shown. Application of the method is particularly meaningful in the case of the development types 2 and 3. In the case of Release Engineering, regular product eliminations and new positioning need to be developed. Therefore, the method presented gives an overview and common language for the development teams.

In the next step, strategies for the actual changes that need to be made to the program will be derived. The portfolio analysis, known as the BCG Matrix [10], assigns the products of a company according to their dimensions Market Growth and Relative Market Share (Figure 6). Stars have high Market

Growth and Share, but typically demand high investment costs to maintain growth. When Market Growth slows down, Stars change to Cash Cows and become an important source of support for other business units. Cash Cows have a high Market Share but slow Market Growth. Cash Cows generate high revenues at typically low investment costs. Question Marks have a high Market Growth but low Market Share. Due to the high growth, they need high investment and even more effort to increase their low Market Share. It often needs to be clarified whether high investment will be performed in the hope of a shift towards the Star-type or whether disinvestment would be the better choice.

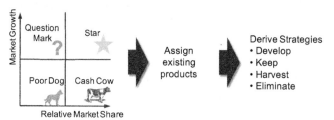

Figure 6. BCG Matrix for deriving product strategies

Once the products of the company are assigned to the BCG Matrix, strategies for structuring the future program can be derived. According to the type of product, i.e. Question Mark or Cash Cow, and the assumed market development, the following strategies for change to product families can be considered [11]:

- *Develop:* This strategy aims to increase the Market Share of the product unit. The strategy should be applied to Question Marks that are promising and are therefore a candidate for investment towards becoming a Star.
- *Keep:* This strategy aims to keep the Market Share of the unit at its current level. The strategy should particularly be applied to successful Cash Cows to maintain good revenue into the future.
- *Harvest:* This strategy sets the focus on short-term revenue of a product without considering long-term development. Application can be meaningful particularly for weak Cash Cows that have less promising market prospects. Question Marks and Poor Dogs can be candidates for this strategy too.
- *Eliminate:* Using this strategy, unsuccessful products or business units are to be sold or closed. The strategy is mostly applied to Poor Dogs or unsuccessful Question Marks.

Application of these strategies will lead to changes in the resulting variety of the product program. The tool shown in Figure 5 represents these planned changes. An advantage of the overall visualisation of the product program is that different scenarios can be compared to each other on the same graphical level, even when changes are made to independent units. Figure 7 gives an example. Based on the current product program (Figure 5) alternatives for the future structure of the program are shown. Questionable in the given example may be Product family 1 (PF1), with relatively low revenues, and Product 2.2 of Product family 2 (PF2), also low revenue and high number of units. A conservative approach could be using "Variety Maintenance" and merging Product 2.2 with the remaining products of PF2. A more radical solution could be the elimination of PF1 and a restructuring of PF2, including a new product in PF2 that replaces some essentials of the eliminated PF1. This comes under the strategy of "Release Engineering".

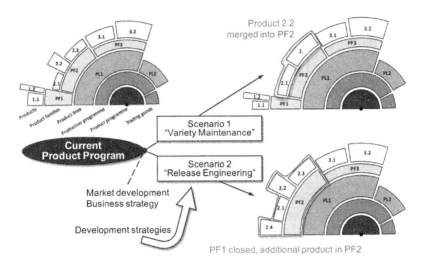

Figure 7. Example for Scenario development using different strategic approaches

Using the visualisation tool, scenarios for the future product program can be developed based on different strategic approaches. In the example given, scenario 2 would be particularly meaningful if the portfolio analysis identified PF1 as a Poor Dog or unpromising Question Mark. Using the same method, profiles and competitor scenarios can be investigated and then mapped towards the company's prospects. This may help to identify market niches or promising segments. The company scenario projections may then incorporate this competitor analysis.

3.3 Evaluation of Scenarios

As in Section 3.2, alternative scenarios for the future structure of the product program have been developed. In the next step, the scenarios need to be assessed to each other. This will be performed by an investigation of the potential of each scenario for supporting strategic platforms. Therefore a different representation is needed. Since the previous step showed the product level without product variants, this step will investigate the variants with respect to potential platform standardisation. Still in the phase of product planning, the economic key figures will be included in this consideration. Figure 8 shows the approach represented by a graph of the products and their assemblies. All products of a scenario, according to Figure 7, are included in a variety analysis in Figure 8. The variety of a product is visualized using differentiation between standard and variant assemblies. The question of how finely the element "assembly" here is understood must be clarified for the individual case. If "assembly" is considered too closely, for example, by singular mechanical elements, the overview will become too complex to handle. However, if "assembly" is too rough, there could be only few options become visible for platform conceptualisation.

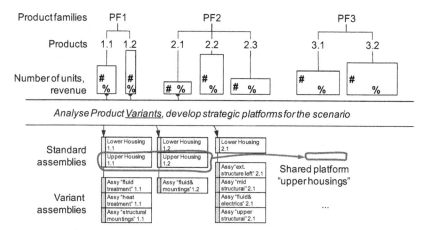

Figure 8. Evaluation approach by conceptualization of strategic platforms

After the assemblies of the products are analysed and noted into the graph, the development team conceptualises possible platform elements for the scenario. For this step, the functions and constraints of the assemblies of different products are compared to each other. The aim is to identify assemblies that commonly have the same function and similar constraints. These common assemblies are candidate for platform modules as exemplary identified in Figure 8. For support, the engineer can perform design modifications such that the assembly matches function and constraints of both products considered. The aim is to design strategic platforms that may be used in products of different product families. If a compromise needs to be made on alternative platform concepts in one scenario, the economic key figures can be assessed to evaluate the benefits of each combination.

Once platform concepts have been developed for the different program scenarios, an assessment needs to be performed to identify the most promising scenario. To support this assessment, the previously analysed economic key figures are used to estimate the economies of scale for each scenario due to the platform concepts.

4 EXAMPLE

This section outlines an example of the application of the proposed method. Figure 9 shows the product program of an aircraft cabin supplier, which was used as the basis for this investigation. For confidentiality reasons, the program is modified and contains fictional numbers. Some products are not shown. The following figures will use the same notations as Figures 5 and 7.

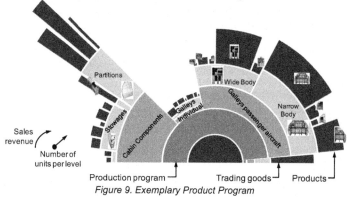

Figure 9. Exemplary Product Program

According to this product program, Figure 10 shows an interpretation of the actual condition: products 7.4 and 8.1 are essential Cash Cows for the company. Both products have high Market Shares and slow Market Development. Product Line 3 (PL3) is the core business of the company since more than the half the revenue is generated by its products. Noteworthy is product 4.3, which shows a high number of units but very low revenue. This product needs to be further investigated. Product Line 2 (PL2) shows a low number of units, but a significant share of the overall revenues. This can be explained by the fact that the company manufactures the products of this line according to individual design orders at low number of units. This causes complexity in the manufacturing division, thus a further investigation would be meaningful. The products 7.1, 7.2 and 7.3 are members of product family 7, which includes the Cash Cow, but show low numbers of units. Investigation of this will also be carried out.

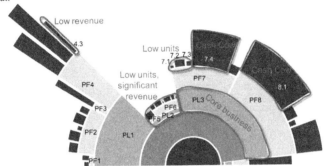

Figure 10. Assumptions regarding the Product program

The next step will be deriving strategies for future scenarios of the structure of the product program. Only one scenario will be given here as an example. In this example, investigation of product 4.3 showed that the induced manufacturing complexity is comparably high and elimination of the product might not be appropriate. Since the structure and function of the product are not of the core business, outsourcing its production was considered. Figure 11 shows the actions performed in the program; in the case of product 4.3 it was moved to the trading goods. For product line 2 (PL2), it was identified that the company offers no products to a certain customer type. A market entry is planned, represented by the new product family 5' (PF5'). Two product families in this line that were rarely sold have been eliminated since no prospects of significant Market Growth could be detected. In the case of product family 7 (PF7), elimination of products was not a choice since they bring in revenue fairly well and are part of the core business. Further investigation of the platform design will be performed, as explained in Figure 8.

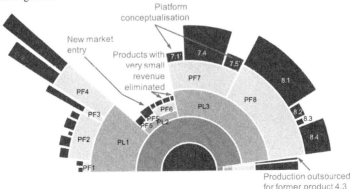

Figure 11. Scenario for changes made to the Product program

This platform conceptualisation, as shown in figure 12, lists the product families, products and their key figures. Next, the assemblies are analysed for their variety and function, as explained in Section 3.3. Using this graph, the engineer uses his product knowledge and creates platform components using assemblies that have the same function and similar constraints. To create a platform, design changes shall also be considered to fulfil the constraints of different products. These design changes can particularly involve concepts of standardised interfaces, harmonisation of geometrical parameters and, if meaningful, over-sizing. As a result, in the current example, housings and structural installations were standardised using interface and minor design changes such that the new platforms 7.1' and 7.5' were introduced.

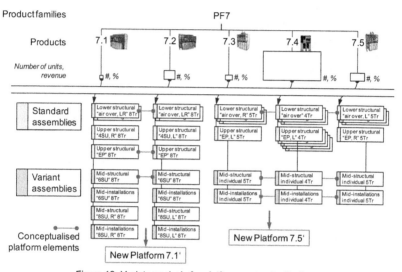

Figure 12. Variety analysis for platform conceptualisation

5 CONCLUSION

A method for strategic planning of modular product families in the context of variety generation was presented. The approach introduces a representation tool of the overall product program, visualising the structure of products and economic key figures. The development type of the company and the market positions of the products were analysed to derive change in the future program. These changes are visualised in the tool presented using different scenarios. In assessing these scenarios, possible platform systems are conceptualised and their benefits estimated. The defined structure of products and variants will be included into the subsequent design phase, where the methods of Design for Variety and Life Phases Modularization will be applied at the product family level.

REFERENCES

[1] Blees, C., Jonas, H., Krause, D., Perspective-based Development of Modular Product Architectures, in *Proc. International Conference on Engineering Design, ICED'09 Vol. 4*, Stanford, August 2009, pp. 95 - 106
[2] Blees, C., Kipp, T., Beckmann, G., Krause, D., Development of Modular Product Families: Integration of Design for Variety and Modularization, in *Proceedings of norddesign2010*, Gothenburg, August 2010, pp. 159 - 168
[3] Erixon, G., *Modular Function Deployment: A Method for Product Modularisation*, 1998 (The Royal Institute of Technology, Department of Manufacturing Systems, Assembly Systems Division, Stockholm)
[4] Göpfert, J., *Modulare Produktentwicklung: zur gemeinsamen Gestaltung von Technik und Organisation*, 1998 (Gabler, Wiesbaden)
[5] Heina, J., *Variantenmanagement*, 1999 (Gabler, Wiesbaden)
[6] Jonas, H., Gumpinger, T., Blees, C., Krause, D., Innovative Design of a Galley Product Platform by applying a new Modularisation Method, in *Proc. 4th International Conference "Supply on the Wings"*, Frankfurt/Main, November 2009
[7] Kipp, T., Krause, D., Design for Variety – Ein Ansatz zur variantengerechten Produktstrukturierung, in *Proc. 6.Gemeinsames Kolloquium Konstruktionstechnik*, Aachen, 2008, pp. 159-168.
[8] Lindemann, U., Maurer, M. & Braun, T. (2009). *Structural Complexity Management*, 2009 (Springer, Berlin)
[9] Pimmler, T.U., & Eppinger, S.D., Integration Analysis of Product Decompositions, in *Proc. 6th Design Theory and Methodology Conference*, New York, 1994, pp. 343-351
[10] Schawel, C., Billing, F., *Top 100 Management Tools*, 3rd edition, 2011 (Gabler, Wiesbaden)
[11] Schmitz, M., *Produkt- und Produktionsprogrammplanung*, 1996 (Gabler, Wiesbaden)
[12] Schuh, G., *Produktkomplexität managen – Strategien, Methoden, Tools*, 2. Auflage, 2005 (Hanser, München)

Contact: Henry Jonas
Hamburg University of Technology
Product Development and Mechanical Engineering Design
Denickestr. 17
21073 Hamburg, Germany
Tel: +49-40-42878-3666, Fax: +49-40-42878-2296
Email: henry.jonas@tuhh.de
URL: http://www.tuhh.de/pkt

Henry is research assistant at the Institute of Product Development and Mechanical Engineering Design of the Hamburg University of Technology. He works on the topics product platform- and program planning as well as modularization, applied mostly in the field of aircraft interiors.

Prof. Dieter Krause is head of the Institute for Product Development and Mechanical Engineering Design at Hamburg University of Technology. He is a member of the Berliner Kreis and the Design Society. The main topics of his research are new design methods for product variety and modularization as well as lightweight design for aircraft interiors.

MODULARITY WITHIN A MATRIX OF FUNCTION AND FUNCTIONALITY (MFF)

Žiga Zadnik[1], Vanja Čok[1], Mirko Karakašić[2], Milan Kljajin[2], Jože Duhovnik[1]
(1) University of Ljubljana, SI (2) University of Osijek, CR

ABSTRACT
The objective of this paper is to present the concept of modularity in the development of a product by means of a descriptive matrix of function and functionality (MFF), based on the generative model and the criteria for describing products, functions and functionalities. The purpose of using the modularity of the descriptive MFF is to improve the initial design process, where only the most basic information is available, such as functions and functionalities, and to use the general functionality method, which is not quite possible with the morphological matrix. The modularity inside the MFF is based on the mutual relation between the function and the functionality, representing the data definition. In relation to the morphological matrix it is built and defined on the basis of a mathematical model and pre-set rules [1], not just on the basis of design intuition. This work represents a method for solving the modularity with regard to the shape and the function. This should facilitate the generation of the functional and shape structures of new and variant products. The developed MFF modularity model was implemented into a prototype web application and confirmed on a concrete product – the Active Lounge Chair 1.

Keywords: Design process, Functional matrix, Functional modelling, Functionality matrix, Modular design

1 INTRODUCTION
Market requirements are the basis for defining basic functional requirements, which in turn represent the initial information on a potential new product [2]. At the beginning of the design process, the functional requirements are usually unarranged, incomplete and sporadically presented, which makes it necessary to arrange, complement and expand them. The product structure can be presented as a functional structure, which at the same time is the basis for defining the shape (physical structure) of the product [3]. In [1, 4, 5] matrix models were developed and presented; they enable the generation of a functional structure of the product, described in matrices. This structure is then the basis for generating a product's shape structure.

In order to reduce the time required for arranging and improving the functional requirements, a modularity model of the matrix of function and functionality (MFF) was later developed. The basic morphological matrix [6] forms the basis for the development of the MFF modularity model. Using a small number of rows and columns, the model can yield a large number of solutions, which often makes them poor and unsuitable. The objective of MFF modularity development is to upgrade and update the deficiencies of the morphological matrix in the following areas: the use of a mathematically based model for creating the links between the function and the functionality, the possibility of the automatic suggestion of solutions, and use of sub-matrices with the modularity.

The MFF modularity model represents a tool that enables the designer to manage the design process faster and better, particularly in the initial concept phases. MFF is a synonym for the tabular presentation of the links between the functions and the functionalities. These functionalities are represented by technical systems [7] or shape models that in part, or in whole, fulfil the required functions. The matrix is used for the development of brand-new products as well as for the development of variant design.

Due to the fact that the MFF in itself connects functions and functionalities, the latter should be carefully and uniquely defined. In [8], the authors approach describing the functions by defining the terminology that is related to the names of the functions, while others describe the functions of technical systems by means of physical laws [9]. With a view to unique identification, rules were defined [1], by means of which the functions, functionalities and products are described. The reference points for designing these rules are those presented in [10]. The functions are described by parameters, based on physical laws, which form the basis for the development of a mathematical model through which the connection with functionalities is established.

Today's market requires ever shorter development times for new products, which triggers the need for a modular architecture of products. Such a modular architecture makes it possible to combine one or several functions in the functional structure with one element that solves them [11]. Such an approach has several advantages; the main one being an increased number of product variants [12]. Erixon [13] developed the Modular Function Deployment method, using the Module Indication Matrix. The established rules (1) also include modularity rules in terms of the function and modularity with regard to the shape. These rules are at the same time implemented into the MFF model and presented on a concrete product, called the Active Lounge Chair 1 – (ALC 1).

Research and development activities within the product-development process have their own characteristic and distinctive features, dominated by unpredictability, creativity, mentality and abstraction. Due to these features it is difficult to thoroughly describe, develop and implement the design process in the initial phases of computer-tools development [3]. From this point of view, we have developed a computer web application, within which the MFF modularity model has been implemented. The application uses a central relational database that includes functionally described technical systems of various complexities, which in turn feeds all the functions, the corresponding parameters and the parameter values of all sorts of products.

2 MFF MODULARITY MODEL

2.1 Modularity with regard to function and shape

Modularity with regard to shape

Modularity with regard to shape is referred to as the appearance of a product in one or more variants (versions). According to the shape-modularity principle [2], products can be pooled into modular assemblies. They are checked in terms of the number of their functions that fulfil individual product variants, i.e., we are determining how many functions are fulfilled by a particular variant. In the case that a product variant includes all the functions of another variant, as well as the functions that other variants do not possess, that variant can replace the other one. A comparatively larger number of functions, fulfilled by a particular variant in comparison to another variant with a smaller number of functions, reflects a greater complexity of the variant. For the final confirmation of the variant with a larger number of functions it is later necessary to upgrade it and carry out an economic analysis, which has not been dealt with in this part because it is too extensive.

Table 1: Modularity with regard to shape

FUNCTION	FUNCTIONALITY			
	Variant 1	Variant 2	Variant 3	Variant 4
Function 1	X	X	X	X
Function 2	X	X	X	X
Function 3	X	X	X	
Function 4	X	X		
Function 5		X		
Function 6				

Table 1 shows that variant 2 entirely replaces variant 1, as it fulfils all the functions that are fulfilled by variant 1. Compared to variant 1, variant 2, in turn, solves some other – additional – functions that the adopted variant does not solve. It can be argued that variant 2 is more sophisticated, compared to variant 1, and that it solves more functions. A back-to-back examination of variants 3 and 4 also reveals that variant 3 entirely replaces variant 4.

Table 2: Modularity with regard to shape

FUNCTION	FUNCTIONALITY			
	Variant 1	Variant 2	Variant 3	Variant 4
Function 1		X	X	X
Function 2	X	X	X	X
Function 3	X	X	X	
Function 4	X		X	X
Function 5		X	X	
Function 6				

If individual variants (1 and 2, for example) have no common function, a new variant (variant 3) should be generated. This variant should fulfil all the functions not common to variants 1 and 2 (Table 2). An economic analysis has not been dealt with at this point because it is too extensive.

Modularity with regard to function:
Within the functional structure, more than one product can have identical or similar functions for performing the same or similar process. For such cases it is necessary to check the technical system overload by introducing modularity with regard to function. The modularity function consequently pools the functions for the larger number of variants. Function pooling represents the introduction of modularity according to the principle of function, where the use of various technical systems for identical functions is protected. For two products with identical functions, and in the case of non-established functions, it is vital to confirm their potential diversity. Only one product should be selected if no additional function has been confirmed for two functionally identical products.

Table 3: Modularity with regard to function:

FUNCTION	FUNCTIONALITY			
	Variant 1	Variant 2	Variant 3	Variant 4
Function 1	X	X	X	X
Function 2	X	X	X	X
Function 3	X	X		
Function 4	X		X	X
Function 5		X	X	
Function 6		X	X	

In Table 3, variants 1 and 2 solve functions 1, 2 and 3. These are two different technical systems that solve their common functions.

2.2 Modularity with regard to MFF

The concept of modularity in the development of a product by means of a descriptive MFF matrix is based on the generative model and criteria for describing products, functions and functionalities. The purpose of using the modularity by shape and function of the descriptive MFF matrix is to improve the initial design process where only the most basic information is available, such as functions and functionalities, and to use the general functionality method. Two aspects are taken into the account: fulfilling as many requirements as possible and fulfilling the requirements as effectively as possible. Both aspects are achieved through consistent combination of solution elements, bindings and modularization. Modularity within the MFF is based on the mutual relation between the function and functionality, which represents the data definition. The presentation is aimed at direct users, developers and researchers of technical systems and recognised technical processes. It is based on the connection between the recognised natural processes in nature and searching for comparable or satisfying technical processes at a certain level of knowledge development.

MFF represents a tabular presentation of the links between the functional requirements and the functionalities. Modularity can be devised if we know the key elements, such as the basic list of functional requirements and a list of functionalities, whose details will be dealt with later. The developed modularity model inside the MFF was created by expanding the matrix of functions and functionalities model, shown in (10), and by examining the functionality as it depends on various functions. The basis for generating and arranging the MFF is the functional structure of a product, which at the same time represents the matrix input. When developing a new product, it is not possible to know and be familiar with a detailed functional structure right at the beginning. Such a structure can be obtained and built only from a rough functional structure, which is subject to constant changes during the design process, as shown in (5).

Within the MFF, functional requirements are introduced into the relation, on one side, and functionalities, on the other, as shown in Figure 1. The functional requirements represent the basic functions, while the functionalities are represented via technical systems. Both functions and technical systems can be either simple or more complex, which depends on the initial description of the individual systems and on the result of a rearrangement. In the theoretical part of the model, simplified and generalised marks for the functions and technical systems will be used for the purpose of showing the MFF modularity. The functions and functional requirement are represented by Fi and are located in the first column, while individual technical systems are represented by TSj and are located in the subsequent columns.

In the MFF model, technical systems are marked with the general marks TS1, TS2,...,TSj $j=1,...,n$, while in the case of implementations and concrete examples the marks are of course replaced by the real names of technical systems. Columns with the names of technical systems, described in the matrix, are placed under the first functionality row. With each new technical system entry, a new column appears in the matrix. Its name matches the newly entered technical system. The same analogy also applies to the presupposed input functions that are to be solved. Hence, if a new modularity by shape or function, or a new functional requirement is added, it is analogously added to the matrix row. All the changes are dynamic and are continuously adapted and updated to the last updated MFF status.

The functions, defined in the MFF matrix, are described in the first column in the table, labelled Function. Each function corresponds to its row. In order to present the model in a simple way, the function names are marked with general marks, such as: F1, F2,..., Fi; $i=1,...,m$, while in the concrete examples within an implementation, they are followed by concrete and real names.

The functions are defined on the basis of the required functional requirements. For systematics and modularity reasons, they are described in the relevant input lists. The MFF vision is that solving the matrix should gradually lead to defining more and more information for a particular functional requirement or function, and that it is solved at the end of the process with a suitable functionality. With a view to fulfilling the function, the differences between particular variants are arranged and the modularity is built.

FUNCTION	FUNCTIONALITY / SOLUTION			
	TS1→	←TS2→	←TS3→	←TSj
Functional requirement - F1 [Suggested solution]	M 88 ↓ ↑ S 49		M 30 ↓ ↑ A 40	
Functional requirement – F2 [Suggested solution] ↑↓		S 85 ↓ ↑ S 45 ↓ ↑ B 23 ↓ ↑ B 15		M 100 ↓ ↑ A 100 ↓ ↑ B 56 ↓ ↑ B 33
Functional requirement – F3 ↑↓				
Functional requirement – F4 [Suggested solution] ↑↓	M 100	M 88 ↓ ↑ S 49		M 88 ↓ ↑ S 49
Functional requirement - Fi [Suggested solution] ↑	M 80			

Figure 1: Modularity model of the matrix of functions and functionality

The result of arranging implies the modularity and/or the growth of the product's complexity. The fulfilment of variants for particular functions should always be ensured. In this case we are determining the number of functions that are fulfilled by a particular variant. This is how to confirm the gradation from the biggest to the smallest possible fulfilment of a function, and to establish a possible connection. We can look for modularity by shape or specifically determine the modularity by functions. This is achieved by providing a fulfilment for a particular variant in one of the adjacent variants. It establishes the modularity by functions, which makes it possible to use different technical systems for identical functions.

For a function that we do not know a lot about at the beginning of the design process, it is possible to determine a suitable solution by means of solving and describing, and to specify in more detail the type of function and all the corresponding parameters, winning parameters, intervals, etc. According to

(1) the functions are divided into four different types of functions: main, supplementary, auxiliary and binding functions. They are all described by parameters, winning parameters and intervals.

The links between the functions and the functionalities that solve them are created by means of the so-called sub-matrices. These sub-matrices in the presented MFF model are coloured and highlighted in grey (Figure 1). Figure 1 shows several different sub-matrices, within which we will explain the solving, arranging, modularity and complexity of solving. As a rule, sub-matrices are not logically distributed at the beginning as their internal distribution is determined by how the design process develops and by the presupposed number of functions and functionalities. Parts of the matrix significantly deviating from the main diagonal are usually evidence that the determined function does not have an accurate basis, that it is specifically oriented and cannot be directly applied in a particular variant. This is a way to determine an unjustified description of function and to develop opportunities for further arranging and modularity. The key feature of the MFF is its arranging ability and the modularity of the sub-matrices, which makes it possible to arrange and sort the whole matrix during the design process. According to the given computer algorithm, the MFF presupposes a hierarchical order at the beginning of the process. It is based on the matching percentage, which can be re-arranged later. Attached is the possibility of two-level-row and one-level-column sorting, representing matrix dynamics within the design process.

The MFF model within the design process always includes all the sub-matrices that are of key importance for the development of further designing. Sub-matrices involving at least one possible solution on at least one function within the presupposed building block or functionality are full and display a partial and complete result for this sub-matrix, while the unsolved sub-matrices are not displayed. The result is displayed in the form of percentage values – numbers in a sub-matrix cell. The value is calculated on the basis of a verbal algorithm of the functional requirement's crossed values and the function on the functionality. Each displayed value corresponds to the informative type of the current function. The function type is based on the description and is determined from the characterised character set M, S, A, B (initial letters for main, supplementary, auxiliary and binding functions).

The number of functions within the sub-matrix is analogous to the number of possible solutions in the functionality column. Results-wise, only the functions with a specific, possible solution are displayed. The functions that are not solving a given situation are not included in the display.

Besides the complete display of possible results between the functional requirements and the functionalities, the MFF model also includes an automatic suggestion for the end solution – the suggested solution in Figure 1 in each row of the first column. It is presupposed that a possible solution is the one that most closely corresponds to the given functional requirement. The end solution is selected on the basis of the individual percentage values; solutions' values, making the end solution the one with the highest calculated percentage value. In the case that there are several solutions with identical percentage values, the higher solution is selected, i.e., the solution ranked higher according to the hierarchical type of function. For example, if there are several identical percentage solutions, the highest-ranking solution on the Main Function type would be selected. The automatically suggested solution is only a suggestion within the design process. In any case, the end and final decisions should always, and in all cases, be taken by the user, i.e., the designer.

The main function's name for each system is shown with a mark (M1). According to the description rules, each building block can only have one function. It can happen that the MFF includes several technical systems with identical names of either the main, supplementary, auxiliary, or binding function. In the prototype model implementation, the real names of the existing status of the description are used for all the functions of the technical system. The supplementary functions' names in the matrix model are shown in Figure 1 and marked with 4, S1; S2,...,Sk; $k = 1,...,p$. A technical system can have more than one supplementary function; it is even possible that a technical system has no supplementary function if it was not planned in the actual descriptions. The auxiliary functions are shown with marks: A1, A2,...,Ak; $k = 1,...,p$. A technical system can have one or more auxiliary functions. In analogy to the explanation above, it can happen that a technical system has no auxiliary function. The binding functions are shown with marks: B1, B2,...,Bk; $k=1,...,p$. In contrast to the supplementary and auxiliary functions, a binding function without a single binding function is not possible because it would make the description incomplete and the technical system would not fulfil the appropriate criteria or rules of the pre-defined rules on describing functions, functionalities and technical systems. A cumulative p value cannot be the same for the supplementary, auxiliary and binding functions. Each technical system can have a different number of functions. For definition and uniqueness reasons, each function of a particular technical system in the MFF matrix is described by parameters, winning parameters and value intervals. However, it is not certain that it will be displayed as it has been mentioned above that it is displayed only when it solves a given functional requirement with a significant probability. Depending on the complexity of the function, it can be described by one or more parameters. In no case can it happen that a function could be left with no parameters, since a function without parameters is no longer a function.

3 IMPLEMENTATION

The MFF modularity model will be presented and implemented on a selected product, called the Active Lounge Chair 1 (ALC 1). The goal of the implementation is not to design a complete Active Lounge Chair 1, but to clearly show and prove the modularity for part of the chair (the arm rest with an exercise mechanism for the hands, Figure 2). Due to the fact that the functional requirements mark the beginning of the design process, the original idea is to direct the designer through the modularity of the shape, function and MFF matrix, and to come to possible new solutions. They will be based on rules and a mathematical model, not only on the basic design intuition.

The Active Lounge Chair 1 represents a product whose basic functions are *sitting*, *resting* and *exercising*. It is aimed at a wide range of users of all ages. The key component parts of the Active Lounge Chair 1 are: the sitting part, the leg/foot rest, the arm/hand rest, the upper body rest, and the hand and foot exercise mechanism, as shown in Figures 2 and 4, where each of the component parts allows and fulfils a precisely defined function. Figures 2 and 4 are composed of several individual pictures that precisely and clearly show the design thinking behind the chair concept, particularly the arm rest with the exercise option.

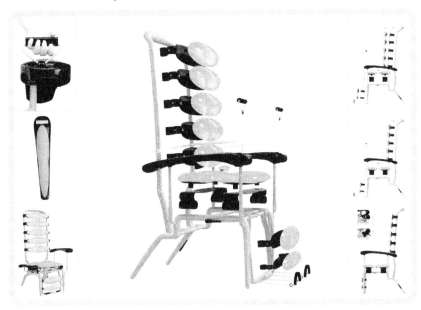

Figure 2: Active Lounge Chair 1 – ALC 1

Before presenting a concrete matrix and modularity within the MFF it should be made clear that the basic, theoretical model of the MFF functioning has been developed and confirmed in a prototype computer web application. The power of managing, running and creating the product data is provided by a central relational database with a relevant database-management system. The model is complemented and upgraded by a number of additional modules, but they will not be dealt with at this point because they are too extensive.

The MFF in Figure 3 represents the real modular matrix *ALC 1* concept. The matrix involves several possible solutions (schematically shown in Figure 4), cross-corresponding to several functions. The main possible functionalities are *Stool*, *Fixed Armchair*, *Variable Armchair* and *Active Lounge Chair 1*, among which it is possible to manipulate the desired functions or functional requirements: *sitting and resting, hand rest, possibility of vertical arm movement, possibility of vertical arm movement independently of lower chair part* and *possibility of exercise*.

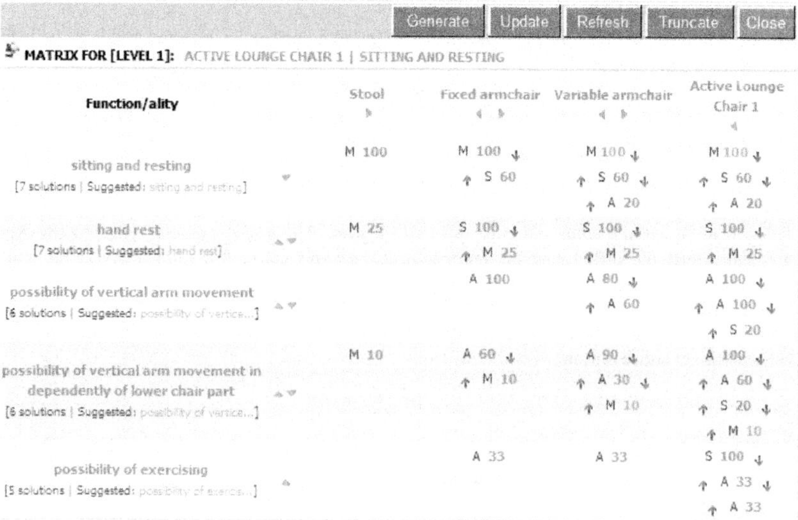

Function/ality	Stool	Fixed armchair	Variable armchair	Active Lounge Chair 1
sitting and resting [7 solutions	Suggested: sitting and resting]	M 100	M 100 ↓ ↑ S 60	M 100 ↓ ↑ S 60 ↓ ↑ A 20	M 100 ↓ ↑ S 60 ↓ ↑ A 20
hand rest [7 solutions	Suggested: hand rest]	M 25	S 100 ↓ ↑ M 25	S 100 ↓ ↑ M 25	S 100 ↓ ↑ M 25
possibility of vertical arm movement [6 solutions	Suggested: possibility of vertica...]		A 100	A 80 ↓ ↑ A 60	A 100 ↓ ↑ A 100 ↓ ↑ S 20
possibility of vertical arm movement independently of lower chair part [6 solutions	Suggested: possibility of vertica...]	M 10	A 60 ↓ ↑ M 10	A 90 ↓ ↑ A 30 ↓ ↑ M 10	A 100 ↓ ↑ A 60 ↓ ↑ S 20 ↓ ↑ M 10
possibility of exercising [5 solutions	Suggested: possibility of exerci...]		A 33	A 33	S 100 ↓ ↑ A 33 ↓ ↑ A 33

Figure 3: Implementation of the matrix of function and functionality on the example of ALC 1

A product can appear in one or more variants, which can be pooled into modular assemblies according to the principle of shape modularity or the principle of function modularity. The basic feature of shape modularity is to establish how many functions are fulfilled by each product variant. For example, Figure 3 reveals that the *Fixed armchair* variant completely replaces the *Stool* variant, as it solves the *Stool's* main function (*sitting and resting*), as well as another function: *hand rest* and the *possibility of vertical movement*, which is by default not fulfilled by the *Stool*. The function solution within the technical system is shown as a percentage value in cells, i.e., cross-intersections in the matrix. The displayed value can be highlighted in various colours, depending on the quality of the sought-after data that can be found within different function types. The probability of a suitable solution hierarchically follows in colours from the most probable green to brown and the least probable grey. Compared to the *Variable armchair* variant, the *Active Lounge Chair 1* variant solves some other, additional functions that the former variant does not solve by default. It can be argued that the *Active Lounge Chair 1* variant, compared to all three other variants, is more sophisticated and fulfils more functions. The *Active Lounge Chair 1* is actually the only modular end solution that fulfils all the set functions according to the shape-modularity principle. It can also be argued that if a product variant includes all the functions of another variant, as well as the functions that the other variant does not possess, that variant can replace and substitute it.

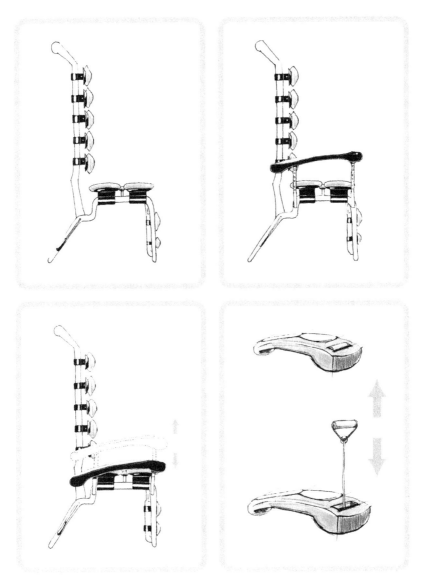

Figure 4: Sketches of possible solutions for the exercise arm rest

On the other hand, modularity by functions can be specifically determined by providing fulfilment for a particular variant in one of the adjacent variants. It provides the possibility of using different technical systems for identical functions, which means that the real function *hand rest* corresponds to all the functionalities, except *Stool*. On the other hand, the functional requirement *sitting and resting* corresponds to all the set solutions. The Active Lounge Chair 1 functional variant completely covers all of the other three variants and so they can be replaced by the said variant. The replacement should be confirmed by an econometric study and technical fulfilment alone is not the only condition.

4 CONCLUDING REMARKS

The goal of the design process is to create new, conceptual product variants. Variants in the process are created by combining functional models and solving technical systems. The MFF modularity model is presented in order to simplify and upgrade the design process. The development of the presented descriptive mathematical model is based on the basic morphological matrix. By means of a developed mathematical model it later enables MFF matrices to create new links between functions and functionalities. Functionalities represent the technical systems of various sophistications, of course described by functions and parameters. On the basis of pre-set rules (description rules, verbal and mathematical rules), the MFF allows the generation of new conceptual variants of products for which we can say that they are not based merely on design intuition.

Due to the increasing competition on the market, enterprises have been increasingly faced with the requirement for a more precise design of the product, adapted to the customers' specific requirements. Enterprises are forced to supply the market with the greatest possible variety of products along with the smallest possible differences between individual variants. The differences are mostly about design, manufacturing and maintenance. For all these reasons, there is an increasing demand for products with an increasingly modular architecture. For this reason, this part presents a theoretical model that implements modularity with regard to shape, function and MFF.

In the case of modularity with regard to shape, products are pooled into modular assemblies. They are checked in terms of the number of functions that fulfil an individual product variant, i.e., we are determining the number of functions that are fulfilled by a particular variant. The variant that includes all the functions of another function plus some new, additional functions, is selected as the end product. The model does not include any econometric analyses that would confirm the economic feasibility of such a selected variant. Using modularity with regard to shape, it is possible to check the complexity of the product variants and the complexity of the process itself by means of the designer's self-checking. The variant that fulfils more functions is more complex and more sophisticated compared to the variant with a smaller number of functions.

In the case of modularity with regard to function, the functions are pooled for the larger number of variants. This ensures the use of various technical systems for identical functions. Two products with identical functions and not yet established functions require confirmation of their potential diversity. Only one product should be selected if no additional function has been confirmed for two functionally identical products.

The MFF model, as well as the modularity model with regard to function and shape, has been included and implemented into a prototype computer web application. By means of a developed central relations database it manages the design data for the development of new conceptual product variants.

The presentation is aimed at direct users, developers and researchers of technical systems and recognised technical processes. It is based on the connection between the recognised natural processes in nature and searching for comparable or satisfying technical processes at a certain level of knowledge development. The mission of the developed models is to contribute to, and find within, the initial design processes the appropriate fundamentals for better and faster design management.

ACKNOWLEDGEMENT

The work presented in this paper is financially supported by the Ministry of Higher Education, Science and Technology of the Republic of Slovenia and the Ministry of Science, Education and Sports of the Republic of Croatia through a bilateral project.

REFERENCES

[1] Zadnik Ž., Karakašić M., Kljajin M., Duhovnik J. Functional and Functionality in the Conceptual Design process. *Strojniški vestnik*, 2009, 55 (7-8), pp. 455-471.
[2] Kušar J., Duhovnik J., Tomaževič R., Starbek M. Finding and Evaluating Customers Needs in the Product-Development Process. *Strojniški vestnik*, 2007, 53 (2), pp. 78-104.
[3] Kurtoglu T. A Computational Approach to Innovative Conceptual Design. *PhD Thesis*, 2007 (The University of Texas at Austin).
[4] Karakašić M., Zadnik Ž., Kljajin M., Duhovnik J. Product Function Matrix and its Request Model. *Strojarstvo*, 2009, 51 (4), pp. 293-301.
[5] Karakašić M., Zadnik Ž., Kljajin M., Duhovnik J. Functional structure generation within multi-structured matrix forms. *Technical Gazette*, 2010, 17 (4), pp.465-473.
[6] Kljajin M., Ivandić Ž., Karakašić M., Galić Z. Conceptual Design in the Solid Fuel Oven Development. In *Proceedings the 4th DAAAM International Conference on Advanced Technologies for Developing Countries*, Slavonski Brod, September 2005, pp. 109-114.
[7] Hubka V., Eder W.E. *Theory of Technical Systems*, 1988 (Springer-Verlag Berlin, Heidelberg)
[8] Hirtz J., Stone R.B., McAdams D.A., Szykman S., Wood K.L. A Functional Basis for Engineering Design: Reconciling and Evolving Previous Efforts. *NIST Technical Note 1447*, 2002 (Department of Commerce United States of America, National Institute of Standards and Technology).
[9] Žavbi R., Duhovnik, J. Conceptual design chains with basic schematics based on an algorithm of conceptual design. *Journal of Engineering Design*, 2001, 12 (2), pp. 131-145.
[10] Duhovnik J., Tavčar J. Product Design Test using the Matrix of Functions and Functionality. In *Proceedings of AEDS 2005 Workshop*, Pilsen-Czech Republic, November 2005.
[11] Pavlić D. Sustav za konfiguriranje proizvoda modularne arhitekture. 2003 (Fakultet strojarstva i brodogradnje u Zagrebu).
[12] O´Grady P., Liang W.Y., Tseng T.L., Huang, C.C., Kusiak A. Remote Collaborative Design With Modules. *Technical Report 97-0*, 1997 (University of Iowa).
[13] Erixon G. Modular Function Deployment-A Method for Product Modularisation. *PhD Thesis*, 1998 (Department of Manufacturing Systems, The Royal Institute of Technology).

Contact: iga Zadnik
Faculty of Mechanical Engineering/University of Ljubljana
LECAD Laboratory
Ašker eva 6
1000, Ljubljana
SI-Slovenia
Phone: +386 1 4771 740
Fax: +386 1 4771 156
E-mail: ziga.zadnik@fs.uni-lj.si
URL: http://www.fs.uni-lj.si

iga Zadnik is an early stage researcher at Faculty of Mechanical Engineering, University of Ljubljana, Slovenia. He received a BSc. in Mechanical Engineering in 2008. His current research interests include design theory, product development, functional and functionality modelling, matrix methods, PDM systems and programming.

Vanja ok is an early stage researcher at Faculty of Mechanical Engineering, University of Ljubljana, Slovenia. She graduated at Academy of fine arts and design in Ljubljana, department for industrial design in 2010. Her current research interests are product design, interactions of emotions and design, Kansei engineering and design for all (user experience).

Mirko Karakašić is a full researcher at the Mechanical Engineering Faculty, J. J. Strossmayer University of Osijek, Croatia. He received his PhD degree in 2010 in the field of design theory, functional modelling and matrix methods. He is author and co-author of many scientific and professional paper published in journals and proceedings of scientific professional conferences.

Milan. Kljajin is a full professor for Machine Design in Mechanical Engineering and Head of Laboratory for Product Development – LECAD Slavonski Brod at the Faculty of Mechanical Engineering in Slavonski Brod, Croatia. His scientific interests refer to theory in mechanical engineering design and its application in practice, recycling, Eco design and diagnostic of failure to improve design.

Jože Duhovnik is a full Professor of computer-aided design at the Faculty of Mechanical Engineering, University of Ljubljana, Slovenia. His pedagogic and research work is oriented towards design theory, development technic, project management, information flow in CAD, and geometric modelling. He is founder and head of the CAD Laboratory at the Faculty of Mechanical Engineering since 1983.

INTERNATIONAL CONFERENCE ON ENGINEERING DESIGN, ICED11
15 - 18 AUGUST 2011, TECHNICAL UNIVERSITY OF DENMARK

PROACTIVE MODELING OF MARKET, PRODUCT AND PRODUCTION ARCHITECTURES

Niels Henrik Mortensen, Christian L. Hansen, Lars Hvam, Mogens Myrup Andreasen
Technical University of Denmark

ABSTRACT

This paper presents an operational model that allows description of market, products and production architectures. The main feature of this model is the ability to describe both structural and functional aspect of architectures. The structural aspect is an answer to the question: What constitutes the architecture, e.g. standard designs, design units and interfaces? The functional aspect is an answer to the question: What is the behaviour or the architecture, what is it able to do, i.e. which products at which performance levels can be derived from the architecture? Among the most important benefits of this model is the explicit ability to describe what the architecture is prepared for, and what it is not prepared for - concerning development of future derivative products. The model has been applied in a large scale global product development project. Among the most important benefits is contribution to:
- Improved preparedness for future launches, e.g. user interface and improved energy efficiency
- Improved synchronization between product- and production development
- Achievement of attractive cost- and technical performance level on all products in the product family
- On time launch of the first generation of the product family

Keywords: product architecture, modeling product architecture, multi product development, production architecture.

1 INTRODUCTION

Many industrial companies are facing serious challenges in maintaining competitive advantages. Among the most often mentioned challenges are:
- There is a need to reduce time to market (and more importantly time to money) for new products and solutions. Some of the companies that have participated in this research have lost 25% of market share in certain business areas during the last year. The reason for this is, they do not have the right products available on the market.
- There is a need to achieve right cost level for global products– Immelt et al. [1] mention that for GE (General Electric) to be cost competitive, the company needs products that are 80% cheaper in China compared to US products.
- The need for localization and customization of products are increasing [2].

There are certainly many approaches to handle the above challenges, which are of organizational-, process-, tool-, and competence nature. The focus in this paper is architectures, i.e. design of product families or product programs based on stable interfaces and standard designs (modules). Implementing an architecture have relations to all of the above aspects, but the overall hypothesis of the research presented in this paper is that in order to improve the design of product families, architectures have to be modeled explicitly and visually.

Many kinds of research projects have been carried out in order to improve the understanding of architecture work. Among the most important contributions are [3], [4] and [5]. So why is there a need for further investigations? One answer is that nearly all definitions of architectures are of structural nature, i.e. what the architecture *is*. This is for obvious reasons very relevant, but equally important are the functional aspects of architecture, i.e. what the architecture able *to do*. For instance the ability to answer the question: Which products can be derived from the architecture? This phenomenon is not very widely understood and described. Furthermore, the links between market, product and production/supply architecture are relevant. This is also not in itself a new recognition, but when it comes to e.g. evaluating the consequence of adding or removing a feature in a product, it is very difficult to model the consequences market- and production wise. It is the ambition to make a model that allows operational linking between the three architectures.

The reason proactive is mentioned in the paper title, is to address that there is a big business potential and necessity for companies to think ahead in product family design, meaning that the next 2, 3 or 4 launches of derivative products have to be taken into consideration explicitly. Architecture wise this means that an architecture shall be able to show the *preparedness* for the launching of future product generations.

The results presented in this paper is based on research in 3 PhD projects, Kvist [6], Harlou [7], and Pedersen [8] within modeling of architectures. The structure of the paper is as follows. Section two will report on some of the findings from observation of architecture work in main Scandinavian companies. Section 3 will identify the relevant modeling aspects to be included for modeling architectures. Section 4 will present state of the art concerning modeling of architectures. After that section 5 will describe a proposal for how to model market, product, and production architectures. In section 6 experience from application of the market, product and production architecture is explained.

2 WHY IS THERE A NEED TO IMPROVE ARCHITECTURE MODELING?

If a product assortment in a company is described by means of a traditional market matrix, it can be shown as below in Figure 1.

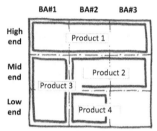

Figure 1. Product mix of a company:
The horizontal axis shows the Business Areas (BA) and the vertical axis describes the performance level of products ranging from low to high performance products

Because many products are designed without conscious decisions concerning the market coverage, poor product family design is carried out. Some of the bad decisions that we have observed in this research are:

One size fits all: In many companies the product architecture is shared from high end to low end products. One consequence of this is that low end products have too high costs and high end products are not sufficiently prepared for future launches. In some companies there is a conception that "stripping" the high end products is a way of developing low end products. There are perhaps examples where this can be done, but is in many cases not possible. In other words, "stripping a Rolls Royce does not bring a Volkswagen into existence".

Dedicated products – future generation products are not addressed: Product families are designed without sufficiently addressing facelifts and next generations. Some examples of this are variants developed on European development sites not prepared for UL approval. The consequence is that US product variants are significantly delayed. Another company is developing a dedicated product for hospitals. This product shall at a later stage also be used in large industrial laboratories. The consequence of developing a dedicated hospital product is a delay of the industrial product family by at least 1 year.

Spaghetti products: Some product families consist of subsystems with very complex interfaces and interactions. The consequence is that development of even small updates becomes very complicated and resource intensive.

Non value adding variety: There are many examples of variety in a product families that does not provide value to customers but only adds complexity cost. A few examples of this phenomenon are: One company is delivering products with actuators that are bolted, welded and glued. This means, that three types of production processes have to mastered, leading to increased cost. Seen from a customer point of view, this variety does not add value. Another company is having pressure tanks certified for

4.1, 4.4 and 5.0 bar. In this case certificates and approvals have to be developed and maintained without adding any extra value to the customers.

The consequences of the above issues are higher costs and reduced ability to launch new products. One of the means to handle the above issues is to develop product families based on explicit architectures. The next section will take a closer look on which phenomena to include in the modeling of architectures.

3 WHICH ELEMENTS SHOULD BE INCLUDED IN MODELING OF ARCHITECTURES

The paper is based on the so-called Product Family Master Plan Framework [7], [9], Theory of Technical Systems [10] and Theory of Domains [11]. Consequently three types of architectures are necessary, i.e. market, product and production/supply chain.

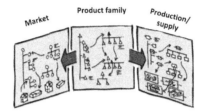

Figure 2. The Product Family Master Plan Framework

There exist many definitions of architectures in literature. Some of the most often quoted are:
"The combination of subsystems and interfaces defines the architecture of any single product. Every product has an architecture; the goal is to make that architecture common across many products", Meyer & Lehnard [5].

"An architecture is a structural description of a product assortment, a product family or a product. The architecture is constituted by standard designs and/or design units. The architecture includes interfaces among units and interfaces to the surroundings", Harlou [7].

"In essence, a PFA (Product Family Architecture) means the underlying architecture of a firm's product platform, within which various product variants can be derived from basic product designs to satisfy a spectrum of customer needs related to various market niches", Jiao & Tseng [12].

All of the above definitions are underlying the importance of interfaces and description of how product families can be described. This is certainly very important, but the above definition is missing the clear distinction between structural and functional aspects of an architecture. Furthermore it does not explain the type of elements that are relevant in the structural and functional definitions. In accordance with Theory of Technical Systems [10] this research will reserve the word structure to how individual products are built up and architecture will be reserved for describing how a product family is built up including the future derivative products.

The next sub sections will explain some of the necessary architecture modeling requirements in market, product and production architectures that this research have identified.

3.1 Market architecture requirements

The overall purpose of the market architecture is to model what the product family shall cover and what it shall not cover. Often this is unclear leading to unfocused product architecture design.

Product properties across the product program: Taking a starting point in properties being obligatory, expected or positioning in the market place, properties can be realized by implementing them as either e.g. basic properties, differentiators or delighters in the product design – depending on the level of fulfillment.

Requirements across individual and all application areas: This is relevant in order to scope the product families, e.g. which areas shall be covered and which shall not be covered. Similarity and differences across application areas is in principle going to drive variety of the elements constituting the product architecture and flexibility of the production architecture.

Product family architecture definitions: This dimension is explaining which product families that shall be developed and how they cover the market grid as previously shown in e.g. Figure 1.

List of features and options: This is an important area since it is often difficult for projects to clarify how many features shall be implemented in high end, medium and low end products. It is also relevant to explicitly specify which features shall be implemented and which ones shall be postponed to later launches or simply omitted from certain market segments.

List of commercial variants that shall be launched to the market: This describes the complete list of individual products and which standard designs and features that goes into each product.

3.2 Product architecture requirements

The overall purpose of the product architecture is to describe the building principles for a product family.

List of structural elements: According to [7] we distinguish between standard designs and design units. The standard designs encapsulate what is reused in several product families, whereas the design units are elements which are not reused. The distinction between standard designs and design units is of importance as their nature is different. Standard designs have to be designed in such a way that they can be used in future products, whereas design units only have the scope of one product. Consequently the application aspects are different for standard designs and design units. A standard design requires a higher degree of documentation, higher degree of maintenance, appointment of responsibility than a design unit, in order to enable reuse in future products.

List of interfaces: This area described the important mechanical, electrical, fluidal and software interfaces between standard designs and design units.

List of product families that can be derived from the architecture: This area describes the functional aspect of an architecture and includes key properties of the individual products that can be developed e.g. cost, energy efficiency, footprint, fault tolerance etc.

3.3 Production/supply chain requirements

The purpose of the production/supply architecture is to describe the building principle for production and how it supports launches of the future derivative products. It means that e.g. flexibility and scalability is of high relevance.

Generic production flows: These flows describe the main production and assembly processes including the necessary production equipment. At the end of the production flows, the types of standard designs that can be produced shall be described. This indicates the flexibility of the production and shows what differentiates each variant and what is common.

List of equipment: This includes the production lines, cells, machinery, tools and fixtures, mapped towards future launches.

3.4 Road mapping – future launches

Future launches: Indicate which products and standard designs to be launched.

Specific product updates: This shall explain which products that shall be launched for each application area.

4 STATE OF THE ART

This section described significant contributions to the modeling of architectures in literature:

Modular Function Deployment: The modular function deployment (MFD) [3] builds largely on the methodology of the QFD method and on the formulation of eight so-called module drivers. The purpose of MFD is to enable cross functional teams (including mainly marketing, development and production personnel) to create a mapping from the physical structure of the products within a family to the functional structure of those products and to ensure that the functional structure corresponds to the demands of the customers. Modular Function Deployment method consists of five consecutive steps. Customer requirements are mapped to functional criteria and subsystem design characteristics and subsequently forming a physical design in which a modular architecture supports a carefully selected set of modularization incentives called module drivers.

Design Structure Matrix: This approach takes a starting point in the decomposition of a product into components/systems and an identification of interfaces/relations among these, Pimmler & Eppinger [13], Höltta-Otto & De Weck [14]. By the use of algorithms, it is possible to encapsulate components into modules or chunks that are closely related to each other from an interaction point of view [15]. This process is referred to as clustering. The outcome of a DSM is a proposal for a future modular product architecture.

Generic Bill of Materials: The generic BOM originate from the assemble-to-order environment [16]. The end-products typically have a number of features for which a number of options are available to choose from. Not many options are required in order to make the number of combinations (i.e. end-products) enormous. The number of end-products can easily become too large to able to define specific BOM's for every single combination. Furthermore, forecasting, BOM-storage and maintenance become unmanageable. The generic BOM is a concept that is introduced to enable creation of a specific manufacturing BOM when the customer places an order. The generic BOM is used to describe related products in one all-embracing model by using generic and specific items.

Decision tree: The decision tree [17] is used by Tiihonen & Soininen [18] as a product configuration model, which basically represents all the valid combinations of the components that can be used to obtain the desired functions for the customer. The decision tree presents the multitude of component variety within a product family and by the use of positive combinatory relationships (e.g. if "engine size"=D13 then "engine power" must be 360 or 420 hp) and/or incompatibility relations (e.g. if "engine size"=D13 then "engine power" cannot be 220 or 700 hp) it defines the possible product configurations.

Value analysis: Value Analysis is a discipline founded at General Electric in the late 1940's [19]. In short, value analysis is a methodology that has as its purpose to relate cost with functions in a product. It is a stepwise methodology in which a product is partitioned into smaller constituents for further analysis – that may be analysis of cost or value. Value is not the same as the Japanese idea of customer value we may see within the lean paradigm. Value is specifically defined as the "worth" relative to cost, i.e. value = worth/cost. Worth in this sense actually resembles the idea of customer value in lean very well. It is a denominator of those aspects, functions and features a customer wants to pay extra for. Some practitioners try to quantify worth and relate it directly to cost. Obviously cost is rather quantitative and measurable in hard currency, while "worth" is a more soft and qualitative size. Whether qualitative or quantitative, value has a focus on identifying value elements from a customer perspective and relate it directly to the functions of the product and thereby indirectly to the way the products are built.

Function structures: The function-based design methods are characterized by establishing either a function model [20] or the schematics of the product [4]. The function structure describes the flow of material, data, and energy through sub-functions of the product using a set of rules (e.g. the rules that are referred to as the functional basis which basically is a common language to describe functional elements. The schematic of the product is somewhat similar to the function model. But where the function model describes the product using functional elements the schematics on the other hand can describe both functional and physical elements, whichever being the most meaningful. Having established an understanding of the functional structure of the product some methods base identification of modules on experience and some simple guidelines, i.e. a rather qualitative approach [4], [13] and [20].Basically, these methods identify potential modules in a way similar to the way the MFD method makes use of the so-called module drivers.

Multi criteria assessment: Otto & Hölttä-Otto [21] presents a technique based on multi-criteria assessment where product architecture concepts are given a score based on a set of different weighted criteria. Although, the method is designed to be used for screening of preliminary product architecture concepts, and not - as it is the focus of this research - analysis and re-design of product families, the method include analysis aspects that should be considered. The method is based on relatively quantitative metricc adapted from the field of modularity, product architecture design, and product development in general (e.g. functional structure, DSM, commonality indices, etc.).

Value stream mapping: Most value stream mapping tools has a focus on information and physical goods passing through the supply chain. The value stream is consequently often perceived as the flow of materials through the value adding processes. There are several value stream mapping tools, e.g. by Womack & Jones [22]. This section describes the "traditional" value stream mapping tool. Other tools or methods re describe in the subsequent sections. A less graphical depiction of the value stream is a process activity map. It is a schematic representation of the critical path of a production. It is basically a matrix containing a mapping between process steps and machines, time consumption and distance along with other factors of choice. This tool may be used in conjunction with the traditional value stream map or as a preparation of that.

Conclusion: It is clear that all the above approaches can play a role in identifying structural aspects of an architecture, but the functional aspects are not explicitly described. Furthermore the structural contents of architectures are not described in terms of different design types, e.g. standard designs and design units. This topic is relevant in order to design flexibility in product architectures. In large projects this plays an important role concerning scoping of the development task.

The next section will present a proposal for the modeling of market, production and production architectures.

5 ARCHITECTURE MODELING

5.1 Market architecture

The purpose of modeling the market architecture is to bring clarity into decision making concerning the choice of which segments to cover or not cover and what properties are needed in order to do so across different business areas with different applications. A clearly defined market architecture is able to guide and control the engineering efforts towards profitability by "smart" product family design.

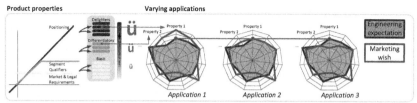

Figure 3. Product properties and their mapping towards varying market applications.

The radar diagrams show the total performance of the product by mapping the properties capable of positioning the product against competitors e.g. by differentiation. During the early phases of product scoping and requirements definition in close cooperation with competencies representing marketing, the mapping can serve as means of matching the wanted product performance from a marketing point of view with the expected product performance from an engineering point of view. Hereby, the explicit mapping can have a brokering function facilitating the meeting between sometimes unrealistic marketing wishes and best guess engineering expectations. If applied to a product family intended to cover different applications in different segments with varying requirements, it is of fundamental interest to map marketing professionals' perception of the spectrum of varying demands. As it is most often impossible to fulfill requirements for all segments, the mapping can help focusing the product architecture towards the most appropriate and favorable segments. To concretize the product properties, features and options can be modeled e.g. by the means of the "customer view" [7] mapped towards the different applications and varying the performance levels (low-, mid- and high end).

Figure 4. Features/options and their mapping towards performance levels in different applications and the identified product architecture(s).

This mapping serves to answer the questions of which product features that are in scope for the development task. Some features are too expensive or simply irrelevant for certain applications and are outside scope. Other features will be outside the standard program for all applications since they may 'pollute' a robust product architecture. Finally, the mapping towards one or more product architectures closes the gap towards engineering and sets the boundaries for the development task.

The detailing of the link between matching product features and identified product architecture, calls for a visualization of the commercial variants. They serve as being the 'contract' between engineering and marketing explicitly identifying the development task. The detailing of this list requires the product development task to be past the early stages, but major value is represented in conducting this modeling as early as possible.

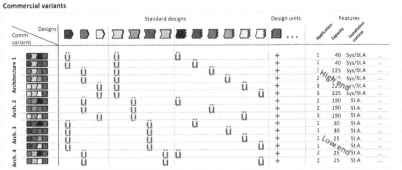

Figure 5. Commercial variants and their utilization of standard design,
design units and associated product features

The modeling will vary according to the application variation, general market aspects etc., however, the models shown in Figure 3, Figure 4 and Figure 5 are made to illustrate the general purposes.

5.2 Product architecture

According to the suggestions presented in this article, the modeling of product architectures encompasses the constitutive structural elements of a product architecture and the behavioral functional abilities. In other words, the aim of this model is not solely to describe what the product architecture *is*, but also what the product architecture is able *to do*.

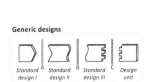

Figure 6. Generic structural elements of
the product architecture:
Standard designs and design units.

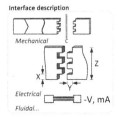

Figure 7. Modeling of interfaces between
standard designs, between standard designs
and design units and/or surroundings.

Equally important to the standard designs and design units, the interfaces capable of maintaining a predictable product structure, must be modeled explicitly as well.

Different standard designs can adopt different roles. Some are closely related to specific functions and/or application, while others are universal to the product architecture. Finally, design units are used for embodying functionalities that vary between individual product variants.

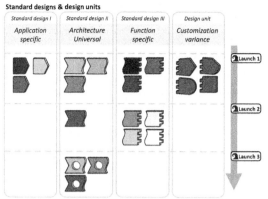

Figure 8. Standard designs & design units.

Figure 8 shows the variance across the different structural elements while incorporating the dimension of future launches: Which designs need to be prepared for which launches? Naturally, it is impossible to plan further than a certain realistic extent in rapidly changing markets, but the higher the detail this modeling can achieve, the better the basis for improving the launch preparedness is.

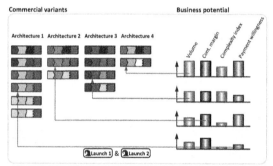

Figure 9. Assessing the business potential of commercial variants.

As described earlier, the explicit modeling of commercial variants early in the development process, act as the explicit link between the market- and product architectures, it is of fundamental commercial importance to map the expected production volumes, contribution margins, and payment willingness from customers – the payment willingness being the quantitative interpretation of the 'worth' phenomenon described earlier. If established, a measure of the complexity induced by different product variants can be included to qualify discussions with industrialization professionals with the task of freezing production architecture aspects. These four measures can help balancing out the product architecture(s), ensuring a leveled variance spectrum composed of "smart" variants with an appealing overall business justification.

5.3 Production architecture

Depending on the size of the product architecture development project, the associated production system will need either an update, a modification or a complete redesign. The production system is designed coherently, as the product architecture matures and passes from concept to detail design.

As basis for the modeling of production architecture is the generic production flow shown in Figure 10. This is capable of showing how and when the product variants are created in the production lines, which elements in the production system that are alike and which elements that differ. The relevant decoupling points (either *variant creation points* or *customer order points*, depending on the context) can be established and fixed. Furthermore, an inclusion of relevant machinery, tools and capacity utilization metrics provides the opportunity of assessing key financial characteristics of the suggested setup.

Generic Production Flow

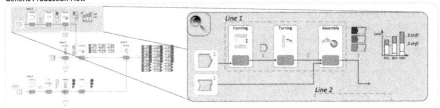

Figure 10. Generic Production Flow: Modeling the flow of all variants in one visual model

Since production equipment can require extensive capital investments, a mini roadmap of the lines, machinery and tools is valuable to map towards the suggested launch rhythm. As shown in Figure 11, the addition of further parts and components intended for launch 3 and beyond, will most likely entail a larger utilization of the production capacity, take up physical space of the production floor and require additional investments in machinery and tools downstream.

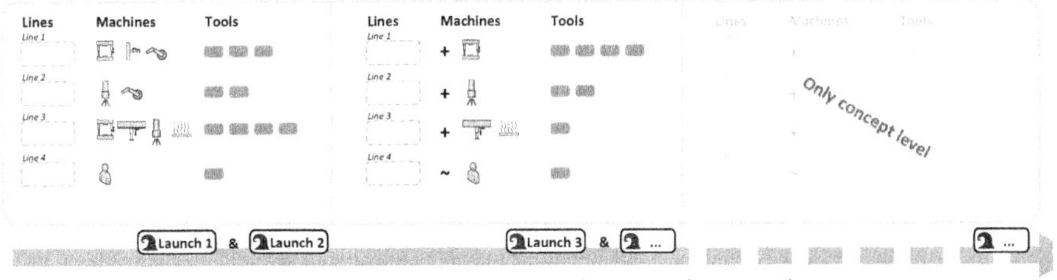

Figure 11. Production equipment needed for 1st launch, 2nd launch, 3rd launch etc.

These are all aspects that are predisposed by the design of the product architecture(s); thus requiring explicit and coherent models.

As marked in Figure 4, certain features will be part of the standard program incorporated in specific commercial variants, while other features will need an individual business case in order to be fulfilled as e.g. customizations. Setting up a global chain of supply and delivery, service levels of standard lead times, degree of local customization possible etc., are also factors predisposed by the architectures of the product and production. Figure 12 shows an example of how a global company could utilize the price of cheap labor in some regions with the local capability of customizing product (and perhaps conduct final assembly) around the world in product/distribution centres.

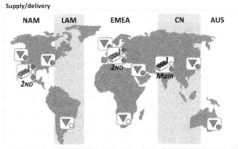

Figure 12. Global supply and delivery capabilities

5.4 Roadmap

The behavioral aspects of the market-, product- and production architecture is considered in the architectures' future launch preparedness. This is a function of the architecture, explaining what the architecture is able *to do*. This ability is modeled by visualizing the launches, derivative products and specific product updates – already planned for.

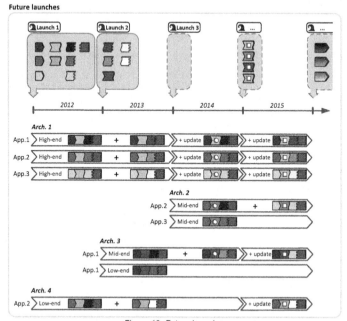

Figure 13. Future launches:
Launch preparedness, launch waves, derivative products and specific product updates.

6 EXPERIENCE FROM APPLICATION

The above architecture modeling approach has been utilized in one large scale product development project. The case company is operating globally and develops industrial products in high volumes. During the 2 year project, approximately 100 designers have been working on developing the product program. The development project has included complex fluid-dynamics, mechanics, materials, software, electronics, solid state mechanics and thermodynamics - architecture wise only mechanics and electronics have been included. In the area of production, complete new facilities have been established in Asia, Europe and the US. Market wise the product program has partly been launched. Sales is taking place through existing sales companies. The application areas include mainly existing well known areas, but also a few new applications are included, e.g. renewable energy.

The PhD students have been working for more than a year and a half, several days a week in utilizing the market, product and production architecture model. During the case study, the architectures has been developed and described by the PhD students in close collaboration with employees in sales, product management, engineering, production and supply chain. Four types of architects have been responsible for the contents of the market, product and production/supply chain architecture. The four types of architects are named market, product, production/supply and cross functional architect. The market architect is based in product management and is responsible for the market architecture and roadmap; the product architect is based in engineering and is responsible for the product architecture; the production architect is based in production and is responsible for the contents of the production architecture; and the cross functional architect is responsible for the alignment of the market, product, production architecture and roadmap. The project manager has acted as the cross functional architect.

Two kinds of meetings have been conducted in the project: They are named architect meeting and cross functional architect meetings. The first year both architect and cross functional architect meeting was carried out each Thursday from 9.00 to 11.00. During the last period the market, product and production architect was held each week, but cross functional architect meeting was held every 2nd week. Participants in the market architect meetings were program management and product management. In the product architecture meetings, senior designers from relevant specialist areas participated. In the production architect meetings new product introduction managers, tool designers and production line designers participated.

Experience from application of the market architecture: The feature/options has enabled an earlier and more explicit definition of what defines a high end, mid end and low end product, i.e. clarification of which features and options that shall go into which variants.

Experience from application of the product architecture: The interfaces have been decided much more conscious compared to previous projects in the company. It means that the next 3 product launches have been explicitly planned in such a way that the architecture is prepared for one new technology, an update of the user interface and more advanced wireless communication.

Experience from application of the production architecture: The project has had the task to establish completely new production lines with three kinds of automation levels, fully automatic, semi automatic and manual production and assembly. Particularly the full automatic production line design have benefitted from the product architecture. It has been possible to order new production and assembly equipment earlier since the product program have been decided earlier and therefore variety of each part have been known earlier. Also the product architecture specification has been beneficial to production design since flexibility and scalability is very important design properties for automatic production equipment.

Experience from cross functional application of the market, product and production architecture: The main benefit of the cross functional review meetings have been continuous scoping of the project, i.e. decisions concerning what shall be developed now and what shall be postponed. Another aspects that have been more consciously considered, is clarification of where the architecture shall be prepared and where is shall not be prepared for future launches. The performance limits concerning cost, energy, foot print and availability have also been clearly defined.

All in all the main benefits of applying the explicit modeling of market-, product-, production architectures (including the roadmap) has been a contribution to:

- Improved preparedness for future launches, e.g. user interface and improved energy efficiency.
- Improved synchronization between product- and production development
- Achievement of attractive cost- and technical performance level on all products in the product family
- On time launch of the generation of the product family

Concerning future application the cross functional architect role has to be reconsidered. With a traditional organisation, one could argue that "no one" or everyone is responsible. No single person or department have all the competencies necessary to handle the cross functional architect role. This will be a topic for further research and case studies. The architecture models are mainly handling technical decisions whereas business decisions are only implicitly addressed. This is another area that obviously should be improved.

7 CONCLUSIONS

The paper has presented an explicit proposal for description of contents of a market, product and production architecture. The main contribution is the distinction between structural and functional contents of architectures. By this distinction it is possible to improve the description of what the architecture is prepared for concerning future launches.

Further work includes test in two other companies. So far only the mechanical and electrical elements are included. It is clear that also software has to be included in the next version of the architecture model. Also other life phases such as service/aftermarket will in many cases be of high importance. A follow up case study is planned in order to study whether the intended preparedness is realized in reality.

REFERENCES
[1] Immelt, J.R, Govindarajan, V. and Trimble, C.: How GE Is Disrupting Itself, *Harvard Business Review*, 2009, 87(10), 56-65
[2] Hvam, L., Mortensen, N. H. and Riis, J., *Product Customization*, 2007 (Springer, Berlin)
[3] Ericsson, A. and Erixon, G., *Controlling design variants – Modular product platforms*, 1999 (Society of Manufacturing Engineers, Michigan, US)
[4] Ulrich, K. T. and Eppinger, S. D., *Product design and development,* 2nd ed., 2000 (McGraw-Hill, Boston, MA, USA)
[5] Meyer, M. H. & Lehnerd, A. P., *The power of product platforms – Building value and cost leadership*, 1997 (The Free Press, New York)
[6] Kvist, M., *Product Family Assessment*, 2009, Dissertation/Thesis (DTU Management, The Technical University of Denmark, Kgs. Lyngby).
[7] Harlou, U., *Developing product families based on architectures - contribution to a theory of product families*, 2006, Dissertation/Thesis (Department of Mechanical Engineering, Technical University of Denmark, Kgs. Lyngby).
[8] Pedersen, R., *Product Platform Modeling*, 2009, Dissertation/Thesis (DTU Management, The Technical University of Denmark, Kgs. Lyngby).
[9] Mortensen, N. H.,Hvam, L.,Haug, A.,Boelskifte, P. and Hansen, C. L. Making Product Customization Profitable, *International journal of industrial engineering*, 2010, 17(1), 25-35.
[10] Hubka, V. and Eder, W.E, *Theory of Technical Systems,* 1988 (Springer, Berlin)
[11] Andreasen, M. M., *Syntesemetoder på systemgrundlag – Bidrag til en konstruktionsteori* (in Danish), 1980, Dissertation/Thesis (Department of Machine Design, Lund Institute of Technology, Lund, Sweden)
[12] Jiao, J., & Tseng, M. M. A methodology of developing product family architecture for mass customization, *Journal of Intelligent Manufacturing,* 1999, 10(1), 3–20.
[13] Pimmler, T. U. & Eppinger, S. D., Integration analysis of product decompositions, *Proceedings of the ASME Design Theory and Methodology Conference*, 1994 (Minneapolis, MN, USA)
[14] Hölltä–Otto, K. & De Weck, O., Degree of Modularity in Engineering Systems and Products with Technical and Business Constraints, *Concurrent Engineering: Research and Applications*, 2007 15(2), 113–126 (Sage Publications, Thousand Oaks, CA, USA)
[15] Steward, D. V., The Design Structure System: A Method for Managing the Design of Complex Systems, *IEEE Transactions on Engineering Management*, 1981
[16] van Veen, E. A. & Wortmann, J. C., Generic bills of material in assemble-to-order manufacturing, *International Journal of Production Research*, 1st International Production Management, Conference, 1987, 25(11), 1645-58
[17] Rea, R. C., Helping a client make up his mind - The use of a "decision tree" helps a client in planning his estate, *The Journal of Accountancy*, 1965, 39-42
[18] Tiihonen, J. & Soininen, T., *Product configurators – Information system support for the configurable product*, 1997 (Product Data Management Group, Helsinki University of Technology, Helsinki, Finland)
[19] Fowler, T. C., *Value Analysis in Design,* 1990 (Van Nostrand Reinhold Company, New York)
[20] Pahl, G. & Beitz, W., *Engineering Design – A systematic approach*, 1996 (Springer, Berlin)
[21] Otto, K. and Hölttä-Otto, K., A multi-criteria assessment tool for screening preliminary product platform concepts, *Journal of Intelligent Manufacturing*, 2007, 18(1),59-75
[22] Womack, J. P. and Jones, D. T., *Lean thinking – Banish waste and create wealth in your corporation*, 2003 (Simon & Schuster UK Ltd., London)

Contact: Niels Henrik Mortensen, Technical University of Denmark, Department of Management Engineering, Building 426, DK-2800 Kgs Lyngby, tel: +45 45 25 62 75, email: nhmo@man.dtu.dk

Niels Henrik is Professor of platform based product development in the Department of Management Engineering at the Technical University of Denmark. He teaches and researches in engineering design and platform based product development.

INTERNATIONAL CONFERENCE ON ENGINEERING DESIGN, ICED11
15 - 18 AUGUST 2011, TECHNICAL UNIVERSITY OF DENMARK

DESIGNING MECHATRONIC SYSTEMS: A MODEL-INTEGRATION APPROACH

Ahsan Qamar[1], Jan Wikander[1] and Carl During[2]
(1) KTH-Royal Institute of Technology, Sweden (2) Micronic Laser Systems, Sweden

ABSTRACT

Development of mechatronic products requires different types of design models in order to support both domain-independent specifications and domain-specific principles. This research aims to find out how *system-level modeling* can support mechatronic design, and how the integration of system-level modeling and domain-specific modeling can be supported during different design phases. A design example of a hospital bed's propulsion system is presented to show firstly the relationship between conceptual design and system-level modeling, and secondly the need for integration of system level and domain specific design models. An integrated modeling and design infrastructure is proposed to support abstraction between mechatronic design models, hence supporting co-evolution of design models. The paper concludes that a mechatronic design problem can be better supported through such an integrated design approach. However, usability of this approach needs to be further supported by more case studies in the future

Keywords: Mechatronics, system design, design infrastructure, model integration

1 INTRODUCTION

Technical systems today have to support a large number of functions at reduced cost. The efficiency and cost effectiveness gained by implementing the required functionality through electronics and computer software has become a major driver towards development of mechatronic products. These products are characterized by increased integration of mechanics, electronics, and computer software, requiring companies to establish cross-disciplinary teams consisting of several domain-experts. Tomiyama et al. [1] state that multi-disciplinary product development (as in mechatronics) introduces difficulties such as the need for an inter-disciplinary design language, how to deal with different stakeholders, and how to deal with inter-disciplinary design problems. Mechatronic design requires a collaborative effort between different domain experts within the cross-disciplinary team. However, each domain contains different collections of design methods and tools. An overall mechatronic design method is yet to be conceived. As engineers do not necessarily possess cross-domain knowledge, it is difficult for an individual to understand the inter-disciplinary problems, especially in the context of complex products. Therefore, communication between involved domains is essential throughout different design phases in order to avoid integration problems in mechatronic product development. In addition to this, technological advancements and increased functionality contribute to high complexity in designing mechatronic products.

Modeling is an important design activity, where modelers try to gain information about consequences of their decisions early in design. Buur and Andreasen [2] state that the success of a mechatronic design project depends especially on the ability of the designers to communicate and visualize their ideas to the rest of the group. Design models permit a designer to describe his or her thoughts for better understanding, both individually and by the group. Depending on the design stage, a design model can be abstract or detailed. A design model should be carefully developed to model only the product properties necessary at the current design stage [2]. This restriction in scope is necessary: firstly, since information about a design problem increases through different phases of design, secondly, because a design model with too many product properties will become unnecessarily complex to serve the purpose for the designer. Hence, product design is based on different design models reproducing different product properties. In a mechatronic design scenario, design models vary between different domains. Some of these models define and describe the product from the domain perspective such as mechanics or electronics; others are used to evaluate product properties within a domain such as dynamic analysis of a mechanical design.

The partitioning between different mechatronic domains is laid very early in product design. By defining function principles to the function structure, the designer allocates different technologies to the product function/sub-functions (*design concept*). However, this often leads to an isolated development within a domain, and optimization of the individual modules, rather than the complete system. Buur et al. [2] and Gausemeier et al. [3] state that specifying a mechatronic *design concept* requires a new design model. Buur et al. [2] define the new model types to support abstract function structure independent of a technology, function principles supported through different technologies, and specification of the interfaces between different technologies. Gausemeier et al. [3] utilize a new modeling language for supporting abstract function structure, domain-spanning function principles, and interfaces.

One of the main ideas behind the research treated in this paper is to utilize design models that are suitable to capture domain-independent specifications. At the same time, the designer can apply multi-technological function principles to the *design concept*. The domain-specific models evolving from these multi-technological function principles can be integrated with the domain-independent model, the other main proposal of this paper.

Since design evolves not only from one design phase to the next, but also in between domains, it is important to support abstraction between models of different engineering domains. Moreover, it is also important to keep the design models consistent with each other if a certain design model is modified at a certain design stage. One approach towards achieving this abstraction is through model transformations. Some examples of integrating design models through model transformations are presented in [4] and [5]. Consistency between mechatronic design models has been discussed in [6]. While these approaches extend the software engineering principles towards model-based development, this paper aims to explain the relationship between design models within a mechatronic design problem, and proposes a solution to multi-domain model integration. The paper utilizes a design study of a servo-propelled hospital bed performed at the Department of Machine Design at the Royal Institute of Technology, Sweden, to answer questions such as:

1. How to establish an initial domain-independent design specification through a design model, and obtain a complete system view required in a mechatronic design problem?
2. How to establish relationships between domain-independent and domain-specific design models?
3. How to integrate different design models developed at different design stages?

The remaining part of the paper is structured to answer the above questions as follows. Section 2 discusses model-based design in relation to engineering design methods. The section proposes a solution for solving the communication problem between domain experts in a mechatronic context (question 1). Section 3 presents the design study of the hospital bed system highlighting the conceptual design phase where domain-specific design models were also utilized. Section 4 is about answering the question: how to integrate models (question 2), and how better design can be achieved through model integration, supported by a small example (question 3). Section 5 concludes the paper, including a discussion on proposed future work.

2 MECHATRONIC DESIGN AND MODEL-BASED DEVELOPMENT

Model-based engineering (MBE) is about elevating models in the engineering process to a central and governing role in the specification, design, integration, validation, and operation of a system [7]. The systematic design approach from Pahl & Beitz [8] shows three main product design phases: conceptual design, embodiment design, and detailed design. During these design phases, models increase in detail with the passage of time, and abstraction between models needs to be supported in order to manage the modeling process as a whole.

Design activities for a mechatronic system are typically performed by a multi-disciplinary team of domain experts. Different design methods (available in different domains) are followed by the domain experts, who are unlikely to possess inter-disciplinary knowledge to get a detailed enough understanding of the whole design problem. Therefore, it is difficult to establish a common mechatronic view; rather different domain views are established and the dependencies in-between are not clear. Hence, it is necessary to establish some means of communication between such views in order to avoid integration problems. Frey et al. [9] classify the communication between two design-domains in terms of communication possibilities between persons, between methods, between models, and between analysis tools. We argue that for a mechatronic design problem, two possibilities can be undertaken to attack the communication problem:

1. A mechatronic design methodology could be followed. VDI2206 [10] introduces such a design methodology. However, it does not cover the management of dependencies between mechatronic design domains. Also, the means to support abstraction between system-design and domain-specific design (per VDI2206) are lacking. A vertical abstraction adds detail to a model or reduces it, while a horizontal abstraction typically takes place between models of the same detail, often in different domains. As highlighted by Buur et al. [2], new mechatronic design models are needed that support both domain-independent and multi-domain modeling capability. Some modeling languages, such as Modelica [11] and MapleSim [12] support the creation of multi-domain design models (spanning different stages of VDI2206). However, domain-independence is also important for the different domain-experts to share a common product view. We conclude here that a single multi-domain modeling language cannot provide a solution for mechatronic design problems, a conclusion also supported by Shah et al. [5]. Such a language is difficult to develop, support, and evolve with time.
2. A framework that supports mechatronic design can be utilized. Here, we can let the domain experts utilize the existing methodologies inside each domain, and provide means of communication between domains through the framework. As suggested by Frey at al. [9], a person-to-person and method-to-method communication is either error-prone, or not directly possible. Even though the communication between analysis tools is possible, however, it is based only on execution of design models and not their development. Therefore, a model-to-model communication is employed during this paper, which allows for both vertical and horizontal abstraction during model-based development.

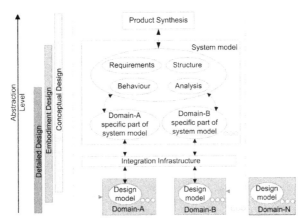

Figure 1. Model—based development in engineering design in the context of an integrated modelling and design infrastructure for mechatronic system design

For communication through multi-domain models in a perspective of engineering design, an infrastructure is required, supporting evolution of design models while design proceeds between different design phases. The infrastructure should support development of domain-independent design models and specification of interfaces between design domains, development of domain-specific design models, and integration between all these design models. Such an infrastructure is vital to integrate multi-domain models, which is a key proposal of this paper, as explained in Section 4.

Conceptual design is a dynamic phase in terms of change in design and interaction of designers. Initial product synthesis – being part of conceptual design – serves as a basis for developing an abstract function structure and corresponding function principles. In a model-based development with model as the primary artifact, it is important to capture the product synthesis information inside models. In agreement with Gausemeier et al. [3], we will use the term *common specification language* to denote a common platform for different domain experts to define and specify a system. Figure 1 shows that a system-model can be developed through a common language, based on the findings of product synthesis phase. Here, the interfaces between chosen *design concepts* can be specified starting in a black-box manner and continuing towards increasing detail. The Systems Modeling Language (SysML) [13] is a general purpose modeling language where a modeler can specify a system to a level that enables its association to other design models [14]. We utilize SysML to represent the multi-domain function principles through generic SysML constructs. The resulting model can be utilized as a common design model, by building relationships with other design models. The relationships can be built by utilizing the extension capability in SysML to create domain-specific parts of the system model (Figure 1), where the parts are built with concepts of a particular domain. The proposed integration infrastructure allows integration of the domain-specific design models with the domain-specific parts of the system model. Hence, as the design progresses through domain-specific design models, the system model also increases in detail. Design iterations continue to take place between different design stages, and design models consistently evolve through the integration infrastructure (Figure 1). Another approach is to first create domain—specific design models based on product synthesis (red arrows in Figure 1). The resulting design models can be related with each other by transforming them into the system model through the integration infrastructure. We do not strictly propose to follow this order or the order starting with the system model, to leave flexibility for future work.

3 DESIGN EXERCISE: A SERVO-PROPELLED HOSPITAL BED

This section is based on a mechatronic design example in order to better reflect on the following questions (rephrased from questions: 1, 2, and 3 in section 1):
- How to establish relationships between domain-independent and domain-specific design models?
- How to integrate different design models developed at different design stages?

By performing a design exercise on the hospital bed example, we try to identify the needs for the integrated modeling and design infrastructure, and determine how dependencies can be managed through the SysML model. The aim with the exercise is to design an active (driven) wheel module which can be utilized on common hospital beds. The design activities performed within the conceptual design phase are explained in the subsequent sections.

3.1. Conceptual design

During conceptual design, a group of designers were provided with a set of requirements for the hospital bed. The design team consisted of six team members with backgrounds in mechanical engineering, electrical engineering, control engineering, system engineering and computer science. The requirements for the design of the wheel module were classified in different categories on a white board, and used as a reference throughout the initial conceptual design phase. Some of the main requirements for hospital bed are presented below; the rest is omitted for space concerns:
- Bed speed of 2 m/sec on a maximum 5° slope (power, driving requirements)
- Being able to turn on the spot and move in any direction (driving configuration, control)
- Wheel solution packaged as one module (sensor, motors, control, wheel module configuration)

Based on these requirements, the design group discussed the following main points:
- Which configuration of driving and steering (number and placement of active wheels: *propulsion system configuration*) will provide the best solution in terms of drivability, steering, and cost effectiveness. Six different configurations were discussed.
- How to provide a modular solution, yet one that works on current hospital beds. This includes decision on sensors (wired/wireless), power required to move maximum load, battery (central or distributed), battery charging scenarios, and mechanical interfacing.
- Configuration of driving and steering within one wheel module assembly (*wheel module configuration*). Two configurations were discussed, knowing available wheels and information about drive and steering actuators. The braking requirement for each wheel was also discussed to be accounted in the solution.
- A central controller is required to control the driving and steering of the bed as a whole, by controlling the available drive and steer actuators in each wheel module assembly.
- Safety considerations in relation to the use of wireless devices, patient safety and comfort requirements.

During the initial conceptual design phase (lasting for 5 hours), an initial decision was made on some of the concepts. However, some concepts needed further analyses before any decision about them could be made. Some of these concepts were:
- Wheel-module configurations
- Analysis of driving, steering and braking actions in relation to bed movement for six different propulsion system configurations, and two different wheel module configurations

- Controller complexity for different wheel module- and propulsion system- configurations
- Which battery is suitable for the bed, i.e. an analysis of power to weight ratio and cost considerations of batteries
- Relationship of distributed battery and central battery in connection to wired and wireless sensors, and the cost difference for each concept

Different propulsion system configurations led to six possible design alternatives, all providing driving and steering capability. These concepts were compared in a weighted decision matrix (omitted here due to space concerns) against a set of basic design criteria. The decision matrix showed that the propulsion system configuration based on either a diagonal configuration or a fully configurable configuration as the most feasible options. The same result is modeled inside a system-model (Figure 3(b)) which is discussed in the following section.

3.2. System design/modeling

After completing the initial conceptual design, the requirements for the wheel module along with the possible solution concepts were known (working structure). A system model was built based on these findings. The suitability of SysML as a modeling language during the conceptual design phase has been documented in [15] and [16]. This section will present a few snapshots of the SysML model, especially its relation to the performed synthesis, and to the creation of domain-specific models.

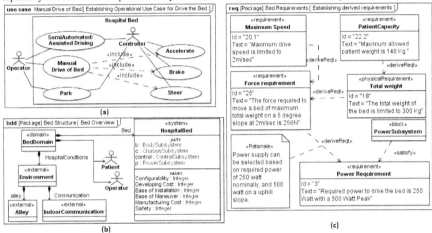

Figure 2. (a) Use cases showing manual and propelled operation of bed. (b) Top-level overview of bed showing main actors. (c) A set of requirements for the power subsystem

Figure 2(a) shows the manual and automated driving use cases performed by the bed operator. The accelerate, brake, and steer use cases are performed by the operator manually. The top-level structure of a hospital bed is shown in Figure 2(b), containing the main actors and the environment in which the bed operates. The *system HospitalBed* contains parts (regardless of technology) mentioned in the *parts compartment*. Figure 2(c) shows some of the derived requirements, in this case for the power subsystem. All models in Figure 2 are developed based on the information obtained through the product synthesis phase of the design exercise. The main structure of the hospital bed propulsion system is shown in Figure 3(a), showing different subsystems. Each subsystem contains a combination of multi-domain components, represented through general purpose semantics. The interfaces between different system components (represented by generic constructs) can also be specified.

The *ChassisSubsystem* in Figure 3(a) contains two to four *WheelModule* blocks. This is based on the conceptual design phase discussion deciding that there will be a minimum of two active wheels on each bed, and each wheel will be enclosed as one complete module with driving and steering capability (*Modular Components Constraint*). The six alternative propulsion system configurations are compared against stated criteria by a weighted objective function (*WheelModuleObjectiveFunction*) used inside the *WheelModulePerformance TradeOff* block (Figure 3(b)).

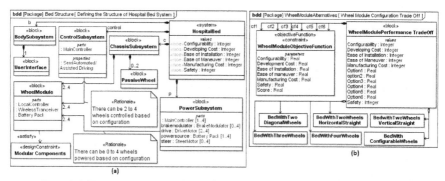

Figure 3. (a) Main Structure of the propulsion system (b) A model of a trade-off study for propulsion system configuration alternatives. The criteria such as configurability etc. are represented as measures of effectiveness (moe) of each alternative.

During the conceptual design phase, a discussion was made about off the shelf wheel-motors, only requiring a steering mechanism to provide a steering and driving capability in one module. Based on the wheel-motor, two configurations for the wheel module assembly were discussed, as represented in Figure 4(a). Each configuration consists of a drive motor-wheel unit and a steering actuator/transmission as constituent parts. Interfaces *Drive* and *Steer* are *provided interfaces* to each configuration, letting the drive motor and steer configuration to interact with the wheel-module configuration. Interface *Sensor* (to measure angle and velocity of wheel) is a *required interface*, letting the wheel configuration send sensor measurements to other blocks. Configuration 1 consists of a gearing with a steer motor to steer the drive motor assembly (Figure 4(b)). Configuration 2 contains freely revolving driving assembly with the drive wheel mounted off the vertical rotational axis of the steering assembly. A brake modulator inside the steer assembly locks the drive assembly at the required angle. Position encoders were selected to measure the steering and driving rotation angles.

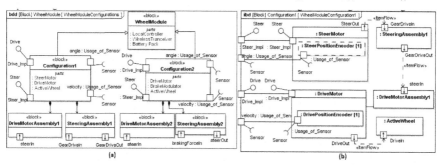

Figure 4. (a) Two alternatives for wheel module configuration. (b) Internal block diagram showing internal connections for configuration 1

In order to make a decision about the wheel module configuration, it was important to realize each configuration in terms of form, and analyze them in terms of behavior. For this purpose, a CAD model and a dynamic analysis model for each configuration were built, as discussed in section 3.3.

For the controller, a centralized architecture and a distributed architecture was considered. Although both the architectures were also dependent upon selection of wireless or wired transmission, it was decided to control the angle and velocity of each wheel through a local controller (Figure 5(b)). If a wireless transmission is considered, a local battery can power the wheel module and the local controller. In this case, a wireless transceiver would send data to the main controller and receive reference commands from it as shown in Figure 5(a). Figure 6 shows the complete distributed control architecture with two wheel modules and the user interface inputs. It can be noted that in case of wired transmission, the interface *WirelessData* and *WirelessCmd* will be replaced by wire connections. In this case, a central battery can also be considered relevant instead of a localized battery.

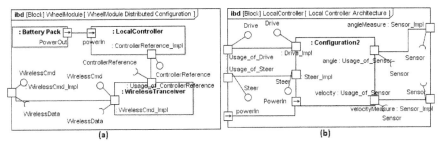

Figure 5. (a) Wheel module in distributed architecture. (b) Internal structure of local controller, and its interface with the wheel module configuration 2

The system-level modeling through SysML proved to capture the information gathered during the synthesis phase effectively, as displayed through different SysML diagrams. Moreover, all SysML diagrams presented here are consistent with each other. This means that introducing a change in one diagram during design leads to relevant changes in other diagrams too. The SysML tool (MagicDraw [20]) supports keeping different SysML diagrams consistent with each other.

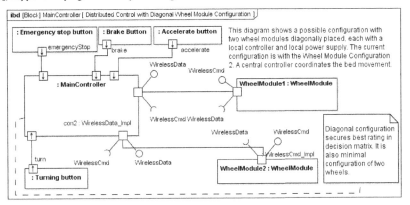

Figure 6. Architecture of a distributed controller with the user interface buttons, and two wheel modules in the diagonal propulsion system configuration

3.3. Domain-specific design
As discussed earlier, the decision about some of the components, configurations, and alternatives needed further analyses through domain specific models. This section provides a limited overview about mechanical design and analysis of the wheel module, in order to make a decision about the wheel module configuration assembly and the relevant components. Other issues such as controller bandwidth, energy supply and communication between wheel modules etc. were also discussed. However they are not presented here due to space concerns.

Wheel module assembly design
The CAD modeling of the two wheel module configurations was performed to get an estimation of the size of the whole module and to know the necessary components. The *wheel-motor* model was utilized to construct the assembly. The two concepts can then be compared based on manufacturing cost, safety, and performance criteria. Figure 7 (a) and (b) show the two configurations.

Analysis of wheel module configurations
In order to analyze how the bed moves with each wheel module configuration, a dynamic analysis model was created. The multi-domain physical modeling and simulation tool MapleSim [12] was used

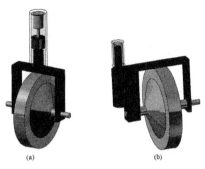

Figure 7. CAD models of wheel module configurations. Both configurations contain the same drive motor and wheel (a) Configuration 1 with steering motor for steering (b) Configuration 2 with brake modulator for steering

to construct an initial simplified model of the two concepts. This analysis also highlights the control complexity of each configuration. Figure 8(a) shows the rigid body model using configuration 1. Figure 8(b) shows the rigid body model using configuration 2 with the brake modulator represented as a clutch that locks or releases the drive wheel. Both configurations gave acceptable performance. However, configuration 2 (i.e. utilizing a brake modulator to provide steering) requires an intelligent control strategy to control steering angle, whereas configuration 1 is a much simpler control problem.

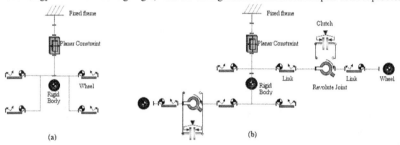

Figure 8. Design models created to analyze the bed movement. (a) Configuration 1 (b) Configuration 2, the clutch mimics the brake modulator. A planar force in each driven wheel mimics the steering and driving actuator in both configurations

The above design and analysis through domain-specific design models was necessary in order to take a decision about wheel-module configuration – a situation which is typical to most mechatronic system developments. This further emphasizes the need to integrate the system-model with these domain specific design models in order to ensure consistency while designing. Section-4 throws further light on this topic.

3.4. Dependencies between domain-models

Through the SysML model, the *design concept* for the propulsion system was specified. Further domain-specific design provided more information in order to make decision about sub-systems, components, and alternatives. The domain-specific models were created based on information obtained from the SysML model. For example, the CAD model in Figure 7(a) was made based on the initial sketches made during conceptual design phase, which led to creation of figure 4(b). The information about wheel, wheel-motor, steering motor and steering configuration was needed before the CAD model could be created; a dependency between the SysML model and the CAD model. Moreover, the dynamic analysis model in Figure 8(a) was created based on the propulsion system configuration, a dependency between CAD model, SysML model and the dynamic analysis model. Since the SysML model contains the wheel module requirements and the complete system structure, it is important to keep this model consistent with other design models. A change in e.g. the CAD model during design iteration should be traced back to the corresponding model element inside the SysML model, a problem of multi-view consistency as explained in [5]. The following section presents our proposed infrastructure for integrating models.

4 INTEGRATING MODELS FOR ABSTRACTIONS IN DESIGN

Several iterations may take place both between the design phases and between the corresponding design models before the design process is complete. Each model contains a certain level of detail. For example, SysML provides a rather abstract system view, whereas a CAD model provides a detailed view. Figure 9(a) shows a domain-independent system-model, and the corresponding domain-specific models that are typically created during design.

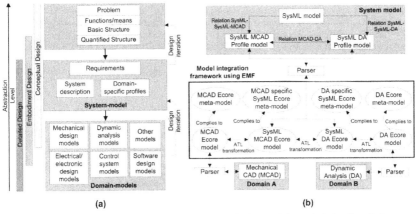

Figure 9. (a) Iterations between multi-domain design models, and abstraction during different design phases (b) Integrated modeling infrastructure based on EMF

Keeping the domain models consistent with each other requires means to manage inter-domain dependencies. A domain-model contains information that has relation to other domain-models (dependencies). It also contains information that is relevant only within the domain-model itself. We utilize a meta-modeling approach to characterize the information relations between domain-models. A meta-model specifies the abstract syntax of a modeling notation [17]. A meta-model for a domain specifies the concepts that exist within that domain [5]. Therefore, different domain-models comply with different meta-models. By defining relationships between these meta-models, it is possible to write transformations between models, which comply with those meta-models. We utilize the Eclipse Modeling Framework (EMF) [18] to define the meta-model of each domain, and we define relations between meta-models through a rule-based language (ATL) [19]. Figure 9(b) shows the integrated modeling infrastructure, in this case involving two domains: mechanical CAD, and dynamic analysis. Inside EMF, a model is represented as an Ecore model. Therefore, each domain is specified through a domain-meta model in Ecore, e.g. MCAD Ecore meta-model (Figure 9(b)).

Since SysML is a general purpose modeling language, the aim of using SysML is to create a system model independent of any domain or technology. However, our proposition to establish relationships between domain-models through SysML requires SysML to support domain-specific concepts. This has been supported in SysML through creation of profiles. A profile extends the SysML meta-model with the needed domain-specific constructs. Figure 9(b) shows two profile blocks, each relating to a domain-specific model. For example, the SysML MCAD profile model contains a model built with MCAD concepts such as: assembly, part, relations etc. An MCAD-specific SysML meta-model can then be created as an Ecore model. For dynamic analysis (DA) models, a DA-specific SysML meta-model can be created. Relationships between the SysML model and the domain-specific part of the SysML model are necessary to establish a link between the system design and the domain-specific design. This can be achieved by allocating (manually) elements/components in SysML model to their counter parts in domain-specific part of the SysML model. In this way, the relationships between different design models can be established (inter-domain dependencies) through the SysML model, by establishing allocation relationships between SysML model and the corresponding domain-specific parts of it. For example, in Figure 9(b), it is possible to establish relationships between the MCAD model and the dynamic analysis model (DA) by establishing relationships between the *SysML model* and the *SysML MCAD Profile Model,* and between *SysML model* and the *SysML DA Profile Model.* These relations can be manually drawn, or can be automated. An automation procedure requires relating a concept in one meta-model to a corresponding concept in the other meta-model and writing transformation rules based on those relations. Figure 9(b) shows an ATL transformation between each domain model and the domain-specific part of SysML model in EMF.

4.1. Integration Example

This section will illustrate an integration example between the MCAD model of the hospital bed wheel unit and the SysML system model. Figure 10(a) shows a generalized meta-model for MCAD, based on constructs such as a*ssembly*, *part*, *variable*, *relation* etc. There are differences between MCAD tools, which can lead to a different meta-model for each tool. However, we propose a generalized meta-model to be adapted for a domain such as MCAD, and aligning different modeling languages within the domain to that meta-model. An important consideration here is that the proposed MCAD meta-model does not contain all MCAD concepts; rather it only contains constructs relating to the type of information that we are interested to obtain from MCAD tools.

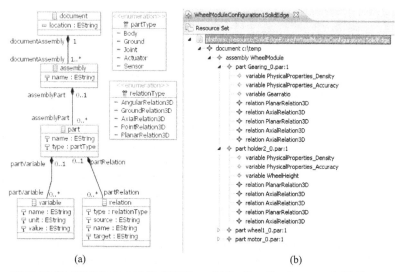

Figure 10. (a) MCAD meta-model in EMF Ecore (b) Configuration 1 represented in Ecore

A model complying with the MCAD meta-model can be created by making API calls to the MCAD tool of choice. We utilize Solid Edge [21] in this example, and populate a Solid Edge Ecore model (representing the MCAD Ecore model) through a developed parser (see Figure 9(b)). The same parser allows us to create a model inside Solid Edge based on an Ecore model. Hence, it is now possible to represent the wheel-module assembly configuration 1 as an Ecore model shown in Figure 10(b).

The SysML profile for MCAD is based on MCAD concepts, which are extended from the SysML meta-model elements. For example, a *Part* extends a SysML *block* etc. Figure 11(a) shows SysML4CAD profile meta-model represented in Ecore. Another parser is used that populates an Ecore model complying with SysML4CAD domain-specific meta-model. The same parser allows creating a *SysML MCAD Profile* model from an Ecore model (see Figure 9(b)). Figure 11(b) shows the wheel module configuration 1 as SysML4MCAD Ecore model.

Declarative ATL rules can be specified for each meta-model construct, to create a target model element based on the source model element. This completes a transformation between MCAD and SysML4CAD. In a similar fashion, transformations will be written between the DA model and SysML4DA model based on their corresponding meta-models as shown in Figure 9(b). At the end, the relationships between SysML4CAD profile model and SysML4DA profile model can be specified, to describe the dependencies between the two domain specific design models.

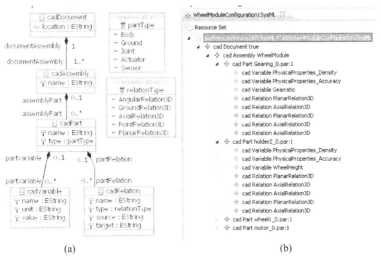

Figure 11. (a) SysML profile for CAD (SysML4CAD) in Ecore (b) Wheel module configuration 1 represented as an instance of SysML4CAD profile

4.2. Mechatronic procedure, domain models under SysML Umbrella

The extended system model allows us to first establish an overall overview of the mechatronic system, to specify the function principles, and to specify the principle solution through a common specification language. Through the model integration framework, domain-specific views are integrated, and dependencies and consistency between design models are maintained via the system model. We believe that establishing abstract system information through the system model and keeping other design models consistent with the system view provide good support for identifying inconsistencies among design models, and avoiding integration failures as a result. The integration infrastructure can in this way support development of better mechatronic design solutions.

5 CONCLUSION

In this paper, a proposal for mechatronic design infrastructure based on the integration of design models is presented. Mechatronic product development requires a common specification language for different domain experts to communicate with each other. It also requires design models that can support multi-domain constructs inside one modeling language to be able to model a mechatronic concept. Though some modeling languages support multi-domain modeling and analysis (such as Modelica), other design models are very domain-specific. Using SysML to establish a domain-independent system model, and establishing relationships and means for automated integration with other design models is the main theme of this paper. The paper presents a step by step construction of a SysML model and some domain-specific models for the design problem of a hospital bed's propulsion system. System level modeling can play a major role in mechatronic product development, thus it has to be supported through all development phases. The proposed integration infrastructure enables us to maintain both the system model and domain-specific design models throughout the product development process. A small integration example between an MCAD model and a SysML model is presented to exemplify our proposal. Future work targets extending the model-integration example towards a more comprehensive integration example between: SysML, MCAD, and dynamic analysis (DA), evaluating the support potential of this approach during mechatronic design phases.

ACKNOWLEDGEMENT

The authors are thankful to Carl-Johan Sjöstedt, Daniel Frede, Daniel Malmquist, Hamid Shahid, and Mohammad Khodabakhshian for their valuable inputs in performing the hospital-bed design-study.

REFERENCES

[1] Tomiyama T., D'Amelio V., Urbanic J. and ElMaraghy W., Complexity of Multi-Disciplinary Design. *Annals of the CIRP*, 2007, 56(1), pp 185-188
[2] Buur J. and Andreasen M. M., Design Models in Mechatronic Product Development. *Design Studies*, 1989: 10(3), pp 155-162
[3] Gausemeier J., Schäfer W., Greenyer J., Kahl S., Pook S. Management of Cross-Doamin Model Consistency During the Development of Advanced Mechatronic Systems. *Proc. Internationl Conference on Engineering Design, ICED'09,* Stanford, California, USA, 2009
[4] Johnson T. A., Paredis C. J. J. and Burkhart R. Integrating Models and Simulations of Continuous Dynamics into SysML. *Proc. 2008 Modelica Conference*, Germany, March 2008.
[5] Shah A. A., Kerzhner A. A., Shaefer D. and Paredis C. J. J. Multi-View Modeling to Support Embedded Systems Engineering in SysML. *Lecture Notes in Computer Science, Graph Transformations and Model-Driven Engineering*, 2010, Volume 5765/2010, pp 580-601
[6] Hehenbrger P., Egyed A. and Zeman K. Consistency Checking of Mechatronic Design Models. *Proc. ASME 2010 International Design Engineering Technical Conferences & Computers and Information in Engineering Conference, IDETC/CIE 2010*, Montreal, Canada, 2010
[7] Estefan A. J. *Survey of Model-Based Systems Engineering Methodologies,* Technical Report, Revision B, Incose Focus Group, 2008
[8] Pahl G., Beitz W., Feldhusen J. And Grote K., *Engineering Design- A Systematic Approach*, 2007, (Springer-Verlag), ISBN 3-540-19917-9
[9] Frey E., Ostrosi E., Roucoules L. and Gomes S. Multi-Domain Product Modelling: From Requirements to CAD and Simulation Tools. *Proc. Internationl Conference on Engineering Design, ICED'09,* Stanford, California, USA, 2009
[10] Association of German Engineers, VDI-guideline 2206, Design Methodology for Mechatronic Systems, Berlin, 2004
[11] Modelica Association, *Modelica Language Specification V3.2*, 2010, https://www.modelica.org/documents/ModelicaSpec32.pdf
[12] Maplesoft, *MapleSim V4.5,* http://www.maplesoft.com/products/maplesim/
[13] Object Management Group, *OMG System Modeling Language Specification V1.2*, 2010, http://www.omg.org/spec/SysML/1.2/PDF/
[14] Friedenthal S., Moore A. and Steiner R. *A Practical Guide to SysML- The Systems Modeling Language*, 2008, (MK/OMG Press), ISBN 978-0-12-374379-4
[15] Wölkl S. and Shea K. A Computational Product Model for Conceptual Design Using SysML. *Proc. ASME 2009 International Design Engineering Technical Conferences & Computers and Information in Engineering Conference, IDETC/CIE 2009*, California, USA, 2009
[16] Follmer M., Hehenberger P., Punz S. and Zeman K. Using SysML in the Product Development Process of Mechatronic Systems. *Proc. International Design Conference, Design2010*, Dubrovnik, Croatia, 2010
[17] Czarnecki K. and Helsen S. Feature-Based Survey of Model Transformation Approaches. *IBM Systems Journal*, 2006, Vol. 45, No. 3, pp. 621-645
[18] Eclipse Foundation, *Eclipse Modeling Framework (EMF)*, 2009. http://www.eclipse.org/modeling/emf/
[19] Eclipse Foundation, *Atlas Transformation language (ATL)*, 2009, http://www.eclipse.org/m2m/atl/
[20] NoMagic, *MagicDraw UML/SysML V16.9*, http://www.magicdraw.com/
[21] Siemens PLM Software, SolidEdge, 2010, http:// www.plm.automation.siemens.com/en_us/products/velocity/solidedge/index.shtml

Contact: Ahsan Qamar
PhD. Candidate
KTH- Royal Institute of Technology,
Division of Mechatronics, Department of Machine Design,
Brinellvägen 83, 10044, Stockholm, Sweden
Email: ahsanq@kth.se, Web: www.md.kth.se/~ahsanq

INTERNATIONAL CONFERENCE ON ENGINEERING DESIGN, ICED11
15 - 18 AUGUST 2011, TECHNICAL UNIVERSITY OF DENMARK

AN APPROACH FOR MORE EFFICIENT VARIANT DESIGN PROCESSES

Sebastian Schubert[1], Arun Nagarajah[1], Jörg Feldhusen[1]
(1) Chair and Institute for Engineering Design, RWTH Aachen University, Germany

ABSTRACT

Today, as a result of a steady increasing pressure to reduce costs for being competitive in the automotive supply industry, the majority of the products are designed by adaption of already existing products to new requirements. In the embodiment design phase, engineers take CAD models as the design base and adapt those to the new requirements. Due to the fact that the same models are used in different generations of product variants, the models are getting more and more complex and unstructured. Thereby the effort for adaptation increases. By using parametric design this effort can be reduced significantly.

In addition to the optimization of the embodiment design process, further capability for cost reduction can be found in the design process. Mandatory analyses, like the FMEA, have to be redone completely within every variant, although the main product remains similar. In order to reduce the effort, the FMEA has to be standardized.

In the following, the presented approach shows how the approximation of parametric models in the embodiment design phase can be enhanced by using function structures and parameters derived from the FMEA execution in order to shorten the time needed for the variant design process.

Keywords: Variant Design Process, FMEA, Embodiment design, CAD

1 INTRODUCTION

In times of globalization, the strategic aim of every supplier has to be the cost leadership without neglecting the product quality and the adherence to delivery dates [1]. In domains, such as body and exterior, economic growth is not expected in the automotive supply industry [2]. Due to the worldwide economic crisis, Original Equipment Manufacturers (OEM) and suppliers are facing a further increasing pressure on development costs. The OEMs are able to deal with this problem by shifting the development effort to the suppliers. [3] Therefore, the suppliers have to improve their development processes in order to reduce the development costs.

In the mentioned domains, the products designed by one supplier are fulfilling the same main function, e.g. "Keep the bonnet closed" for a bonnet locking system. The applied physical principles realizing the functions remain similar as well. Just the geometric layout of functional elements and the shape of those are different. Therefore, in the automotive supply industry, the products are designed by variant design [4]. From projects in this field, it is known that the variant design process is based on adaptation of already existing product variants. Normally the engineer selects the product variant with which he is most familiar; therefore, the most recent product is the common choice as initial point of engineering design. As the engineer uses this design model repeatedly, the CAD models are getting more complex with every new product variant development. The part's features which are getting dispensable during the advancement of the design process often remain in the CAD model. These features still exist in later product generations. As a result, the manufacturing process of the parts is more expensive. The increasing number of elements in sketches complicates the handling of the models, leading to longer development times.

The first aim is to make the handling of the CAD models easier and therefore to reduce the time spent for development. In the 80s, in the Automotive Industry the customer orientation increased. In order to detect problems early in the design process and to take the right measures, the Failure Mode and Effects Analysis (FMEA) was introduced [5]. Today, the FMEA execution accompanies the whole design process [6]. For every new product variant, the FMEA has to be redone completely, even though the product itself basically remains the same.

The second aim is to decrease the effort spent for the execution of the FMEA in order to reduce the expenses in the design process.
This paper describes how the time spend in the embodiment design phase of a product variant can be reduced by using parametric and direct modeling techniques in the CAD system. Further, it is shown how the design process is improved by introducing standardized system elements in the FMEA as well as reusing data out of the FMEA in the embodiment design phase. An introduction into the application of FMEA in automotive industry is given first.

2 BASICS

2.1 Bonnet locking systems

As mentioned before, the approach presented in the following is applicable for products developed by variant design. In chapter 4, the bonnet locking system is used as a case study for presenting the application of this approach. In chapter 2 and 3, examples are shown using the bonnet locking system for explanation purposes. The bonnet locking system consists of the lower part shown in Figure 1 and a striker which engages in the catch bolt. Bonnet locking systems are fulfilling the function of keeping the bonnet in a secure position. The function is always fulfilled with the identical mechanical solution principle: The assembly's core consists of a ratchet brace and a catch bolt mounted in a housing (Figure 1).

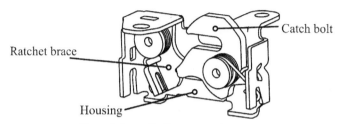

Figure 1. Bonnet locking system, lower part

2.2 Execution of System-FMEA-Product

The System-FMEA-Product is applied throughout the whole design process. As a preventive reliability method, possible system failures can already be detected in the design process. In this context, the system is defined as each technical entity, which can be divided into system elements. [7] According to the guideline provided by the German Association of the Automotive Industry (VDA), the System-FMEA-Product is executed in the following five steps:
1. System Elements and system structure
2. Functions and Function structure
3. Failure Analysis
4. Risk Assessment
5. Optimization. [6]

Due to convenience in the following the term "System-FMEA-Product" is set equal to "FMEA".

2.2.1 System Elements and system structure

In step 1 of the FMEA procedure the interfaces of the system are determined before the structure of the whole system is worked out. The top element represents the considered system. On the lower levels the system elements can either be sub-assemblies or parts. The number of levels for setting up this system structure is arbitrary. [6]

2.2.2 Functions and Function structure

Functions describe the relationship between the inputs and the outputs of the system [8]. In the next step, the functions of all system elements are determined beginning with the complete system. The system implements the main function; the functions realized by the sub-assemblies contribute to the main function (Figure 2) [7]. Taking the functions into account exclusively, a hierarchical function structure is derived. Considering the collocation to the system elements as well, a product architecture is set up. The product architecture is the scheme by which the functions of the product are collocated to physical elements [9], e.g. sub-assemblies, parts or features.

2.2.3 Failure analysis

Based on the functions, a failure analysis is carried out for each system element. By doing this failure analysis all failure functions are detected, which lead to an insufficient fulfillment of the associated function and the superior function as corollary (Figure 2). For abstract functions such as the top function, failure functions can be determined by negation of the function itself. Failure functions of components are defined as physical failure modes such as fracture, wear-out, jams, clamps [7], and insufficient manufacturing.

In the majority of cases the superior failure function is caused by a failure of subordinated system element. Therefore the failure analysis has to be carried out until the lowest system element is reached. Insufficient manufactured parts lead to failure of a function or an insufficient fulfillment of the selfsame function. Therefore, a systematic search of failure functions on part level is carried out by checking every relevant geometric parameter of the examined part.

Geometric parameters in this context can be either thicknesses or angles but also distances between surfaces (chapter 2.6). For example, a failure of a geometric parameter can be a wrong wire diameter of a spring leading to a spring stiffness being too strong which causes a malfunction of the whole system. The FMEA requires a systematic analysis of the product. By doing this analysis, a complete list of geometric parameters is derived. These geometric parameters are reused in the embodiment design phase.

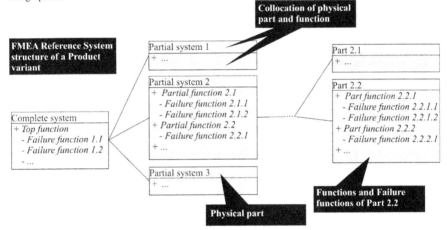

Figure 2. FMEA of product

2.2.4 Risk assessment and Optimization

After the definition of the failure functions the risk assessment is executed. For every failure function severity rating, occurrence rating and detection rating are determined and the risk priority number (RPN) is calculated by multiplication. Based on the RPN, a decision is made whether an optimization is necessary. [5]

2.3 Standardization of FMEA

In variant design the effort necessary for execution of the FMEA is easily reducible by standardization. Due to a similar system structure using the same physical principles fulfilling the same functions, the structure of the risk assessment remains constant among the product variants. Based on already existing product variants, standardized elements for each single system element are introduced in the FMEA. Each system element contains a reference set of functions, failure functions. The reference set concludes all functions and failure functions being realized in least one product variant.

In case of a new product variant development it is checked which functions are part of the new variant and which ones are omitted, based on the reference variant of the FMEA (Figure 3). If a new function has to be implemented, it is added to the reference set.

When implementing a new function in the product variant, it has to be checked, which system elements must be changed for this new function. For these system elements steps 1-5 have to be redone. The other system elements only need step 4 and 5 to be redone, leading to a significant effort reduction in the FMEA execution. In this manner it is reasonable to implement standardized elements early.

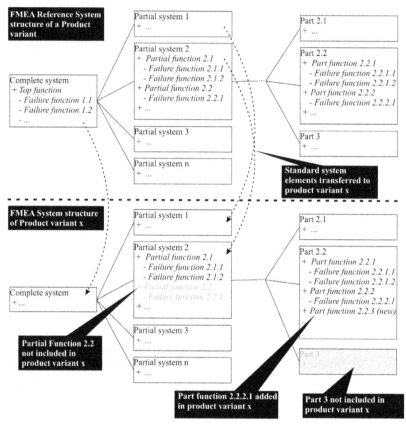

Figure 3. FMEA in variant product design

2.4 Embodiment design

2.4.1 Feature

Features are physical elements of a part fulfilling part functions, e.g. the upper arm or the boring. While the layout of the assembly is identical, the layout of the parts features might vary, such as the position of the stop in the opened state (Figure 4).

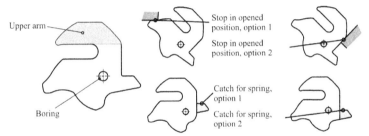

Figure 4. Examples for features of the catch bolt

2.4.2 Geometric Elements and Skeleton model

Geometric Elements are planes, axes, points, or coordinate systems in the CAD model. A skeleton model is a part-file which contains only geometric elements. Each geometric element represents an item of the product. For example planes are representing surfaces and axes are representing axes of rotation (Figure 5). The positions of those geometric elements are defined by parameters.

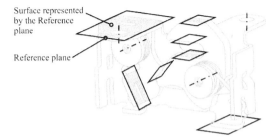

Figure 5. Skeleton model (The product model is not part of the skeleton model.)

2.4.3 Parameter

Geometric parameters are user-defined variables to adjust the CAD model. Those variables represent dimensions like a length or an angle. The embodiment design process is normally initiated by defining the main geometrical sizes of the product, either directly given by the specifications or determined by the engineer. The aim is to define the dimensions in the product model only once. When the dimensions are modified, the model is adjusted according to these parameters. Examples for dimensions being parameters are shown in Figure 6.

The different layouts of the features (Figure 4) are also determinable by parameters. In that case the parameter to be determined is "Stop of catch bolt in opened position" and the settable value is either "option 1" or "option 2".

While an adaptation of a geometric parameter causes no new failure analysis during the FMEA execution (chapter 2.2, steps 1 to 3), a new layout of the features requires the system structure, the functions, and the failure functions to be checked. Risk assessment and the optimization have to be revised anyway.

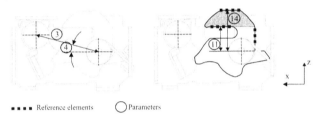

Figure 6. Extract of Parameters

3 APPROACH

In the Automotive Industry, the effort spend for developing products by variant design can be significantly reduced without reducing the quality of the product, if a continuous use of product data is established. Due to a standardization of the FMEA, a reference function structure and a reference set of geometric parameters exist. In case a reference function structure or reference set of geometric parameters have to be established, a similarity analysis has to be carried out. The establishment is done similar to the creation of a reference product structure [10].

In case of a new order, the system FMEA is executed. Based on the reference set of functions, those functions are selected which need to be realized in this particular variant. The functions are determined by the engineer based on the requirements list provided by the OEM. An approach to how the engineer can be supported in the selection of the required functions is worked out in the PONNGA project [11].

The solution presented here is based on three layers (Figure 7). The first layer contains the function structure derived out of the FMEA. Based on the selection of the functions, parameters and geometric elements related to these functions are activated or deactivated. Therefore, in the skeleton model only those geometric elements are provided which are needed to design this particular product variant, eliminating features related to the deactivated geometric elements. This clearly arranges the CAD model and reduces the complexity of the product.

Layer 2 represents the skeleton model including the geometric elements and the parameters. Each parameter is either collocated to one dimension determining the position of a geometric element or it manipulates a part directly, e.g. the thickness definition of a sheet metal part.

The features and parts situated in layer 3 are basically defined by the parameters and geometric elements of the skeleton model. The geometric elements not being determined by the parameters are modeled directly in the CAD part by adding further elements, such as fillets and rounds.

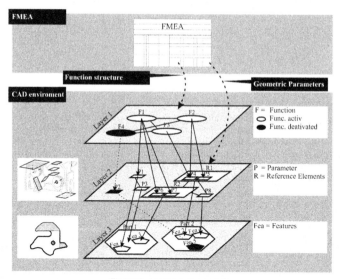

Figure 7. Layer representation

3.1 Setting up the approach

The first step to realize this approach is to set up a product architecture, consisting of the function structure and the feature structure. The product architecture is elaborated based on all existing product variants. By going through all existing product variants, each part is examined for realized features. Each feature being realized in at least one product version is collocated to the part function out of the reference system structure of the FMEA. This collocation of function to feature represents this product architecture (Figure 8). Out of the reference system structure, the set of geometric parameters is taken and collocated to the features which are determined before.

In the application of the approach, the required partial functions are selected; the no longer required partial functions are deselected (layer 1). Automatically, the related geometric elements and parameters are deactivated (layer 2), disabling the embodiment design of the related features (layer 3).

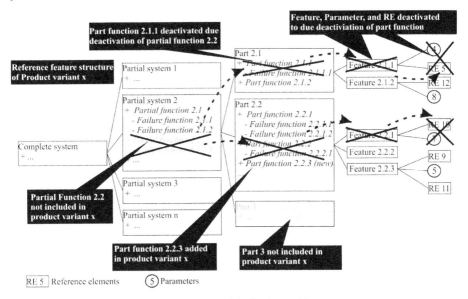

Figure 8. Establishment of the Product architecture

3.2 Part modeling

In the CAD-system, the parts are modeled using the selected geometric elements. All geometric elements belonging to one particular part are transferred from the skeleton model to a separate part file. The basic geometry is designed using the geometric elements. The final shape of the part is created by using the direct modeling techniques (Figure 9).

The linkage between the geometric elements in the skeleton model and the part-files still exists. When an update of the assembly's geometry is necessary, the update will be made in the skeleton model. The part files are automatically adjusted.

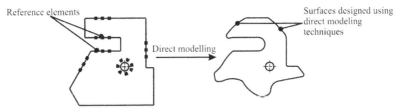

Figure 9. Part modeling

4 CASE STUDY

In the Automotive industry, OEMs order a new system by providing the supplier specifications containing requirements. Normally the specifications are copied from a previous version and edited afterwards. Basically the requirements remain mostly similar. An example is shown in Figure 10.
Based on the provided specifications the development is initiated. The system FMEA is executed continuously in the design phase. Due to the similar requirements, the System structure and the

functions of each system element remain similar, too. In order to reduce the effort for the FMEA, standardized system elements are elaborated containing a reference set of functions and failure functions. The engineer is able to decide which functions are needed to fulfill the requirements of the new order. All unnecessary functions are omitted; new functions have to be added. (Figure 10) An example is shown in Figure 11. Certain bonnet locking systems require a high stiffness. The engineer is able to determine whether the housing requires additional elements to achieve the higher stiffness (Figure 11). If the regarding of the stiffness is not required, the related function of the housing is deactivated.

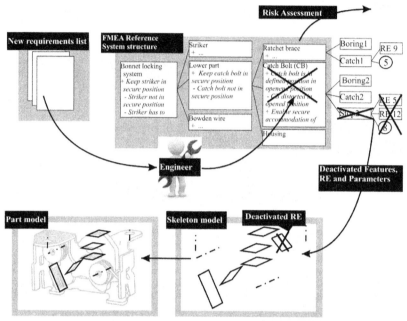

Figure 10. Application of the approach

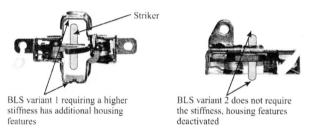

BLS variant 1 requiring a higher stiffness has additional housing features

BLS variant 2 does not require the stiffness, housing features deactivated

Figure 11. Bonnet locking system variants with different requirements

The set of functions is exported into an MS Excel-file. The data base contains the same system element as the FMEA software. All not required functions are deselected, setting the related parameters and REs as "deactivated". (Figure 12) Thus the initializing parameters for the skeleton model are provided, the start of the part modeling is enabled. The Excel-file directly controls the skeleton model. CATIA provides an interface for Excel-files.

System Element	Function	Subordinated SE	Features	Parameter	Geometric Element	Activated?
Bonnet locking system		Striker; Lower part; bowden wire ...				true
	Keep striker in secure position					true
Striker						true
Lower part		Ratchet Brace; Catch bolt; housing, Catch bolt				true
	Keep catch bolt in secure position					
Catch bolt						
	Catch bolt is pivotable		Boring2	12		true
				13		true
				14		true
	CB is defined position in the opened state		Stop2		5	false
					12	false
				8		false
				15		false
				16		false

Figure 12. Excerpt of the data base

In the embodiment design phase in CATIA, the engineer creates a new set of files including an assembly file, a skeleton model, and the required part files. All parameters of the active geometric elements can be set in the skeleton model. Having set these main boundary conditions, the part design can be started.

In CATIA, geometric elements and parameters are transferred to the part-files in three steps: publishing, copying, and pasting. The publishing enables the elements to be copied from the skeleton model and to be pasted into the part-files. (Figure 13)

Figure 13. Publishing Reference Elements to enable part design

The advantage of this procedure is the remaining linkage between the geometric elements in the skeleton model and the part-files. Adaptations that have an impact on the assembly are made in the skeleton model. All part-files related to this change are updated, giving the CAD model an unambiguous state.

As a result, the model just contains the features being needed. Complex models arising by reason of using the same CAD model over generations of product variants are avoided. The time needed for adapting the whole model is significantly reduced.

5 SUMMARY AND OUTLOOK

In this paper, an approach is presented how the design process for products designed by variant design can be improved. Data elaborated by executing the FMEA is reused in the embodiment design phase, in order to reduce time need. Further, the costs are reduced and the quality of the product is improved.

In the Automotive industry the execution of an FMEA is mandatory for all companies developing and manufacturing products. Today in supply industry the FMEA is completely redone, even though most of the products remain similar in functions and physical effects, changing only the geometrical dimensions. In order to reduce the effort spend for the FMEA, standardized elements are elaborated for the system items, which are reused. Doing this, three of five steps during the execution of the FMEA are avoided.

As a part of the FMEA, a function structure and the geometric parameters of the parts are determined. Both are reused in the embodiment design phase. The function structure is used to select features of the product to be developed. Doing this selection a skeleton model, used as a starting model, is adjusted. Only those required geometric elements are provided which are necessary to develop the required features of this model. The mentioned geometric elements are manipulated using the geometrical parameters. The use of the geometric elements supports the engineer, because they provide the main constraints for the embodiment design. Necessary changes having an impact on more than one part are done by adjustment of parameters in the skeleton model. The part models are adjusted automatically. Applying this approach, the time needed for the development of a new variant is significantly reduced without reducing the quality of the product. Due to this the costs are reduced.

This approach is applicable for mechanical parts like the shown bonnet locking system. It has to be checked whether this approach is also appropriate for systems having electrical components. Further, the use of two separate software systems, one for the FMEA, one for the control of the skeleton model has to be improved, either by using just one system or by the establishment of a standardized interface between the software systems.

REFERENCES

[1] Maggioni, S., Thiele, R., Rivard, S. and Turner, H. *Leadership in the Automotive Industry*, 2006 (Spencer Stuart, http://content.spencerstuart.com/sswebsite/pdf/lib/Automotive_Study_March_06.pdf)

[2] Bernhardt, W., Dressler, N. and Tóth, A. *Mastering Engineering Service Outsourcing in the automotive industry*, 2010 (Roland Berger Consultants, Munich)

[3] Wallentowitz H., Freialdenhoven, A. and Olschewski, I. *Strategien in der Automobilindustrie*, 2009 (Vieweg + Teubner, Wiesbaden)

[4] Franke, H.-J. *Variantenmanagement in der Einzel- und Kleinserienfertigung*, 2002 (Hanser, Munich)

[5] Tietjen, T., Müller, D. *FMEA-Praxis*, 2003 (Hanser, Munich)

[6] VDA. *Sicherung der Qualität vor Serieneinsatz, Vol. 2*, 1996 (Henrich GmbH, Frankfurt)

[7] Bertsche, B. *Reliability in Automotive and Mechanical Engineering*, 2008 (Springer, Berlin)

[8] Pahl, G., Beitz, W., Feldhusen, J., Grote, K.-H., Wallace, K. and Blessing, L. T. *Engineering design – A systematic approach*, 2007 (Springer, London)

[9] Ulrich, K. T. and Eppinger, S. D. *Product design and development*, 2008 (McGraw-Hill/Irwin, Boston)

[10] Nurcahya, E. *Configuration instead of New Design using Reference Product structures, CIRP Design Conference,* Berlin, March 2007, pp. 1-10 (Springer, Berlin)

[11] Feldhusen, J., Nagarajah, A. and Schubert, S. *A Data Mining Method for selecting the suitable existing product variant as a development base for a new order, DESIGN 2010,* Vol.2, Dubrovnik, May 2010, pp.895-903 (Faculty of Mechanical Engineering and Naval Architecture, University of Zagreb, Croatia)

Contact:
Dipl.-Ing. Sebastian Schubert
Email: schubert@ikt.rwth-aachen.de
Prof. Dr.-Ing. Jörg Feldhusen
Email: feldhusen@ikt.rwth-aachen.de

RWTH Aachen University
Chair and Institute for Engineering Design
Steinbachstrasse 54B
52074 Aachen
Germany
Tel: +49 241 80 27341
Fax: +49 241 80 22286
www.ikt.rwth-aachen.de

Short CV of the presenting author:
Sebastian Schubert is a full time research and teaching assistant at the Chair and Institute for Engineering Design, RWTH Aachen University since May 2008. The main themes of his research include variant design processes, engineering design and methods. He obtained a diploma degree in mechanical engineering from RWTH Aachen University, Germany in 2008.

INTERNATIONAL CONFERENCE ON ENGINEERING DESIGN, ICED11
15 - 18 AUGUST 2011, TECHNICAL UNIVERSITY OF DENMARK

THE PROCESS OF OPTIMIZING MECHANICAL SOUND QUALITY IN PRODUCT DESIGN

Nielsen, Thomas Holst; Eriksen, Kaare Riise

ABSTRACT

The research field concerning optimizing product sound quality is a relatively unexplored area, and may become difficult for designers to operate in. In some degree, sound is a highly subjective parameter, which is normally targeted sound specialists, if the sound has significant quality meaning to its context of use. This paper describes the theoretical and practical background of managing a process of optimizing the mechanical sound quality in a product design by using simple tools and workshops systematically. The procedure is illustrated by exploring a case study regarding a computer navigation tool (computer mouse or mouse). The process is divided into 4 phases, which clarify the importance of product sound, defining perceptive demands identified by users, and, finally, how to suggest mechanical principles for modification of an existing sound design. The optimized mechanical sound design is followed by tests on users of the product in its use context. The result of this article is a tangible, systematic process, which has the possibility of enhancing the knowledge about sound design in products and its cause and effect.

Keywords: Sound quality, process, mechanical sound.

1 INTRODUCTION

This article is a pilot study regarding the development process of a product design, with focus on how mechanical produced action sounds can be optimized in quality by focusing on users' subjective perceptions. The research project, called MechanicalSound+, is a starting point for further exploration and discussion within this research field. MechanicalSound+ tests the hypothesis regarding; "with use of a systematic approach, it is possible to optimize the mechanical sound quality of existing products by focusing on the sound". The theoretical and practical research is conducted via a case study and is targeted one specific product and demographical target group. Further research and exploration will verify the validity of the process in practical use in design practice.

In the area of product sound, the terminology is defined into four categories by Bernsen [1]:
1. Operational sounds
2. Action sounds
3. Signal sounds
4. Passive sounds

The field within designed product sound is a relatively new design research area. Studies have mainly been concerning re-produced operational sound in digital form, with use of trained listening panels in a controlled environment [2] [3] [4]. In these research projects, sound characteristics (also known as attributes) are used to define and rate the subjective perception of the operational sound in relation to objective, measurable characteristics. None of these projects deal with un-trained listening panels, and how to provide tangible guidance for product designers who have interest in optimizing the mechanical sound quality.

The methodology draws inspiration from DELTA's Filter model [5]. DELTA is a part of the Danish research and technology organization named GTS/ATG (Advanced Technology Group), and consults private and public businesses both locally and internationally. The Filter model describes the relationship between objective and subjective measurements when using a listening panel as a tool of verification. In the Filter model, objective measurements are conducted in the beginning of the process.

In MechanicalSound+, objective measurements are delimitated due to the purpose of conducting fast research using subjective parameters for systematic product sound optimization. Therefore, the focus is put upon the process of facilitating workshops with a tangible outcome for Product Development Departments (PDD). The outcome is perceptual demands identified by users of the product using a user-oriented methodology approach.

1.1. Market opportunities

Many product designers are only focusing on feeling and sight when developing products, and marketing strategists define it as 2D-branding [6]. By focusing on more senses, it is possible to create a differential advantage with increased sales as outcome [6]. This tendency of 5D-branding (branding for all senses) is a combination of service and product design. However, the 5D-branding aspect is not exclusively implemented into the product design alone, but it is normally a part of a system design solution [6]. Manufactures and developers can suffer economical consequences by ignoring the importance of the sound [7]. For instance, the Danish actuator manufacture, LINAK A/S, suffered mayor economical consequences due to noise in their hospital elevation products [7].

As a part of the screening process for MechanicalSound+, a survey [8] was conducted among 120 practicing Industrial Designers in Denmark. The response rate was poor (25 responded), and could be characterized by a lack of interest in sound design. Therefore, the survey was seen as a starting point containing inspiring data. The data indicated a gap between how designers work with sound in product designs, and in what degree they felt the sound parameter can be optimized in practice. 15 out of 25 considered sound design as a competing parameter within their business. 73 percent out of the 15 respondents asked felt that they "in some degree" or "in a strong degree" could optimize the sound in their product designs [8].

1.2. The designer's environment

Within product development, the designer has significant power to focus the product design towards a specific goal. It is commonly known that in a structured development process, a Design Brief or list of demands controls this process [9], and can be seen as a contract between PDD and client. It prevents the designer from creating solutions which become irrelevant and not beneficial for the client. Typically, the designer's role is to visualize a better solution than the existing situation, plan it, execute it, and, in some cases, facilitate the final stages of implementation [10].

MechanicalSound+ conducted research via a four-phased process, which is described in the following sections:
- Phase 1: Identifying the importance of sound
- Phase 2: Defining attributes
- Phase 3: Rating attributes
- Phase 4: Creating mechanical modifications and tests

2 MANAGING THE PROCESS

The process of sound quality optimization proclaims demands to the product developer regarding basic theoretical knowledge about sound and product development. MechanicalSound+'s outcome was a handbook with guidance for product developers to facilitate and execute workshops resulting in improved knowledge about optimization of mechanical action sounds in a specific product targeted a specific demographical user group. It is meant to be an inspirational platform and tangible toolbox for freely use and modification. However, it is not seen as a rigid tool, but more as an ongoing iterative tool. A case study concerning the optimization of a computer mouse's action sound was used throughout MechanicalSound+ to conduct results for later data analysis and discussion.

2.1. Identifying the importance of sound

The starting point of the research was identifying target users, its context of use, and hereby the demographical importance of sound in the specific product. Numerous methodologies are able to gather this information, and fast situated interviews were proven being a beneficial tool. As Lyon states [11], it was found crucial to rate the importance of the sound parameter in the beginning of a project due to decisions regarding design and construction all have influence on the sound design [11]. Furthermore, analysis of the product structure in relation to its surrounding components was preferable to consider at this stage of the process. The mechanical sound can be caused by isolated settings of components (named generator) [11], or several combinations and relations of components.

It concluded a tendency of four users in average rated the product sound of their mouse to 3.75 on a scale from 1 to 5. The mechanical sound was important but not essential because other design parameters were rated higher. Due to the purpose of MechanicalSound+, the aim was to differentiate the mouse by its mechanical sound. The aim of the sound parameter was hereby rated 5 out of 5, which can be seen in figure 1.

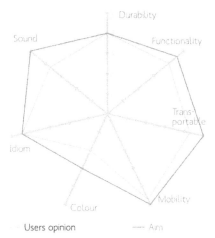

Figure 1. The rating of design parameters in MechanicalSound+.

2.2. Defining attributes
Designers can focus their development by defining various demands to the product (or product specifications) [9]. As a result, it is possible to verify the outcome (product) before final implementation to the market. A user-driven approach was used in MechanicalSound+ to define attributes describing the perceived sound characteristics. Afterwards, the attributes were rated and used as demands for subsequent mechanical modifications. The attributes are conducted in a five-people workshop situated in the product's context of use.

It has been proven beneficial to use a setup of tools prepared by the facilitator before executing the workshop. It was mainly writing tools, post-its, A1 paper map, and most important of all, products with diversity in mechanical action sound. In this experiment, it was positive that the facilitator was a part of the PDD for further documentation, development, and final tests.

With instruction from the facilitator, the workshop began. Here, it became important to constantly notice the dynamics of the listening panel, and observe how well users group and categorize the products. Non-artistic users may have difficulty in describing subjective parameters due to its intangible characteristics. Video material and pictures might become beneficial for later data analysis and documentation.

Figure 2. Users grouping computer mice in accordance to its sound characteristics (attributes).

The workshop resulted in four groups of attributes describing sound characteristics. The first group was named "sharp, fast, and stress" which was evaluated in direct contrast to the other defined attribute "lazy". Another grouping was named "round and smooth", which also has an opposite attribute named "hollow". The attributes "lazy" and "hollow" were eliminated due to consensus of opinion in the listening panel of its direct contrast. Afterwards, the listening panel rated the products on an undefined scale exclusively in relation to the sound attributes. The results were used in selecting products with highly diversity in sound perception for the following workshop.

The attributes named "sharp, fast, round, and smooth" are all perceptual descriptors of sounds in accordance to DELTA's word classifications [5]. "Stress" was the only attribute whose characteristic was described as affective responses to sound [5]. Many attributes are localized in the same word classes, but cannot be generalized into one group of sound descriptors in this type of product.

2.3. Rating of attributes

The following workshop had the purpose of rating the attributes defined with use of individual blindfolded listening tests exclusively focusing on the sound. Before conducting the results, the users needed to be validated as a useful human measurement tool. An individual blindfolded test needed to be carried out to eliminate all senses besides the ability of hearing. After validation, a scaling of the most accepted value, named X+, was located by the users on a similar undefined scale. The rating of the most accepted value was executed with the sound design of five other products as reference. This rating indicates a trend of what this specific user group accepts the most.

Each user rated the two attribute categories; "sharp, fast and stress" and "round and smooth" blindfolded but with the ability to;
1. Generate the stimuli (sound).
2. Scale the attribute value for each product.
3. Scale the most accepted value X+.

The two tests (each attribute category) were executed twice in order to validate the users' responses. When facilitating the second round, the facilitator reorganized the position of the products. In figure 3, all users' average values are represented (the ratings of both tests divided by 2) including the most accepted value X+ in the last column.

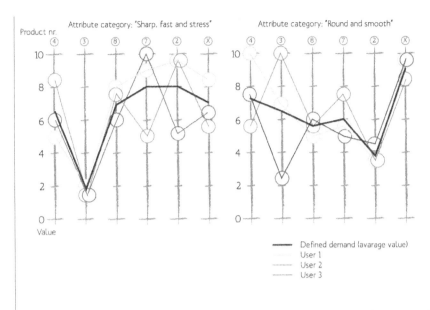

Figure 3. The ratings of 2 attribute categories by 3 users. The answers by users are the average value conducted from 2 similar tests. The thick purple lines indicate the overall average value from all users participating in the test.

Three out of four users did not have major variations in rating the attributes categories. One user had difficulty in making consistent answers and was excluded from further data analysis. The rating was segmented via photos and going from the lowest value of 0 to a maximum scale of 10. Here, the average value form first and second round of tests was put into a scheme containing all user responses. The rating of the most accepted value indicated an average of 7 in the attribute category "sharp, fast and stress" and 9 in "round and smooth".

2.4. Creating mechanical modifications and tests

The two defined attribute categories with ratings were used as guidelines for targeting improved sound quality in the product design. In accordance with Lyon [2], the developer needs to identify the source of the sound and localize the mapping between design and perceptive parameters before making sound design modifications. This was carried out by analyzing the product, its components, and the relationship in between. Lyon [11] categorizes the passage of sound throughout a product into three groups:

- Generator (source of noise)
- Transmission path
- Surface reflection

Several products were detached and analyzed regarding the localization of the sound source. In this type of product, a micro switch was identified as the generator of the sound and of most importance at all. Many different types of micro switches were identified in other computer mice with high diversity in sound characteristic. The product volume, internal wall placement, tightening of overall assemblies/joints, and assembly of the PCB-board with the bottom part of the mouse all had influence in the transmission path. By decreasing the volume with fabrics, it was possible to create a sharper sound. Dampers in between the PCB-board and the bottom part reduced the sharpness of the sound, and lowered the sound level together with sound absorbing material inside the mouse. Surfaces decorated with absorbing material had minor effect on the sound. All relationships found can be seen in figure 4.

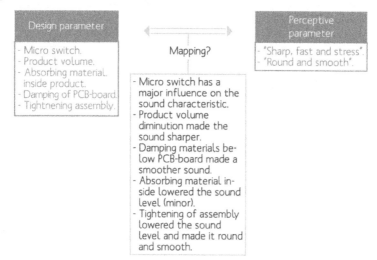

Figure 4. The mapping of design and perceptive parameters of computer mice, and what kind of influence these parameters have on the sound design.

Five similar products were modified, and the sound designs of two products were directly aimed at the users' identified ratings of the two attribute categories. The only differences in these two products were the loudness of the action sound. The remaining products were targeted a higher value in "sharp, fast, and stress", and another product a higher value in "round and smooth". The remaining mouse was not modified at all, and all modifications can be seen in figure 5.

Figure 5. The sound designs of five similar products were modified. A: filled with material reducing the product's volume. B: tightening the overall assembly using tape in all joints. C: replaced the micro switch. D: replaced the micro switch, dampener below the PCB-board and reduced the product's volume. E: no modifications.

The modified products were tested by 82 design students, and the results can be seen in figure 6:

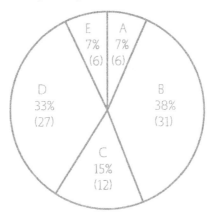

Figure 6. Test of five different sound designs with same product base (identical products from the beginning). 82 design students participated, and evaluated which product sound was the most acceptable in the given context of use (group rooms at Aalborg University).

As seen in figure 6, product B and D are evaluated most acceptable by the users. With knowledge from the conducted tests, it would be preferable to adjust the following design parameters in order to optimize the mechanical sound design in the computer mouse used in the test:

- Use the existing micro switch with which the product is manufactured.
- Tightening the overall assembly and joints using thermoplastic elastomeric or other rubber materials.
- Integrate dampers between the PCB-board and the bottom part of the mouse.
- Coating the inside of the top part with absorbing material.

These principles need to be tested using a prototype in further research for final validation.

3 CONCLUSIONS

A four-phased model for mechanical sound quality optimization has been explored throughout a case study concerning the action sound of a computer mouse. The process included a user-oriented methodology approach by including users identifying and rating the defined attributes categories followed by evaluation of the final sound design by design students. It can be concluded that it is seen possible for developers to optimize the existing sound design in a computer mouse by executing workshops systematically throughout the product development process. The sound of the micro switch was found highly important in modifying the mechanical produced action sound. Furthermore, the internal volume, the damping of the PCB board, and the tightening of the overall assembly and joints were found relevant in this specific product researched in MechanicalSound+.

It is seen in other research projects regarding sound quality optimization that sound descriptors are exclusively in each product category and very much context and user dependent [12]. Another research project explored how to predict user responses to sound quality by setting up metrics and Acoustical Sensory Profiles [2]. The conclusion points out that it is very much product, user, and context dependent, and it is very difficult to generalize to other products, segments, and markets.

MechanicalSound+'s research results are based on minor statistically responses, and further research can, with a greater number of respondents and resources, conduct more valid data. Also, it would be beneficial to explore other methodologies and more complex product structures.

The outcome is seen as a possibility for improved knowledge and awareness within the PDD about how to identify which parameters that have influence on the mechanical sound design. In a product development process, the PDD needs to identify the importance of i.e. a mechanical produced action sound at an early stage of the process. If the sound has major importance to the users' quality perception, workshops and simple tests can be incorporated into the product development process.

The workshops and tests can become a time, resource, and economical demanding post in the project's overall structure. It is, therefore, important to evaluate the importance of the product sound contrary plausible outcome of quality improvement at the beginning of a project. If not, major economical consequences are plausible [7]. In the conducted case study, the mechanical produced action sound was exclusively targeted eliminating other important senses like seeing or/and feeling when using the mouse. Further studies can explore this relationship in between targeted senses and adjust the process presented in this case study towards a more holistic sensorial approach. The overall quality assessment is a combination of all senses affected by the product in-use, and hereby a possibility of greater quality improvement in the product design.

Sound is not the only subjective parameter at which designers target their products. It is a fine balance between all five human senses. The importance of other subjective parameters can be beneficial for the PDD to classify at an early stage of the process. A product's idiom and design is also a very subjective parameter, whereas a PDD can - with use of mood or style boards - create a platform of communication between designers and users to target users' emotional demands regarding i.e. aesthetic preferences. This case study differs from mood or style boards by creating tangible tools for designers to target their mechanical produced action sound towards with optimized quality perception as outcome by users.

ACKNOWLEDGEMENTS
The author would like to give thanks to Søren Bolvig for his inputs and knowledge about design research and strategy. Moreover, it is proper to give thanks to students participating in surveys, interviews, workshops, and tests.

REFERENCES
[1]   Jens Bernsen, Danish Design Centre. "Lyd i Design. Sound in Design", 1999.
[2]   Richard H. Lyon, RH Lyon Corp. "Product Sound Quality - from Perception to Design", 2003.
[3]   David L. Bowen, Acentech Incorporated. "Correlating Sound Quality Metrics and Jury Ratings", 2008.
[4]   Flemming Christensen, Geoff Martin, Pauli Minnaar, Woo-Keun Song, Benjamin Pedersen, and Morten Lydolf, Sound Quality Research Unit at AAU. "A Listening Test System for Automotive Audio - Part 1: System Description", 2005.
[5]   Torben Holm Pedersen, DELTA. "The Semantic Space of Sounds, Lexicon of Sound-Describing Words - Version 1", 2008, pp. 92-93.
[6]   Lindstrøm M. BRAND sense. "Branding for alle sanser: føle, smage, lugte, se og høre", 2005 (Børsens Forlag A/S).
[7]   Anders Berner and Kenneth Munck, ing.dk. "Produktlyd: Aktuator-producent på vej mod lyddesign", 2001.
[8]   Thomas Holst, Working Report. "Spørgeskemaundersøgelse blandt designere i praksis", 2010.
[9]   Karl T. Ulrich and Steven D. Eppinger. "Product Design and Development", 2008.
[10]   Ken Friedman, Department of Knowledge Management, Norwegian School of Management. " Creating design knowledge: from research into practice", 2000.
[11]   Richard H. Lyon, Marcel Dekker, Inc. "Designing for Product Sound Quality", 2000.
[12]   Statement from Flemming Christiansen, Associate Professor at Centre for Acoustics at Aalborg University and former staff member of Sound Quality Research Unit at AAU.

INTERNATIONAL CONFERENCE ON ENGINEERING DESIGN, ICED11
15 - 18 AUGUST 2011, TECHNICAL UNIVERSITY OF DENMARK

DEVELOPMENT OF MODULAR PRODUCTS UNDER CONSIDERATION OF LIGHTWEIGHT DESIGN

Thomas Gumpinger, Dieter Krause
Hamburg University of Technology

ABSTRACT
Whether it is reduction of complexity during development or individual configuration for the consumer, modular products have many benefits throughout their product lifecycle. It is not surprising that many products are based on this principle. Along with modularisation, the tendency toward lighter, more efficient products is growing. Lightweight design of moving masses is crucial, particularly in the transportation sector. However, design conflicts between the two principles are hard to resolve. Modularised products tend to be heavier than non-modular products. Overcoming this conflict is an important step in serving the individual consumer and meeting environmental responsibilities.
The effects of modularisation on lightweight design are outlined. A strategy to handle the identified impacts is then presented.

Keywords: Modularisation, Lightweight Design, DSM

1 INTRODUCTION
In 2008 a modularisation approach for an aircraft galley was designed to reduce the inner variability while maintaining outer variability [1]. The results were promising but with one major drawback: weight. Several lightweight design optimizations [2] where made to reduce galley weight. Some weight drawbacks in the modularisation concept could not be assigned to specific parts nor actioned. In a small comparison with a few similar modularised and non-modularised galleys of competitors, the modularised concepts always fell short in the weight category. In the aviation industry, this extra weight is unfavourable in the purchase decision of the airline. This is obvious when the impact of additional weight in the aircraft is examined. Because an aircraft is designed to withstand high acceleration in an emergency, weight increase of a component always has secondary effects on the structure. The additional weight increases the load, which has to be absorbed by the structure. The structure may need to be reinforced to sustain this, creating more weight gain. The propagation of weight in aircrafts is typically around a factor of 4 [3], for example, if an aircraft has to carry 100kg additional weight, this leads to a weight increase of 400kg in total.
A modularisation has to be at least weight neutral or the costumer will not accept it. However, to capitalise on the modularisation concept, the additional effort required for the necessary weight optimisation in engineering and production has to be kept at a minimum. The question of how modularised products can be efficiently developed using lightweight design therefore arises.
Based on a literature review, the interdependencies between modularisation and the lightweight design concept are outlined and the main conflicts between them are identified. A supporting system model is then presented as the basis for modular lightweight design optimization. The strategy is trialled on the aircraft galley and then discussed.

2 MODULARITY AND LIGHTWEIGHT DESIGN
The section begins with a critical reflection of the similarities and differences between modularisation and lightweight design. Based on a literature review by Salvador [4], the characteristics of the modular concept and its transformation into the product are compared to lightweight design principles. There are multiple definitions of modularity as a result of the many perspectives on the term. Salvador summarises them and divides the perspectives into component commonality, component combinability, function binding, interface standardization and loose coupling [4].
The effects on lightweight design of the different perspectives may be similar; the main drivers for weight increase and the positive effects of modularisation are outlined.

2.1 Component commonality and lightweight design

Commonality of components in modularity is most often seen as the reuse of standardized components in different parts of the product and/or different product variants. This commonality view of modules may define them as an independent functional entity rather than just reused parts. The use of components in different applications (even if they are only slightly different) causes over-sizing of these parts [5] because they have different requirements in each configuration. This is in clear contrast to lightweight design, where over-sizing should be kept at a minimum [6]. Within a product, the over-sizing of a reused component may be traceable. However, the consequences in additional weight across a product family are hardly tangible. Over-sizing of modular products is often inevitable and, due to the complex interaction of modules in different product variants, it is often not quantifiable and hard to optimize.

To illustrate this effect, Figure 1 shows the variants of a cantilever. The variants consist of the same set of modules. Accordingly, the modules have to fit all the requirements across the variants. Variant V 1 is the initial state of the cantilever. All modules are designed with a minimum of over-sizing. In Variant V 2, the load F increases so the modules have to be adapted to withstand the additional load. With the load increase in module M 4, the weight propagation due to structural reinforcement occurs throughout all modules. It ends up with significantly more weight than the module set. In variant V 3, the position of module M 4 and M 3 is changed; due to this different position the load induced by M 3 has to be conveyed through to M 4 in the remaining modules. Therefore, M 4 in particular has to be reinforced and so gets heavier. This is just the first round of adaptation. The cycle repeats until a stable state is reached. In many variants, it ends with an undesirable over-sizing of modules.

Figure 1 shows a fractional weight propagation tree for module 4. The cycle begins in Step 2 of the propagation because M 4 indirectly influences itself. With an increasing number of variants and modules, such trees branch out rapidly, visualising the complexity.

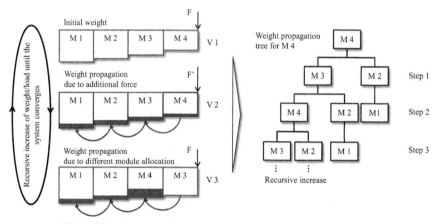

Figure 1 Weight propagation throughout the variants in modularised designs

2.2 Component combinability and lightweight design

Component combinability is building product variants out of the combination of a given set of modules. This combinability of components to maximize the variety of created products to fulfil customer needs is a widely understood meaning of modularity. This combinatorial freedom has to be reflected within the components. Thus, the universal application of modules leads to higher requirements, which result in functional and structural over-sizing. Variant 3 of Figure 1 shows the combinability of a module set. Module 3 and module 4 are both combined with module 2. This leads to higher requirements for module 4, which results in more weight.

2.3 Function binding and lightweight design

The concept of function binding is a specific functionality linked to one module, enabling the costumer to combine functions for their desired use. This is the opposite to integration of functions, which is a key concept of system lightweight design. Rather than separating functions into different modules, functions should be integrally combined in the product. With integration of function, the load-carrying structure has additional functionality, consequently, components and weight can be reduced.

For example, a set of modules should be able to build a motorized aircraft and a glider. The plane would need a module for fuselage, wings, engine and tank. For the glider, the engine and tank module could be spared. This has the disadvantage that the wing and tank must be split into separate modules. Modern aircrafts store fuel in the wing, thus wing and tank are combined to reduce weight. A modularised product family could not benefit from this lightweight design principle.

Independent functions may lead to redundant sub-functions within the modules [7], causing additional weight.

2.4 Interface standardisation and lightweight design

According to this view, interfaces should be standardised so that variants can be combined via a universal interface. This leads to even more design restrictions: over-sizing of the interface is the result. These interfaces may be designed for easy reversibility, e.g. for maintenance reasons or re-configurability in product system modularity. This reversibility requirement prohibits the use of lightweight interface connections, such as adhesive bonding, soldering and welding, because they are non-reversible.

Standardised interfaces are often restricted in their dimensions, because they have to fit in multiple designs. The size restriction, in combination with possible high loads of a variant, results in a high stress peak. This stress peak should be avoided in lightweight designs; instead, an evenly distributed load application should be favoured [6].

2.5 Loose coupling and lightweight design

In this concept, the cohesiveness of interactions within the module is stronger than external ones. Due to a reduction in interactions between modules, a specific module may have to fulfil certain functions itself. Another module may need this functionality, but each module has to provide this function individually due to the loose coupling. This redundancy leads to functional over-sizing.

Göpfert mentions that a strong focus on the modules runs the risk of losing track of the overall product [7]. In lightweight design, interaction of the whole system to reduce the weight is a significant factor [8]. Loose coupling also allows building a multitude of variants, which leads to a complexity and requirement increase.

2.6 Summary

In the literature review, no specific advantage to modularisation for lightweight design could be found. However, lightweight design could benefit from the overall positive effects of modularisation. Dividing the product into small, hierarchically-structured parts leads to better understanding of the system, so the concept and design phases profit. System interrelation can be more easily identified due to clear separation. The integration of function approach can be applied at a module level [2]. The modularisation leads to multiple reuse of a part, thus effort for development and production enhancements per module can be increased. These scale factors can be used to reduce the module's weight.

As different perspectives of modularity are compared against lightweight design, the same conflicts occur. The major drawbacks of modularisation arise in three categories: weight increase due to interfaces, over-sizing, and complex design interactions.

Göpfert describes a way to set weight targets for functions or corresponding modules in his tool METUS. This reduces the weight, but does not include the effect on the system. It becomes clear that modularity and lightweight design are in some ways opponents. Thus, a strategy to overcome the conflicts and benefit from both is needed. A strategy for efficient reduction is presented in the next section.

3 OPTIMIZATION STRATEGY FOR MODULARISED LIGHWEIGHT DESIGN PRODUCTS

Figure 2 shows the consecutive steps of the approach for modularised products using lightweight design. Starting from the Module Interface Graph (MIG) visualisation of the variants of a modularised product family, the data is converted into a node-link representation. In the next step, these node-link diagrams are transferred into Design Structure Matrices (DSMs). All modules are globally defined to address all variants simultaneously, hence the DSMs are interlinked. With this system model, the optimization can address the previously stated conflicts. The optimization includes 3 steps to meet the weight target. First the interfaces of the modules are evaluated. Secondly, the over-sizing of the modules is reduced by matching module requirements across the variants. Last, for efficient lightweight design, weight-sensitive modules are traced and weight optimized. The three weight reduction steps have an increasing level of effort and costs. The degree of modularisation and the robustness against new variants or changes are reduced. Therefore, as soon as the target weight is reached optimization can be stopped to ensure a cost efficient and robust design.

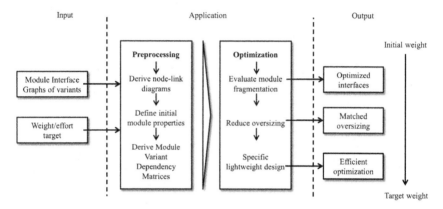

Figure 2 Overview of steps

3.1 A system model to support modular lightweight design

The aim of the system model is to provide the basis for the optimization of the modularised product family. It consists of an abstracted model of the product variants and their interrelations. It only contains information needed for the optimization to reduce complexity. The three steps are as follows:

Derive node-link diagrams

The output of the previous modularisation defines the modules, their interactions and, if applicable, the variants. The Module Interface Graph visualises this output and shows the modularisation in a transparent and intuitive way. This is used as input for the system model. The MIG 2D shapes of the modules are transferred to nodes in the node-link diagram. For this optimization, unnecessary complexity drivers are removed, hence the interactions between the modules, such as electrical power flow and media flow, are taken out, leaving only the relevant interface load interactions.

Therefore, interface loads link the nodes (modules) together. Figure 3 shows the MIG and the corresponding node-link diagram representation of a modularised product.

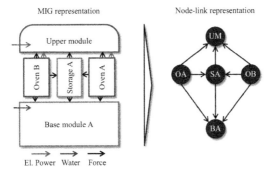

Figure 3 Derive node-link diagram from MIG

The load exchanged between modules represents interactions between the nodes. Depending on the level of detail of the product development phase, this information can be estimated, calculated, simulated or measured from a prototype test. To include weight propagation, a first estimation of a modules interface load consists of the forces of the accelerated module mass and the enforced load on the acting module. For modules with multiple outputs, the forces are split up, represented by a factor . In the example of Figure 4, the interface conveys load from module 2 to module 1. The load increases with an increase in the mass or the interface load of module 2.

$$F_{1\leftarrow 2} = m_{M2} \cdot a \cdot \alpha_1 + (F_{IM2'} + F_{IM2''}) \cdot \alpha_2$$

with $F_{IM2'}$ and $F_{IM2''}$ at M2.

Figure 4 Interrelation of modules through distributed load

For a modularised product family, all variants of interest need to be converted into node-link diagrams. Figure 5 shows three different product variants and the corresponding node-link diagrams in different views. It is possible to stack these diagrams on each other. With a collapsed view of the node-link diagrams, the multitude of relations is visible. The variants consist of the same set modules. Therefore, the requirements of a module are defined by the variant in which the module occurs. This forms a connection between the node-link diagrams. If the collapsed view is rotated, the network of relations expands and the module connections are visible. The result is a multi-layered network that represents the modularisation of the product family.

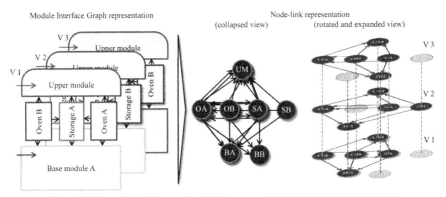

Figure 5 MIGs and multi-layered node-link representation of three product variants

Define initial module properties

The different modules are derived from the MIG and the essential data for the optimization process is added. To simulate the behaviour of the network, the following properties of the modules are needed:
- Payload: Not structure supporting masses.
- Weight of the structure: Weight of the load-conveying structure.
- Weight of interfaces: Estimated weight of additional interfaces due to previous modularisation.
- Weight/load correlation: Increase or decrease of the structural weight, due to additional or reduced loads. For simple components, such as slender struts, the load-dependent weight can be calculated analytically [9]. Depending on experience and development status, this factor can be estimated or calculated. Discrete steps of weight and maximum load can often be identified, for example, additional layers of Carbon Fibre Reinforced Plastic in a sandwich panel rise in discrete steps in weight and maximum load.
- Max. structural load: The module's maximum structural load from the corresponding variants. This is a requirement for the module weight/load correlation.
- Max. acceleration: Maximum acceleration of the module is necessary as we consider lightweight design of moving masses.

Derive Module Variant Dependency Matrices

Node-link diagrams and matrices are both visualisations for graphs. They can be transferred from one to another. In this case, the modules are the nodes and are represented by a row and column in the matrix. An entry in the matrix links two modules together. The entry value is identical to the link value from the node link diagram. It describes the conveyed load between two modules. The total load output is summarised for each module.

The previously described properties of the modules are entered into a matrix. The maximum load value per module is derived from the complete set of variants (Figure 6). This guarantees that the highest requirements are included in the development of the modules.

As the maximum load has an impact on the structural weight, a higher load increases the module's weight. The interface loads in the variants are module weight sensitive; hence, a higher weight increases the interface load. Because of this correlation, a circular dependency between the matrices occurs. For a stable system, this circular reference has to converge. In most cases, the initial modularisation is over-sized, therefore all modules fulfil the maximum requirement and the system is valid. If this is not the case, the modules that do not fulfil the requirements can be traced back in the system model and adapted accordingly.

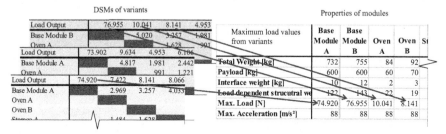

Figure 6 Matrix representations of the variants and modules.

The matrices represent the weight and load dependencies between the modules across the variants. This Module Variant Dependency Matrices is the basis for the following optimization.

3.2 Optimization of the module variant matching

A good balance between decision criteria needs to be found for a good optimization. The system model of the modularised product family must provide evaluable information to make these decisions. As per the conflicts outlined above, the interfaces have to be evaluated whether they are needed or not, the oversizing of the modules has to be reduced, and an efficient lightweight design has to target product weight-sensitive modules. Execution of the steps is described below.

Evaluate module fragmentation

Pahl and Beitz portray the risk of unnecessary increases in interfaces with unnecessary maximisation of modularity [10]. The additional weight of the interfaces in a modular design has an adverse impact on the lightweight design of a product. To reduce this effect, the interfaces are critically evaluated. First, the necessity of an interface is reconsidered, then the weight increase impacts of the remaining interfaces are evaluated, and, if required, appropriate actions are applied.

To achieve this, a search in the system model for modules that are directly connected with each other throughout all variants is conducted. If such a combination of modules is found, the modules can be merged together. In this context, Salvador refers to a "weak product system modularity" concept [4]; this combination of modules lacks the requirement of interface reversibility. The adverse effect of the interface on product weight is eliminated, but the degree of modularity decreases. The remaining interfaces have to be critically evaluated in a decision matrix. Therefore, the weight/load factor of the interface is rated against its importance. The interface importance of the module is taken from a review from the former perspective-based modularisation. The weight/load factor is taken from the interface weight and the conveyed load. The decision matrix indicates whether to weight-optimize, realize, remove or decide on the interface (Figure 7).

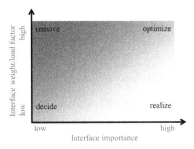

Figure 7 Decision matrix for interfaces

Reduce over-sizing

The requirements across the variants and the resulting weight propagation cause the over-sizing of a modularised product family. For a minimum of over-sizing, the requirements of the variants have to be matched. Due to the complexity of the module variants network, this is not really achievable by hand. The system model can be used to match this over-sizing. The first step is an incremental decrease in the weight of one module where the maximum load is below the module's load capacity. With the decrease in weight, the load capacity of the module converges to the necessary one. Secondly, the load conveyed to the connected modules also drops because less weight is accelerated and so less force is conveyed. Hence, the connected module's weight can be reduced as well. This is incrementally done until a minimum is reached. The over-sizing of each module is minimized and matches the variants overall. If this iterative decrease in requirements, corresponding load and weight only reaches a local minimum, enhanced algorithms can be used to find a global minimum. Figure 8 shows the weight reduction propagation of module OA, from the example above. The structural weight of the modules influenced reduces with a snowballing effect.

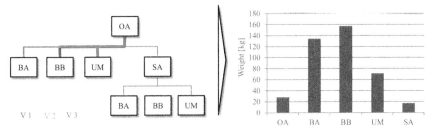

Figure 8 Weight reduction propagation of module OA and resulting effects on other modules

Module-specific lightweight design optimization

The previous two optimizations may have relatively low costs because they do not alter the basic design of the modules or apply costly lightweight design optimizations. However, if the weight target of the product family is not reached, additional lightweight design optimization steps have to be taken. To achieve this, the weight/load factor of specific modules is optimized (minimized): the module's load capacity is increased while maintaining the weight or the load capacity is kept while reducing the module's weight.

For optimum efficiency, the goal is to select modules that have a big impact on the product family's overall weight. The weight sensitivity of a module describes how the overall system reacts to a weight change in the module. Therefore, weight change of the whole product family is set in relation to the weight change of the module.

$$\Delta W_{PF}/\Delta W_M \qquad (1)$$

If this is correlated with the effort (cost) of reducing the weight of the module, an optimization efficiency plot can be drawn (Figure 9). The modules to optimize can be determined with this efficiency plot. The maximum reasonable number of modules is limited by modules in the upper right side of the plot. An evolutionary algorithm (EA) is then run on the module definition matrix. The EA has the goal of decreasing the weight/load to minimize the product family weight until the target weight is reached. The EA has the following boundary conditions:

- Change modules with high weight sensitivity ( ) and high ratio of module weight reduction to effort
- Keep a specific minimum weight per module
- Keep over-sizing of optimized modules equal
- Keep a specific minimum over-sizing

As the algorithm tries to reduce the weight, the secondary weight reduction of modules, as described above, is also included. This ensures an efficient weight reduction that incorporates the whole system rather than isolated modules. The percentage of weight decrease after each optimization step is presented on the right side of Figure 9.

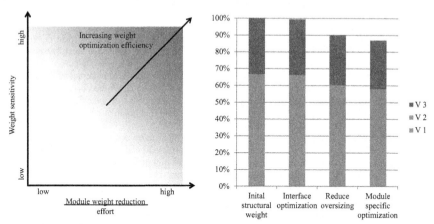

Figure 9 Optimization efficiency plot and weight reduction after each optimization step

The result of the EA provides the input for the development of the modules. The requirements are all matched for an efficient lightweight design of the modularised product family. With this matching of modularisation and lightweight design, additional information for efficient development is created. As the development of the product family matures, the information for the system model gets more exact, hence the prediction and optimization becomes more accurate.

4 EXAMPLE

The aircraft galley is used as an example. Previous projects, a modularisation concept [1] and its CAD realisation of the aircraft galley [9], are the input. The CAD model showed that the modularised concept, despite other optimizations [2], led to a too-heavy structure, hence, this approach was created. The MIG visualisation of the previous modularisation concept [1] was the starting point for validating the approach. The MIG was converted into a node-link diagram using a customized software tool (Figure 10). Module boundaries and components are both visualized in the MIG and the software tool. The pre-processing was carried out for ten different configurations of the galley family.

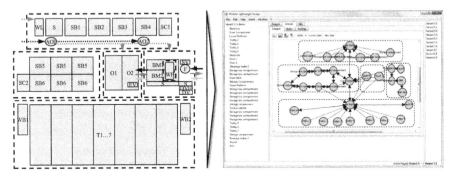

Figure 10 MIG of Galley (according to [1]) converted into a node-link diagram

Following this step, the information for the modules was gathered. The CAD model (Figure 11) enabled accurate aggregation of the initial structural weight, payload, and interface weight. The maximum acceleration is also specified to 9 g. From a former FEM analysis, some interface loads could be calculated and serve as reference points. The interface loads were then interpolated. In the first step, the weight/load correlation was estimated with a high, medium and low strength design per module. The correlation was then interpolated using these points.

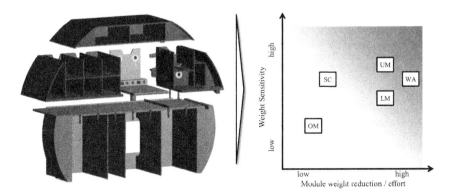

Figure 11 Modularised Galley (according to [11]) and optimization efficiency plot

The product family was optimized according to the steps above (Figure 11). The lightweight design matching of modules and variants provides the basis for further development of the modularised product family.

5 CONCLUSION

This approach attempts to add more transparency to the development process in modularised lightweight design products. A good balance between the decisions of which interface to reduce, which maximum load is required and which module has to be optimized is required to even out the conflicts between modularisation and lightweight design. An initial validation using the example of an aircraft galley proved the substantial benefits of the outlined approach.

Undoubtedly more work needs to be done. More in-depth research of the relationship between modularisation and lightweight design is needed to ensure that all important aspects are covered. The possibility of interlinking the system model with enhanced CAx tools, such as FEM and topology optimization, should be investigated to more accurately predict weight propagation. Nevertheless, the need and opportunities for efficient modularised lightweight design for product families is clear.

REFERENCES

[1] Blees C., Jonas, H. and Krause, D. Perspective-based Development of Modular Product Architectures. In *Proceedings of the 17th International Conference on Engineering Design (ICED)*, Vol. 4, Stanford, August 2009, pp.95-106.
[2] Gumpinger T., Jonas H. and Krause D. New Approach for Lightweight Design: From Differential Design to Integration of Function. In *Proceedings of the 17th International Conference on Engineering Design (ICED)*, Vol. 6, Stanford, August 2009, pp.201 – 210.
[3] Hertel H. *Leichtbau,* 1980 (Springer, Berlin).
[4] Salvador F. Toward a Product System Modularity Construct: Literature Review and Reconceptualization. *IEEE Transactions on Engineering Management 54*, 2007, pp. 219-240.
[5] Piller F. T. *Modularisierung in der Automobilindustrie,* 1999 (Shaker, Aachen).
[6] Klein B. *Leichtbau-Konstruktion,* 2007 (Vieweg, Wiesbaden).
[7] Göpfert J. *Modulare Produktentwicklung: zur gemeinsamen Gestaltung von Technik und Organisation,* 1998 (Gabler, Wiesbaden).
[8] Wiedemann J. *Leichtbau,* 2007 (Springer, Berlin).
[9] Rees D. W. A. *Mechanics of optimal structural design,* 2009 (Wiley, Chichester).
[10] Pahl G. and Beitz W. *Engineering Design,* 2007 (Springer, Berlin).
[11] Jonas H., Gumpinger T., Blees C. and Krause, D., Innovative Design of a Galley Product Platform by applying a new Modularisation Method, in *Proc. 4th International Conference "Supply on the Wings"*, Frankfurt/Main, November 2009.

Contact: Thomas Gumpinger
Hamburg University of Technology
Product Development and Mechanical Engineering Design
Denickestr. 17
21073 Hamburg
Germany
Tel: +49 40 42878 2148
Fax: +49 40 42878 2296
Email: gumpinger@tuhh.de
URL: http://www.tu-harburg.de/pkt

Thomas studied mechanical engineering at the Technische Universität München. Since 2007, he has worked as a scientific assistant at the Hamburg University of Technology. He works on lightweight design, integration of function, and testing, mostly in the field of aircraft interiors.

Prof. Dieter Krause is head of the Institute Product Development and Mechanical Engineering Design and dean of mechanical engineering at Hamburg University of Technology. He is a member of the Berliner Kreis and the Design Society. The main topics of his research are new design methods for product variety and modularization, as well as lightweight design for aircraft interiors.

INTERNATIONAL CONFERENCE ON ENGINEERING DESIGN, ICED11
15 - 18 AUGUST 2011, TECHNICAL UNIVERSITY OF DENMARK

THE INVESTIGATION AND COMPUTER MODELLING OF HUMANS WITH DISABILITIES

A. J. Medland[1] and S.D. Gooch[2]
[1] University of Bath, United Kingdom
[2] University of Canterbury, New Zealand

ABSTRACT
Aids for the invalid or infirmed are often created simply by modifying those used by the able-bodied, with little care taken as to their individual needs and limitations. This study is aimed at determining their actual requirements through both modelling their anthropomorphic conditions, and measuring their physical capabilities.
The subjects are evaluated in an experimental rig where, for example, the appropriate force data is collected. The physical limitations of the skeleton are also recorded and entered into a manikin model incorporated within a constraint environment. Together the manikin models are used to evaluate the disability aid under consideration.
This approach has been employed in the study of wheelchairs for people with spinal injuries. Here the positions at which the maximum pushing capability of the subject can be determined and the chair modified, or redesigned, to allow this to be achieved. A similar approach can be applied to other invalid aids and medical equipment.
A procedure is now being developed that can be applied to the collection of this data which can handle a range of problems for the creation of more effective aids for the elderly and infirmed.

Keywords: anthropomorphic, human measurements, constraint modelling, invalid aids

1 OBJECTIVE OF STUDY
In order to design aids for invalids and the infirm it is important to fully understand their capabilities and needs. Many of the current devices have been provided through the modification, or adaptation, of existing designs used by the able-bodied user. Very little work has been carried out into the fundamental needs of this group of subjects and their ergonomic capabilities. Similarly, the majority of human modelling techniques offer only the capabilities of the able-bodied and ignore those with extreme and critical disabilities.
The current research study has set out to provide this information within a combined investigation that includes both experimental and analytical techniques [1]. The capabilities of the individual subjects are studied through experimental procedures, and computer modelling used to allow chosen tasks to be determined and resolved, which allow realistic body postures to be developed [2]. The resulting posture and strengths then provide the basis for the device redesigns.

2 EXPERIMENTAL PROGRAMME
The major experimental investigation has been (and continues to be) conducted in collaboration with the University of Canterbury, in New Zealand. Within the Department of Mechanical Engineering, they have been creating experimental techniques and rigs to allow postures and forces to be measured on the handicapped [3]. This is currently directed towards the study of tetraplegics through Burwood Hospital, which is the national centre for spinal injuries (and through which they obtain ethical approval). This work is undertaken with real subjects and through this process, of modelling their problems, improvements are sought in their living conditions.
The effort to date has centred on mapping the strength capabilities of wheelchair users. Here an experimental rig has been developed in which the forces created, whilst pushing, can be measured for different sitting and hand positions (figure 1.).

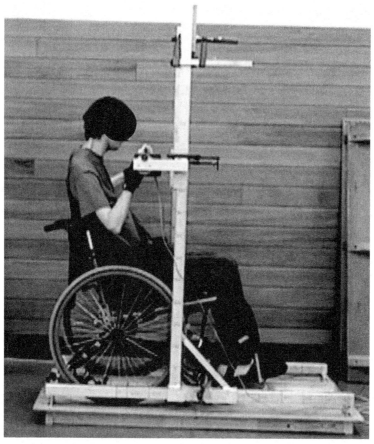

Figure 1. Experimental wheelchair pushing rig.

Investigations undertaken on a range of subjects have shown that their capabilities are highly dependent on both the level of spinal injury and the position at which they are able to apply the force. The position of this point of application, and its movement across the drive wheel rim, results in considerably different forces over small changes in pushing position, as can be seen in the force mapping in Figure 2.

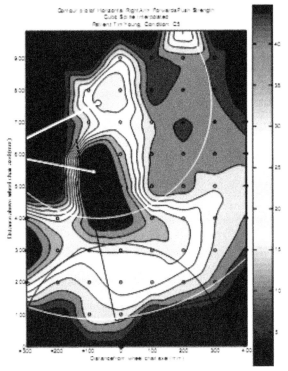

Figure 2. . Force map created from experimental data for an individual subject, showing the variation in forces with reach capability.

The rig has been designed to allow it to be adapted to provide data on other studies that can range from working across a surface to that of direct lifting by the subject. As problems are proposed and investigated, a number of force mapping experiments will be conducted for a range of individuals. This will provide the basis of a library of mappings that can be used to compare capabilities and to provide information upon which other devices can be considered and redesigned.

3 ERGONOMIC DATA

In such studies of human posture it is necessary to understand and incorporate the ergonomic limitations of the human body. This has been achieved through collaborative work with Professor Johan Molenbroek at the Technical University of Delft [4]. Within the Department of Ergonomics and Design, data has been collected over the last twenty years on a large group of humans, covering ages from a few months old, to males and females in their eighties. This range of data has been made available to the research together with their ADAPS anthropomorphic computer representation [5]. Since being incorporated into the University of Bath research programme the Delft manikin has been extensively modified and adapted to allow closer human movements to be represented that are necessary for the study of invalids (these include, amongst other things, the ability to balance, a more complex three segment spine model and the modelling of hands with the ability to grip [6]).

4 COMPUTER REPRESENTATION

The resolution of a desired posture or task can present a difficult and uncertain problem. The variables involved in the solution of any task can vary from simply the movement of, say, the eyes in looking at an object, through to, perhaps, 50 or more when the same looking activity is required and the object to be 'seen' is perhaps behind the manikin, thus requiring complete body motion and the selection of a new posture.

The approach adopted in this study has been based upon that of a constraint resolution approach developed by the Bath research group over the past fifteen years [7]. Here the problem to be solved is described by a collection of 'rules' that are tested against the total truth of all the rules of the problem. Direct search techniques are used to search for a combination of variable values that make the problem true. When used in the design of mechanical systems (such as packaging machines) all of the rules can be determined and the variables known [8]. Within the modelling of humans, it was necessary to create an approach using sensitivity analysis in which the variables influencing the solution could be found and ranked [9]. This allowed the variables to be increased and changed during the course of the investigation.

5 APPLICATION OF APPROACH

During the development of these techniques a number of human studies have been undertaken both to provide a greater understanding of the type of human problems that can be addressed, and to provide an increasing confidence on the experimental and computational techniques being created.

These studies have concentrated on differing aspect of the problem and have included the following:
1. Individual studies of human actions such as balancing, pointing, looking and gripping (both individually and in combinations).
2. Stair climbing.
3. Working at an ironing board.
4. Sitting in, and propelling, a wheelchair.

With further development and refinement of the approach, the method can be applied to a wider range of problems.

6 MODELLING CAPABILITY

The capability to model a wide range of human problems and limitations depends on the creation of a generic modelling environment with high flexibility and control.

The constraint modeller has been created with a hierarchical modelling structure that allows embedded capabilities. Any desired level of space can be embedded within another to provide both simple embedding and pivoting. This, together with the built-in nine degrees-of-freedom of the individual spaces (3 of translation, 3 of rotation and additionally 3 of scale) provide a very large number of freedoms that can be applied to the solution.

Such numbers of freedom are often beyond the normal requirement, and so within the environment each can be either 'fix' or 'freed'. The fixing/freeing can be either defined in the definition of the problem, or the effect of the variables can be determined dynamically, during the problem progression, by sensitivity analysis. Those having no, or very little, influence on the problem can be 'fixed' during individual activities and reinstated later.

The range of values allowed in each freedom of human motion is not unlimited, as joints can not continuously rotate, as in machines. The direct search technique used incorporates limit checking techniques. If the search tries to move the motion beyond a limit that is specified then the chosen freedom is reset to the defined limit and the search continued without that variable.

The range of limits, both upper and lower, can be set and selected from a number of data files. They can represent the extremes of human movement, the normal range, the 'socially acceptable', or even tailored to the capability of an individual.

6.1 Obtaining Individual Capabilities

With this wide range of modelling capability the human model can be adapted to represent a subject that extends from that of a full range of human capabilities through to one of severe disabilities. In order to capture such a range a program has been developed that can be employed by the occupational clinician to observe and record approximate data of postures and deformability. Here the proposal is to follow their normal approach of estimating the subject's postures in chosen position, and passing this data directly into the graphic generating programme. This can then be compared to the subject from different viewing positions and when the different postures are adopted.

The modelling environment commences with normal human geometry and limitations (figure 3). These can then be modified, using constraint rules to provide a representation, such as a paraplegic sitting in a wheelchair (figure 4) .Here rules provide, and determine, the sitting posture that results from the needs to contact the various parts of the chair and to undertake other actions such a looking and reaching the drive wheels. In previous paraplegic studies[10] the postures taken by different spinal injuries were seen to be quite different due to the level of the injuries as can be seen in Figure 5. and required additional rules to represent each condition.

Figure3. Manikin in normal standing posture in front of wheelchair.

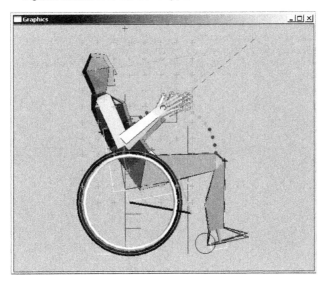

Figure 4. Manikin complying to rules of sitting in wheelchair, looking and operating it.

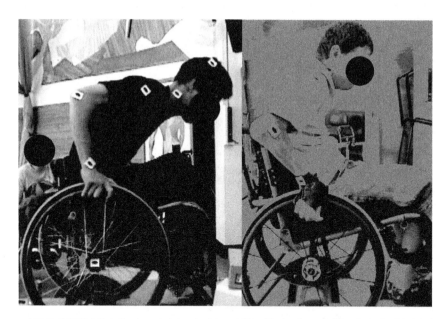

Figure 5. Different postures undertaken by L1 and C6 subjects when of sitting in a wheelchair.

The capability of the constraint modelling environment allows the standard manikin to be readily modified. At the first level the model can be parametrically changed by the insertion of different geometric values to represent the skeleton. With the ADAPS programs, obtained from Delft, at least twelve body sizes were provided that ranged from a 3 month old baby to 80 year old males and females. Individual models can also be made of chosen subjects.

In the study of invalids it is necessary to model both truncated and lost limbs. These are achieved by manipulation of the embedded scale parameters available in each model space, which can be reflected through the embedding function down the complete limb hierarchy. Figures 6 and 7 show representations of both the loss of a right leg below the knee and a person with stunted lower legs.

Figure 6. Model with the loss of a lower right leg

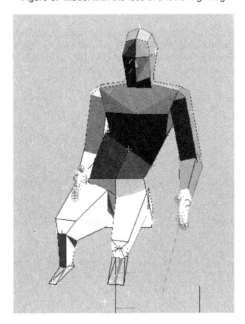

Figure 7. A representation of a subject with stunted lower legs and twisted spine

Further manipulation of the body geometry allows the distortions resulting from a range of illnesses to be modelled, as with the twisted spine (also shown in Figure 7). Here the manikin model has been reduced to a three segment spine as the full spine representation is difficult to create and for which it is not easy to obtain accurate data for all segments by observation.

7 DESIGN MODELLING APPROACH

The purpose of this study is to create an integrated environment in which it is possible to collect the data necessary to both analyse and design invalid aids, and support devices, for the handicapped. The approach is centred on the integration of determining the force capabilities of the individual in the desired posture necessary to undertake the task required. It is also necessary to determine what modifications will be made to those postures as the result of their illness. Once these are understood then designs can then be proposed and evaluated, as has already been undertaken in the repositioning of the driving force actions in the wheelchair, to align with the subject's maximum force profile.

7.1. Experimental data rig

The experimental rig needs to be very flexible to allow force measurements to be taken in any desired position and for any posture. This must include from the normal sitting positions through to those of standing, lying down and, possibly, of working at a desk. For these reasons the rig has been generated as a space frame in which the attitudes and activities can be set up and undertaken. The force data is collected at selected points throughout the space and transferred into the manikin modelling program, where it is mapped into the appropriate surfaces.

7.2. Individual anthropomorphic data collection

The limitation of the individual human subject needs to be collected and evaluated by professional medical staff. The approach being adopted is to create a modelling environment that allows them to follow, as closely as possible, their normal procedures in evaluating a subject. This approach uses a simple assessment form on which all limitations are recorded. If a posture is being assessed the angles and distortions observed are recorded on the sheet (either as estimated angle values or as '+' or '-'). These are then combined and postures estimated in further analysis.

The approach currently being investigated in the prototype system is illustrated in Figure 8. Here the various data collected is used to manipulate the manikin model directly. The effect of these distortions can be readily observed and corrections made, if necessary. The representation of the complete human subject can then be recorded and filed upon the computer system.

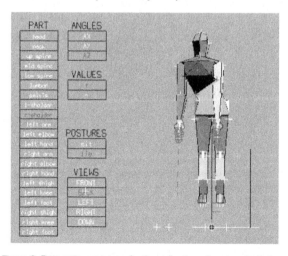

Figure 8. Data entry prototype for the collection of posture limitations

This data collection can then be repeated, if necessary, with the subject in other postures, such as sitting or lying. The results can then be compared and conflicts between observations resolved. In Figure 9 a resulting sitting posture is represented which can be checked and modified by observing it from different view positions.

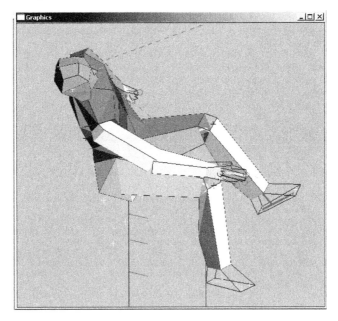

Figure 9.Resulting sitting posture of subject.

7.3. Design studies

This information, on both force capability and limitations in posture, thus provides the basis for any design study. Any successful design must not exceed the force and physical capabilities of the subject who is going to use it. The data thus collected can be used in the evaluation of proposed designs and alternatives sought. Within the constraint modelling environment optimisation and direct search procedures can be used to allow the parameters of a proposed design to be manipulated is search of a solution that satisfies all the desired conditions that need to be satisfied.

8 FUTURE WORK

This paper provides the background to the preliminary study undertaken of the approach being adopted for the design of invalid aids. The work so far has shown its feasibility, and has led the authors to investigate the possibilities of funding a full research project into this subject. The success of the full research activity is now dependent upon undertaking further case studies, such as the current wheelchair redesign, which will provide the evidence and support for such an approach.

ACKNOWLEDGEMENTS

The authors wish to acknowledge the support and help of staff and colleagues who contributed to the experimental investigations and to also recognize the support of the Erskine Visiting Fellowship of the University of Canterbury, New Zealand, which has provided financial support for exchange visits and the attendance of conferences on five separate occasions.

REFERENCES

[1] Floris L., Dif C. and MA Le Mouel. "Chapter 5 - The tetraplegic patient and the environment" Surgical rehabilitation of the upper limb in tetraplegia. Ed Hentz, V. R., ISBN 0702022713 / 9780702022715 Saunders May 2002

[2] Singh B., Hicks B. J., Medland A. J., Mullineux G., Molenbroek J. F. M., 2010. A constraint based human model for simulating and predicting postures. In: Horvath, I., Mandorli, F., Rusak, Z., eds. Tools and Methods of Competitive Engineering. Delft: Delft University of Technology, pp. 221-229.

[3] Gooch S. D., Woodfield T., Hollingsworth L., Rothwell A. G., Medland A. J. and Yao F. (2008), On the Design of Manual Wheelchairs for People with Spinal Cord Injuries, Proceedings of the 10th International Design Conference, Design 2008, pp 387-394, May 19 – 22 2008, Dubrovnik, Croatia.

[4] ADAPS., Human Modelling System (Ergonomics software), Section of Applied Ergonomics, Faculty of Industrial Design Engineering, The Technical University of Delft, The Netherlands, 1990.

[5] Molenbroek J.F.M. and Medland A.J., "The application of constraint processes for the manipulation of human models to address ergonomic design problems", Proceedings of Tools and Methods of Competitive Engineering Delft, The Netherlands pp. 2000, 827-835.

[6] Medland A.J., "Modelling the movements of the human hand", Fourth Asian Pacific Conference on Biomechanics, University of Canterbury, Christchurch, New Zealand 2009.

[7] Medland A.J., Matthews J. & Mullineux G., " A constraint–net approach to the resolution of conflicts in a product with multi-technology requirements", International Journal of Computer Integrated Manufacture, 2009, DOI:10.1080/09511920802372286.

[8] Hicks B.J., Mullineux G. & Medland A.J. "The representation and handling of constraints for the design, analysis and optimisation of high speed machinery", AIEDAM Journal, Artificial Intelligence for Engineering Design, Analysis and Manufacture, Special Issue Constraints and Design, 2006, Vol. 20, No. 4, pp. 313-328, ISSN 0890-0606. DOI: 10.1017/S0890060406060239

[9] Medland A.J. & Matthews J., "The implementation of a direct search approach for the resolution of complex and changing rule-based problems", Engineering with Computers 2009. DOI 10.1007/s00366-009-0148-z

[10] Gooch S.D., Hollingsworth L. & Medland A.J. "The Study of the interaction of humans with wheelchairs to improve the design", Proc. Of ASME2009, DETC2009-87475, August 2009, San Diego ,USA.

INTERNATIONAL CONFERENCE ON ENGINEERING DESIGN, ICED11
15 - 18 AUGUST 2011, TECHNICAL UNIVERSITY OF DENMARK

EVALUATION OF AN AUTOMATED DESIGN AND OPTIMIZATION FRAMEWORK FOR MODULAR ROBOTS USING A PHYSICAL PROTOTYPE

Vaheed Nezhadali[1], Omer Khaleeq Kayani[1], Hannan Razzaq[1] and Mehdi Tarkian[1]
(1) Linköping University, Sweden

ABSTRACT
This paper presents an automated design and evaluation framework, by integrating design tools from various engineering domains for rapid evaluation of design alternatives. The presented framework enables engineers to perform simulation based optimizations. As a proof of concept a seven degree of freedom modular robot is designed and optimized using the automated framework. The designed robot is then manufactured to evaluate the framework using preliminary tests.

Keywords: Automated design, simulation-based optimization, multidisciplinary design, modular robot, CAD automation

1 INTRODUCTION
The first generation of industrial robots was introduced in 1950s, and at that time it was expected that in a near future they will be widely used. However, due to long development times, high initial costs and complexity involved in the design process, use of robots has been mainly limited to specialized industrial task. Conventionally robot design is application-specific, implying that a given robot cannot be modified for diverse applications, reducing its reusability and forcing manufactures to incur high initial costs for new applications.

Evidently, a new design methodology is required to overcome the mentioned problems faced during robot design process. A modular design approach addresses the shortcomings of conventional design method by sharing and reusing modules, leading to a higher customization level.

Modular robots, like conventional robots, typically consist of subsystems which belong to multiple engineering domains, such as mechanics, electronics, and control. These subsystems are designed and evaluated using various engineering tools. Design modification of a component belonging to a certain domain will consequently require an update in other engineering domains. Propagation of such modifications across the engineering tools is generally done manually, which is a time-consuming process. To propagate these modifications in a time efficient manner, an automated interaction between these engineering tools is preferred. An automated design framework integrates the engineering tools and eliminates the need for manual modifications. Moreover, in contrast to a manual design evaluation process, more design iterations can be performed for a given development time. As a larger design space can be explored, the likelihood of finding an improved design increases. Additionally, an automated design framework would lead to reduction in time to market of the product.

1.1 Related Work
A robot design process mainly involves geometry generation, kinematic analysis and preferably optimization. The benefits of generating flexible and robust geometries for automated design have been demonstrated in various research groups and disciplines. The aircraft research domain has made efforts to describe methods for the automatic generation of geometries. This has been effectively demonstrated by Jouannet et al. [1], for micro-UAV design, Tarkian et al. [2] for civil aircraft design and La Rocca et al. [3] in the analysis for specific aircraft feature. The advantages of automated geometric modeling have also been illustrated for industrial robots, Tarkian et al. [4], as well as modeling airfoil shapes for use in wind turbine design, as presented by Cooper et al. [5]. Tarkian et al. in [6] discuss the dynamic models for industrial robots. Dynamic models for aircraft design analysis have also been effectively utilized by Johanson et al. [7]. Ölvander et al. have presented an optimization framework for aircraft design in [8], and a study of modular robots drive train

optimization is carried out by Pettersson and Ölvander [9]. In [10] an optimization procedure for the selection of actuators and mechatronic components is presented.

An automated design framework for industrial robots has been presented in [4] and [6]; however, no validation has been performed for this framework. These works also lack an automatic selection of optimized actuators which is an important part of the robot design process.

In this paper a new automated design and evaluation framework for modular robots is presented. The proposed framework enables the designer to create and evaluate different concepts rapidly, and facilitates the implementation of a simulation-based optimization for the selection of actuators. The presented optimization procedure, however, can also be extended to other robot design parameters. Additionally, a physical prototype, designed using this framework, is manufactured to ensure the reliability of the presented framework.

1.2 Modular Design

Modular design is an approach that subdivides a system into smaller parts (modules) which can be independently created and then used in different systems to drive multiple functionalities [4]. Modular robots refer to the idea of a family of robots which share modules within the family. Figure 1 shows an example of a modular robot family sharing modules among each other. The concept of modular robots has been discussed intensely since late 80s, e.g. Krenn et al. [11] and Paredis et al. [12]. Recently these robots are becoming increasingly attractive for robot manufacturers, such as Motoman [13] and Nachi [14]. Studies regarding some aspects involved in the design process of modular industrial robots have been carried out at Linköping University; Petterson et al. [15] and Safavi et al. [16].

The paper is divided into four sections. After this first introduction, the second section presents different components of the automated design framework in detail. In the third section, a design example is presented in which the automated design framework is utilized to design a seven degree of freedom (DOF) modular robot. An evaluation of the framework is also presented in this section. Finally section four comprises the conclusions and way forward.

Figure 1. Example of a modular robot family

2 AUTOMATED DESIGN AND EVALUATION FRAMEWORK

The automated design and evaluation framework is implemented by integrating design tools as shown in Figure 2. In the framework, Microsoft Excel serves as common platform for data exchange between various tools. It also acts as an interface through which users can interact with the framework. The framework can be broken down into an automated CAD framework, automated dynamic simulation framework and an optimization routine.

2.1 Automated CAD Framework

For simulation, evaluation and verification of the properties of any given product, a geometric CAD model is preferred [17]. In the beginning of the geometric modeling phase, simplified geometries are used which can result in inaccurate representation of the product. However, in order to define a sufficiently accurate geometric CAD model, frequent re-modeling has to be performed to introduce more details in the model.

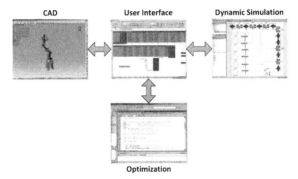

Figure 2. Design tool integration

Updating and generating geometric CAD models is a time demanding process. An automated CAD framework (ACF) allows for the rapid generation of geometric CAD model, hence, reducing the time to reach a sufficiently accurate geometric representation of the product.

Commercial CAD tools available in the market are becoming increasingly suitable to generate automated geometries for multi-disciplinary optimization (MDO) and design. CAD tools such as CATIA, Solid Works, Pro Engineer and NX6 all offer parametric design with varying functionalities. Geometric models created using these CAD tools can be very flexible and robust in the sense that both shape and number of geometric objects of the model can be parametrically defined. This parameterization is accomplished by defining templates and context manuals. These manuals contain complete construction procedures of the template objects. The template objects can be parametrically modified to obtain different sizes and shapes, and these construction procedures enable the template objects to be instantiated into different contexts, consequently, increasing reusability of created geometries in the CAD model. The geometric complexity of the CAD model dictates the accuracy of the model, whereas, degree of parameterization defines its flexibility. In general, the geometry should be flexible enough to allow generation of all conceivable configurations or concepts of interest. However, to define a generic geometry a large number of parameters have to be introduced, so a compromise has to be made between the two.

2.1.1 Automated generation of geometric CAD model for modular robots

The mechanical structure of a modular robot consists of a base followed by a series of modules. Each module mainly consists of a link, actuator, shaft and bearings. The choices made while designing these modules to make them modular are not trivial. Considering the fact that in the beginning, limited knowledge is at hand about the properties of the finished product, some design parameters should be kept flexible to be easily modified during the entire design phase. Hence, the possibility to remedy the shortcomings of certain choices made early in the design cycle is vital. For example, the choice of actuator dictates the shape of the module and in particular the link length. Changing the actuator type can result in re-modeling of entire geometry of the robot. Such changes can be verified rapidly using an automated CAD framework. Moreover, properties like mass, centre of gravity and moment of inertia; required for dynamic simulation and optimization can also be obtained.

The CAD tool selected for automated geometry generation of the modular robot is CATIA V5 which offers great flexibility for automated CAD generation. The parameters for changing the shape and number of different geometric parts in the robot CAD model are controlled from a user interface.

The main parameter which defines the number of instances of the modules in the robot CAD model is the DOF. The CAD model hierarchy in CATIA consists of a main assembly containing sub-assemblies. The first sub-assembly represents the base, and the remaining sub-assemblies, depending on DOF, correspond to the modules of the robot.

As mentioned earlier, templates are essential for automatic generation of CAD models. These templates are parametrically modeled using CATIA and stored in a library folder. Depending on the geometry of the CAD model being generated, CAD files are copied from the library folder to a working folder. The parts from the working directory are instantiated and constrained to each other in their respective sub-assemblies. Once all the sub-assemblies or modules are ready, they are inserted into the main assembly of the modular robot. All the instantiations are parametrically done using Visual Basic Script (VB Script). The above discussed approach forms the automated CAD framework for the modular robots which is shown in Figure 3.

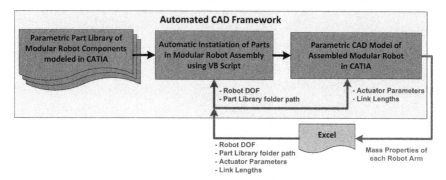

Figure 3. Automated CAD framework for modular robots

2.2 Automated Dynamic Simulation Framework

To investigate the dynamic aspects of the geometric CAD model created by the ACF, a dynamic model is required. These models are based on differential equations and algebraic equations. In the process of designing a robot, such models are often used to select suitable joint actuators by calculating the maximum torque required to rotate a joint at a certain rotational speed. Moreover, such models are also used to study the workspace of a robot and can provide a life time estimate of different components making up the drive train as demonstrated in [18].

To perform these analyses rapidly, an automated dynamic simulation framework, similar to ACF, is developed with a dynamic model at its core which is shown in Figure 4. This framework automatically retrieves mass and geometric properties of the current geometric CAD model, applies them to the dynamic model, performs a dynamic simulation and finally outputs the results of the simulation to the user interface for evaluation by a user or an optimization routine.

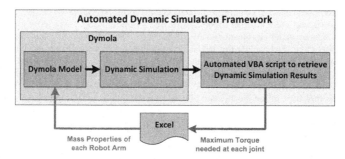

Figure 4. Automated dynamic simulation framework for modular robots

The degree of flexibility of an automated design framework is defined by the flexibility of its subcomponents. Introducing a non-parametric dynamic model in the framework would make the whole framework rigid. On the other hand a dynamic model offering more flexibility than the geometric model is also redundant. Commercially available softwares like Visual Nastran, ADAMS and Dymola offer the possibility of developing a parametric dynamic model, to different extents. For the presented work the dynamic model is developed using Modelica in Dymola.

Modelica is an object-oriented modeling language used largely for physical modeling of interconnected multidisciplinary systems. The hierarchical and object-oriented nature of Modelica makes it a prime candidate for modular design. Component models from the Modelica standard library [19] are used to develop the dynamic model.

2.3 Simulation-Based Optimization

In serial robots, the mass properties of each actuator affect the required torque in all previous actuators; therefore, a sequential approach is not suitable for selection of the actuators. Moreover, actuators of any industrial robot can highly affect weight, performance and cost of the robot. Therefore, it is important to ensure that a reliable procedure is implemented to select the actuators.

In this work, an optimization routine is utilized to choose the best suited actuators among available choices from a library which is stored in a Microsoft Excel sheet. During the optimization process, values of the various objectives are calculated by the use of different softwares and transferred to Microsoft Excel which acts as an interface between all framework components.

3 FRAMEWORK VALIDATION

To validate the automated framework, first, the CAD model of a seven DOF modular robot is generated. Afterwards, maximum torques required at each joint of the robot is calculated through a dynamic simulation of the most torque-demanding trajectories. Using an optimization process, the best suited actuators are selected based on the torques calculated in the previous step. Finally, a physical prototype of the designed modular robot is manufactured to evaluate the framework.

3.1 Automated CAD Model Generation

Figure 5 shows the procedure followed by the ACF to generate a seven DOF modular robot. As it is shown, first, components of modules are assembled and then the modules are placed in the robot structure. At the end of robot generation process in CATIA, the mass properties of each module which would be needed in dynamic simulation are transferred to the Excel interface.

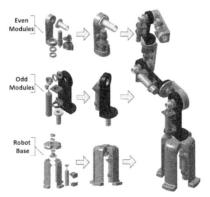

Figure 5. CAD model generation procedure for a 7 DOF modular robot

3.2 Automated Dynamic Simulation

To evaluate the dynamic properties of the generated CAD model a dynamic simulation is performed by following a predefined trajectory. The maximum required torque at each joint obtained from the dynamic simulation is passed to the Excel interface automatically. The torque curves obtained from the dynamic simulation for each joint are shown in Figure 6.

3.3 Optimization: Selection of Joint Actuators

When design problems are characterized by discontinuous and non-convex design spaces, nonlinear programming techniques are inefficient, computationally expensive, and, in most cases, find the relative optimum which is closest to the starting point. Genetic algorithm (GA) is applicable for the solution of such problems. GA is based on the principles of natural genetics, and is a stochastic method that can find the global minimum of problems with a high probability [20]. In this work, GA is used as the optimization algorithm.

The CAD model shown in Figure 5 contains randomly selected actuators from the actuator's library. In the following section, it is illustrated how the optimized actuator at each joint of a 7 DOF modular robot is selected by the optimization algorithm.

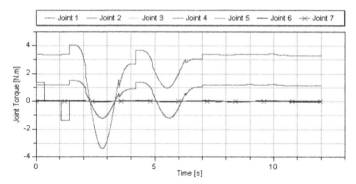

Figure 6. Torque requirement at different joints

3.3.1 Problem Formulation

In an optimization problem, the problem formulation is of a high importance as it would directly affect the results of the optimization. Choosing the lightest actuator for each joint in a serial robot would minimize the total weight of the structure and is an objective which is subjected to some constraints. The length of actuators is the most important constraint since it should be smaller than the link length so that the actuator would be mountable on the structure. The next constraint is the delivered torque by each actuator which should obviously be more than the required torque calculated in the dynamic simulation.

Briefly, the problem of selecting the best actuators for each joint can be summarized as follows:

Objective (F): Minimizing the weight of robot structure by selecting the lightest actuators

Subjected to:

Constraint 1 (g_1): Length of the actuators should be less than the link length

Constraint 2 (g_2): Required torque at each joint should be delivered by the actuator

Weighted sum method is chosen to formulate the abovementioned problem. The constraints are reformulated as penalty functions and multiplied by suitable weight factors according to their level of importance. Subsequently, the penalty functions are added to the objective function according.

The predefined reach of the robot limits the length of each link in the structure. Thus, as mentioned before, the length of each actuator could not exceed a certain value. The formulation of the length constraint as a penalty function, $p_1(x)$, would look like:

$$L_i = \begin{cases} q_1, & \text{Length of Link (i) } - \text{Length of Actuator (i) } < 40 \, mm \\ 0, & \text{Otherwise} \end{cases} \quad (1)$$

$$p_1(x) = \sum L_i, \, i = 1, 2, \ldots, DOF$$

Reformulation of the second constraint as a penalty function, $p_2(x)$, is:

$$T_i = \begin{cases} 0, & \text{Calculated torque } < \text{Torque delivered by the actuator} \\ q_2, & \text{Otherwise} \end{cases} \quad (2)$$

$$p_2(x) = \sum T_i, \, i = 1, 2, \ldots, DOF$$

Large values are selected for q_1 and q_2 in (2) and (3), to avoid the selection of unsuitable actuators. The formulation of f(x) is according to (3):

$$\begin{aligned} M_i &= \text{Mass of each module containing all component} \\ M(x) &= \sum M_i, \, i = 1, 2, \ldots, DOF \end{aligned} \quad (3)$$

As a result, the primary optimization problem which had two constraints can now be rewritten as the following optimization problem without any constraint.

$$Min\ f(x) = M(x) + \sum w_j p_j (x_i) \qquad (4)$$

$$\sum w_j = 1$$

$$g_k(x) = \sum w_j p_j (x_i)$$

$$x_i \in \{1, 2, 3, \ldots, 30\}$$

$$i = 1, 2, \ldots, DOF;\ j = 1, 2;\ k = 1, 2$$

According to the importance of the constraints, the weight factors are selected to be $w_1 = 0.6$, $w_2 = 0.4$.

3.3.2 Optimization workflow

Actuators are represented as parametric CAD models. Depending on the DOF, corresponding number of actuator models are instantiated in the robot structure. In each iteration, the dimensions and mass of all selected actuators are sent to CATIA. After updating the CAD models, the mass properties of each module are sent back to the Excel interface. Figure 7 shows the workflow of the optimization process.

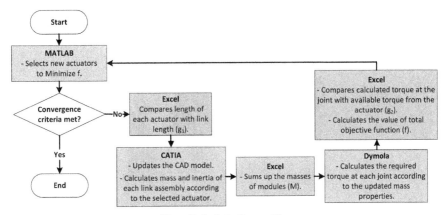

Figure 7. Optimization workflow

3.3.3 Optimization results

The optimization process took nearly 13 hours on a computer with 3.2 GHz microprocessor and 2 GB of RAM. Initial settings for the GA are as follows:

Number of generations = 150
Number of individuals = 20
Generation Gap = 0.9
Number of design variables = 7 (The following results are for a 7 DOF robot)

Each individual x is an array of 7 integers which correspond to the number of actuators. Since the problem is formulated as a minimization problem, the better individuals are the ones with lower objective values. According to Figure 8 and Figure 9, one can see how the objective values of individuals and the average of objective function values have changed during the optimization process.

3.4 Framework Evaluation Using Physical Prototype

Finally, a physical prototype is manufactured using a 3D printer to further evaluate the framework results using some preliminary tests. The physical prototype of the robot along with its CAD model is shown in Figure 10.

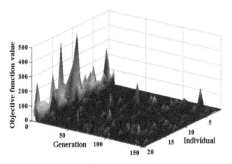
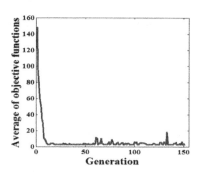

Figure 8. Objective function values for all individuals during the optimization process

Figure 9. Average objective function value in each generation

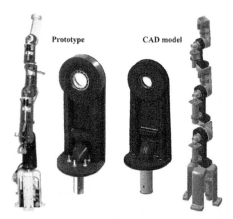

Figure 10. Prototype vs. CAD model

In Dymola, the torque calculated at each joint is dependent on the mass properties calculated in CATIA. To ensure the validity of the calculated torques, a comparison between the mass of manufactured components and the calculated mass of corresponding CAD model should be made. This comparison is shown in Table 1.

Table 1. Mass comparison between CAD model and physical prototype of a 7 DOF modular robot

Mass	Module 1	Module 2	Module 3	Module 4	Module 5	Module 6	Total
CAD Model [kg]	0.954	0.535	0.935	0.441	0.506	0.427	3.798
Prototype [kg]	0.869	0.503	0.886	0.399	0.465	0.389	3.511
Difference (%)	8.9	5.9	5.2	9.5	8.1	8.9	7.6

According to the comparison, shown in Table 1, the calculated mass of each module is higher than the measured mass of the respective module in the prototype. This is the result of assigning higher density values to the CAD models than that of the material used during manufacturing. This ensures that the torque calculation at each joint has a certain amount of safety factor.

As mentioned in Section 3.3, an optimization routine is employed to select the actuators. To evaluate this selection of actuators, the physical prototype is set to follow different trajectories including the one used in dynamic simulation during optimization. These trajectories for the physical prototype are generated in joint space using a trajectory planner implemented in Dymola, which are passed to the corresponding joint actuator in real-time as shown in Figure 11. The trajectory generated in Dymola is transferred to LabVIEW using a MATLAB (Simulink) interface. A motion controller and a servo amplifier are used to generate and amplify the required command signals for each actuator. A control system is also implemented by using position feedback from an encoder mounted on the actuator and velocity feed-forward.

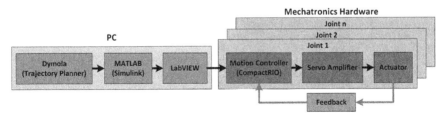

Figure 11. Robot motion-control procedure

The actuators were able to follow their target trajectories with an acceptable amount of positional error. As an example, the performance of an actuator whilst following one of the trajectories is shown in Figure 12. This test confirms that the selection of actuators by the optimization routine is valid.
These tests, though preliminary in nature, help in increasing the confidence of the designer in the presented automated framework.

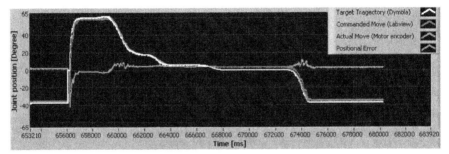

Figure 12. Trajectory tracking performance of an actuator

4 CONCLUSIONS, DISCUSSION AND FUTURE WORK

In this paper an automated framework for design and evaluation of modular robots is presented. A design example focusing on the design of a seven DOF robot using the presented framework is also discussed. An optimization is performed to select the best suited actuators to the robot structure. Finally, a prototype is manufactured to validate the results of the framework.
During the design process, using the automated framework the changes were incorporated in the robot structure rapidly. Using an automated dynamic simulation framework, the dynamic analysis of the robot was done effortlessly, which enabled for a simulation-based optimization to be carried out. Moreover, the framework proved to be useful during the prototyping/manufacturing stage as last-minute changes could not only be incorporated but also were evaluated before production which would have been impossible to perform manually in the same time.
As a future work, optimization of other design variables like link lengths, payload, sharing of modules within a robot family, etc can be studied. For the optimization process, GA was used. As a future work, other algorithms, like Complex algorithm, can be implemented to confirm the optimality of the optimization results. Furthermore, a finite element analysis of the robot structure will be included in the framework to further enhance its functionality. This will allow the designer to optimize the morphology of the robot module for a given payload. For further evaluation of the framework, a full-scale robot based on the manufactured prototype, using industry standard components can be designed and manufactured.

ACKNOWLEDGMENTS
The authors wish to thank Professor Johan Ölvander and Edris Safavi of Linköping University and Dr. Xialong Feng of ABB Corporate Research for their valuable guidance and constructive comments. We would also like to express our gratitude to National Instruments for their sponsorship and technical support.

REFERENCES
[1] Jouannet, C., Lundström, D., Amadori, K. and Berry, P. Design of a Very Light Jet and a Dynamically Scaled Demonstrator. In *46th AIAA Aerospace Sciences Meeting and Exhibit*, Reno, NV, USA, January 2008.
[2] Tarkian, M. and Zaldivar, F. Aircraft Parametric 3D Modeling and Panel Code Analysis for Conceptual Design. In *26th ICAS*, Anchorage, USA, September 2008.
[3] La Rocca, G., van Tooren, M.J.L. Enabling distributed multi-disciplinary design of complex products: a knowledge based engineering approach. *Journal of Design Research*, Vol. 5, No. 3, 2007, pp.333-352.
[4] Tarkian, M. Ölvander, J., Feng X. and Pettersson M. Design Automation of Modular Industrial Robots. In *ASME CIE'09*, San Diego, USA, September 2009.
[5] Cooper D. and La Rocca G. Knowledge-based Techniques for Developing Engineering Applications in the 21st Century. In *Proceedings of 7th AIAA Aviation Technology, Integration and Operations Conference*, Belfast, Northern Ireland, September 2007.
[6] Takian M., Lunden B. and Ölvander J. Integration of parametric CAD and dynamic models for industrial robot design and optimization. In *ASME CIE'08*, New York, USA, August 2008.
[7] Johanson B., Jouannet C. and Krus P. Distributed aircraft analysis using web service technology. In *Proceedings of World Aviation Congress*, Montreal, Canada, September 2003.
[8] Ölvander J., Lunden B. and Gavel H. A Computerized Optimization Framework for Morphological Matrix Applied to Aircraft Conceptual Design. *Computer-Aided Design*, Vol. 41, No. 3, 2009, pp.187-196.
[9] Pettersson M. and Ölvander J. Drive train optimization of industrial robots. *IEEE Transactions on Robotics*, Vol. 25, No. 6, 2009, pp.2047-2052.
[10] Roos F. *Towards a methodology for integrated design of mechatronic servo systems*, Doctoral dissertation, Dept. Machine Design, Royal Institute of Technology, Stockholm, Sweden, 2007.
[11] Krenn, R., Schäfer, B. and Hirzinger G. Dynamics Simulation and Assembly Environment for Rapid Manipulator Design. In *7th ESA Workshop on Advanced Space Technologies for Robotics and Automation*, ESTEC, Noordwijk, Netherlands, November 2002.
[12] Paredis, C.J.J., Brown, H.B. and Khosla, P.K. A rapidly deployable manipulator system. In *Proceedings of IEEE International Conference on Robotics and Automation*, Vol. 2, Minneapolis, MN, USA, April 1996, pp. 1434-1439.
[13] Motoman SIA20, http://www.motoman.com
[14] Nachi MR20, http://www.nachirobotics.com
[15] Petterson M., Andersson J. and Krus P. Methods for Discrete Design Optimization. In *Proceedings of ASME DETC'05, Design Automation Conference*, Long Beach, California, USA, September 2005.
[16] Safavi E., Tarkian M. and Ölvander J. Rapid Concept Realization for Conceptual Design of Modular Industrial Robots. In *NordDesign 2010*, Göteborg, Sweden, August 2010.
[17] Mäntylä M. A modeling system for top-down design of assembled products. *IBM Journal of Research and Development*. Vol. 34, No. 5, 1990, pp.636-659.
[18] Tarkian M, Ölvander J and Lundén B. Integration of Parametric CAD and Dynamic Models for Industrial Robot Design and Optimization. In *2008 ASME International Design Engineering Technical Conferences (IDETC) and Computers and Information in Engineering Conference (CIE)*, New York City, New York, USA, August 2008, pp. 761-769.
[19] Modelica Standard Library, https://www.modelica.org/libraries/Modelica/
[20] Rao S.S. *Engineering Optimization Theory and Practice, 4th Edition: Modern Methods of Optimization*, 2009 (John Wiley & Sons).

INTERNATIONAL CONFERENCE ON ENGINEERING DESIGN, ICED11
15 - 18 AUGUST 2011, TECHNICAL UNIVERSITY OF DENMARK

DESIGNING CONSISTENT STRUCTURAL ANALYSIS SCENARIOS

Wieland Biedermann[1] and Udo Lindemann[1]
[1]Technische Universität München

ABSTRACT
Companies face challenges due to rising complexity through shorter market lifecycles, manifold costumer requirements, additional solutions options and discipline-spanning cooperation. Efficient tools for analyzing and assessing solutions and processes are necessary during the development. Structural considerations are an established approach, which can be used in early phases of the innovation process. Manifold structural analysis criteria such as cycles and clusters are applicable in complexity management. The criteria are interconnected. Their interrelations cause redundant analyses. Developers must choose appropriate criteria combinations to gain significant results efficiently. Researchers have to develop consistent, non-redundant structural analysis scenarios. In this paper we present a model of the interrelations of structural analysis criteria. We propose a procedure for the development of structural analysis scenarios and show its application in one case study. Researchers get a tool for the systematic creation of structural analysis scenarios. Industrial applicators get efficient tools for structural complexity management.

Keywords: Structural complexity management, graph theory, structural analysis, design structure matrix, multiple-domain matrix

1 INTRODUCTION
Companies face challenges due to rising external complexity in engineering design. Reasons are shorter product life cycles, manifold costumer requirements, more solution options due to technological advances and combinations of products and services. Companies react by offering more products and introducing discipline-spanning collaboration. This increases internal complexity. If complexity is not managed successfully it leads to longer development times, cost overruns and wrong decisions with highly detrimental and long-term consequences [1-3].
Structural considerations are an established approach to manage complexity. One of the most used methods in engineering design is the design structure matrix (DSM) [4]. It has been applied to products, organizations, processes and parameters [5]. Its analytical capabilities have been supplemented by graph theory [1] and network analysis [6]. Its modeling capabilities have supplemented by the domain mapping matrix [7] and the multiple-domain matrix [1]. Maurer has proposed a structural approach to deal with complexity in technical systems [1,2].
Manifold structural analysis criteria have been proposed in complex systems research. They are from graph theory [8], network analysis [9], matrix theory [2] and motif analysis [10]. The criteria comprise properties of entire structures like planarity or connectedness, subsets of structures like cycles or clusters, metrics like degree or relational density and visualizations like matrices, graphs or portfolios. Maurer [1] and Kreimeyer [3] have proposed collections of structural criteria. Especially, the introduction of motif analysis has led to an almost infinite variety of structural criteria. The need for careful selection of analysis criteria arises. Developers must choose appropriate criteria combinations to gain significant results efficiently. Kreimeyer sets up a collection of about 50 metrics to evaluate engineering design processes which he models from six viewpoints. Kreimeyer has used an approach based on the goal-question-matrix to guide applicators in choosing the right criterion [3]. His approach neglects the internal dependencies of the criteria. Therefore, many criteria are redundant in at least one view. However, they are consistent as the complete set of metrics has been thoroughly discussed in workshops. This approach to consistency checks is rather tedious and still produces redundant criteria. A more efficient approach is needed which produces non-redundant analysis scenarios.
Researchers have to develop consistent, non-redundant structural analysis scenarios. The scenarios describe how and for which purpose structural criteria are applied. They may comprise multiple steps

of refinement of the analysis results. They tell applicators which criteria can be used at the same time to produce non-redundant results.

Following research questions are addressed in this paper:
- How can consistent structural analysis scenarios be developed?
- How do structural criteria interdepend?
- Which criteria are unique in terms of significance?
- Which criteria refine others?

In this paper we present a model of the interrelations of structural analysis criteria. We propose a procedure for the development of structural analysis scenarios and show its application in one case study. We show how the model can be used for systematic consistency checks and for deriving non-redundant analysis scenarios. Figure 1 shows the context of this work in structural complexity management. The focus is structural analysis. We do not address the applicability of structural criteria (see [11] for a detailed treatise). We focus on the combination of structural analysis criteria.

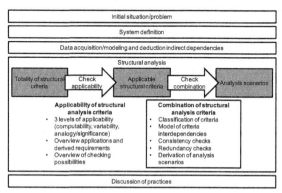

Figure 1: Integration of this work into the general structural complexity management process (based on [1])

The paper is structural as follows. In the next section we present a collection of structural criteria and their interdependencies. In section 3 we describe a procedure to design consistent structural analysis scenarios. In section 4 a case study dealing with networked requirements is presented. In section 5 and 6 the results are discussed and conclusions for structural complexity management are drawn.

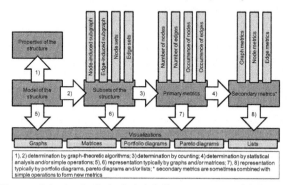

Figure 2: Classes of structural criteria, their definition processes and their representations (partially based on [3])

2 STRUCTURAL CRITERIA AND THEIR RELATIONS

In this section we present the model of the relations among structural criteria. First, we introduce the taxonomy of the criteria. Then, we present the meta-model of the relations. Finally, we present the model itself.

2.1 Taxonomy of structural criteria

Figure 2 shows the classes of the structural criteria. We follow loosely the taxonomies proposed in [1] and [3]. The root criterion is the model of the system structure. Properties and subsets are derived by graph-theoretic algorithms. The subsets subdivide into node- and edge-induced subgraphs and node and edge sets. Subgraphs are partial graphs of the complete structure. They differ in the carrier of information. In edge-induced subgraphs knowing the edges suffices to reconstruct the whole subgraph including its nodes. The node and edge sets do not include edges or nodes respectively. The primary metrics are derived from the subsets by counting the nodes or edges in the subset or by counting how often a node or edge occurs in a type of subset. The secondary metrics are combinations of primary metrics. One way to derive them is to compute mean or extreme values of primary metrics. Another way to derive them is combination by algebraic operations. The criteria are visualized by matrices, graphs, diagrams and lists. In the remaining paper we omit visualizations.

Table 1 shows the taxonomy of structural criteria we use in this paper. It is not exhaustive but easily extensible. We focus on criteria originating in graph theory and network theory. The taxonomy does not contain most of the metrics discussed in [3] as many of them require parameterized or labeled graphs. It does not contain criteria introduced by motif analysis. The definitions of the criteria are available in [1-3,8]

Table 1: Taxonomy of structural criteria (partially based on [1] and [3])

Main category	Sub-category	Structural criteria and references
Sub-sets	Node sets	independent set, vertex cover, adjacency set, active adjacency set, passive adjacency set, reachable set, active reachable set, passive reachable set, separating set
	Edge sets	feedback arc set, edge cover, incidence set, active incidence set, passive incidence set, cut set
	Node-induced subgraphs	connected component, strong component, k-connected component, block, clique, biclique, start node, end node, leaf node, transit node, articulation node, isolated node
	Edge-induced subgraphs	open sequence, closed sequence, path, shortest path, cycle, triangle, elementary cycles, tree, spanning tree, bridge edge
Primary metrics	Number of nodes	Order, order of clique, order of separating set, order of independent set, order of vertex cover, degree, active degree, passive degree, reachability, active reachability, passive reachability
	Number of edges	Size, size of cut set, size of edge cover, distance, path length, cycle length
	Occurrence of nodes	No. of cycles per node, no. of cliques per node, no. of triangles per node, no of shortest paths per node
	Occurrence of edges	number of cycles per edge, number of cliques per edge
Secondary metrics	Graph metrics	average path length, average degree, relational density, diameter, girth, cyclomatic number, vertex connectivity, edge connectivity, independence number, clique number, vertex covering number, edge covering number, degree distribution
	Node metrics	degree centrality, betweenness centrality, closeness centrality, snowball factor, forerun factor, clustering coefficient, activity, criticality
	Edge metrics	Karatkevich number

2.2 Taxonomy of relations among structural criteria

The types of relationship were derived from the description of the criteria and the model shown in figure 2. We differentiate three types: inheritance, composition and derivation. Inheritance and composition only occur among subsets and the structural model. Inheritance means that one criterion is the parent of the other. The child criterion has all properties of the parent and may have additional constraints and properties. Our model allows for multiple parents. Composition means that one criterion is a subset of the other. The part criterion partially defines the properties of the composition criterion. The part criterion has more constraints. Our model allows for multiple parents. Derivation occurs between subsets and primary metrics, between primary and secondary metrics and among secondary metrics. Derivation means that one criterion is used to compute the other. Our model allows for multiple derivation paths but not for their distinction. Table 2 shows the taxonomy of relations among structural criteria.

Table 2: Taxonomy of relations among structural criteria

Relation	Definition	Example
Inheritance	One criterion is a more specific kind of the other	Each triangle is a clique consisting of three nodes.
Composition	One criterion is a subset of the other.	Each cycle is part of one strong component.
Derivation	One criterion is used to derive or compute the other.	The degree is derived from the incidence set of a node.

The taxonomy is incomplete as it omits relations which result from the type of model and the application context. This includes coexistence, correlation and exclusion relations.

2.3 Model of structural criteria and their relations

Figure 3 and figure 4 show the complete model of interdependencies among the criteria listed in table 1. Figure 3 shows the inheritance relations. Figure 4 shows the composition and derivation relations. The figures show the direct relations. We omit indirect relations for the sake of simplicity. The inheritance and composition relations are transitive. For example isolated nodes inherit all properties of leaf nodes, block and k-connected components.

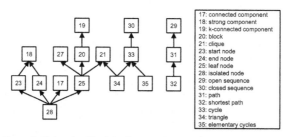

Figure 3: Network of the inheritance relations of the structural criteria

The four inheritance relations of the isolated nodes result from the rigorous interpretation of the criteria definitions. In practice they do not play a prominent role. The exposed position of the connected components in the composition network results from the fact, that most subset definitions require connected graphs as reference system. The prominent positions of degree and order in the derivation network result from the wide application as reference and/or norming metric.

3 DESIGN OF STRUCTURAL ANALYSIS SCENARIOS

In this section we present the theoretical foundations of our approach and a procedure to create structural analysis scenarios.

3.1 Implications of the relations among structural criteria for their significance

The relations shown in table 2 imply constraints for significances of the connected structural criteria. Figure 5 shows the rationale of the constraints. If one criterion is a subset of the other its significance must be more specific and contribute to the significance of the composition criterion. If one criterion is a child of the other its significance must be same but may contain more specific aspects. If one criterion is derived from the other its significance must be more general and may highlight partial aspects. Table 3 shows the implications for the three relations in our model.

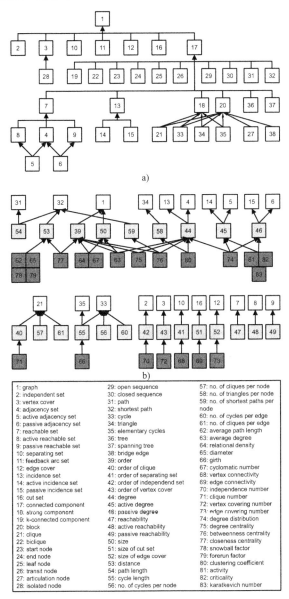

Figure 4: Networks of the composition a) and derivation b) relations of the structural criteria

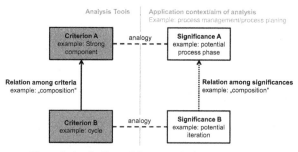

Figure 5: Applications of the network of structural criteria

Table 3: Implications of the relations among structural criteria

Relation	Implication
Inheritance	The child criterion has the same significance as its parent. As the child is more specific and fulfills more conditions its significance may be a special case of the parent's.
Composition	The significance of the composition criterion is an aggregation of the significance of its parts. The significances must not contradict each other. Part criteria of the same composition may not be related.
Derivation	The derived criterion is either a property of a subset of the network or an aggregation of metrics. Its significance is more general than the original criterion's or highlights the original's significance partially.

3.2 Procedure to design structural analysis scenarios

Based on the rationale shown in figure 5 we propose a procedure for designing consistent structural analysis scenarios. They depend on the analysis context and the structural model. The analysis context defines the scope and aim of the analysis. The structural model defines the types of elements and relations. Together, they impose requirements for the applicability of structural criteria (see [11] for a detailed treatise). The requirements reduce the totality of the criteria to applicable ones. By considering their interdependencies the applicable criteria can be reduced and structured to form analysis scenarios. For each combination of analysis context and structural model a new analysis scenario has to be designed. The proposed procedure is shown in figure 6.

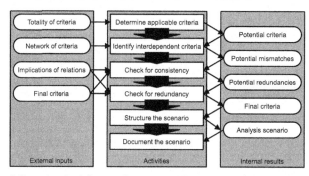

Figure 6: Procedure to define consistent, non-redundant structural analysis scenarios

Determine applicable criteria – As shown in [11] the criteria have to fulfill three criteria: computability, distribution and significance. Computability and distribution impose hardly any limitations. Significance is hard to test and quantify [11]. This step results in a list of potential criteria for the scenario.

Identify interdependent criteria – The interdependency model is reduced to the applicable criteria. All relations in the reduced model have to be for consistency and redundancy. This step results in list of pairs of criteria which represent potential inconsistencies and redundancies.

Check interdependent criteria for consistency – Based on the constraints in table 3 the pairs are tested for consistency. Usually, the significances should be consistent. If they are not the applicability of the connected criteria has to be retested. If a test is not possible one or both criteria have to be omitted. This step results in a list of consistent criteria and a list of potential redundancies.

Check interdependent criteria for redundancy – The remaining pairs are tested for redundancy. The criteria are redundant if they have the same significance. If a pair is redundant one of the criteria can be omitted. Usually the more specific criteria should be omitted to avoid unnecessary computations. This step results in the final list of criteria for the scenario.

Structure the criteria – The criteria are assigned to the analysis aims base on their significance. One criterion may be assigned to multiple aims. The criteria in each group are ordered to form incremental steps of analysis. The ordering can be done by partitioning [4] the criteria network. The most general or most aggregated criteria are placed first. More specific criteria are assigned to subsequent analysis steps as they allow for in depth analysis if necessary. This step results in an ordered scenario.

Document the scenario – The documentation contains a description of all criteria including their significance and computation and the structure of the criteria.

The presented approach is straight forward as it guides the discussion about consistency and redundancy towards the criteria which interdepend. One critical step is to determine the applicability of the criteria. We omit the discussion of applicability for the sake of brevity and refer to [11] for a thorough discussion. All subsequent use the model of criteria interdependencies for consistency checks, redundancy checks and structuring of the critera.

4 A STRUCTURAL ANALYSIS SCENARIO FOR NETWORKED REQUIREMENT MODELS

We use the results of Eben and Lindemann [12] in this case study.

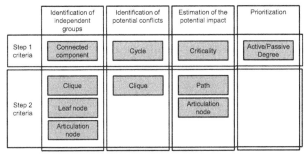

Figure 7: Structural analysis scenario for networked requirements

4.1 Applicable structural analysis criteria

Eben and Lindemann present a collection of 16 structural criteria to analyze requirement networks. The aims of the application are:
- Identification of independent groups of requirements
- Identification of potential conflicts among requirements
- Estimation of the potential impact of changing a requirement
- Prioritization of requirement

In the remaining section we omit all criteria which are not depended on other criteria for the sake of simplicity.

4.2 Interdependencies, consistency and redundancy

Table 4 shows the interdependencies of the criteria. Based on their significance the interdependencies are tested for consistency and redundancy. The test results are shown in table 4. All 13 pairs of criteria are consistent. One pair is redundant. Four pairs are partially redundant. The remaining eight pairs are non-redundant.

Table 4: Combined criteria in requirement models and their significance (based on [12])

Criterion with significance	Criterion with significance	Consistency Redundancy
Composition relations – first column comprises second column		
Connected Component – A subset having no influence on other subsets. It can be regarded separately.	**Clique** – Requirements forming a clique may belong to the same class, and be highly interdependent.	Consistent, non-redundant
	Leaf node – The requirement is influenced by one other directly. Not necessarily the whole requirements structure is affected.	Consistent, partially redundant
	Articulation node - It links subsets of requirements. It may represent an interface or interaction in the system.	Consistent, partially redundant
	Path – Requirements connected via a path to a requirement can be affected by a change of the latter.	Consistent, non-redundant
	Cycle – Requirements connected in a cycle might form a conflict.	Consistent, non-redundant
	Tree – Requirements of a lower hierarchy level may inherit the priority of higher level ones.	Consistent, non-redundant
Inheritance relations – first column inherits from second column		
Isolated node – The requirement can be regarded on its own.	**Connected Component** – A subset having no influence on other subsets. It can be regarded separately.	Consistent, redundant
	Clique – Requirements forming a clique may belong to the same class, and be highly interdependent.	Consistent, non-redundant
	Articulation node - It links otherwise independent subsets of requirements. It may represent an interface or interaction in the system.	Consistent, non-redundant
Leaf node – The requirement is influenced by one other directly. Not necessarily the whole requirements structure is affected.	**Clique** – Requirements forming a clique may belong to the same class, and be highly interdependent.	Consistent, non-redundant
	Articulation node - It links subsets of requirements. It may represent an interface or interaction in the system.	Consistent, non-redundant
Derivation relations – first column is derived from second column		
Criticality – A requirement with a high criticality affects and is affected by a large number of other requirements. It should be given high priority	**Active degree** – Stands for the intensity of the requirement's influence on other requirements.	Consistent, partially redundant
	Passive degree – Passive requirements are affected by many others. It might be a source of uncertainty.	Consistent, partially redundant

4.3 Structural analysis scenario for networked requirements

Figure 7 shows the structural analysis scenario. It comprises four analysis aims, two steps and nine analysis criteria. The criterion isolated node was removed as it is redundant to connected components in the analysis context. Next, we describe each aim and the corresponding criteria in detail.

Identification of independent groups of requirements – The primary criterion is the connected component. It represents groups of requirements which are mutually independent. For more detailed analyses the scenario proposes three criteria: clique, leaf node and articulation node. Cliques represent highly-interconnected requirements which cannot be separated. Leaf nodes represent side requirements which are only loosely connected to the rest of the structure. Articulation nodes represent integrative requirements which have the potential for separating larger groups.

Identification of potential conflicts among requirements – The primary criterion is the cycle. It represents connected requirements which form a loop. For more detailed analyses the scenario proposes cliques. They represent highly-interconnected requirements which cannot be separated.

Estimation of the potential impact of changing a requirement – The primary criterion is criticality. It measures the local connectivity and impact of the requirements. For more detailed analyses the scenario proposes two criteria: path and articulation node. Paths represent modes of impact on the requirements. Articulation nodes represent integrative requirements which have the potential for separating larger groups.

Prioritization of requirement – The two primary criteria are active and passive degree. Active degree measures the intensity of the requirement's influence on other requirements. The passive measures the intensity of the influence on the requirement by other requirements. Passive requirements might be a source of uncertainty. The scenario proposes no criteria for more detailed analyses.

The original paper [12] gave a set of nine criteria for analyzing requirement networks. It showed the applicability of the criteria. In this case study we extended the original approach by checking the consistency and redundancy of the criteria. We showed that all criteria are consistent. One criterion is redundant and therefore removed from consideration. The final scenario comprises eight criteria. Two of them are applicable to two aims.

5 DISCUSSION OF THE RESULTS

We presented a model of the interdependencies of 83 structural criteria, a procedure to define consistent structural analysis scenarios and a case study. The criteria interdepend in three types of relations: inheritance, composition and derivation. Each relation imposes consistency constraints onto the criteria and their significance. The procedure comprises six steps and uses the model for systematic consistency and redundancy checks. The application of the procedure in the case study results in a two-step analysis scenario with only five out of ten applicable criteria in the first step. One applicable criterion was eliminated from the scenario as its significance is redundant. For three criteria subsequent criteria are available which support the refinement of the analyses. This supports incremental analysis approaches which allows for better planning and more efficient work.

Our approach to designing the analyses scenarios is more efficient and goal-oriented than previous guidance approaches such as the goal-question-matrix [3]. The GQM approach requires pairwise comparison of the criterion concerning their significances. In the case study this requires $(n^2-n)/2=(16^2-16)/2=120$ comparisons. In our approach only 13 comparisons are necessary. This corresponds to a time saving of about 90%. Moreover the new approach provides consistency requirements for each type of relation among the criteria. This leads to more savings compared to the GQM approach, where the requirements have to be worked out for each pair anew.

6 CONCLUSION

Our results allow for the first time the systematic creation of structural analysis scenarios under consideration of the inherent complexity of the analysis criteria and their interdependencies. They provide researchers with a tool for structuring and guiding their work. Industrial applicators get efficient tools for structural complexity management. The scenarios guide the planning and application of structural analysis criteria. They give an overview of the applicable, non-redundant criteria. They allow for efficient access to the criteria via the application context and aims. The handling of complex systems becomes more efficient.

Our results are not comprehensive. The network model neglects the existence of relations based on coexistence, correlation and exclusion. To include them is a task in future research. The taxonomy neglects criteria, which require parameterized or labeled graphs or were introduced by motif analysis. These need to be included to cover the complete spectrum of structural analysis criteria. However, there is no consensus in the research community, which criteria are developed. Through recent developments the amount of available criteria has reached the manageable limit. We think that our approach helps in focusing, guiding and structuring the work with structural analysis criteria.

ACKNOWLEDGEMENTS

This research was made possible through the generous funding by the German Research Foundation (DFG) in the project A2 ("Analysis of discipline-spanning changes in product development") within the Collaborative Research Centre SFB 768 ("Managing cycles in innovation processes").

REFERENCES

[1] Maurer M. Structural Awareness in Complex Product Design, 2007 (Dr.-Hut, Munich).
[2] Lindemann U., Maurer M. and Braun T. Structural Complexity Management - An Approach for the Field of Product Design, 2009 (Springer, Berlin).
[3] Kreimeyer K. A Structural Measurement System for Engineering Design Processes, 2010 (Dr.-Hut, Munich).
[4] Steward D.V. Design Structure System: A Method for Managing the Design of Complex Systems. IEEE Transactions on Engineering Management, 1981, 28(3), 71-74.
[5] Browning T.R. Applying the design structure matrix to system decomposition and integration problems: a review and new directions. IEEE Transactions on Engineering Management, 2001, 48(3), 292-306.
[6] Collins S.T., Yassine A.A. and Borgatti S.P. Development Systems Using Network Analysis. Systems Engineering, 2008, 12(1), 55-68.
[7] Danilovic M. and Browning T.R. Managing Complex Product Development Projects with Design Structure Matrices and Domain Mapping Matrices. International Journal of Project Management, 2007, 25(3), 300-314.
[8] Gross J.L. and Yellen J. Graph Theory and Its Applications, 2005 (CRC Press, Boca Raton).
[9] Cami A. and Deo N. Techniques for Analyzing Dynamic Random Graph Models of Web-Like Networks: An Overview. Networks, 2008, 51(4), 211-255.
[10] Milo R., Shen-Orr S., Itzkovitz S., Kashtan N., Chklovskii D. and Alon U. Network motifs: simple building blocks of complex networks. Science, 2002, 298(5594), 824-827.
[11] Biedermann W. and Lindemann U. On the Applicability of Structural Criteria in Complexity Management. In 18th International Conference on Engineering Design (ICED11), Copenhagen, August 2011. (Design Society) (paper no. 191 accepted on 26th Mar 2011).
[12] Eben K.G.M. and Lindemann U. Structural Analysis of Requirements – Interpretation of Structural Criterions. In Proceedings of the 12th International DSM Conference, Cambridge, July 2010, pp. 249-261 (Hanser, Munich).

Contact:
Wieland Biedermann
Technische Universität München, Institute of Product Development
Boltzmannstr. 15, D-85748 Garching, Germany
Phone +49 89 289-15129
Fax +49 89 289-15129
biedermann@pe.mw.tum.de
http://www.pe.mw.tum.de

Wieland Biedermann is a scientific assistant at the Technische Universität München, Germany, and has been working at the Institute of Product Development since 2007. He has published several papers in the area of structural complexity management.
Udo Lindemann is a full professor at the Technische Universität München, Germany, and has been the head of the Institute of Product Development since 1995, having published several books and papers on engineering design. He is committed in multiple institutions, among others as Vice President of the Design Society and as an active member of the German Academy of Science and Engineering.

INTERNATIONAL CONFERENCE ON ENGINEERING DESIGN, ICED11
15 - 18 AUGUST 2011, TECHNICAL UNIVERSITY OF DENMARK

DESIGN OF AN UPPER LIMB INDEPENDENCE-SUPPORTING DEVICE USING A PNEUMATIC CYLINDER

Norihiko Saga, Koichi Kirihara, and Naoki Sugahara

ABSTRACT
This paper describes the design of a device to support a patient's upper limb motion. For safety, light weight, and flexibility, it uses a pneumatic cylinder for which the optimum arrangement is presented. The independence-supporting device has two modes corresponding to livelihood support and rehabilitation. A compliance control system and a position control system are designed for those modes. We evaluate the independence-support mode's effectiveness through some experimentation.

Keywords: Medical systems, Pneumatic systems, Actuators, Human-machine interface, Quality of work life

1. INTRODUCTION
Restriction of motion of a joint's range is called contracture. A joint's range of motion exercise is effective for preventing contracture. However, if the exercise is performed by a physiotherapist, then a joint's range of motion will improve, but if the time not to exercise is long, then contracture will progress again. Then, a rehabilitation instrument is necessary for motion exercise to be performed after exercise with the physiotherapist.

Some continuous passive motion (CPM) devices are used as rehabilitation instruments for the maintenance or restoration of joint's range of motion (ROM). During CPM therapy, the joint area is secured to the CPM device, which then moves the affected joint through a prescribed arc of motion for an extended period of time. In fact, CPM devices are available for numerous joints such as the knee, ankle, jaw, hip, elbow, shoulder, and finger.

Nevertheless, most instruments use motors to provide high power. For that reason, they are heavy and large. Installation and movement of instruments at facilities are difficult. It is also difficult to use such devices freely at home. Additionally, it is not possible to use such a device at rest, which is the most effective time for rehabilitation training.

We therefore specifically examine a rehabilitation instrument that is small, light, and easily put on and taken off. According to an annual report on the aging society in Japan, a super-aging society is expected in the near future [1]. A simultaneous increase of patients and a decrease of medical workers is feared. In addition to aging, the number of handicapped people is expected to increase because of sickness and injury, and for other reasons. Because of many people's physical handicaps, activities of daily life (ADL) will become difficult. Moreover, the burdens for those giving treatment will increase.

Handicapped people require training for rehabilitation to recover their ability to use upper limbs. Additionally, impairment of abilities is known to be can recoverable through rehabilitative training. In a clinical scene of rehabilitation, the patient and the occupational therapist (O.T.) train together. The O.T. demonstrates and facilitates motions that give a constant load to patient's limbs and which move their limbs, slowly repeating flexion and extension. The machine can often substitute for the O.T.'s motion during rehabilitation. Some rehabilitation devices have been developed [2]–[8]. In a clinical scene, such a device should be a simple mechanism with a simple control system that is easy to use.

Therefore, we developed an upper limb rehabilitation support device with a wide operating range. It is compact and has a link mechanism. Welfare apparatus, such as the rehabilitation support device that we developed, must be safe, flexible, and lightweight because this device must have contact with humans during operation. A DC motor and a hydraulic actuator are used for industrial robots. However, if we use these actuators for welfare apparatus, the system would become complex and bulky, which is undesirable. The necessary functions increase when a target patient extends the device. Thereby the rehabilitation device becomes ever larger, and its operation becomes increasingly complicated. Therefore, we used a pneumatic cylinder to drive the device because the shock can be absorbed using air compressibility: it has a simple structure with a high power–weight ratio.

For this study, the target patients are few and the rehabilitation instrument can be designed to have only two modes with two control systems, which many patients find necessary. Furthermore, a position control system is applied on the livelihood support device and a compliance control system is applied on the device instead of an O.T.'s motion of rehabilitation training. Some experiments were performed to evaluate the device and its control system.

2. DESIGN OF SUPPORT DEVICE

Fig. 1 depicts the upper limb independent support device. This device has five degrees of freedom by virtue of its link mechanism. It consists of joint 1, joint 2, and joint 3. Joint 1 reciprocates on the y-axis by a linear guide to supports the upper limb for the reach action, as depicted in Fig. 2(a). Joint 3, with an attached gas spring (Y0061, Tokico; Hitachi Ltd.), rotates around the x-axis, as depicted in Fig. 2(b). Joint 2, with an attached a pneumatic cylinder, rotates around the x-axis to support arm flexion and extension, as depicted in Fig. 2(c). Joint 1 and joint 3, with attached rotation joints, can rotate around the z-axis, as shown in Figs. 2(d) and 2(e). Joint 2 is operated actively by a pneumatic cylinder, but the other joints are operated individually by the patient.

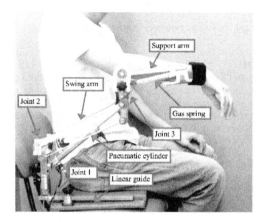

Figure 1. Independent support device

(a) Motion of joint 1
by linear guide

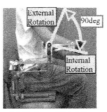
(b) Motion of joint 3
by gas spring

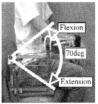
(c) Motion of joint 2
by pneumatic cylinder

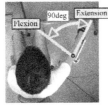
(d) Motion of joint 3
by rotation joint

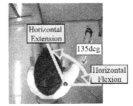
(e) Motion of joint 2
by rotation joint

Figure 2. Motion for the livelihood and the rehabilitation

A simple link mechanism is used with the device. However, it has a wide operating range. Consequently, using the device, the patient can operate the upper limb without unpleasantness. Additionally, the weight of the device is about 4 kg. Therefore, it is possible to do training without choosing a particular place because the device is portable

3. INDEPENDENT SUPPORT FUNCTION

We assume that patients with paralysis who remain independent will use the device, as will patients with decreased muscular power attributable to an accident or aging. The independent support device has two support functions that correspond to livelihood support and rehabilitation contents. By undergoing rehabilitation with a device, it is expected that the treated person's load is decreased, and that a patient can therefore train at home.

3.1 Livelihood support function "Mode A"

"Mode A" is a function to recover practical function of an upper limbs. The device supports training that operates the upper limb on the desk, as portrayed in Fig. 3(a). In this function, a position control (on rotation angle of joint 2) is applied to support an upper limb's vertical motion (i.e. shoulder flexion and extension). According to this function, a patient who has trouble operating the arm to resist gravity can train easily on a desk.

3.2 Rehabilitation support function "Mode B"

The device supports the patient's upper limb flexion and extension motion for rehabilitation, as portrayed in Fig. 3(b); because of this function, the patient's muscular power recovery and movable region of expansion are expected. In a clinical scene, the O.T. adjusts training considering the level of the patient's trouble. In this "Mode B", compliance control was applied to operate with the device as an occupational therapist. The patient can conduct ergo-therapy corresponding to the level of the patient's muscular power.

(a) Mode A (b) Mode B

Figure 3. Support Function

4. CONTROL SYSTEM

Fig. 4 depicts the device control system. The electropneumatic regulator (ETR200-1; Koganei Corp.) regulates the pneumatic cylinder's (T-DA20_100; Koganei Corp.) inner pressure. A rod in the pneumatic cylinder expands and contracts when the pneumatic cylinder's inner pressure changes. The swing arm rotates around the y-axis. The rotation angle of joint 3 is measured using the rotary position sensor. The load cell (LMA-A-100N; Kyowa Electronic Instruments Co. Ltd.), installed in a stand for the elbow, measures the force that the patient is adding.

A compliance control system for "Mode B" is applied to change the joint 3 stiffness. The compliance control equation is written as

$$\tau = K(\theta_d - \theta) \quad (1)$$

Therein, $\_d$ stands for the desired angle, $\_$ signifies the measured angle, $\_$ denotes the torque of the joint 3, and K represents the constant of stiffness. In addition, $d\_$ is defined as the difference between the desired angle and the measured angle $(\_d - \_)$.

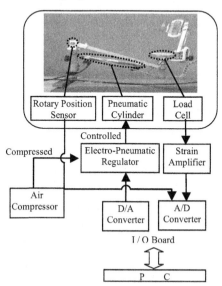

Figure 4. Control system of rehabilitation support device

5. EXPERIMENTS

In this section, we describe compliance control for Mode A and position control for Mode B. Furthermore, we evaluate the effectiveness of the rehabilitation support mode through experimentation.

5.1 Position control "Mode A"

This experiment is performed with and without a load (wrist part, 1 kg; elbow part, 1.8 kg), which assumes the weight of a human arm. The loads of the wrist and elbow part were estimated using the ratio of the weight of each part to the weight of a human. Moreover, the target value was given from 110 deg to 90 deg in the ramp input, which was assumed to represent the arm extension (shoulder joint).

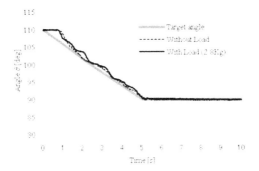

Figure 5. Experimental results of position control

Fig. 5 portrays the experimentally obtained results of position control. The rotation angle smoothly followed the target value without overshooting. It converged to the target angle (90 deg). Therefore, when the device is used for assistance of rehabilitation training on a desk, the patient's arm can be moved to the position that the patient desires. The device is useful safely, without giving discomfort to the patient.

5.2 Compliance Control "Mode B"

The rehabilitation support device is fixed with a jig so that the rotation angle _ might be 90 deg. We measured the $d\_$ and generated torque _.

Fig. 6 presents the experimentally obtained results of compliance control. The solid line represents the theoretical value of the generated torque from eq. (1). The gray solid line shows torque according to the weight of the arm of a typical adult male (65.7 kg body weight; arm weight 3.2 kg). Comparison of experimentally obtained results and theoretical values presents a strong correlation. Figure 6 shows the generated torque as 18 Nm; the torque by the arm weight is 8.9 Nm, as depicted by the gray solid line. Sufficient margins exist from the torque by the weight of the arm to the limit of the generation torque. Therefore, the patient can add force from the state to put the arm on the device.

We confirmed that the joint 3 stiffness rose by increasing the constant of the stiffness through this experiment. When actually using the device for rehabilitation, we assume that the constant of stiffness is set low for a patient with weak muscles, and that the constant of stiffness is set high for patients with strong muscles, presumably those in advanced stages of recovery.

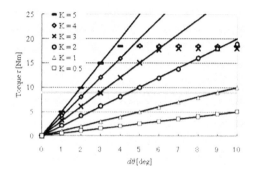

Figure 6. Experimental results of compliance control

6. EVALUATION OF EMG

Evaluation of the upper limb rehabilitation device measured the EMG of the body. The measurement part is a greater pectoral muscle, a broadest muscle of back, and a deltoid muscle front part in each of Mode A and Mode B. Furthermore, Fig. 7 and Fig. 8 present measurement results of EMG. The EMG that was effective for the dorsal flexion was confirmed.

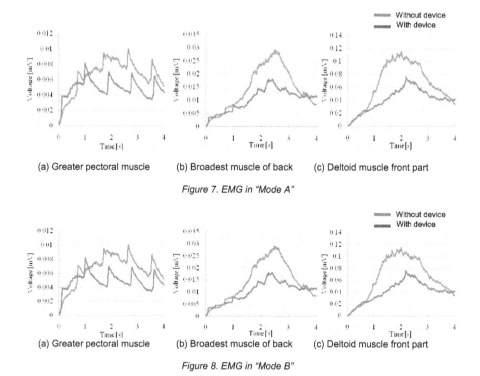

(a) Greater pectoral muscle   (b) Broadest muscle of back   (c) Deltoid muscle front part

Figure 7. EMG in "Mode A"

(a) Greater pectoral muscle   (b) Broadest muscle of back   (c) Deltoid muscle front part

Figure 8. EMG in "Mode B"

7. CONCLUSION

For this study, we developed an upper limb independence support device using a pneumatic cylinder. A summary of the obtained results is presented as follows.

- By arranging the pneumatic cylinder optimally, the device is compact. Nevertheless, it provides widely various movements.
- The device has two support modes corresponding to livelihood support and rehabilitation contents. A position control system was applied in Mode A to support recovery of a patient's practical function of the upper limb. In Mode B, a compliance control system was applied to support a patient's muscular power. In Mode B, to support recovery of a patient's practical recovery and movable region expansion, a compliance control system was applied.
- The position control performance for Mode A was verified experimentally. The results confirm that the rotation angle of joint 3 followed the target angle smoothly.
- A compliance control performance for Mode B was verified experimentally, revealing high correlation with measured values and theoretical values of torque of joint 3.

These results confirmed that the device that we developed can support a patient's training activities.

REFERENCES

[1] Government of Japan Cabinet office, Annual report on the Aging, 2006
[2] Kazuo Kiguchi, Takakazu Tanaka, Keigo Watanabe and Toshio Fukuda, Design and Control of an Exoskeleton System for Human Upper-Limb Motion Assist, IEEE/ASME International Conference on Advanced Intelligent Mechatronics , pp.926--931, 2003
[3] Keijiro Yamamoto, Kazuhito Hyodo, Mineo Ishii and Takashi Matsuno. Development of Power Assisting Suit for Assisting Nurse Labor, JSME International Journal Series B, Vol.45, No.3, pp.703—711, 2002
[4] Norihiko Saga, Naoki Saito, Seiji Chonan, Development of a Support Arm System Using Artificial Muscle Actuator and Gas spring, 2nd Frontires in Biomedical Devices Conference, 2007
[5] Norihiko Saga, Takashi Saikawa and Hideharu Okano, Flexor Mechanism of Robot Arm Using Pneumatic Muscle Actuators, IEEE International Conference on Mechatronics & Automation, pp.1261—1266, 2005
[6] Toshiro Noritsugu, Fuminori Ando and Takashi Yamanaka, Rehabilitation Robot Using Rubber Artificial Muscle (1st Report Realization of Exercise Motion with Impedance control), Journal of RSJ, Vol.13, No.1, pp.141-148, 1995
[7] Zeungman Bien, Dae-Jin Kim, Myung-Jin Chung, Dong-Soo Kwon and Pyoung Hun Chang, Development of a Wheelchair-based Rehabilitation Robotic System (KARESII) with Various Human-Robot Interaction Interfaces for the Disabled, Advanced Intelligent Mechatronics , pp.902—907, 2003
[8] Robert Richardson, Michael Brown, Bipin Bhakta and Martin Levesley : Impedance control for a pneumatic robot-based around pole-placement, joint space controllers. ELSEVIER Control Engineering Practice 13, pp.291-303, 2004

INTERNATIONAL CONFERENCE ON ENGINEERING DESIGN, ICED11
15 - 18 AUGUST 2011, TECHNICAL UNIVERSITY OF DENMARK

ACTUATION PRINCIPLE SELECTION – AN EXAMPLE OF TRADE-OFF ASSESSMENT BY CPM-APPROACH

Torsten Erbe[1], Kristin Paetzold[2], Christian Weber[1]
(1) Technische Universität Ilmenau (2) Universität der Bundeswehr München Germany

ABSTRACT
The generation of motion is a task of many technical systems. Customized drive systems constitute a challenge to the selection of a suitable actuator during the development of drive systems. Common approaches and tools for actuator selection are limited to the selection of known actuators from a database. However, especially in cutting edge technology conflicts of requirements complicate the selection of a suitable actuator or even actuation principle.
This paper uses the CPM / PDD approach to describe a concept of visualizing the properties and characteristics of actuator-principles in order to identify potential for an influence by the designer. Based on the context of precision engineering, measures to meet conflicting objectives and to identify convenient characteristics for adaption as well as limitations of the proposed approach are discussed.

Keywords: decision-aid, requirement management, CPM/PDD, actuator selection

1   INTRODUCTION
Motion generation is an important function in many technical systems. Various environmental conditions and the variety of the required movements result in a trend towards automation and an increasing number of tailor-made drive systems. By using a suitable control system state-of-the-art gear-less-drives allow almost every motion pattern without using further transmission components, e.g. gears or mechanisms. Therefore, the task of selecting or designing an actuator, which is optimally adapted to the given requirements, is an important task for engineers developing drive systems.
Although all relevant actuation principles are scientifically well characterised and rule sets are available for their design, in practice early ad-hoc decisions for this central component are often made purely based on preferred options and/or experience from previous cases; a systematic search through the entire range of potential solutions is usually not done. This is despite the fact that requirements of the actuator can vary considerably from one case to the next, depending on the application (i.e. motion range, environmental restrictions) and that some requirements may be difficult to formulate in a general manner at all (cost, reliability of supplier, etc.). A careful actuator selection is of particular interest in cutting edge technology, such as low cost automation, high precision engineering or micro systems engineering.
This paper refers to drive systems in the field of precision engineering, which are based on electrical and electromagnetic actuators. In this application area, the achievement of a specific resolution or repeatability is often more important than cost reduction. Trade-off assessment is of particular interest since the additional implementation of other requirements (holding force, performance in high magnetic fields, self-heating, aspect ratios ...) is often decisive. The required parameters are properties of the actuator system and can only be determined by the development engineer via changing the material and/or geometrical characteristics of the potential actuator.
The distinction between properties, which can only be influenced implicitly, and characteristics, which can be affected directly by the designer, is the foundation of the CPM/PDD-approach [1]. The existing (as said: generally well investigated) descriptions allow mapping the characteristics of an actuator to the its properties, i.e. they model the relations between the two, which, according to the CPM/PDD approach, are crucial for the development process..
The intention of this paper is to integrate existing knowledge on actuator principles into the CPM/PDD framework in order to make actuator evaluation and selection (and development???) better structured. This approach does not address actuator developers but design engineers who intend to evaluate and select an actuator principle for tailor-made drive systems –who are expected to have some overview knowledge of (some) actuators and actuation principles. The purpose of the approach is to provide

these engineers with a method for the qualitative comparison and assessment of actuation principles referring to conflicting requirements during conceptual design stages. With regard to the context of precision engineering, the mapping concept will be presented with a view to the selection of actuation principles. The difference between actuator and actuation principle selection will be highlighted and limits as well as conclusions referring to the concept of actuation principle selection will be discussed in this paper.

2 ACTUATION PRINCIPLE SELECTION VS ACTUATOR SELECTION

In the process of designing a drive system, the engineer has to decide between using an *actuator*, which is available on the market, or developing one independently (often with the assistance of a specialised supplier), based on a common *actuation principle*. Actuators are usually selected from a convenient database – i.e. a paper or software-based product catalogue.

There are different approaches and tools to support actuator selection. A practical solution for this problem is the one proposed by HUBER [2]. He adapted the concept of the *material selector* software [3] and applied it to actuator selection. This resulted in a number of particular ratios/indicators of actuator properties, which are called *performance indices*. Working ranges of various actuation principles can be visualised according to the required quantitative performance indices.

Proposals for software-based implementations are delivered by [4] and supplemented by a *Skyline/Pareto* based approach of visualisation in [5].

The main advantage of the *material selector* based approach is a small amount of required data sets (suitably distributed) to give a representation of the working space of different actuator principles. However, the range of properties in the application area of actuators is larger and depends on the actuator principle, which also makes it more heterogeneous than the one of materials. The dependencies between the individual parameters are only given implicitly. In addition, two-dimensional graphic representations lead to the assumption of a direct one-to-one dependency, which does not exist in that manner. The number and complexity of inter-dependencies between actuator properties vary according to the actuation principle. Furthermore actuators, which are capable of working in the same operating area, are in many cases characterised by distinct and not directly comparable qualitative properties.

Another approach was described by EGBUNA [6], addressing the actuator selection in the context of low-cost automation. He proposed a qualitative actuator selection based on a subtractive rule set. The actuator selection support is focused on cost comparison, for low-cost automation on the basis of lifetime-relative costs. Although low-cost automation and precision engineering are different application fields, the approach is, in principle, also applicable for precision engineering tasks.

Both approaches focus on the selection of existing actuators, leading top a sort of interactive product catalogue. That is why those are of limited use for the estimation of trade-offs in the development of new actuators.

While EGBUNA's approach enables the selection of actuators with respect to qualitative and quantitative parameters, it is impossible to search beyond the solution space of given data sets. By contrast, HUBER's approach allows interpreting the distances between the required quantitative parameters and the "working area spots" of known actuators.

In both approaches described, the actuator selection is conducted by providing an overview of many different actuators represented by numerous (ideally arbitrarily selectable, but not yet fully implemented [5]) properties. For actuation principle selection, however, it is necessary to represent the influencing options and their influence on the required properties. This involves an at least qualitative model of the dependencies of the properties for different actuation principles.

3 CHARACTERISTCS AND PROPERTIES SUBJECTED TO ACTIVE PRINCIPLES

The CPM/PDD approach proposed by WEBER (e.g. [1], [7]) addresses two fundamental aspects of the theory of technical systems:

- *Characteristics-Properties Modelling* (CPM) as an approach for the modelling of technical products using its properties and characteristics, and
- *Property-Driven Development* (PDD) as a process model for the development of technical products based on their properties and characteristics.

The CPM/PDD-approach is still subject of ongoing research. Therefore, different distinctions between the terms properties and characteristics as well as between implicit/indirect/dependent and explicit/direct/independent can be found in articles of various authors (cf. Tab.1).

Table 1. Selection of different definitions for the terms "property" and "characteristic" with respect to the influence of the designer.

Author/Reference	can be		Comments to the term definition
	directly influenced by the designer	not directly influenced by the designer	
WEBER [6], [8]	Characteristics	Properties	Characteristics as like properties are generally defined and not limited to physical products
HUBKA [9], [10]	Elementary Designproperties	External properties	Structure and behaviour are considered as properties, attribute and property are alternative terms and considered as a subset of characteristics
EEKELS/ ROOZENBURG [11], [12]	intensive properties as the sum of physio-chemical form and geometric shape	Extensive Properties	Extensive properties are often associated with a "property pattern". This "pattern" sums up both the properties and their dependencies. (see also [15])
EHRLENSPIEL/PONN/ LINDEMANN e.g. [13]	Direct Properties Consistance Characteristics Relational Characteristics	Indirect Properties Functional Characteristics	„Distinguishing" properties are called characteristics, they have a meaning (quality) and an expression (quantity)
SUH [14]	„design parameters" basically – to a limited extend – as well as the dependencies of the "design matrix"	„functional requirements"	Proposed by SUH the "axiomatic design" approach is not restricted to the domain of mechanical engineering resulting in application dependent „parameters" and „requirements"
BIRKHOFER/ WÄLDELE [18]	Independent properties	Dependent properties	The difference between independent and dependent properties is used to discuss product models, development processes an supporting tools

Essential for WEBER is the distinction between properties (only indirectly assignable by the designer) and characteristics (directly assigned by the designer). Furthermore, he describes the behaviour of a technical system as the sum of its properties.

For further discussion, the term distinction by WEBER [1] is used. According to the CPM/PDD-approach and this distinction the analysis of a product can be understood as the determination of its properties (P_i) by the characteristics (C_j), and its synthesis as the determination of its characteristics by the properties (Fig. 1) [1].

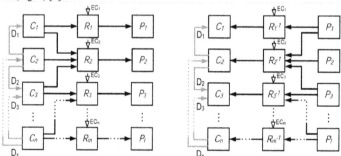

Figure 1. Types of relations in the CPM/PDD-approach. The arrows represent the determination of properties by characteristics in the case of the analysis (on the left) and vice versa (on the right) in the case of the synthesis of technical systems (cf. e.g. [1])

The determination can be displayed by relations (R_k, R_k^{-1}) and is, besides the characteristics and the properties, dependent on external conditions (EC_k). In addition, there are dependencies (D_x) between the characteristics themselves.

The essential challenge of the theoretically powerful CPM/PDD-approach is the identification of the relations (R_k^{-1}) in the process of the synthesis. The main reason is that the properties, which the designer attempts to achieve, are the effects of the characteristics. Since it is impossible to conclude from the effect to the cause unequivocally, the synthesis process is undetermined. Moreover, usually only a few properties of the ones required of a product are given from the beginning. Specifying only a few properties – requirements and definition of other "means" of the product in general – results in a reduced model of the intended behaviour (this again can be understood as the purpose function, cf. e.g. [12], [15], [16]).

Another difficulty is to create a manageable CP-model. The attempt to create as unambiguous relations (R_k, R_k^{-1}) as possible would result in an impractical and complex map of physical interactions. Thus, the set of relationships needs to be limited to the required ones instead of expanding it.

The consideration of actuator principles, however, allows building different models of properties/ characteristics-relations of each actuator principle on the basis of existing descriptions.

4 CONFLICT OF OBJECTIVES DURING ACTUATOR DESIGN

Transforming non-mechanical (usually electrical) energy into mechanical energy of motion is the purpose of a drive system and its basic function is precisely defined prior to the selection or design of an actuator. The parameters listed below represent the basic requirements (properties) for a general actuator system and are predefined in the beginning of its design process:
- mode of motion (limited or unlimited),
- direction of motion (reversible or non-reversible),
- degree of freedom (DOF) of the actuator (between 1 and 5),
- type of motion (rotational or translatory),
- ability to maintain position without actuator energy supply and
- (effective) power output

In particular cases all these requirements have to be fulfilled, so that they cannot be a differentiating criterion for the actuator or actuation principle selection.

The first step during drive system design is to search for existing actuators (e.g. via the actuator selection tools). If no existing actuator meets the requirements, the cause could be that (cf. change request types in CPM/PDD, [17]):
- a different set of required properties,
- different external conditions or
- different (direct) restrictions of characteristics

are given in this particular case.

The option of the designer is developing an existing actuator into the region (cf. "working area spots") of the new required properties/external conditions. As stated before, this can only be accomplished by modifying the *Gestalt*-characteristics – the sum of the geometry, material and their state (e.g. magnetisation, hardness, tolerances …).

However, if properties/external conditions cannot be met in this process, conflicts will occur. According to PDD a "conflict" exists, if in order to achieve a required property (or properties) one or more characteristic(s) are modified and this change leads to a deterioration of one or more other required property/properties, which in turn can not be compensated by changing other characteristics (cf. [17]).

If there are several actuation principles that can be developed to meet the given requirements, this may lead to different conflicts in each case; the task is then to identify the most promising actuator principle without having to detail all options. What *promising* means can be rated on the basis of:
- the type of characteristics eligible for an adaption (size, material or condition of the latter)
- the relative magnitude of adaption of eligible characteristics (also dependent on the type of relation, e.g. linear, logarithmic, cubic, …)
- the absolute magnitude of the eligible characteristic itself (e.g. hardness, modulus of elasticity but also the grain size with respect to magnetisation are physically limited).

for each potential actuation principle. The assessment itself depends on the set and types of requirements; besides that, physical boundaries, technological capabilities, lot sizes, technical periphery etc. have to be taken into consideration.

5 USE CASE – EXAMPLES AND OPPORTUNITIES TO SOLVE TRADE-OFFS

For the purpose of a simple representation, the CPM model was rearranged. The aim was the separation into input and output parameters, as it is common for actuator representations. The parameters are separated with respect to the forms of energy, as shown in Fig. 2 and Fig. 3 for electro-magneto-mechanical and piezoelectric actuators.

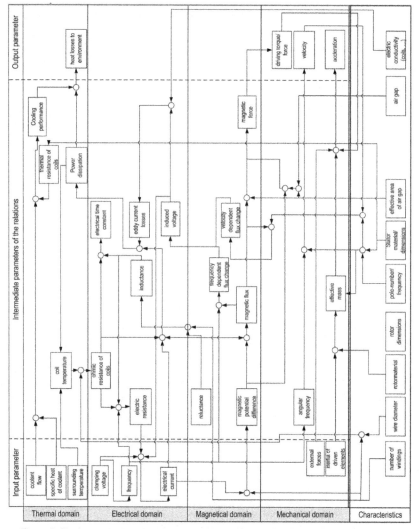

Figure 2. Electro-magnetic actuator-model arranged in order to distinguish input and output parameters as well as characteristics

Because of the different energy domains the models of electromagnetic actuators (Fig. 2???) appear to be more complex than the ones of other actors which internally do not have further energy conversions (apart from the basic one between input and output energy and omnipresent heat losses).
The task is to identify for which potential actuation principle(s) it is, in order to satisfy particular requirements (required properties), easier to make changes to the characteristics without compromising other properties.. The proposed model helps to illustrate the dependencies qualitatively

and visualises the required changes of characteristics to achieve the desired properties and - vice versa - the qualitative impacts of changes of characteristics on other properties.

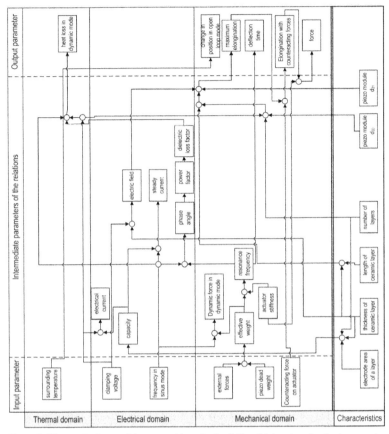

Figure 3. Piezo-actuator-model arranged in order to distinguish input and output parameters as well as characteristics

Based on the property or characteristic that has to be changed, the effects can be mapped comparably to a cascade or chain of opportunities. By using fuzzy (+ +, +, 0, -, --) correlations, a rough estimate of the impact can be made (Fig.4).

The necessary changes to achieve the particular (set of) requirements are dependent on the actuation principle.

Because relations between characteristics and properties are fixed by the actuation principles the conflicting objectives can only be resolved by:
- changing dependencies (resulting in a new actuator configuration; e.g. piezo, piezo-stack, inch-worm drive, …),
- redistributing requirements of the drive system (e.g. using a coarse-fine drive concept instead of a direct drive or to use yet a gear),
- changing characteristics (by extending the scale of the parameter set – e.g. change materials; ferrit to neodymium) or
- weakening the requirements of the particular actuator (appears at first sight preposterous, but is sometimes the last resort).

An alternative strategy could be considering the above listed measures starting from the bottom: In most practical cases, the development of a new actuator configuration is, due to the complexity of the required special knowledge, the least preferred option and usually not viable for an unexperienced designer.

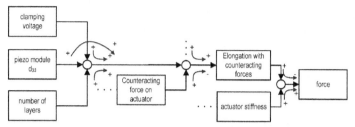

Figure 4. Qualitative correlations for the estimation of the impact of changes of characteristics; exemplified on the piezo actuator-model (reduced)

Whether and when requirements can be redistributed or weakened depends on the particular case. As the decision on the suitability of an actuator can be reduced to:
- the applicability of the type of energy conversion itself (magnetism, heat, etc.) or
- ratios between properties and characteristics (specific sizes, densities, stiffness, aspect ratios etc.)

the starting points for the evaluation of the suitability of an actuator principle are physical or technological limitations. Changing characteristics is (physically) limited, which must be considered. For example, the density of functional materials (e.g. transformer plate metal, piezo ceramic) can not be reduced easily to half of the original value. Also, air gaps in the nm range are technically hard to achieve.

6 DISCUSSION AND CONLUSIONS

The intention of the approach presented in this paper was to give the designer the opportunity to identify his/her options of influence on a known actuator design with respect to new requirements and not to give design rules for developing new actuators from scratch. It provides a supplement to the existing actuator selection tools and approaches. Approach and measures how to meet conflicting objectives were discussed. However, general rules for suitable trade-offs are hard to formulate due to the variety of possible sets of requirements.

The proposed model represents the relations between characteristics and properties for a special type of technical solutions - actuators. This was possible because the relations are well described for most of the actuation principles. Basically a comparable approach is conceivable for other application areas where a large number of potential solutions exists and the solutions are well understood and described.

The presented approach seems to be similar to a design structure matrix, but differs in purpose and use. The purpose is the assessment of the accessibility of outputs by quantitative changes of characteristics in the early design stages. For this purpose, either the characteristics influencing the desired outputs are determined or the impact of the changes of characteristics on the output values can be estimated. Changing the structure - if possible at all - is, however, reserved for specialised actuator developers.

The reasoning shown allows the designer to identify convenient characteristics and to assess them on a qualitative basis. A quantification of individual dependencies by using appropriate literature is possible. Ongoing research will be focused on software implementation and interlinking of the models of different actuation principles and how to link them to (existing) actuator/design catalogues. The implementation of mathematical equations for the relations to support the decision making process is also subject of current research. Linked to that, an important question is which benefits the calculation can offer, using the values of parameters known in the early stages of the design process.

In the future the systematisation, the study of physical and - if possible - technological limitations as decision-making criteria is of particular interest. In this context, recent investigations suggest that the consideration of scaling effects is very important, too.

Physical and technological limitations (e.g. ceramic grain sizes, magnetic/Weiss domains, heat losses ...) are non-linear relationships and can affect the decision model significantly, especially in the design of microsystems. Although the models presented here are still valid, the extent of the influence of these limitations will be subject to further investigations.

ACKNOWLEDGEMENTS
The authors would like to thank the members of the Collaborative Research Centre 622 "Nanomeasuring and Nanopositioning Machines" and the German Research Foundation (DFG) for their support.

REFERENCES
[1] Weber Chr. CPM/PDD - An Extended Theoretical Approach to Modelling Products and Product Development Processes, In: *Proceedings of the 2nd German-Israeli Symposium on Advances in Methods and Systems for Development of Products and Processes)*, Stuttgart 2005, pp. 159-179. Fraunhofer-IRB-Verlag
[2] Huber J. E. and Fleck N. A. and Ashby M. F. The Selection of Mechanical Actuators Based on Performance Indices, In: *Proceedings - Royal Society. Mathematical, physical and engineering sciences*, 1997, vol. 453, pp. 2185-2205
[3] Ashby M.F. Materials Selection in Mechanical Design, 3rd Edition. 2004. Elsevier Butterworth-Heinemann
[4] Zupan M. and Ashby M.F. And Fleck N.A. Actuator Classification and Selection-The Development of a Database, In: *Advanced Engineering Materials*, Volume 4, Issue 12, December 2002, pp. 933–940
[5] Erbe T. and Stroehla T. and Theska R. and Weber Chr. Decision-aid for actuator selection, In: *Proceedings of the 11th International Design Conference DESIGN 2010*, Dubrovnik, 2010, pp.1503-1512
[6] Egbuna C.C. and Basson A.H. Electric Actuator Selection Design Aid for Low Cost Automation, In: *Proceedings of the 17th International Conference on Engineering Design (ICED'09)*, Vol. 6, Stanford, August 2009, pp.43-54
[7] Weber, C.: Looking at "DFX" and "Product Maturity" from the Perspective of a New Approach to Modelling Product and Product Development Processes. In: *Proceedings of the 17th CIRP Design Conference*, pp. 85-104, Springer, Berlin, 2007
[8] Weber, C.: How to Derive Application-Specific Design Methodologies. In: *Proceedings of DESIGN 2008*, Vol. 1, pp. 69-80, Faculty of Mechanical Engineering and Naval Architecture, University of Zagreb, 2008.
[9] Hubka V. Theorie der Maschinensysteme. Berlin-Heidelberg : Springer-Verlag, 1973
[10] Hubka V. Theorie der technischer Systeme. 2nd Edition. Berlin-Heidelberg. Springer-Verlag, 1984
[11] Chakrabarti A. Engineering design synthesis. Berlin-Heidelberg. Springer-Verlag, 2002
[12] Roozenburg F. M. N. and Eekels J. Product design: fundamentals and methods. Wiley, 1995
[13] Ponn J. and Lindemann U.: Konzeptentwicklung und Gestaltung technischer Produkte. Berlin-Heidelberg. Springer 2008
[14] Suh N.P. Axiomatic Design. Oxford University Press, 2001
[15] Reitmeier J. and Paetzold K. Property and behavior based product description – component for holistic and sustainable development process. In: *Proceedings of the 11th International Design Conference DESIGN 2010*, Dubrovnik, 2010, pp.1673-1680
[16] Paetzold K. Ansätze für eine funktionale Repräsentation multidisziplinärer Produkte. In: *17. Symposium „Design for X"*. Neukirchen. October 2006
[17] Deubel T. and Conrad J. and Köhler Chr. And Wanke S. and Weber, Chr. Change impact and risk analysis (CIRA): combining the CPM/PDD theory and FMEA-methodology for an improved engineering change management. *ICED '07 - Paris, 16th International Conference on Engineering Design.* . Paris, France. 2007, pp. 9-10 [Executive Summary]
[18] Birkhofer H. and Wäldele, M. The Concept of Product Properties and its Value for Research and Practice in Design. In: *Proceedings of the 17th International Conference on Engineering Design (ICED'09)*, Vol. 2, Stanford, August 2009, , pp. 227-238

Contact:
Dipl.-Ing. Torsten Erbe
Research Assistant
Ilmenau University of Technology, Department of Engineering Design
PO-Box 100565, 98684 Ilmenau, Germany
torsten.erbe@tu-ilmenau.de

AUTOMATED USER BEHAVIOR MONITORING SYSTEM FOR DYNAMIC WORK ENVIRONMENTS

Yeeun Choi, Minsun Jang, Yong Se Kim*, Seongil Lee
Dept. of Industrial Engineering, Sungkyunkwan University, Korea
*Dept. of Mechanical Engineering, Sungkyunkwan University, Korea

ABSTRACT

The aim of the study is to improve existing methodology for user observation to evaluate user performance in a dynamic and complex work environment. We developed a new user behavior monitoring system and used the system to analyze user behaviors in a complex and dynamic work environment. The proposed monitoring system is composed of an object tracking module using RFID tags on objects and a RFID reader on user's hands, and video cameras for recording user behavior. We also designed a smart floor with embedded pressure sensors for monitoring user movement through variation of pressure sensor signals. The system is installed in an observation room where a model take-out coffee shop was simulated for verifying the utility of the proposed observation system. The system would provide valuable insights to improve user performance and work environment redesign.

Keywords: User behavior monitoring, RFID, Smart Floor

1 INTRODUCTION

Since today's technology improvement and manufacturing advancement can affect the user-centered design, the application of new technology would give a new experience to users and provide improving products and services. Traditionally usability tests are conducted in order to obtain user needs. Most of current usability evaluating protocols currently were designed to observe only one user at a time and also got limited position as well. According to Hsu et al. (2006), they experimented using a variety of sensors to analyze the behavior of users. They monitored with combined video and RFID data in the reading room environment [1]. A 'sensing room' was created by Mori et al. (2006), and the location and lifestyle of users were analyzed in the room. Pressure sensors and RFID tags attached to the floor and the furniture were used in constructing the room.

Typical usability evaluating systems observe only one user using a specified target objects in a specially pre-set laboratory. A different monitoring system from the traditional laboratory environment that can record continuous activities of multiple users in a more real-life setting could be used to evaluate the usability of not just a product but of spaces and environmental design. This type of system can take advantage of high technologies such as wireless communication modules and various sensors. Symonds et al. (2007) used RFID tags in a house, bedroom and to identify the movement of certain goods which are easy to be lost in traditional tracking system using just cameras. They were trying to solve the problems what the elderly and disabled people are faced in real life [2]. Pulson and Hammond (2008) observed the behavior of static users who work in the same place without moving in an office environment. They attached RFID tags on the desk, phone or and cups, with a small RF reader attached on the user's wrist. Through RFID tag data they tracked the use of stationaries, based on the user's arm movement [3]. M. Buettner et al. (2009) observed the behavior of users using a RFID reader and antenna which were installed on the landmarks inside apartment for daily life. In their study, small items were moved inside an apartment with attached RFID tags. The RFID tag was detected based on location and 95% of user behaviors were correctly identified from the data [4]. These systems, however, could only monitor each user's movement path and observe whether a certain product was used. They could not analyze user intention and behaviors reflecting contextual situation.

In this study, we have developed an automated user behavior monitoring system that can observe multiple users engaged in dynamic tasks and can provide detailed information regarding the individual user behavior in the working environment, using RFID tags attached on the machines and objects in a model shop and a wireless camera mounted on a cap worn by the user.

2 SYSTEM DESIGN

2.1 Hardware Development

The suggested system is composed of three modules - a video-based observation system, WPAN (Wireless Personal Area Network)-based measurement system, and a Smart floor-based measurement system. We built a typical model take-out coffee shop for applying the suggested system for the purpose of user behavior monitoring and usability evaluation of the current work setting. Each objects, tools, and machines in the model shop, including an espresso machine, a coffee grinder, a toaster, bottles of syrups, and cups are attached with RFID tags, ISO 15693 - 13.56 MHz. Some have two or more tags attached to differentiate the parts and orientation to see which side and part user grasps, while using the objects. There are approximately 150 RFID tags attached on a total of 27 different machines and objects in the model shop.

For building a video observation system, we used two types of camera to monitor user movements and user traffic in the work area. The video observation module records all the users' movements and behaviors using three cameras. Two PTZ(Pan, Tilt and Zoom) type dome cameras were positioned at the both ends of the ceiling, and a small wireless camera was attached under the brim of user's cap that user would be able to put on during the experiment, so that we can track user's eye gazes and monitor the current status of objects used.

Figure 1. (a) PTZ dome camera (b) wireless camera

A WPAN-based measurement system was also developed by using a 13.56 MHz RFID package. User puts a small wrist guards' band with RFID readers attached with a small antenna (40x20 mm) on the ring finger of both hands. Since the RFID tags are attached on all of the objects in the model shop workshop environment, the RFID reader on the wrist band recognizes the corresponding tag and leaves a log data on the server any time user grabs an object for performing tasks. The RFID reader was connected to a Bluetooth communication module to transfer the tag data to a server.

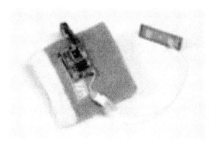 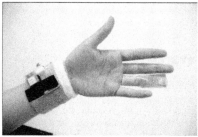

Figure 2. The wrist band with a RFID tag and a reader

The smart floor-based measurement system was developed for monitoring user movement and work traffic, to complement the video-based observation system. The user location and the change of location can be identified in real-time using flexi-force pressure sensors installed underneath of the floor tiles.

For connecting all the pressure sensors to collect the data, the 1019-Phidget Interface Kit Input/Output board was used. A total of 135 pressure sensors and 20 Phidget modules were installed to build a smart floor, with a size of 5m x 1.5m. Each hardware module consists of 6 to 7 flexi-force pressure sensors connected with a regular pattern. All the flexi-force pressure sensors calibrated and checked for maintenance before every measurement. A small bump at the end of the pressure sensor was attached to amplify the pressure signals to increase the recognition rate. Each hardware module was connected to a USB communication module to transfer the pressure sensor data to a server.

Figure 3. (a) Phidgets board (b) Smart floor development

2.2 Software Development

An observation module was developed for automated user behavior monitoring. The observation software monitors the user behaviors and analyzes all of video and RFID tag data from video cameras, RFID readers, and 135 flexi-force pressure sensors. The observation software is divided into two parts – an RFID monitoring module and a smart-floor monitoring module.

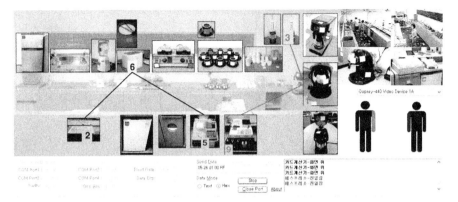

Figure 4. RFID Monitoring Software

The RFID monitoring module (Figure 4) performs two functions. First, this monitoring module logs the RFID tag data to the database server for further data analysis. The system records data such as log time, RFID tag IDs, object name, RFID reader IDs and video frames. Second, the monitoring module can differentiate RFID readers used by each user and hand, and provide the number of times how many each tag was detected on the screen in real time. Also, the module presents the path of user's movement in conjunction with the objects with which the user was engaged in for work. The path can be composed by connecting the RFID data from the objects user touched in time sequence. All the data including the video file from the cameras are synchronized in time and saved on the server.

The smart-floor monitoring system (Figure 5) was developed for monitoring user movement by analyzing the pressure sensor data. The system basically supports and double checks the RFID monitoring module for user movement path calculation. The system saves the pressure sensor data from the smart-floor monitoring module to the server for data analysis. To enhance recognition rate of the user movement on the smart floor, a total of 135 pressure sensors provide signal data in real-time. Faulty pressure sensor could be detected in real-time to allow easy maintenance. The smart-floor monitoring module can present the configuration of sensor arrangement on the floor and can also show movement path of each on the screen. The path is composed by connecting the pressure sensor point on the smart floor whose signal data exceeds certain threshold big enough to be considered as a user's step in time sequence. All the signal data are recorded on the server being synchronized in time with the data from the RFID monitoring module.

3. MONITORING EXPERIMENT

3.1 Beverage Serving Experiment

We tested the user monitoring system in the model coffee shop to find out if the system can be used to provide any improvement in service design. Two users, an experienced server and a novice server, respectively, wear a cap with a small wireless camera attached and wrist bands on both hands with the RFID reader.

Three experimental scenarios were provided for serving coffee, toasts, and iced tea. The users were asked to proceed to serve customers without restrictions on time or service orders. With two users working on the three scenarios independently, the data were collected from the RFID monitoring module and the smart-floor monitoring module and saved in a server wirelessly connected to the modules.

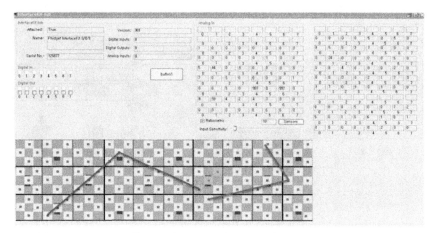

Figure 5. Smart Floor Monitoring Software

Data from the two modules were synchronized in time when and integrated into one. Recorded video data from four video cameras were also synchronized with the data from the two modules. Figure 6 shows a user performing coffee-making services to a customer according to a scenario. Table 1 shows the machines and objects required for use to perform the services and the use frequencies according to the task scenario.

Two servers participated in the experiment: one experienced server and one inexperienced server. Each server takes the same order, and was asked to produce the item and serve the customer as fast as possible. Beyond taking simple observation of the working process, we have observed differences in quality of product and services.

The experienced participant was a male in his twenties and had had 3 months experiences in a coffee shop. The novice participant was also a male in his twenties and knew how to make beverages basically but had no previous experiences. Both were introduced to the experimental setup and provided with the scenarios. The experiment in each trial started with greeting customers, and proceeded to taking orders for an iced-tea, or an Americano, and ends up with serving the customer with what s/he ordered. Ten trials have been taken for each type of beverage. Adequate rest was afforded between each trial to avoid participant's fatigue. For monitoring convenience, we assumed that only one customer comes and place an order in a trial for each server.

Table 2 shows sample rules to configure automated analysis of user behaviors from the RFID tag data.

3.2 Results and Analysis

In this paper, only the result for the 1st scenario would be analyzed and presented. Figure 7 shows an example result of activity flows for the tasks performed by each server in a diagram. Figure 7a shows how the experienced server made an Americano coffee using the machines and objects provided in sequence and presents the number of times visited for each machine/object with the size of nodes. The activity flow displays the amount of time to be taken in previous node to next node, the time spent to perform sub tasks and total task, intuitively.

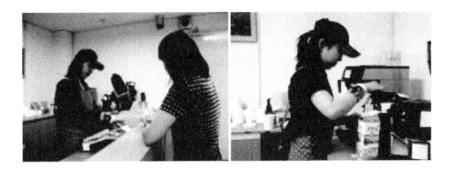

Figure 6. Coffee-serving tasks by a user wearing a camera-mounted cap and wrist bands with RFID reader

Table 1. Dimensions of the DPG Matrix

performance		Tracking Objects (process & frequency)					
Scenario 1	User A americano	Card counter (2)	Grinder (12)	Espresso machine (20)	Cup (4)	Hazelnut syrup (1)	Take out place (5)
	User B iced-tea	Cup (2)	Ice-tea (7)	Water purifier (12)	Tea-spoon (2)	Take out place (4)	
Scenario 2	User A toast	Card counter (3)	Refrigerator (7)	Toast dish (4)	Toaster (17)	Refrigerator (3)	Take out place (7)
	User B iced-tea	Cup (3)	Ice-tea (6)	Tea-spoon (2)	Water purifier (11)	Take out place (3)	
Scenario 3	User A fruit juice	Cash counter (8)	Refrigerator (4)	Blender jar (2)			
	User B fruit juice	Cup (2)	Vanilla syrup (1)	Blender (5)	Blender jar (7)	Cup (2)	Take out place (5)

Table 2. Behavior tracking rule base example

R1: IF(Checked credit card counter =True) THEN Credit payment

R2: IF(Checked grinder =True & Checked Espresso machine = True & Checked a water purifier =True), THEN Americano

R3: IF(Checked refrigerator=True AND Checked blender=True AND Checked syrup bottle=True), THEN Fruit juice

Server's movement path for service was comprehended by extracting a trace from the pressure sensor data logged through the Smart-floor module. The RFID-module synchronized with the smart-floor module was used to compare the task processes between the novice and experienced server.

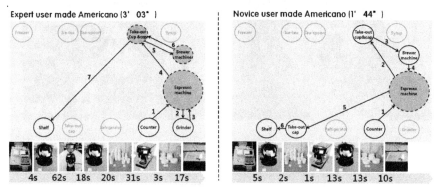

Figure7. (a) Experienced server activity flow (b) Novice server activity flow

The experienced server made an Americano coffee in 3 minutes and 3 seconds. There was no repeated or omitted visit to each machine/object, and seemed to perform the optimized procedures in each trial. On the other hand, the novice server followed the procedures diagramed in figure (b) to make an Americano coffee. Even though there was no waste in movement path flow, comes the same machine (Espresso machine) repeatedly, and skipped a grinder machine to grind the coffee bean into powder. It was observed that the novice server used the grinder machine once in two trials, while grinding 2-serving amount of beans a time to save the powder in a potter for the next service. This made the total task time of the novice (1' 44") shorter than that of the experienced server (3' 03'').

A further user performance was evaluated for both of the novice and experienced servers based on the quality of service provided to the customer. Time to take for serving was a key performance evaluation metric, while the service quality in terms of taste was another though it was hard to be quantitatively measured. While there was a missing procedure for the novice server, the quality of beverage served by the novice server also lacked the required quality in terms of freshness. The problem here is that there should be a clear statement or a goal to achieve in terms of quality of service so that certain procedures are required to take place by any means to stick with the goal. In this context, the quality of product would be the more important factor than shorter task time. By setting the priority in service requirement, servers would never omit any procedure in working process to their convenience and the quality of service can be maintained at a certain level

4. CONCLUSION

This study presents a new system to monitor user behaviors in a work environment carrying out relatively dynamic works more efficiently and accurately using sensor-based technologies and wireless communication technology. Automated data collection was possible through the automated user behavior monitoring system. The data analysis could provide misbehaviors of the server in providing a quality service. The system can be used as a guide for training novice servers and checking for improvement in work behaviors for maintaining a required service quality.

To determine the reliability of the system, the analysis result from the system was manually compared with the recorded video data. It was found that the automated monitoring system provides as valid and reliable analysis result with the RFID tag data producing an object tracking diagram, and the pressure sensor data from the smart-floor producing the movement path diagram as any current manual monitoring system such as video surveillance systems, making the system a comprehensive analysis tool for service analysis and re-design. The average success rate of the system for object recognition for RFID tag was measured to be 95.7 percent. We found that the recognition rate could be enhanced with modification of the direction of the RFID antenna on the wrist band, and the RFID tag attached on the objects with metal surface results in poor recognition.

To improve the recognition rate of the RFID tag data, a variety of sizes and types of the RFID tag needs to be further tested. The data from the smart-floor with pressure sensors was measured to be correctly extracted at 95% based on the comparison from the video To improve the recognition rate of RFID tagging would be the key to improve the overall system performances in accuracy and reliability.

Future study will be focused at various contexts of work using the developed system. For example, the cases in which many orders are placed to create a service queue and customers are waiting to be served, and the cases in which each server would have specific roles to cover and there are more than two servers would provide a valuable insight for the use of the automated monitoring system and for the service re-design. We can further design a usability evaluation toolkit using the current automated monitoring system that can provide more detailed information on server's using each machine and tool in the workshop. Furthermore a different model shop would be applied for more generic application of the system to be possible.

The system can be extended to a multi-user real-life usability evaluation system once more detailed analysis modules can be developed in the future. The ultimate purpose of the system would be to construct a usability evaluation system that not only monitors user behavior but also provides detailed analysis and prediction on user performance and recommends the desired change in the work environment design.

ACKNOWLEDGEMENT
This work has been supported in part by the Research Fund from the Korean Ministry of Knowledge Economy.

REFERENCES
[1] Hsu, H.H., Cheng, Z., Huang, T., Han, Q., *Behavior Analysis with Combined RFID and Video Information,* Proc. of Ubiquitous Intelligence and Computing, LNCS 4159, pp176-181, 2006
[2] Symonds, J., Parry, D., Briggs, J., *An RFID-based System for Assisted Living Challenges and Solutions,* The Journal on Information Technology in Healthcare, 5(6), pp387-398, 2007
[3] Paulson, B., Hammond, T., *Office Activity Recognition using Hand Posture Cues,* Proc. of British CHI Group Annual Conference on HCI 2008, pp75-78, 2008
[4] Buettner, M., Prasad, R., Philipose, M., Wetherall, D., *Recognizing Daily Activities with RFID-Base Sensors,* Proc. of Ubicomp, pp51-60, 2009

Ye Eun Choi received a B.S. degree from Sungkyunkwan University, Suwon, Korea, in 2010. She is currently working toward the M.S. degree at Sungkyunkwan University, Korea. Her research interests include human factors and ergonomics, Human-Computer Interaction, user experience design, social network service.

Min Sun Jang received a B.S. degree from Sungkyunkwan University, Suwon, Korea, in 2010. She is currently working toward the M.S. degree at Sungkyunkwan University, Korea. Her research interests include human factors and ergonomics, universal design, Human-Computer Interaction, user experience design.

Yong Se Kim is the Director of the Creative Design Institute, and a professor of Mechanical Engineering at Sungkyunkwan University. He received PhD degree in Mechanical Engineering with CS minor from the Design Division of Stanford in 1990. His research interests include design cognition and informatics, product-service systems design, experience and service design, and design learning.

Seongil Lee is a professor of the department of Systems Management Engineering at Sungkyunkwan University, Suwon, Korea. He received the PhD degree from the University of Wisconsin-Madison in 1995 with works on haptic interfaces for people with disabilities. His research interests include Human-Computer Interaction, human factors and ergonomics, universal design, accessible computing, and interaction design in mobile and ubiquitous systems.

INTERNATIONAL CONFERENCE ON ENGINEERING DESIGN, ICED11
15 - 18 AUGUST 2011, TECHNICAL UNIVERSITY OF DENMARK

ON THE DESIGN OF DEVICES FOR PEOPLE WITH TETRAPLEGIA

S.D. Gooch[1], A. J. Medland[2], A. R. Rothwell[3], J.A. Dunn[4], M.J.Falconer[1]

[1] University of Canterbury, New Zealand
[2] University of Bath, United Kingdom
[3] Christchurch Medical School, University of Otago, Christchurch, New Zealand
[4] Burwood Spinal Unit, Christchurch, New Zealand

ABSTRACT

People with complete tetraplegia are required to work at or near their physical limits in performing daily activities. Hence, subtle improvements to the design of assistive devices can have life changing consequences. This paper establishes a new procedure for characterizing the strength of people with tetraplegia. The data obtained along with the specifications of assistive devices are implemented in the Bath Constraint Modeller and then predictions made of a subjects ability to use the assistive device. This paper shows how improvements in wheelchair propulsion ability can be made within the constraints of normal wheelchair adjustment. From the characteristic strength maps produced in this study, it is predicted that more marked improvements can be obtained by changing the position of the applied propulsion force. The study proposes a new design concept involving an offset push rim which is expected to improve wheelchair propulsion ability for people with tetraplegia. More generally, the results of this study pose new opportunities for improvements to assistive devices for people while seated.

Keywords: design of assistive devices, tetraplegia, human strength

1. INTRODUCTION

People with Spinal Cord Injuries (SCI's) face a daily struggle with everyday tasks. For many, independent living is an unrealistic expectation. The most common level of SCI [1] is in the cervical spine. A complete break in the spinal cord at the cervical level results in total paralysis from the neck down. The upper limbs have varying degrees of motor and sensory function depending on the exact location of the injury in relation to the cervical nerves.

Each year a significant number of people are affected by SCI's. For example, the annual incidence spinal cord injury in the United States is approximately 40 per million population, equating to 12,000 new cases per year [2]. Estimates of the total prevalence of SCI's in the US have ranged from 127 080 to 300 938 persons. Tetraplegia has constituted 52.4% of all new SCI's since 2000, with 34.1% of new SCI's incomplete tetraplegia and 18.3% complete tetraplegia.

Surgical procedures have been evolved to improve quality of life and independence for people with SCI's. One of authors has performed or supervised around 100 posterior deltoid to triceps transfer (TROIDS) procedures. One of the benefits of this surgery is that it improves a person's ability to perform basic activities such as feeding themselves, brushing teeth as well as manual wheelchair propulsion [3]. A benefit of this study is that it provides both a visual and quantitative measure of human strength in the sagital plane. The method can also be used to demonstrate the effectiveness of the surgical procedures such as TRIODS.

The purpose of this study is to provide information for the design of effective assistive devices for people with tetraplegia. In this paper we establish a procedure for characterising human strength using able bodied subjects and we obtain characteristics for three tetraplegic individuals. This information will be used to help better design and prescribe assistive devices for people with tetraplegia. The differences found will also be used to modify and validate mathematical human movement models for people with disabilities which aid in streamlining the design process. Seven people voluntarily participated in this study, four able bodied and three with tetraplegia, the criteria for the later three was that they had complete SCI.

2. BACKGROUND

Pervious studies have mainly concentrated on the voluntary strength of various upper body articulations, particularly shoulder and elbow articulations within an able bodied population. This study which investigates the upper body strength in the combined articulation task of pushing has previously only been studied over a much coarser grid spacing in one dimension. In one particular study, Kumar [4] performed 2-handed tests from a standing position, at 350mm, 1m and 1.5m above the ground and established that the mid-level height was the strongest position.

There is also little data on the upper-body strengths of people with SCIs mainly due to the difficulty in testing this particular population group. The most comprehensive study of the strength of persons with SCIs on the sagittal plane is that of Das and Forde [5] which measured the seated right handed isometric push-up and push-down arm strength in 24 positions for subjects with C4-T11 SCIs. The study found that the push-up strengths of the candidates were only 30% of that of able bodied forces measured by Hunsicker [6] and the pull down forces were approximately 50%. Das and Forde did not, however, distinguish between candidates with different SCIs and they did not include subjects with higher level tetraplegia.

In this paper we consider individuals with tetraplegia. Each person had a SCI as the result of physical trauma. In each case these injuries affected vertebrae of the cervical spine at different levels resulting in complete paralysis of the lower body and varying degrees of sensory and motor loss to the arms and hands. A summary of the subjects, injury levels and resulting sensory and motor control is given in Table 1.

Table 1. SCI level and function (adapted from Floris et al. [7])

No. Subjects	SCI Level	Sensory and Motor Control
1	Cervical injuries (C5-C6)	• Preservation of shoulder abduction + external rotation • Preservation of elbow flexion + variable wrist extension • Little/no voluntary control of elbow extension • No hand function
1	C6 'TROIDS'	• Limited elbow extension
1	Cervical injuries (C7)	• Elbow extension /Wrist extension • Finger extension, no grasp
4	Able body function	• Participants were told they were unable to use there legs for posture support

In a previous study, Gooch et al [8] investigated the difference in wheelchair propulsion ability between people with various levels of tetraplegia. There were found to be distinctly different wheelchair propulsion characteristics between three groups of people with tetraplegia. The three groups were: people without arm extension ability, people with arm extension ability and people with TROIDS for arm extension. These three groups are represented by the three people with tetraplegia measured in this study.

3. EXPERIMENTAL PROCEDURE AND OBSERVATIONS

Each subject was loaded onto the test rig shown in Figure 1. The wheelchair is anchored to the base platform using four ties so that the central axel of the back wheels is located at a reference datum point in the superior/inferior and anterior/posterior directions.

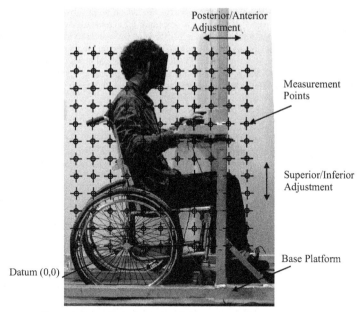

Figure 1. Upper body strength test rig showing measurement points

The subjects horizontal push force is measured using two calibrated LPX-50kg compression load cells, one for each arm, which are fixed to aluminium posts. The position of the applied force can be adjusted in the sagittal plane in the superior/inferior and anterior/posterior directions as indicated in Figure 1. The measurements are recorded over a grid spacing of 100mm. While the subject's strength can be measured at any point and in any direction, the grid positions used for measurements in this study are listed in Table 2.

Table 2. Grid positions for strength test measurements

Direction	Distance from the centre of the wheel (mm)											
Anterior/Posterior	-400	-300	-200	-100	0	100	200	300	400	500	600	
Superior/inferior	0	100	200	300	400	500	600	700	800	900	1000	1100

While there are 120 possible grid positions, some positions are physically unreachable depending on the person's flexibility, stature, sensory function and motor control. In each position the subject pushes on the centre of a hemispherical hand support which transfers the force onto the LPX-50kg compression load cell as shown in Figure 2. During the strength test, the subject is free to adopt the posture they think will allow them to produce the maximum push force. Subsequently, the maximum force is recorded and if the position is deemed physically unreachable a force reading of zero is recorded.

At the beginning of the test the frame is positioned so that the hemispherical hand support is at a point close to the centre of the user's range of motion. Once data has been recorded at each vertical position, the post is moved in either the anterior or posterior directions by a distance of at least 200mm and a new set of data captured. This minimum distance is used to help mitigate the effects of fatigue from the prolonged use of one muscle group.

The effect of fatigue is also reduced by allowing the candidate to have sufficient rest between each push measurement. Using the Rohmert fatigue model [9] the recommended rest time necessary to eliminate fatigue with a force application of one second at a maximum voluntary contraction of 0.9 was predicted to be 48.1 seconds. Depending on the difficulty in adjusting the load cells to the new position the approximate rest time between positions was found to be 40 – 60 seconds. The effect of fatigue was checked periodically throughout the testing procedure by moving the vertical posts back to the starting position in anterior/posterior direction and re-measuring. A review of strength-training literature indicates that there is a direct relationship between reps-to-fatigue and the percentage of maximal load. As the percentage of maximal load increases, the number of repetitions decreases in a linear fashion [10]. The average difference in these three repeated measurements can then be used to estimate a percentage loss in strength per repetition. Therefore each recorded strength measurement is multiplied by this loss factor and the number of reps since the commencement of the test to give an equivalent un-fatigued strength measurement. If fatigue is noticed and is consistent in the superior/inferior direction then the effects of fatigue can be considered in the analysis.

Figure 2. Load cell attachment brackets

Force maps were created using the force measurement results obtained from measuring four subjects with normal motor and sensory control. These subjects were 22 to 60 year old males and their force maps are shown in Figure 3. The results obtained for the four people with normal motor and sensory control show that people seated have a maximum forward push force in their lap area.

The results of strength measurements for the subject with C5/C6 tetraplegia are shown in Figure 4. The subject had slightly better motor and sensory control on his left side and a moderate strength region exists around the -200,300 position. From these results it is evident that the subject would be expected to have better wheelchair propulsion ability if his seat height was lowered by approximately 100mm.

The results of strength measurements for the subject with C6 tetraplegia (TROIDS), Figure 5, also illustrate an asymmetric strength profile. This result is consistent with anecdotal evidence suggesting that the subject had a markedly more successful outcome of the TROIDS procedure on his right side than on his left side. Comparing Figures 4 and 5, the C6 (TROIDS) subject is approximately twice as strong as the C5/C6 subject. His strength map indicates more function in the lap area which is consistent with having improved arm extension ability.

The results of strength measurements for the subject with C7 tetraplegia are shown in Figure 6. While the subject with C7 tetraplegia had a similar force map to the people tested with normal arms, his strength was approximately 20% of that measured from the subjects with normal motor and sensory control.

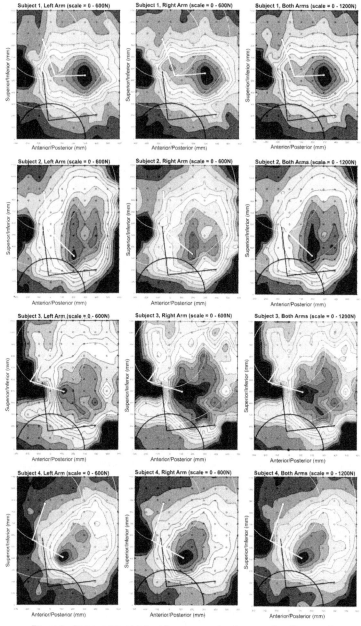

Figure 3. Force maps obtained from measuring four people with normal motor and sensory control.

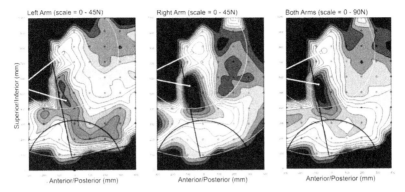

Figure 4. Force maps obtained for the subject with C5/C6 tetraplegia.

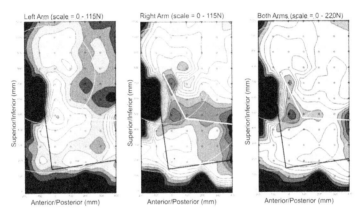

Figure 5. Force maps obtained for the subject with C6 tetraplegia (TROIDS).

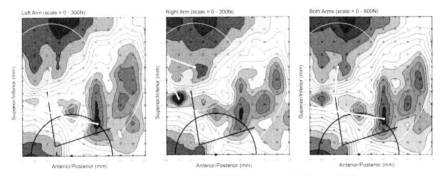

Figure 6. Force maps obtained for the subject with C7 tetraplegia.

The people in each group tested have particular motor and sensory function and this is illustrated in the distinctly different force maps. Given that the people with normal motor and sensory function have similar characteristics, Figure 3, it is likely that people in the three tetraplegic groups may be representative of other people with similar injuries.

4. IMPLICATIONS FOR DESIGN

This study shows that the ability to provide a force for tasks such as wheelchair propulsion will vary widely between subjects with different SCI's, between their arms and throughout the range of motion in the sagittal plane. As the aim of this study is to provide support for the design of effective assistive devices, such variations have to be understood and incorporated in the approach.

The research programme has thus concentrated upon the integration of three major issues. Firstly the modelling of a manikin in which such human variations can be represented, secondly to be able to incorporate realistic data, as has been illustrated in the previous sections, and finally to be able resolve and optimise the tasks.

All of this research has led to the construction of a constraint-based approach built upon the Bath Constraint Modeller which incorporated the ADAPS human model from the Technical University of Delft [11], [12], [13]. This uses rules to define the explicit tasks and requires the implicit rules necessary to create an articulated and life-like manikin. Within the manikin model are in excess of 22 body parts, related hierarchically or linked by greater than 52 degrees-of-freedom. All of these are themselves limited by defined boundary conditions. Such complexities has required the creation of new direct search approaches incorporating sensitivity analysis [14], [15], [16]. These research activities have led to a study of wheelchair sitting postures [17] that provide the basic models for humans interacting with a wheelchair.

The experimental studies, presented in this paper, are now to be incorporated into the manikin environment in order to allow the potential forces provided by the individual subjects to be assessed. The forces obtained from experimental measurements are read in and forces across the push rim of the wheel calculated. Within the constraint modelling environment design variables, such as positions and sizes of the wheels, can be selected and used in the optimisation of the design. With this approach the maximum stroke and forces can be obtained for individual cases.

As all wheelchairs are represented fully parametrically within the environment, various parameters can be changed and there effects observed. The position and size of the drive wheels have a considerable effect on both the social usability and balance of the chair, as well as changing the force profile that the user can provide. Furthermore, due to the difference in force mapping observed in the experimental studies, the modelling analysis predicts that different subjects will be able to achieve significantly improved force profiles, often with only subtle changes in chair geometry and sitting position. For people with tetraplegia, a small increase in capability may be enough to make the difference that allows an individual to use a manual wheelchair.

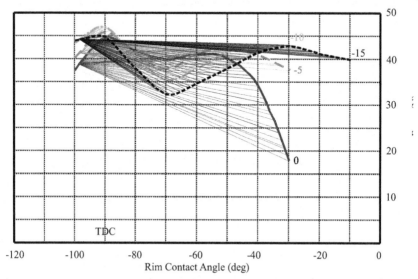

Figure 7. Showing the effect of lowering the seating position (in mm) for an individual.

In Figure 7, the effect of lowering the users sitting position on the force profile across the rim is easily seen. The effectiveness of this for the user is not only in the ability to gain higher forces throughout but selecting the regions in which greater forces can be provided. For example at –70 degrees the higher force is obtained at the original height but of pushing is required forward of –40 degrees then a lowering by -5 or –10 will provide a vastly improved capability.

5. STUDY OF PUSH RIM DESIGN

While it has been demonstrated that the model predicts an increase in propulsion ability for a small change in seat height, Figure 7, the force maps created from our experimental procedure indicate that a more marked improvements could be achieved with greater changes to push rim position. These potential improvements are, however, not within the limits of normal wheelchair adjustment. Consequently a number of design configurations are being investigated to allow such push rims to be practically achievable.

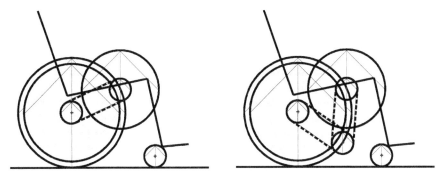

Figure 8. Showing principles to designs providing rim offset.

The simplest arrangement (shown in the first sketch in Figure 8) is to provide a chain drive between the driving wheels and a higher mounted push rim wheel. Such a solution could be implemented on an existing wheel chair but its position would need to be chosen to provide the improved forces required. This approach would result in the push rim being permanently fixed in position offset from the central wheel axis, which could be an inconvenience for other wheelchair activities.

For this reason a more advanced scheme is also to be investigated that provides both flexibility in push rim positioning and allows it to be lowered back to its original position at the drive wheel centre (where it can be used normally). This is shown in the second sketch and consists of a double chain (or gear) drive mounted on a v-linkage arrangement. The lower linkage mounted on the drive wheel centre can be locked in a forward mounted position, while the second linkage, mounted on the first is rotated to the preferred position for the rim wheel. Such a mechanism can be arranged to allow the rim wheel to be lowered down to align with the driving wheel centre. This design can thus be moved through a range of positions from the wheel centre position up to a higher force region selected by an investigation of the experimental data produced for a chosen individual.

In the study shown in Figure 9, the subject's capability before and after the TROIDS surgery is compared with the predicted improvement possible through raising the push rim above the driving wheel. Whilst little is achieved beyond the –40 degree push position as the user starts to push from the –100 position nearly twice the force is obtained.

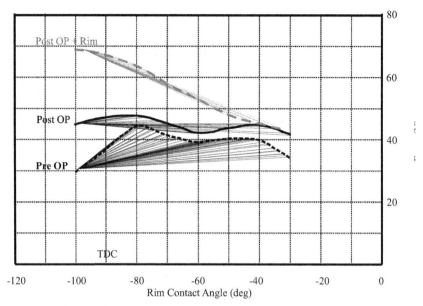

Figure 9. Force across push rim showing improvement due to rim offset.

6. CONCLUSIONS

This paper presents a novel method of characterizing human strength capability. It has been demonstrated that this method characterises the arm strength capability for able bodied subjects. Force maps have also been obtained for three subjects with tetraplegia. These three subjects had distinctly different motor and sensory upper limb control and where found to have distinctly different characteristic force maps. From the results of the able bodied participants it is likely that the tetraplegic subjects will be representative of people with similar injuries. Further work has commenced that involves measuring a large cohort of tetraplegic subjects in the three groups tested in this study, namely: those with no tricep function; those with TROIDS for arm extension; and people with C7 tetraplegia who have functioning triceps. The individual maps obtained will be a useful tool in establishing design parameters for the design of assistive devices in the future.

The constraint-based approach demonstrates the improvements in rim force that can be achieved with only subtle changes in wheelchair set up. From the force maps and the analysis using the Bath Constraint Modeller, it is likely that more marked improvements to wheelchair propulsion ability can be achieved by allowing the wheelchair user to apply the propulsion force closer to their area of maximum force. The design of a wheelchair with an offset push rim is now being evolved.

ACKNOWLEDGEMENTS

We thank all the participants in this study for generously offering their time and assistance. This research is supported by Industrial Research Limited in New Zealand. The authors also wish to thank the Department of Mechanical Engineering and the IdMRC at University of Bath and the Department of Mechanical Engineering at the University of Canterbury. In particular the assistance provided by Mr Julian Phillips and Mr Ron Tinker at the University of Canterbury is gratefully acknowledged.

REFERENCES

[1] O'Connor, P., *Spinal Cord Injury, Australia 1999–00*, Australian Institute of Health and Welfare, Canberra, AIHW cat. no. INJCAT 40, December 2001, ISBN 1 74024 172 X

[2] National Spinal Cord Injury Statistical Center. *Spinal cord injury: facts and figures at a glance*. The University of Alabama at Birmingham, USA.

[3] Gooch S.D, Rothwell A.G., Yao F.,Dunn J. & Woodfield T., *A procedure for measuring manual wheelchair propulsion ability for people with spinal cord injuries*, International meeting of upper limbs in tetraplegia conference, September 2007,Shiriners Hospitals for Children, Philadelphia, Pennsylvania, 2007.

[4] Kumar, S., Narayan, Y., and Bacchus, C. (1995). *Symmetric and asymmetric two-handed pull-push strength of young adults*. Human factors, 37(4).

[5] Das B and Forde M, 1999, *Isometric push-up and pull-down strengths of paraplegics in the workspace: 2. Statistical analysis of spatial factors*, Journal of Occupational Rehabilitation, Vol.9, No.4, 293-299.

[6] Hunsiker, P.A. (1955). *Arm strength at selected degrees of elbow flexion*. Wright Air Development Center Report, pages 54-548.

[7] Floris, L., Dif, C. and MA Le Mouel. *Chapter 5 - The tetraplegic patient and the environment Surgical rehabilitation of the upper limb in tetraplegia*. Ed Hentz, V. R., ISBN 0702022713 / 9780702022715
Saunders   May 2002

[8] Gooch, S. D., Woodfield, T., Hollingsworth, L., Rothwell, A. G., Medland, A. J. and Yao F. (2008), *On the Design of Manual Wheelchairs for People with Spinal Cord Injuries*, Proceedings of the 10th International Design Conference, Design 2008, pp 387-394, May 19 – 22 2008, Dubrovnik, Croatia.

[9] Rohmert, W (1973). *Problems in determining rest allowances part 1: use of modern methods to evaluate stress and strain in static muscular work*. Applied ergonomics, 4(2):91

[10] Sale, D., & MacDougall, D. (1981), *Specificity in strength training: A review for the coach and athlete*. Canadian Journal of Applied Sport Sciences, 6, 87-92.

[11] ADAPS., *Human Modelling System (Ergonomics software)*, Section of Applied Ergonomics, Faculty of Industrial Design Engineering, The Technical University of Delft, The Netherlands, 1990.

[12] Hicks B.J., Mullineux G. & Medland A.J. *The representation and handling of constraints for the design, analysis and optimization of high speed machinery*, AIEDAM Journal, Artificial Intelligence for Engineering Design, Analysis and Manufacture, Special Issue Constraints and Design, 2006, Vol. 20, No. 4, pp. 313-328, ISSN 0890-0606. DOI: 10.1017/S0890060406060239

[13] Molenbroek J.F.M. and Medland A.J., *The application of constraint processes for the manipulation of human models to address ergonomic design problems*, Proceedings of Tools and Methods of Competitive Engineering Delft, The Netherlands pp. 2000, 827-835.

[14] Medland, A.J. & Mullineux, G. *A decomposition strategy for conceptual design*, J. Eng. Design, 11, (1),2000, 3-16.

[15] Medland A.J. & Matthews J., *The implementation of a direct search approach for the resolution of complex and changing rule-based problems*, Engineering with Computers 2009. DOI 10.1007/s00366-009-0148-z

[16] Medland A.J., Matthews J. & Mullineux G., *A constraint–net approach to the resolution of conflicts in a product with multi-technology requirements*, International Journal of Computer Integrated Manufacture, 2009, DOI:10.1080/09511920802372286.

[17] Medland, A.J. and Gooch S.D., (2010) *An Improved Human Model for use in the study of Sitting Postures*, Proceedings of the 10th International Design Conference, Design 2010, May 17 – 20 2010, Dubrovnik, Croatia.

INTERNATIONAL CONFERENCE ON ENGINEERING DESIGN, ICED11
15 - 18 AUGUST 2011, TECHNICAL UNIVERSITY OF DENMARK

BROWNFIELD PROCESS FOR DEVELOPING OF PRODUCT FAMILIES

Timo Lehtonen[1], Jarkko Pakkanen[1], Jukka Järvenpää[2], Minna Lanz[1], Reijo Tuokko[1]
(1) Tampere University of Technology, FIN (2) Finn-Power Oy, FIN

ABSTRACT

This paper represents a development process of product families in a case where already available designs are emphasized. This can be called a brownfield process. Tools, which support the individual steps of brownfield development projects, do exist. In this paper it is described how these tools, methods and procedures can be used to cover a whole development process of a product family. The development of a product family was divided into five steps: setting of goals, developing of a generic element model, analyzing the customer requirements, analyzing the minimum variation and describing the resulted product structure. In the first four steps existing tools were used. In the fifth step a new description method, Product Structuring Blue Print (PSBP), for describing a product structure was represented. PSBP shows how items are related in assemblies, how modules include assemblies, how modules are realized, and what customer requirement is connected to each module. PSBP helps in creating the view of the significance of the product structure solution principles. PSPB gives also a response to how product structuring decisions have to be made.

Keywords: Product Structuring Blue Print, variability, product family

1 INTRODUCTION

Many of the product development processes are focused to new product development. There are less product development processes that focus on brownfield processes, where product range and markets are available. The term "brownfield" is used in our paper as it is used in the building industry and in modernisation projects of process facilities. The brownfield process stands for the re-using of available assets and it includes notions that there are limitations to designing and solutions because of existing structures. Old product solutions, product structures or customer requirements limit designing of new products. Because of this, the brownfield process is not the preferred solution from the designer point of view. Old solutions can include waste that has to be cleaned away before the rest of the solutions are useful. In developing product families this means that for example the quantity of parts might be unnecessarily high or that product solutions are not matching against current customer requirements. In the incremental development process the object is developed step by step. One example of incremental development process has been represented in Oja's dissertation [1]. In many industrial cases though making good use of available products is needed.

New product development or greenfield process (which does not include constraints for development work like brownfield process), has higher risks. Markets usually have dominant designs, which affect the customer behaviour. When a new product has been developed, there is a risk that the customers do not accept it. Investments to infrastructure of the company and existing resources have an effect on the selection of whether to develop current products to higher level or to develop completely new product. Design re-use is one of the most important things that motivate the utilization of brownfield processes. A new product usually includes a set of new requirements to other downstream phases such as manufacturing, maintenance, and sales. These are just a couple of instances why it is unfeasible to start from the scratch.

Tools, which support the individual steps of brownfield development projects, do exist. In this paper it is described how these tools, methods and procedures can be used to cover a whole development process of a product family. Experiences of these tools applied into the case are also discussed. In addition, the tools that were used in this case are analysed in relation to other tools known in the area of developing of product families.

In brownfield processes, it is known that there exist certain product solutions that the customers buy for a specific purpose. The information about the configuration rules is often more unclear. The main challenge in realising product family, which supports customer requirement variation is to develop partition logic. The purpose of the partition logic is to provide rules for selecting the product elements for needed customer variants. The partition logic has to take into account the procurement of elements, and that the elements are suitable for production. In this paper an approach to describe module structures is represented.

The case where product family tools were tested is introduced in Chapter 2. The steps of the brownfield process are explained in Chapter 3. Chapter 4 represents a documenting approach for the product structure. Issues regarding to information management is included in Chapter 5. The results of the research are concluded in Chapter 6. Chapter 7 includes discussion.

2 FINN-POWER OY – CASE COMPANY

The company manufactures production equipment for sheet metal processing. Figure 1 represents an example product of the company. Devices of material management are for example loading equipments, portal robots, tables, carriages and conveyors. Some of these have been illustrated in Figure 2.

Over the years demand has diversified and a quite a selection of different devices for material handling have been developed. The quantity of different alternatives and options has become a challenge for the management over time. Many of the designs have been projects for specific customers. This has led to situations such as that for example the development of certain robot models has gone inevitably on their own paths. Matters have been thought about in different devices in the same ways but requirements have been examined mainly one device at a time.

Figure 1. Finn-Power portal robot equipped with a material carriage and a conveyor.

In the year 2006 there was a modularisation project of carriages in the company. In this project, the relation between customer requirements and product properties were managed successfully. The whole product range has not been analysed in regard to modularisation aspects. Commonalities in the product range have been noticed, but the matter has not been effectively improved. There have been different projects, but all of them have not been successful. On the other hand, old projects have served as good learning points for this project. To conclude the description of the current situation, now it was right time to start a consolidating project on the whole product range. The main idea is in the realisation of customer requirements in the whole concept level, not just in one individual product.

This project has many benefits for in the company:
- Comprehensive analysis of customer needs and above all directing of them to the products in a controlled way.
- Managing of variety without losing the management of the whole concept.
- Utilizing the commonality of the product family.
- Speeding up the material management in production.
- Enabling of variation during the production.
- Speeding up the order-delivery process.
- Scaling of the scheme on the whole product concept, not just on single product.
- Simplifying of the product range and elimination of unnecessary combinations.

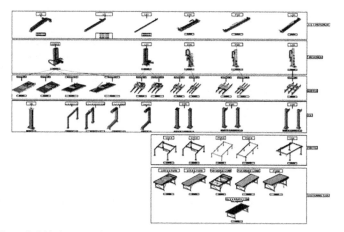

Figure 2. Old elements of the product family. Do we really need all these variations?

Existing products are utilised in the development project although there has been also thoughts that new elements are needed to be developed. Devices to be examined were loading devices, portal robots and discharge equipment where applicable at the beginning of the project.

3 STEPS OF THE BROWNFIELD PROCESS

In this case the brownfield process for the product family development is based on five elementary steps.
1. Defining (business) targets.
2. Drafting the proposed module architecture using mainly old solutions and components.
3. Updating and rationalising the market and customer requirements.
4. Creating module architecture with minimum scale of variation. Defining the amount of new design needed.
5. Documenting the reasoning behind the selected module architecture.

At first the targets of the development process were analysed. The purpose of the first step was to give a clear picture of why the development work should be done and what benefits the results could bring. After analysing the drivers to the development process it was time to model the entity of products and to sketch the generic element model in step two. The generic element model describes the design rationale and the intent of product realisation. The element structure alone is not sufficient enough for developing of product variants. The product variants have to match customer needs. The customer needs were examined in step three. The minimum scale of variation was analysed in the fourth step. These steps are explained in more details in the following subsections. The fifth and the last step includes description of the formed product structure. Actually a sixth step is also needed, because the cost and income effect of the selected solution must be validated against the original business targets. However this step must be taken in any development project. Because it is not specific for brownfield process, it is left outside the scope of this paper.

3.1 Target setting

At first the business goals for the development project have to be discussed. In this case the target setting was done based on the benefits of systematically variable products, which are illustrated in Figure 3. This figure helped to find answers to the question "Why to design variety with commonality for a technical system?" The primary goal was detected; the cost reduction of operating expenses. At this point, the pilot product was chosen. It was discussed that the suitable approaches to facilitate this goal would be 1) form a common architecture for product family and 2) the development of the elements (i.e. modules) as different streams, explained in "Dynamic modularisation concept", see Figure 4 and references [2], [3].

The case specific targets for the product structuring were:
1. Common architecture: the product variants should have a common architecture, as large part as possible.

2. Common modules: the architecture should be made in such a way, that as many as possible of the modules could be shared between the variant customer products in the product family.
3. Elimination of unnecessary variants: the variations, which do not add customer value or facilitate the production or long term development of the product, should be eliminated.

These goals leads to a modular product family, that is able to fulfil the customer needs without extraneous quality and is made out of a minimum inventory of different modules. Naturally, the modules must be real modules with unbreakable and manageable interfaces in mechanical, electrical and information domains [4].

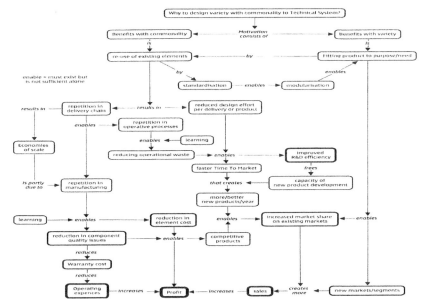

Figure 3. Cause-effect chain of benefits using commonality and variability [5].

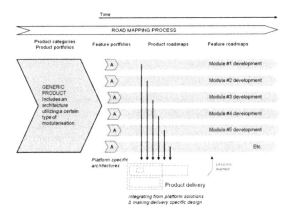

Figure 4. Dynamic Modularisation concept is a R&D method, in which everything is being developed as modules that fit into the common product range architecture. All the products are assembled from those modules. [6].

3.2 Generic element model

The formulation process of a generic element model includes several steps starting with the brainstorming session having participants from various functions such as sales, product development, mechanical, electrical and software engineering.

In this case, the individual work was done first in a brainstorming. During the common session, the participants formed pairs and tried to find a common understanding on terms and connectivities in this particular design. The session resulted in five proposals for a general element model candidate of the future product family. The proposals were all analysed based on the groups of cross-disciplinary expertise.

Due to the fact that there were experts from different areas of the company; the proposals differed a great deal from each other. This was seen extremely beneficial. During the presentation and discussions, the facilitator made a collection of the proposed elements on the white board. After all there were 37 elements on the white board. In addition to written elements there were more than one element, that should be divided further. The software elements were still excluded at this point and there was a separate proposal including a dozen of software elements.

The previous product structuring division could not fulfil case targets completely and so invention of something new was also reasonable. Again brainstorming session was held. In this session everyone was asked to describe his idea of the elements that will be building blocks in the new product architecture. In the beginning of this stage the proposals for the quantity of elements were seven or fourteen and so on. However later on it was discovered to be a too small number. In the end there were several tens of elements. As the final conclusion a combined generic element proposal was accepted by all parties as a work draft for developing the modular architecture.

3.3 Customer requirements

The brownfield processes have existing designs in the background which means that old products have been manufactured for some customer requirements. These requirements can include out of date information and traditions, which are based on old technological realisations. The challenge is to grab the customer requirements according to real customer needs, not according to old products. The validity of customer requirements was analysed using the Gripen method. This is a method inspired by the rationale how the product structure and configuration in Scania trucks were realised [7],[8]. The starting point in this method is the process of customer - what is customer doing. Process can be for example forming of sheet metal products. A handful of master questions, the questions that are most important in for the process of customer are found out. When the customer's preferred ways of working and processes are understood they are segmented in to specific groups and solutions. These groups and solutions have to meet the requirements of a certain customer segment. The most important questions are those that determine in which segments the customer belongs.

One policy of the Gripen method is that instead of selling individual components, larger assemblies or "solutions" are offered, since it is easier to be sure of the compatibility and correct functioning of limited amount of layout variants than compatibility or correct functioning of remarkable amount of single pieces. The Gripen method includes also a thought that individual components are sold only when they are important elements for the customer alone.

In the discussion about the Gripen method, mainly catching of the understanding of customer needs, both pros and cons were considered. Generally the approach was highly regarded as a rational approach to this case and it was decided to give it a try. However there is one point that should be noted. It is possible to develop unnecessary variation. A care should be keep that the level of variation is sufficient enough but not too wide. This is due the fact that it is usually easier to define two solutions that one that fits on the requirement. The issue of minimizing the level of variation has been discussed in the next step.

The focus in customer requirement mapping is to get out of the product orientation and to move on to analysing variability from customer perspective. Giving up on product names and naming products based on configurations can be one helpful approach to this because this helps to get out of a feeling of product specific parts. This has been done for example in Scania 4-series models [7],[8].

3.4 Minimum scale of variation

In the fourth step the suggested element groups, which can form modules later on, of the product are compared against the customer requirements. The objective is to find the minimum quantity of variation that fulfils the customer requirements.

Product Family Master Plan (PFMP) [9] was used for analysing the minimum variation of product family. PFMP includes customer, engineering and part view of the product family.

At first the elements of the generic model were written in the middle of the white board on purpose. Now, there was room for drawing of empty domains and relations on the both sides of the elements. On the left hand side there was the domain of customer need. On the right there was the domain of parts.

On the side of customer needs the "use case" needs were listed. Relations were drawn between customer needs and generic elements. The relations formed visible route from customer needs to existing parts. This method illustrates the formalisation of the product family described above. The formalisation gives possibilities to see certain solutions as well as problems and enables the participants to ask important questions relating to specific relations.

For example in the part domain, the real parts of the product family were arranged according the elements in which they were supposed to be part of. This enables designers to see clearly the feasibility of the elements as modules in the sense of commonality. If an element has parts set, which could be formed with small amount of standard bill-of-materials (BOM), it is a good candidate for a module, where variation is achieved by selecting a suitable module. Similarly if major amount (weighted with a price more than a item number) of the parts of the element could be standardized and only minor part is varying, it could be a configurable element. Naturally there must exist a remarkably large "base unit", which is a standard module. Figure 5 represents examples of PFMP workshop.

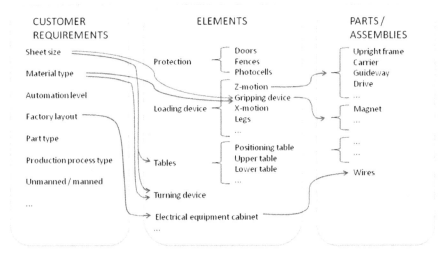

Figure 5. Example of relations between customer, element and part view in PFMP.

In examination of proposed structure the goal was to find or to create a commonality between product variants. This addresses the need for standardisation. Standardisation forms the bedrock for modularisation and thus there must be commonality between X and Y or the goals will never be achieved.

If any kind of standardized BOMs for an element are not generated, element as a future module is evidently challenging to be realised. In this case the suggestions would have been:
- Divide it further.
- Change the element division.
- Consider of changing the technical solution.
- ...and if nothing helps, consider if this is an feature which should not be part of this product family (divide the product to families, remove this element from standard product).

With the use of PFMP it is also possible to discover unnecessary internal variation. Every variant part/module should have a connection to a specific customer need that tells why there has to be variation. Only exception to this rule is that there may be a really good reason for variation coming from own processes of company. It is also possible to check, whether there is possibility to use the same parts or assemblies in elements that as a whole serve different functions.

Configuration knowledge was clarified with the help of the K- & V-Matrix method [10]. In this case we emphasised the use of K-Matrix that included analysis of customer needs against the proposed element structure. This matrix tool helped in ensuring that the elements the configurable product consists would be feasible in production and delivery network. The original K- & V-Matrix method does not consider variation strategies. This missing knowledge was added to solution principles of product variation next to proposed elements of the products to the matrix. These solution principles described that how the variation is done in each element. For example the element could have three options to certain customer requirements.

4 DOCUMENTING THE PRODUCT STRUCTURE FOR DESIGNERS

After four steps that have been demonstrated in Chapter 3, the plan of the product structure exists. This product structure is based on the existing solutions mostly but matches with minimum scale of variation and in reasonable way to the customer variation. The next step was to design elements according to the new product structure. This development needed suitable documentation that represents partition logic of the product and objectives the solutions will provide. In this case these matters were represented with the help of graphs, which were called the Product Structure Blue Prints (PSBP).

The idea in representing module structures using PSBP description was to make the blueprint drawing to show what elements have to be developed or exist and which customer requirement is connected to that element.

The syntax of the method is represented in Figure 6. The left most side includes name of the element (from generic bill of materials) in the product family. Next to it (when moving towards the right hand side) is a description of what kind of modules the element consist of. In this case the generic modules are mostly functional modules that are linked to solution principles of product variation. The solution principles of product variation describe the structure of items in the final assembly. Five different strategies of product structuring can be seen behind the solution principles.

1. Standard component that is included every time.
2. Module that is interchangeable without layout change.
3. Module that is interchangeable with layout alternatives.
4. Module with parametric variation.
5. Element that requires delivery specific design (these should be avoided).

These strategies have been marked with numbers in the generic example represented in Figure 6. Effects of the product structuring strategies on value creation (value chain), procurement and production can be estimated and with the help of this the product structure can be validated. The right hand side of the Figure 6 represents the relation between customer requirements and actual modules (solution principles of product variation).

The Figure 6 shows, for example, that the Beta module includes solution principle Eta to product variation and that the Eta is solved with standardisation. The Gamma module answers to the Customer requirement #9 through the Eta solution principle of product variation.

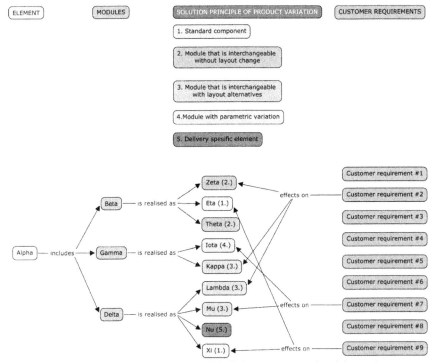

Figure 6. Product Structuring Blue Print (PSBP) for representing of module structures.

5 DIGITAL PRODUCT PROCESS REQUIREMENTS

In today's product development environments the data and information related to a product and its processes are managed in many different systems, often with little integration and with a great deal of data redundancy. The modelling of complex products and their production processes and systems is typically done by multi-disciplinary, often multicultural and geographically distributed teams, working in collaboration. The cooperation required between the teams and each individual member of the team is heterogeneous because of the different platforms and each member's specialization and tasks. Therefore a great number of different types of CAx systems are used throughout the whole life cycle of the product. This heterogeneity affects the communication both between humans and information systems [11].

For example, engineering drawings and product models may be maintained in a proprietary CAD system format, whereas the information on materials, surface finishing, packaging, electrical connections, assembly processes and sequences, resources, and so on are contained in various dispersed documents and stored in a variety of formats. What is common for these documentation formats is the lack of computer interpretable meaning of data [11].

The use of multiple proprietary formats forces the experts to serve as a manual human-machine interface between different systems, causing a vast amount of manual work and the possibility of human errors. After each life cycle phase of the product the different design teams create the models again from their own perspective. Also during these remakes the model gets filtered, because the current phase does not need all the information that was needed in the previous development phase, or because the transfer format does not support all the data types of the original model. The filtering leads to the "snapshot" approach where the product information is reduced to a screen-captured picture in the end forming a cartoon like design documentation [11].

A commercial solution for capturing the meaning of models and documentation does not exist. In other words "the meaning of" is not implemented into the IT-systems. In academic circles some approaches have been utilized for adding the computer interpretable semantics into the models. These technologies include use of logical representation such as first-order-logic and description logic, light-weight ontologies and semantic web technologies. Unfortunately, these have not yet been accepted into the solutions of software vendors.

As the technology rarely solves the underlying problem, it has been proposed that the companies should concentrate on the knowledge preservation through daily interactions. In this case the interactions would be the cooperative design procedures and definition where the information comes from and where the knowledge is used later on. It is a known fact that once the designers know the meaning of their work and its impact to colleagues work, the amount of errors and short-cuts tends to drop [11].

6 CONCLUSIONS

Development process in the case of a product family can be identified to be an business-oriented process. It means that it is most important to think of the business profits that the results could realise first. This is why the process does not give any modularity type until the goals have been set.

The product family concept has been defined in the company. The next steps are in designing each section. Design work is done in mechanical, electrical and software disciplines. At this stage, it has been estimated that the project would last about one year. Also a technical innovation will be tried in the form of a prototype. If results of the testing are positive and it looks promising to be developed further, the innovation will be adapted to the whole concept.

The PFMP is a powerful although a little laborious tool. However, a lot of information can be extracted from the company's IT-systems. Use of only PFMP can result in products that are not configurable or configuration rules are not unambiguous. Because of this, other tools are needed in addition for detecting the configurable parts from which the product is assembled.

K- & V-Matrix method is not as visual method as PSBP. And the deficiency in the former is that it does not include the production point of view.

Both pros and cons can be recognised about using the PSBP description approach. The final element structure of the products can be hard to piece together from table tools like MS Excel. Standardised elements are viewed fairly well in customer requirement - module relation. The idea is that the product does not include anything that has not been modelled. One question is that is the quantity of variants sufficient?

If modules of the product are explicit, they have been validated and all the customer requirements can be fulfilled with them, realistic product structuring plan of the item is ready. The developer group can be given a drawn PSBP as a work instruction in designing of the part. If in the developing work solutions have to be changed. It can be seen from PSBP immediately where the changes have effects and what estimates must be redone. One challenge is the fact that the method is new and not yet established among the industry. The PSBP can be therefore be drawn how it is wanted, not necessarily how it is supposed to be drawn in the first place. To improve this, information systems should have support for this kind of a description method.

When modules are known from the product structuring strategy point of view, the basis for validation exists. It is possible to estimate or to calculate what advantages in the processes and networks of the company this module structure brings to us. In our example case, the expected benefits were calculated using activity based costing approach. The cost savings and added value differed highly from activity to activity. Because the product was not completely re-engineered, the expected savings on material and components were moderate. But the effects on company operational costs were in this example remarkable high. In certain cost topics they came up to near 40 %, which could be considered a good result even for Greenfield project.

7 DISCUSSION

Other methods and approaches could have been used instead of using methods represented in Chapter 3. Selection of product structuring solutions like modularity without evaluation of business objectives is possible but then the outcome is not very likely an optimal from the company effectiveness point of view.

The research group of Ishii [12] has developed calculation models for optimizing the quantity of product variants. There is evidence that these models can result in the most optimal solution but this means also that all of the existing variants need to be developed again. Main challenges of these kinds of greenfield processes were discussed in Chapter 1.

Erixon [13] has represented Module Indication Matrix (MIM) where module drivers are checked against function carriers. This method starts from existing components or modules but does not consider is it possible to reduce amount of modules by standardisation. In spite of this the tool can be considered as a possible tool variant also in a brownfield process.

There are many methods that say that modules should be treated as functional ones [14]. Adapting of function based module structure leads to good results only if recognised variation is only functional. The Gripen method that we used helps to identify the importance of functionalities in variations.

Quality Function Deployment (QFD) [15] does not include alternatives of product structuring strategies but concentrates on handling of product properties. The essential contribution of this paper stays outside of QFD studies. QFD is not a tool which is sufficiently design oriented. Currently the mere control of requirements does not help. It must be possible to design a product family which meets the customer requirements and the own requirements of the developing of the operation.

Brownfield method offers a process and a formalised way to rationalise existing product range. This kind of an approach is useful because in projects where it is not possible to develop new products, it might be hard to decide a good starting point and direction for the development work.

ACKNOWLEDGEMENTS
Authors like to thank The Finnish Funding Agency for Technology and Innovation (Tekes) and KIPPcolla project (Knowledge Intensive Product and Production Management From Concept to Recycle in Virtual Collaborative Environment) consortium and partners for support and feedback.

REFERENCES
[1] Oja, H, *Incremental Innovation Method for Technical Concept Development with Multidisciplinary Product*. 2010, Dissertation, Publication 868, 126 p (Tampere University of Technology, Tampere)
[2] Riitahuhta, A. and Andreasen, M.M, Configuration by Modularisation. 1998, *Proc.of NordDesign 98*, pp167-176 (KungligaTekniska Högskolan, Stockholm)
[3] Lehtonen, T., Juuti, T., Pulkkinen, A. and Riitahuhta, A, Dynamic Modularisation – a challenge for design process and product architecture, in *Proc. International Conference on Engineering Design, ICED'03*, 2003 (KungligaTekniska Högskolan, Stockholm)
[4] Borowski, K-H, *Das Baukastensystem in der Technik*, 1961 (Springer-Verlag)
[5] Juuti, T, *Design Management of Products with Variability and Commonality*, 2008, Dissertation, Publication 789, 155 p (Tampere University of Technology, Tampere)
[6] Lehtonen, T., Taneli, H., Martikainen, A-M. and Riitahuhta, A, *Laivatoimituksen tehostaminen joustavan vakioinnin ja moduloinnin keinoin, Final report of MERIMO project*, 2007 (Tampere University of Technology, Tampere)
[7] Lehtonen, T, *Designing Modular Product Architecture in the New Product Development*, 2007, Dissertation, Publication 713, 220p (Tampere University of Technology, Tampere)
[8] Scania, *Scania brochure 1592341 fiFI*
[9] Harlou, U, *Developing product families based on architectures – Contribution to a theory of product families*. 2006, Dissertation, 173 p (Technical University of Denmark, Denmark)
[10] Bongulielmi, L, *Die Konfigurations- & Verträglichkeitsmatrix als Beitrag zur Darstellung konfigurationsrelevanter Aspekte im Produktentstehungsprozess*, 2002, Dissertation, 218p (ETH, Zentrum für Produktentwicklung, Zürich)
[11] Järvenpää, E., Lanz, M., Mela, J. and Tuokko, R, Studying the Information Sources and Flows in a Company – Support for the Development of New Intelligent Systems, in *Flexible Automation and Intelligent Manufacturing, FAIM2010*, 2010
[12] Ishii, K. and Eubanks, C.F, Design for Product Variety. Key to Product Line Structuring. *ASMEDesign Technical Conference* Vol. 2, 1995, pp499-506 (Boston)
[13] Erixon, G, *Modular Function Deployment - A Method for Product Modularisation*. 1998 (KungligaTekniska Högskolan, Stockholm)
[14] Otto, K. and Wood, K, *Product Design*. 2001 (Prentice Hall, New Jersey)
[15] Taguchi, G, *Introduction to Quality engineering: Designing into products*, 1986 (Asian Productivity Organisation)

INTERNATIONAL CONFERENCE ON ENGINEERING DESIGN, ICED11
15 - 18 AUGUST 2011, TECHNICAL UNIVERSITY OF DENMARK

APPROACH FOR THE CREATION OF MECHATRONIC SYSTEM MODELS

Martin Follmer[1], Peter Hehenberger[1], Stefan Punz[1], Roland Rosen[2] and Klaus Zeman[1]
(1) Johannes Kepler University, AT (2) Siemens AG Corporate Technology, DE

ABSTRACT
One of the major challenges in developing mechatronic products is the increasing complexity of the products themselves. The defining feature of mechatronic products is the interplay between various engineering disciplines such as mechanics, electronics, and software. There is a critical lack of methods and tools supporting the interdisciplinary aspects of the development process of mechatronic products, especially in the conceptual design phase. These deficiencies make it difficult to overview the interdependencies of the involved engineering disciplines. Mechatronic System Models (MSM) can improve this unsatisfactory situation and allow for a holistic view on complex mechatronic systems. This affords a validation of design concepts and a qualified comparison of different concepts in early phases. MSM should at least be able to manage existing data and to illustrate the most important relations. Additionally, they should provide the possibility to execute several simulations of load cases, thus allowing specific "global" system properties to be evaluated. Typically these simulations at the system-level differ from those at the discipline-level.

Keywords: process model, product development process, system-level modelling, system-level models, system models, mechatronics, mechatronic systems, systems-of-systems

1 INTRODUCTION
The defining feature of mechatronic products is the interplay between various engineering disciplines such as mechanics, electronics, and software [1]. Due to that and to the increasing complexity of the products themselves, the need of methods and models supporting the overview of the most relevant system data is present from the very beginning of design processes. However, there is a critical lack of methods and tools supporting these interdisciplinary aspects of the development process of mechatronic products as well as the communication between developers from various disciplines, especially in the conceptual design phase [2]. As quoted in [3] these deficiencies make it difficult to overview the interdependencies of the involved engineering disciplines. The level of abstraction at which the whole product is investigated and discipline-overall models are used is hereafter referred to as "system-level".
Mechatronic System Models (MSM) can improve this unsatisfactory situation and allow for a holistic view on complex mechatronic systems which is generated by various single views such as the consideration of requirements, structure, behaviour, function, parametrics etc. MSM should at least be able to manage existing data and to illustrate both the relationships inside a system (between sub-systems) and between a system and its environment [4]. Additionally, they should provide the possibility to execute several simulations of load cases (test cases), thus allowing specific "global" system properties to be evaluated. Typically these simulations at the system-level differ from those at the discipline-level. Since simulations at the discipline-level are usually conducted by highly skilled and specialized engineers who use specialized, discipline-specific software tools, the simulations at the discipline-level can normally not be replaced by simulations at the system-level. Furthermore, simulations at the discipline-level are used to evaluate "local" properties [5].
The basis for simulations in the various phases of the product development process is a shared database. This database also provides interfaces to discipline-specific software tools and their related models, and must ensure data consistency. Further discussions of the database are beyond the scope of this paper.
This article discusses possible process models for the creation of MSM, and is structured as follows: The subsequent section is dedicated to two process models. The first one shows the working steps necessary for the creation of a MSM whereas the second one discusses the simulation-based design

process as shown in [5]. Next, an integrated process model for the creation of a MSM is presented. A conclusion summarizes the main aspects of this article and addresses future activities.

2 RELATED WORKS

VDI Guideline 2206 [6] addresses in particular the design methodology of mechatronic systems and proposes the "V-model" (Figure 1) for the development process. After analyzing all requirements on the total system, the sub-functions and sub-systems are defined (left branch of the V-model). They are to be developed simultaneously by discipline-specific development teams working in collaboration. After verification of the sub-functions and testing the sub-systems, these are integrated stepwise (right branch of the V-model) into the "overall system". Then the performance of this integrated system is checked by analysis and evaluation in order to assure its properties. If the system must be improved, the initial operation phase is repeated (iterative process).

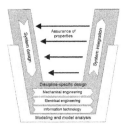

Figure 1. V-model according to VDI 2206 [6]

In [7] and [8] the authors presented an approach for the holistic description of a multidisciplinary system with the consideration of the essential operating modes and the desired behaviour. Therefore aspects such as the environment, application scenarios, requirements, the system of objectives, functions, etc. should be considered in a specific specification technique. Furthermore, a procedure model for the conceptual design phase (which includes four sub phases) was developed. The research group also developed the software tool "Mechatronic Modeller" that is based on the specification technique for modelling mechatronic systems.

A concept for a software prototype supporting the development of mechatronic systems was presented in [9]. The software prototype called "Connection-Modeller" should allow various views on the system under design, e.g., requirements, functions, structure. These views are called partial models and can be developed using proprietary software-tools. The Connection-Modeller provides means to define cross-discipline connections between various partial models which e.g. can be used for the propagation of design changes.

A system-level (high-level) model of a multidisciplinary system based on a functional description was introduced in [10]. The architecture of the system is primarily determined by the main functions which are already known in the conceptual phase of design. In later design phases general functions can be decomposed into more concrete ones which lead to a more detailed functional structure of the system under design. The system model should provide a better overview of the system and should also connect abstract with more concrete models. A verification of specific parameters against requirements can also be supported by the functional structure. The same research group presented a function modelling approach in [11] to connect different views on different levels of abstraction to support the designer in the conceptual design phase. The authors discussed the following three industrial problems: (i) design traceability, (ii) design understanding and (iii) system decomposition.

In [12] a communication framework supporting the communication between the model based system engineering efforts and the discipline specific software tools was introduced. In this paper an approach for an integrated design environment is presented in which mappings between SysML and Matlab/Simulink models are built.

In [13] means to use structural, functional or behavioural relations between entities as cues for engineering change prediction were discussed. The authors mentioned a lack of methods for generating plausible estimates of how changes propagate. In [14] the hierarchical structuring as a proper technique for representing complex systems was proposed. The authors also suggested several guidelines which should support designers for attaining a meaningful decomposition.

3 MECHATRONIC SYSTEM MODEL (MSM)

3.1 Process model I: Creation of a MSM

3.1.1 Generic Approach

Mechatronic systems usually consist of several sub-systems and system-elements on different hierarchical levels (also referred to as levels of abstraction). Therein, the terms system, sub-system and system-element have a relative meaning, hence, the allocation of a specific level to the different system-elements depends on the definition and view of the system under consideration and is thus a matter of definition and view.

A Mechatronic System Model (MSM) should represent the overall mechatronic system under consideration (original) and should include all its relevant properties. As the structure of the mechatronic system may be regarded as a significant property, at least this structure has to be mapped to the model, too. Additional, maybe different, structures with various abstraction levels may arise from other views of the system (e.g. requirements, functions, modelling aspects), leading to a multiple-structured model. The MSM covers the highest abstraction level considered, and may include sub-models and model-elements on levels below. The terms model, sub-model and model-element again have a relative meaning and are a matter of definition and view.

Figure 2 shows the generic approach for the creation of a Mechatronic System Model (MSM). This approach describes the necessary steps and their chronological order to create a MSM both for top down and bottom up modelling. In Figure 2, the different levels are depicted as rectangles representing sub-models of the MSM which can be used to structure the MSM with respect to various views. The grey area of each sub-model of the MSM accounts for interfaces and communication between the connected sub- or discipline-specific models by transmitting input and output parameters (depicted as blue circles). Representatively for all levels of the MSM, the granularity of models and respective parameters is depicted only at the system-level of the MSM by circles with different diameters inside the grey areas. The green rectangles in Figure 2 represent discipline-specific models.

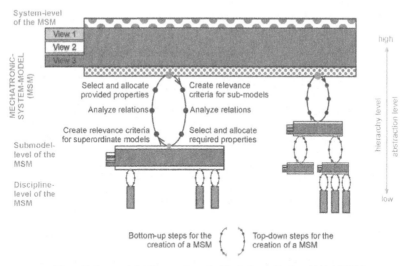

Figure 2. Approach for the creation of a Mechatronic System Model (MSM)

Furthermore, various sub-models of the MSM as well as relations between system- and discipline-level are shown in Figure 2. The interactions between sub-models of different levels as well as between sub-models and discipline-specific models should exhibit the same pattern of steps (depicted as ellipses). To change the level of abstraction, the following steps are required:

- Create relevance criteria (e.g. with respect to the system under consideration, modelling aspects):
  The relevance criteria are used to determine those model-elements which should be included in the MSM on a specific level of abstraction.
- Analyze relations:
  The relations between model-elements on different levels of abstraction must be established, analyzed and documented. These relations connect models of adjacent hierarchy levels and allow for the propagation of changes across hierarchy levels in the model and therefore in the original as well.
- Select and allocate properties (e.g. of the original, of the model):
  Significant properties have to be chosen and allocated to each sub-model and model-element. To all properties, characteristic parameters have to be assigned which are able to quantify the chosen properties. If a parameter is fixed or changed on a certain level, it has to be transferred to all adjacent hierarchy levels on which possible contradictions are examined.

These steps are important for a top-down as well as a bottom-up approach.

3.1.2 Relevance criteria for the inclusion of model-elements into the MSM

As already mentioned, relevance criteria are used to determine which model-elements should be included into the MSM on a specific hierarchy level (system level of the MSM, sub-model- or discipline-level). These criteria cannot be understood as a rigid set of rules but have to be tailored to the specific design or analysis task and to the corresponding questions to be treated by the MSM. As not all model-elements are of the same significance to the MSM and an inclusion of each and every model-element into the MSM would lead to an information overflow in the MSM, detailed models (e.g., of components that are comparably simple and well-understood such as standard bearings or electrical resistors) should be "condensed" to significant models on a higher level.

Therefore it is necessary to specify clear criteria for those model-elements and sub-models which should be included into the MSM. Table 1 shows various general relevance criteria pro and contra modelling on the system-level.

Table 1. General relevance criteria

	General relevance criteria pro modelling on the system-level
GP1	Models of system-elements which are relevant to the understanding of the overall system
GP2	Models of system-elements which are relevant to the behaviour of the overall system
GP3	Models of system-elements which are relevant to the structure of the overall system
GP4	Models of system-elements with interdisciplinary ("global") relevance
	General relevance criteria contra modelling on the system-level
GC1	Models of system-elements with intra-disciplinary relevance

GP1 ... General Pro 1, resp. GC1 ... General Contra 1

Table 2 shows various system-related relevance criteria pro and contra modelling on the system-level.

Table 2. System-related relevance criteria

	System-related relevance criteria pro modelling on the system-level
SP1	Models of system-elements for which GP1 to GP4 are not applicable, but which are relevant due to other criteria (e.g. resulting costs)
	System-related relevance criteria contra modelling on the system-level
SC1	Models of system-elements with interdisciplinary relevance but with low challenge to be mastered (e.g. standard electrical drive for a component with minor importance)

SP1 ... System-related Pro 1, resp. SC1 ... System-related Contra 1

3.1.3 Analysis of relations

The relations modelled in the MSM comprise not only relations of the system under consideration but also additional relations (e.g. inside and between different views).

A suitable representation of a MSM should include only models of those system-elements and sub-systems which have significant importance to the system-level according to the chosen relevance criteria. In the course of the assurance of system properties, a further essential task for the MSM is to model the relations of the system under consideration (under design or analysis). Modelling these relations should contribute to a better understanding of the system and should allow a comprehensive analysis of its internal and external relations. Thereby, the representation of relations between all characteristic system parameters enables both a complete overview of the most relevant system data and the identification of the main parameters. Furthermore, the representation of the relations can be used to structure the model of the overall system and to decompose it into sub-models as well as to trace the impact of design changes. The relations may be represented, e.g., by graphs or standardized modeling languages such as SysML, see also [4] and [15]. In general, various kinds of relations between system-elements are possible e.g., physical, geometrical as well as topological relations. In the following, a (certainly incomplete) list of possible relations is mentioned.

Classification of relations

Table 3 shows a classification into external and internal relations as well as corresponding examples. External relations describe the relation between the system and its environment, whereas internal relations characterize system-inherent relations.

Table 3. Classification of relations

External relations
Interfaces to the system environment (energy, material and signal flows)
Stakeholder requirements
Internal relations
Interdisciplinary relations
Intradisciplinary relations
Geometrical relations

Relations between model-elements of the MSM

It is obvious that a lot of relations between MSM-elements belonging to the same view exist (e.g. relations between requirements). Furthermore, various relations may also arise between MSM-elements belonging to different views. Several possible views of a MSM are listed in the following:
- Requirements
- Structure
- Functions
- Behaviour
- Parametrics
- Models (implemented CAx-Models)

One example is the relation between a specific function and the corresponding requirement that should be fulfilled. The correspondence between a 3D CAD model of an assembly and the related bill of material, may serve as an example for relations between models.

Relations due to flows of energy, material and signals

Various system-elements are influenced by the flows of energy, material and signals inside the system as well as to and from the system. As an example, the change of the energy supply of the system (e.g. from hydraulic to electric) has significant influence on the (already existing) relations of the system, as hydraulic devices always need components for pressure generation that are not required when using electrical devices.

Representation of relations

For the representation of relations of the MSM, manifold possibilities are available, for example:
- Text-based
- Matrix-based (e.g. DSM, [16])
- Graphs
- Standardized Modelling Language (see e.g. [15]).

In our investigations a standardized modelling language is preferred.

Description of relations
Various possibilities are available for the description of relations, as well, for example:
- Text-based
- Arithmetic relations
- Geometric relations
- Logical relations.

A text-based documentation of relations is very simple to create but can quickly become too confusing. Several other possibilities, e.g., as usual in CAD-tools are known.

3.1.4 Selection and allocation of properties

As already mentioned, relevance criteria are used to define the level of abstraction for the elements in the MSM. In the next step, significant properties have to be chosen and allocated to each sub-model and model-element. During the design process the granularity of properties is getting finer and finer which corresponds to an increasing level of detail of system-elements as well as their properties. Corresponding to the design process, the allocation process of properties is iterative as well. Here it is essential to distinguish between global and local properties. Global properties are system-specific properties that cannot be evaluated at a discipline-specific level, whereas local properties are to be evaluated at the discipline-level.

3.2 Process model II: Simulation-based design process

A general approach to a simulation-based design process for mechatronic systems, especially for the early phases of design, was presented in [5]. This approach consists of six phases based on VDI Guideline 2221 [17] and aims at integrating simulation techniques into the design process from the very beginning in order to evaluate the properties of a system under design as far, and as early, as possible within each design stage. Thus, the comparison between actual and desired system properties can be drawn faster, better, and easier, thereby improving the design process itself. Figure 3 shows a depiction of the detailed process model on the left side and a simplified (condensed) representation on the right side.

The input to the process model is a specific "development task". The process model consists of six design phases, whereas only phases 1 to 5 are the focus of the present investigation. The design phases (depicted as large rhombuses) include specific working steps (depicted as rectangles) and corresponding working results (depicted as small rhombuses). Each design phase concludes with a query: Are the requirements reachable? If the requirements are attainable, the process continues with the next design phase; otherwise, an "external" iteration is necessary, or the process must be terminated. The step "Validation/Evaluation" represents an "internal" iteration step (inside the actual design phase) at the end of each design phase. Phase 6 and the output ("further realization and documentation") are beyond the scope of this paper.

In the first design steps, requirements and functions can be simulated; in the principle and architectural design, first mathematical models can be executed. System-level simulations are possible in each phase of the design process, whereas discipline-level simulations are feasible only in later phases of the design process (see the markers on the right side in Figure 3) when the information about the system containing the necessary level of detail becomes available.

4 PROCESS MODEL FOR MODEL BASED MECHATRONIC DESIGN

4.1 Integration of the presented process models

Both process models presented extend the common product development processes by additional investigations regarding simulation-based modelling on the system-level. The following step tries to integrate both process models into one holistic approach for the creation of a MSM.

Figure 4 shows the integration of process model II into process model I as well as the specific working steps depicted as ellipses and "diamonds" leading to an integrated process model for model based mechatronic design. Process model I is specifying the steps that are necessary for changing the abstraction level, whereas process model II is representing the steps for the evaluation of global properties by applying system-level simulations at various hierarchy levels. The relevance of the various design steps depends on the phases of the product development process as well as the specific abstraction level of the system under consideration and the corresponding MSM.

According to Figure 4 simulations are possible on each abstraction level of the MSM. As already mentioned, simulations at the system-level differ from those at the discipline-level and should contribute to a better understanding of the overall system by evaluating system-specific (global) properties that cannot be evaluated at a discipline-specific level.

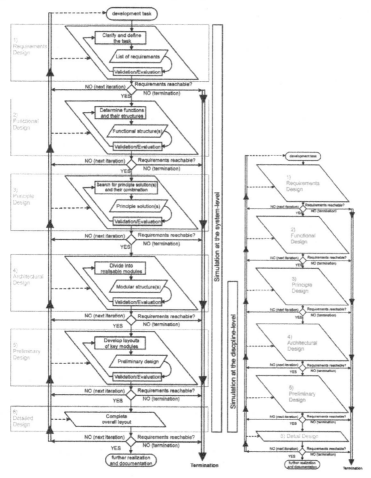

Figure 3. Design process model according to [5]

As explained above, the grey areas in the sub-models are essential for the communication inside the MSM. In the integrated process model according to Figure 4, these areas also include the necessary "Validation/Evaluation" steps for the several system-level simulations depicted in Figure 3.

4.2 MSM for "new-design" and "re-design"

Figure 4 shows the process model for MSM in the distinct cases of "new-design" and "re-design" (see the green arrows). A very significant difference between these two cases is the amount of available information. Normally, the information base for new-design is significantly lower than that for re-design. Depending on the mentioned cases, the design phases from "Requirements Design" to "Preliminary Design" may have different meanings and importance on the distinct levels of the MSM. The diamond shaped arrangement of the design phases in the process model for model based mechatronic design (Figure 4) allows for a free selection of their sequence.

In general, a new-design starts at the highest hierarchy level with a tiny and often uncertain information base. The subsequent phases of the design process contribute to an enlargement of the information base which in turn is requisite for a suitable division into sub-models on a lower level of abstraction.

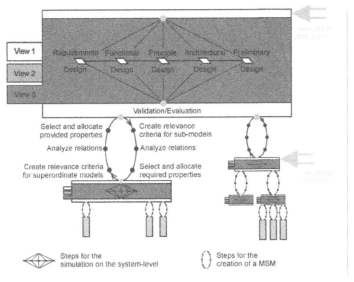

Figure 4. Process model for model based mechatronic design

Re-designs of already existing systems, however, are usually induced by the modification of sub-systems which implies that normally a broader information base is available. Changes caused by re-designing sub-systems must be transmitted to the connected sub-models and discipline-specific models using the MSM-specific working steps such as creation of relevance criteria, analysis of relations and allocation of properties.

5 INTEGRATION INTO COMMON DESIGN MODELS

Figure 1 shows the standard V-model according to VDI 2206 [2]. A complex mechatronic system is, however, usually not developed in the course of a single macrocycle of the V-model. Therefore, it is necessary to iterate through several macrocycles, as shown in Figure 5.

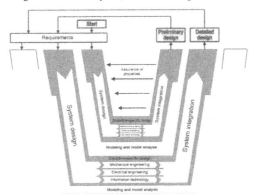

Figure 5. Pass through several macrocycles of the V-model

In addition, the V-model in Figure 5 presents modifications in the phases "System design" and "System integration", which are depicted in more detail in Figure 6 (the inner V-model of Figure 5). Normally, these two phases (depicted as arrows) do not illustrate the separation (or integration) of system and involved disciplines.

Figure 6 shows the integration of the phases of the design process model (presented in Figure 3) into the V-model. Usually, phases 1 to 3 occur at the system-level. Phases 4 and 5, however, interact with both the discipline- and the system-level. The previously discussed external iteration in the design process model can be done in the course of the "Assurance of properties".

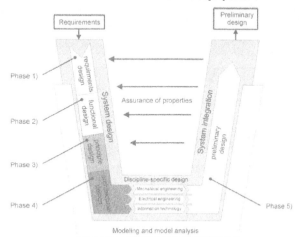

Figure 6. Integration of the different design phases into the V-model

These investigations conclude that even the steps "separation into the disciplines" and "integration of the disciplines" have particular significance in the product development process. It is essential that the interaction between system-level and discipline-level is managed in these steps.

6 CONCLUSION AND FURTHER ACTIVITIES

It is evident that there still exists a considerable lack of methods as well as software tools that support design engineers in executing simulations in the early phases of the product development process and provide a holistic view of the system under consideration. This paper wants to contribute to the remedy of this deficiency by developing an integrated process model for model based mechatronic design which allows also for the evaluation of global system properties. The next steps of the research work will focus on the creation of further views to be covered by the MSM in order to enable a broader range of system-level simulations.

ACKNOWLEDGEMENTS

This work was kindly supported by the Austrian Center of Competence in Mechatronics (ACCM), a K2-Center of the COMET/K2 program, which is aided by funds of the Republic of Austria and the Provincial Government of Upper Austria. The authors thank all involved partners for their support.

REFERENCES

[1] De Silva C. W., *Mechatronics – an integrated approach*, 2005 (CRC Press Boca Raton, London, New York, Washington DC).
[2] Vajna S., Weber C., Bley H., Zeman K. and Hehenberger P., *CAx für Ingenieure - Eine praxisbezogene Einführung*, 2009 (Springer Verlag, Berlin Heidelberg Germany).
[3] Aberdeen Group, *System Design: New Product Development for Mechatronics*, 2008.
[4] Follmer M., Hehenberger P., Punz S. and Zeman K., *Using SysML in the product development process of mechatronic systems*, in Proceedings of the Design 2010 11th International Design Conference, Vol. 3, 2010, pp. 1513-1522.

[5] Follmer M., Hehenberger P. and Zeman K., *Model-based approach for the reliability prediction of mechatronic systems on the system-level*, EUROCAST 2011, 2001, accepted for publication in Springer Lecture Notes in Computer Science.
[6] VDI Verein Deutscher Ingenieure, The Association of German Engineers, *VDI 2206 - Design Methodology for Mechatronic Systems*, VDI Guideline, 2003 (Beuth Verlag Germany).
[7] Gausemeier J., Dorociak R., Pook S., Nyßen A. and Terfloth A., *Computer-aided cross domain modeling of mechatronic systems*, Proceedings of the Design 2010 11th International Design Conference, Vol. 2, 2010, pp. 723-732.
[8] Gausemeier J., Dorociak R. and Kaiser L., *Computer-aided modeling of the principle solution of mechatronic systems: A domain-spanning methodology for the conceptual design of mechatronic systems*, Proceedings of the ASME 2010 International Design Engineering Technical Conferences & Computers and Information in Engineering Conference, 2010.
[9] Stark R., Beier G., Wöhler T. and Figge A., *Cross-Domain Dependency Modelling – How to achieve consistent System Models with Tool Support*, 7th European Systems Engineering Conference, EuSEC 2010, 2010.
[10] Alvarez Cabrera A. A., Erden M. S., Foeken M. J. and Tomiyama T., *High Level Model Integration for Design of Mechatronic Systems*, Proceedings of IEEE/ASME International Conference on Mechatronic and Embedded Systems and Applications, 2008, pp. 387-392.
[11] van Beek T. and Tomiyama T., *Connecting Views in Mechatronic Systems Design, Function Modeling Approach*, Proceedings of IEEE/ASME International Conference on Mechatronic and Embedded Systems and Applications (MESA), 2008, pp. 164–169.
[12] Qamar A., During C. and Wikander J., *Designing Mechatronic Systems, a Model-based Perspective, an attempt to achieve SysML-Matlab/Simulink Model Integration*, IEEE/ASME International Conference on Advanced Intelligent Mechatronics, 2009.
[13] Ariyo O. O., Eckert C. M. and Clarkson P. J., *On the use of functions, behaviour and structural relations as cues for engineering change prediction*, Proceedings of the Design 2006 9th International Design Conference, 2006, pp. 773-781.
[14] Ariyo O. O., Eckert C. M. and Clarkson P. J., *Hierarchical decompositions for complex product representation*, Proceedings of the Design 2008 10th International Design Conference, 2008, pp. 737-744.
[15] OMG, *Systems Modeling Language - SysML*, Version 1.2, June 2010.
[16] Lindemann U., Maurer M. and Braun T., *Structural Complexity Management An Approach for the Field of Product Design*, 2008 (Springer Verlag, Berlin Heidelberg Germany).
[17] VDI Verein Deutscher Ingenieure, The Association of German Engineers, *VDI 2221 - Systematic approach to the development and design of technical systems and products*, 1993 (VDI Guideline, Beuth Verlag Germany).

Contact: Martin Follmer
Johannes Kepler University
Institute of Computer-Aided Methods in Mechanical Engineering
Altenberger Street 69
4040 Linz
Austria
Phone: +43 732 2468 6555
Fax: +43 732 2468 6542
E-mail Address: martin.follmer@jku.at
URL: http://came.mechatronik.uni-linz.ac.at

Martin Follmer is a doctoral student at the Institute of computer-aided methods in mechanical engineering (Head: Prof. Klaus Zeman) at the Johannes Kepler University (JKU) in Linz, Austria. As an undergraduate he studied Mechatronic at the JKU, specializing in Mechatronic Design.

INTERNATIONAL CONFERENCE ON ENGINEERING DESIGN, ICED11
15-18AUGUST2011,TECHNICAL UNIVERSITY OF DENMARK

MODELING AND DESIGN OF CONTACTS IN ELECTRICAL CONNECTORS

Albert Albers, Paul Martin and Benoit Lorentz
Karlsruhe Institute of Technology (KIT)

ABSTRACT
The presented paper focuses on modeling and simulation of electrical connector contacts' behavior and associated design. New solutions are developed, based on the Contact & Channel Approach validated by simulation and experiment.
Primary parameters such as contact resistance, tribological and thermal behavior, contact force, material and connector size strongly influence electrical connector's properties. Therefore, a great deal of experience or effort is needed to design application specific solutions mastering preceding interrelated parameters. However, many state of the art electrical connectors are, especially for high currents, designed by trial and error processes. In order to increase efficiency and effectiveness of the design process, appropriate models are needed. To generate new design solutions, models of a certain level of abstraction are required. In addition to this, holistic computer-aided models enable the prediction of connector's electrical and mechanical performance. Here, design solutions are developed systematically based on the Contact & Channel Approach. At the same time a Finite Element Model is built in order to investigate the behavior of designed connector's prototypes.

Keywords: electrical connector, modeling, finite element method, Contact- and Channel Modeling Approach (C&C-A), product design

1 INTRODUCTION
The general tendency towards modularity and increasing number of high current applications especially in the areas of manufacturing engineering and automotive design cause an increasing need for efficient and reliable electrical connection of independent modules. At the same time connectors have to follow the trend of miniaturization. The majority of connectors' electrical contacts are to be regarded as critical in two extreme working conditions: At very low currents and voltages conducting contact is difficult to ensure whereas at high currents and voltages effects of heat generation and electrical arcing are critical. With increasing number of high current carrying connectors the significance of their performance for the overall system efficiency increases correspondingly. High contact resistance leads to high temperature rise synonym to high power loss.
Taking into account the upcoming technologic developments in the context of Electro- and Hybrid-mobility, the number of high current connectors being particularly problematic with respect to power losses will increase accordingly. Hauck [1] states that vehicles' electrical systems will more and more have to fulfill the tasks of power trains with electric power of 200kW at continuous currents of around 400A instead of only transmitting signals and comparably low energies. Especially in the automotive area Himmel [2] is predicting new demands for high current capacities and reliability, which are going to be combined with general requirements, such as low production costs, small installation space and ergonomic handling properties, i.e. moderate insertion force.
Hence, the task is to design reliable connectors with highest current handling capacity and with resistivity as low as possible for highest efficiency. Efficiency in this context means energy efficiency i.e. the quotient of power output to power input, as well as efficiency of installation space, i.e. the quotient of transferable power to required volume of the connector.
Named requirements have strong interconnections as a low contact resistance causes less power losses and lower heat generation in the contact which in turn increases connector's reliability and lifetime. Furthermore a lower heat generation allows a more compact design and thus decreases the connector's overall volume. To design connectors according to those needs, deepened knowledge and understanding of actual occurring effects is essential. In order to support a corresponding design

process appropriate models are necessary. The major objective is simulation of contact performance based on these models.

2 MODELS OF ELECTRICAL CONTACTS

2.1 Theoretical background

The contact resistance is defined by Rieder [3] as the difference between the resistance of a closed contact and the one of a homogenous conductor of the same shape and dimension. Hence, any additional contact connecting current carrying conductors cause additional electric resistance, which heats the current path locally.

The reason for this contact resistance is not, as often expected, a transition resistance from one material into another. An interface between two metallic surfaces does not necessarily mean a higher resistance for the electron current, as any grain boundary of metallic structure would be. If metallically clean contacts were in real contact over the entire apparent contact area, there would theoretically be no remarkable resistance. Rather according to Rieder [3], resistance originates due to

- mechanical contact area, which takes the contact load, is always smaller than the apparent surface because of unavoidable macro- and microscopic bumps and asperities
- electrically conducting contact area might be smaller than the mechanical contact area due to impurities of almost insulating property
- conducting areas often are covered by non-metallic conducting layers (with higher specific resistivity).

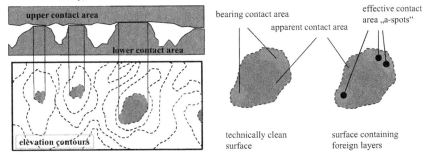

Figure 1: Differentiation between apparent contact area, bearing contact area and actual conducting contact area.

Electric resistance between two surfaces is caused by constriction of the current flow to small conducting areas, so called a-spots (Figure 1), and by the higher resistivity of foreign layers. This constriction is a consequence of the fact that only a number of a-spots conductively connect the contact partners.

Being exposed environmental air, on all materials not being extremely noble, there grow foreign layers. They are made of oxides, sulfides or other compounds and cannot be tunneled through. [3]

Greenwood [4] and Slade [5] state that the constriction resistance of a contact R_C is controlled by following equation:

$$R_c = \rho(\frac{1}{2na} + \frac{1}{2a}) \tag{1}$$

with n as number and a as radius of contact spots as well as as resistivity. Respective distances between single contact spots, which also impact contact resistance, are described by a radius of micro spot clusters.

The resistance defined as /2a for a single contact area is only valid as long as the constriction effect on heat generation is sufficiently small (maximum supertemperature of 3 °C [5]). As under conditions of real high current applications significant joule heat might be produced within the constriction, this assumption cannot necessarily be used in a model assessing new unproven design solutions. Slade [5] deduces from potential theory that the crowding of current lines within an a-spot causes a thermal gradient normal to that constriction.

Parallelization of n contact areas theoretically decreases the overall resistance by $1/n$. The overall area available though stays constant. Since each of the parallel switched contacting surfaces is limited in its (nominal) contact area by this division, contact radius decreases by the factor $1/\sqrt{n}$ accordingly. Hence, each single contact resistance theoretically increases by the factor $\sqrt{n}$. Combining those two effects, a multiplication of contact points by division of a constant overall contact area should decrease the overall resistance by $1/\sqrt{n}$.

2.2 Approaches in modeling electrical and mechanical contact performance

Contact surfaces of the mating partners in interaction with contact force and applied materials mainly control the conducting contact area and thus contact resistance. Difficulties in modeling connectors' performance are mainly driven by the additional aspects of real contact behavior, i.e. surface micro-topography in association with tribological effects, temperature rise and foreign layers on contact surfaces which are very difficult to model and thus to predict and assess.

Most published approaches deal with analysis and modeling of single aspects of contact behavior, such as Maul and McBride [6] who focus intermittency phenomena, fretting, influence of mutual positions of individual contacts on resistance and stiffness studied by Ervin and Sovostianov [7], Schoft and Kindersberger [8] investigating resistance of randomly rough surfaces, and other detail specific views.

There are also some approaches of building models and simulations integrating various mechanical and electrical properties of electrical contacts [9]. Validation is mostly restricted to assumptions like "crossed-rods" models or "sphere - flat-body" combinations. This is due to the fact that the computational models mostly are restricted to very small areas because high resolutions are required for accurate calculations. Another problem concerns the fact that the boundary conditions of connector systems in real applications are mostly very dynamic. For instance, impacts and environment's influences, such as ambient temperature, concentration of corrosive gases and vibrations can vary and scatter largely.

Leidner et al. [9, 11] generate surface topographies based on real measured surface data. They model elastic plastic contact between multi-layered bodies subjected to pressure and shear traction. Basing on the modeled contact, they calculate the resulting constriction resistance and voltage drop. Their models set the actual state of the art in modeling electro-mechanical contacts. However, their approach is focused on contacts transferring low energies at comparably low currents and therefore is not considering temperature effects. For modeling high current contacts a model neglecting temperature rise is insufficient. Simulation focuses only on small areas making an application for generalized design of entire connector systems more difficult. These difficulties reside especially in the objective to unite impacts of micro-scale modifications and macro-scale performance.

The objective is not to model only single a-spot clusters but entire contact areas over several square millimeters with modified surface topographies based on measured surface data. Especially the consideration of temperature rise caused by the electric current is expected to increase the significance of modeled high current contacts.

For creation of new design solutions by analysis and synthesis micro-geometrical models based on mathematical models are not suitable, as they require comprehensive information about design-form and boundary conditions. In order to design new connector principles instead of only varying material, plating, contact force or manufacturing method, the Contact & Channel Modeling-Approach (C&C-A), developed by Albers [10] and latest being developed further by Alink [11], is a suitable method to support abstract modeling and synthesis.

3 A NEW DESIGN APPROACH BASED ON C&C-A FOR ELECTRICAL CONTACTS OF A CONNECTOR

3.1 Analysis, abstraction and modeling

The following deliberations will be formulated using the notations defined by the Contact &Channel Approach [10] as the according model is used for analysis as well as for synthesis.

In order to achieve lowest contact resistance which causes lowest heat generation and thus allows for highest current handling capacity, perfectly smooth and clean contact surfaces would be necessary. Leidner et al. [12] confirmed that smoother surfaces show higher numbers of a-spots then rougher ones. However, this kind of perfection is impossible; production of technically smooth surfaces is very expensive and in this case not even sufficient. This would mean an idealized Working Surface Pair WSP 1.3 in Figure 2, Figure 3 and Figure 4. As described above, between real contact-surfaces there are surface asperities and insulating layers which are to be penetrated or broken mechanically in order to permit electric conductivity. Mechanical penetration could be managed by higher contact forces. However there are narrow limitations for those due to corresponding forces for insertion and withdrawal of the connector as well as resulting mechanical relaxation mechanisms.

Several manufacturers of high current connectors use additional components to decrease contact resistance and increase reliability of the connection. These components aim to provide multitudes of parallelized individual contacts with independent contact forces. It is realized as a lamellae packet or wire cage between pin and socket or also between flat contacting bodies, which creates defined contact areas at high contact pressures and thereby increases reliability and decreases resistance. Another principally related approach is to directly contact woven copper wires with a conventional pin [13]

Figure 2 shows a schematic cross-section of plug connector housing with one exemplary enlarged pin and socket connection. The waved lines illustrate Channel and Support Structures (CSS) transferring energy and the straight lines at their interfaces mark the according Working Surface Pairs (WSP). Interfaces and interrelation to surrounding environmental influences and boundary conditions are represented by so called connectors, marked by a "C" on the outer WSP. The advantage of the element modeling approach C&C-A is that there is a clear focus on the Working Surface Pairs and the accordant Channel and Support Structures fulfilling the system's functions. This abstraction helps to identify the actual task-fulfilling elements, locations of occurring effects and respective properties, as it was shown by Thau an Alink [14].In this case the focus is set on the actual interface of two power lines which are to be connected conductively, i.e. WSP 1.3 (dashed line rectangle in Figure 2).

The very fact that there is electric contact between parts of plug and socket surfaces is not a function but only an effect (Figure 3, top), whereas the transfer of electric current from one wire to another is the actual function of the connector (Figure 2).

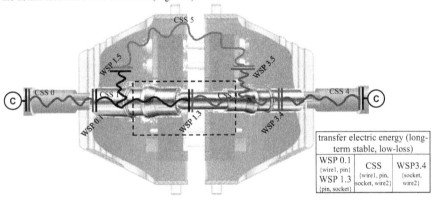

Figure 2: CSS and WSPs marked in a schematic connector with an enlarged view of one pin and socket connection. The tabular representation assigns the connector's main function to single WSP and CSS.

As it was described in the first section, all real surfaces not being extremely noble are partially covered by insulating foreign layers. This can be modeled in an abstract way as a parallel circuit of conducting areas leading the current paths and insulating areas constricting the current paths. The insulating areas are represented by a Channel and Support Structure of unintentional property (CSS 8 in Figure 3 and Figure 4) with the function *locally prevent electric contact*. This CSS can hardly be prevented and consists of the entity of non-conducting media such as foreign layers, impurities or air between the surfaces. Hence it is the design approach of choice, to provide for a multiple parallelization of the effect *conducting electrically* in WSP 1.3. This leads to the design of multiple defined current carrying areas. As mentioned above, this design solution is already realized by various versions of multi-contacting elements (e.g. lamellae cage) between the two initial contact partners. This evolution of thoughts can be illustrated by the Contact and Channel-Approach as shown in Figure 3.

Main function of a connector – detachable stationary connector– is to transfer electric energy as lossless and long-term stable as possible. This main function can be divided in two sub-functions *make and release contact* and *remain electrical contact at low resistance*. The latter can in turn be divided into the sub-functions *uniformly distribute the current flow* and *dissipate heat*. Especially the sub function *uniformly distribute the current flow* is a key function, as its fulfillment also shows effects of decreasing resistance, thereby decreasing heat generation which in turn lowers the importance of dissipating heat and increases the possibility to remain at defined contact area.

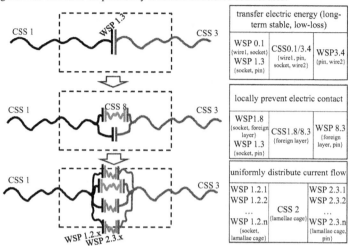

Figure 3: Steps of more detailed analysis of WSP between pin and socket.

The function *uniformly distribute the current flow* is fulfilled by parallelization of a multitude of contact lamellae. In combination with adequate contact forces respectively pressures the contact lamellae also fulfill the tasks of overcoming foreign layers at making contact, as well as hindering growth of foreign layers due to the defined WSP.

On the other hand the additional component between pin and socket provides two serial WSP, i.e. contacts, namely pin – lamellae packet (WSP 1.2) and lamellae packet – socket (WSP 2.3), which is an undesirable side effect. Additional serial contacts mean additional contact resistance and potential cause for malfunctions, as well.

3.2 Synthesis

The idea of uniformly distributing the paths of current flow is carried on onto the micro scale. Creating multitudes of individual contact spots each of very small dimensions equally distributed over the nominal contact area should be able to further decrease resistance and thus increase current handling capacity. Additionally Langhoff and Graesle [15] state that for high current contacts it is particularly important to keep reserve of contact areas in order to enable self-healing processes and prevent overheating. The resulting design approach is illustrated accordingly in Figure 4 (top and middle).

According to the principle of embodiment design *division of tasks* [16], for high production quantities it is favorable to choose the design principle of *integration of functions* rather than the *separation of functions*. Integration of functions generally results in a smaller number of parts or components fulfilling several functions at the same time. Whereas separation of functions consequently aims to a solution in which every part or component fulfills just the one function, which it is perfectly designed for. As every additional component creates double number of serial contacts and high production quantities can be assumed *integration of functions* is chosen.

In Terms of the C&C-A, the principles of embodiment design described above can result in different solutions following the design rules *integration of additional CSS* and/or *division of existing WSPs*.

Application of both of those design rules leads to the solution of integrating the CSS 2 *{lamellae cage}* and of dividing the macro WSPs 1.2 and 2.3 into multitudes of micro WSPs directly on the initial contact partners surfaces cf. Figure 4 (bottom).

This design solution, as a result of the synthesis by C&C-A complies with results of comparable approaches, e.g. application of Axiomatic Design [13].

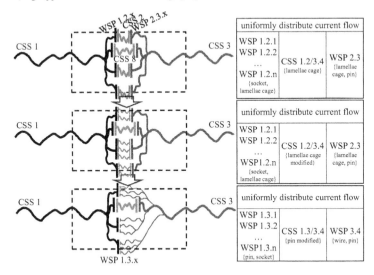

Figure 4: Steps of Synthesis towards new design for connector's contact area.

For synthesis of an improved plug connector system the abstract C&C-model is again linked to real shape design according to the respective requirements. The approach to create multitudes of individual micro contact spots can either be implemented on one of the nominal smooth contact partners or can additionally be applied on existing multi-contact-components. During development and test phase, this design is realized by physical vapor deposition at high vacuum of electrically conductive materials (high purity copper) on substrate specimen which represent the initial contact surface.

During deposition processes masks with different structure images are positioned on the substrate specimens so that various patterns of micro structured topography are designed (Figure 5).

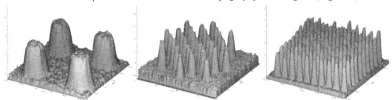

Figure 5: Different structure patterns taken by means of white light interferometry. 4, 29 and 100 contact spots per array (from left to right).

Every surface of micro structured specimens is studied by means of white light interferometry before test. There are three identical arrays of micro-structures in every studied contact to ensure actual contact of every array. Figure 5 shows one array each of different number of contact spots.

A further challenging task is to implement the synthesized C&C-model into a FEM-model of sufficient detail depth. As well, experimental data has to be generated in order to verify models and according simulation. Validated models can be used to investigate concept variants and only optimized ones have to be proved experimentally.

4 FINITE ELEMENT MODEL

4.1 Model development

A three dimensional model of a layer has been modeled, representing a contact between two discs (Figure 6). The upper one has a flat surface whereas the lower one takes into account the surface roughness of the three spot regions (right on Figure 6).

Figure 6: FEM model of a contact pair (left) composed of three spot –regions (right)

Each of the three structure pattern comprises four contact spots (12 for the whole model, as displayed in the left of Figure 5). The spots are made of copper whereas the rest of the plate is structural steel. As a consequence, the model is composed of 5 solids in contact: 3 rough spot layers (copper) and 2 flat discs (steel). The accordingly required simulation process is displayed in Figure 7.

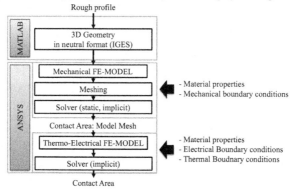

Figure 7: Modeling process

The first step is to import the spot topography into the finite element software. To achieve this, a Matlab script has been developed to generate B-Splines [17] with numerical data coming from the measuring facility in ASCII form. These data are a numerical discretization of the real topography.

Once generated, the three dimensional surface with five additional flat surfaces features a solid in a format neutral file (IGES) in order to avoid any finite element solver dependency. This solid generation is processed three times, as there are three spot regions to be generated. The rough structures are then imported into finite element modeler in order to complete the geometry by creating both steel discs. The three rough bodies are then fixed on the lower disc (Figure 6).

The simulation process composed of two main phases begins after the geometry is imported. The first phase is used to build the contact between both discs with comparable loads a used during experiments (≈ 3.5 N). The second one consists in applying electrical and thermal loads in order to measure the electrical resistance of the system. In order to achieve this, the application of adequate boundary conditions is necessary which is explained in the next subsection.

4.2 Applied boundary conditions and contact configuration

The initial state is an open contact as there is no real possibility to close all of the 12 potential contacts because of their non-planarity (cf. section 2). In order to generate a contact between spot regions and upper disc, a displacement is applied to the first structure whereas the second one is fixed.

As contact properties, the Augmented Lagrange Method is employed. The reason for this is a higher accuracy and better ability to avoid unrealistic high stresses and penetration depth. The algorithm is highly sensible on the contact pinball radius and contact stiffness. As a consequence using a very small pinball radius ($\approx 8 \cdot 10^{-10}$ m) is necessary regarding to the structure finesse. Furthermore, contact stiffness is actualized every substep of the calculation to avoid unrealistic high contact pressure. Results of calculated contact pressure are displayed in Figure 9.

In the second phase, the mechanical contact properties are directly taken from the final state of the first phase and used for electro-thermal calculation. Electrical load consists in applying a current of 10 A flowing through both discs. Additionally, as thermal load, a convection flow (occurred by the surrounding air) is applied on both discs with a convective coefficient of 28 $W \cdot m^{-2} \cdot K^{-1}$.

Thermal contact properties are also dependent from the pinball region, as well as from the contact conductivity linked with the physical state of concerned surfaces regarding oxidation and impurities.

5 RESULTS AND DISCUSSION

The target effect of decreased contact resistance by multiplied number of individual micro contact areas could be verified by experiments (Figure 8). As nominal contact area the cross section of created micro spots is accounted. Every data point in Figure 8 represents the arithmetic mean of hundreds of thousands measured values each series taken over periods of 10 to 50 hours.

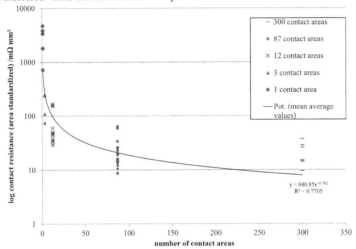

Figure 8: Area-standardized contact resistance over number of individual contact areas.

However, simulation results revealed the fact that the artificially created micro-spots in turn show varying numbers of actual contact spots (Figure 9) as well. Some areas show multitudes of a-spots, others show none. It is reasonable to expect this being real behavior.

This is one reason for a deviation between computational model and experimental data. Other reasons for this are the actual missing consideration of foreign layers in the model as well as the sum of remaining general sources of deviation. Future models are supposed to consider effects from Foreign/insulating layers, Temperature rise, Elastic/plastic deformation, Relaxation effects and Vibration/friction/wear. A main objective is to extend the system boundaries towards a model considering the whole connector, even including its housing.

In numerical simulations, various tests confirmed the significant impact of the normal load on the contact area. Figure 9 points out the differences between high loads (on the left) inducing larger contact areas and also more contacts in general and lower loads here leading to only one spot providing contact areas (on the right). Contact area strongly influences the electrical resistance of the modeled contact as explained before. Still, for higher loads contact pressures are remarkably higher (approx.500 MPa) than for low loaded surfaces (on the right) where the pressure is around 200 MPa.

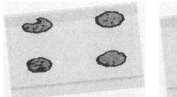
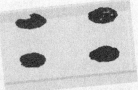

Figure 9: Contact pressure at one of the three spot regions with high load (left) and low load (right).

The thermal-electrical simulation was done with both preceding contact conditions. The applied current and convection gives realistic results but still with deviation compared to experimental measures. The temperature field displayed in Figure 10 corresponds to the contact zones calculated before. High loads induce better temperature homogeneity (on the left) in comparison with lower loads (on the right).

In addition to the calculations discussed before, the electrical resistance of the contact is simulated. Calculated resistance is of around 70 m according to a voltage of approx. 700 mV for the low loaded version. The calculated temperature reaches about 50°C in the contact zones (Figure 10, dark). High loaded contacts show lower temperatures due to larger contact area and higher number of contacts resulting in lower contact resistance.

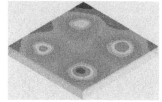

Figure 10: Temperature field in the contact, high load (left) and low load (right).

Comparison with experiments revealed a difference of voltage drop of around 500 mV between simulation and experiment. An explanation for this is found in the conductivity of the boundary layers as well as in size and number of contact areas depending on the contact tolerances of the model. These tolerances define contact identification (opened/closed). Nevertheless, the global trend that higher loads increase the number of contact points inducing a lower resistance was confirmed.

6 CONCLUSION AND OUTLOOK

The design solution of parallelization of electrically conducting contacts by means of micro structuring principally provides potential for high current connectors with higher energy efficiency and thus higher current handling capacity. It is not clear though, if this is valid scale-independent, what is planned to be studied closer during further investigations. Generally, it is a promising approach to overcome the conflict of objectives between connector's performance and its efficiency (energy, connector's size). Hence, the Contact & Channel Modeling Approach (C&C-A) was applied successfully to support focused analysis as well as target-oriented synthesis.

Next steps will be further development of the FEM-model including the influences of thermal conditions and generation of foreign layers. Objectives for design will continue to be decreasing contact resistance, leading to less heat generation and thus higher current handling capability. Finally one holistic model is planned to be used for generating and validating further potential innovative solutions for connector design. Holistic here means considering the entity of actuating variables on the entire connector system.

Only consequent union and comparison of simulation and experiment will allow generating holistic multi-physic models. With Physical Vapor Deposition a very flexible technology is used to create various surface topographies. All resulting designs have to be assessed regarding not only resistance and current handling capacity but also subsequent regarding manufacturing possibilities, wear and impact on actuating force.

REFERENCES

[1] Hauck, U. Requirements of Components for Connectivity and High Voltage Security. In *13th Kooperationsforum Bordnetze, Vol. 1, Ingolstadt, November 2010, section 8.*
[2] Himmel, J. Kriterien fuer die Konfektionierung von Hochvoltleitungssaetzen. In *13th Kooperationsforum Bordnetze, Vol..1, Ingolstadt, November 2010, section 6.*
[3] Rieder, W. Elektrische Kontakte – EineEinfuehrung in ihre Physik und Technik, 2000 (VDE Verlag GmbH, Berlin)
[4] Greenwood, J.A. Constriction resistance and the real area of contact. *British Journal of Applied Physics,* 1966, Vol. 17, 1621-1632
[5] Slade, P. Electrical Contacts – principles and applications.1999 (Marcel Dekker, Basel).
[6] Maul, C. and McBride, J.W. A Model to describe Intermittency Phenomena in Electrical Connectors. In *48th IEEE Holm Conference on Electrical Contacts*, 2002, pp. 165-174
[7] Ervin, J. andSovostianov, I. Effect of Mutual Positions of Individual Contacts on the Overall Resistance and Elastic Stiffness of a Cluster of Contacts. In *Letters in Fracture and Micromechanics*, 2009, pp.101-108
[8] Schoft, S. andKindersberger, J. Joint Resistance of Busbar-Joints with Randomly Rough Surfaces. In *21s International Conference on Electrical Contacts*, Zurich, 2002, pp. 230-237
[9] Leidner, M. and Schmidt, H. A numerical method to predict the stick/slip zone of contacting, nonconforming, layered rough surfaces subjected to shear friction. In *55th IEEE Holm Conference on Electrical Contacts,* Vancouver, 2009, pp. 35-40
[10] Albers, A., Matthiesen, S., & Ohmer, M. An innovative new basic model in design methodology for analysis and synthesis of technical systems. In: *Proceedings of the 14th International Conference on Engineering Design ICED,* 2003, Stockholm
[11] Alink, T. Meaning and notation of function for solving design problems with the C&C-Approach. *Dissertation,* Karlsruhe Institute of Technology, 2010
[12] Leidner, M. and Schmidt, H. A new simulation approach to characterizing the mechanical and electrical qualities of a connector contact. In *European Physical Journal – Applied Physics*, Vol. 49, 2, February 2010, p.9
[13] Axiomatic Desgin, MIT, Axiomatic Design Trounces Tradtitional Methods of Reducing Friction and Wear. In *Axiomatic Design Success Story*, 2004 (www.axiomaticdesign.com)
[14] Albers, A., Alink, T. Matthiesen, S. and Thau, S. Support of System Analyses and Improvement in industrial design through C&C-M. In *International Design Conference – DESIGN,*2008, Dubrovnik, Croatia
[15] Langhoff, W. and Graesle, H. Steckverbinder fuer hohe Bordstroeme. In *Automotive*, 11(12), 2004, pp. 46-48
[16] Pahl, G. Pahl/Beitz Konstruktionslehre. Grundlagen erfolgreicher Produktentwicklung Methoden und Anwendung. *Vol. 6*, 2005 (Springer)
[17] Lee, E. T. Y. A Simplified B-Spline Computation Routine, *Computing,*Springer, Vol. 29, Sec. 4, pp. 365–371, December 1982

Contact:
Prof. Dr.-Ing. Dr. h.c. Albert Albers
Karlsruhe Institute of Technology (KIT)
Institute of Product Engineering (IPEK)
Kaiserstr. 10, 76131 Karlsruhe, Germany
Tel: +49 721 608 42371
Fax: +49 721 608 46051
Email: albert.albers@kit.edu
URL: http://ipek.kit.edu

Albert Albers is head of the IPEK - Institute of Product Engineering at the Karlsruhe Institute of Technology (KIT), Germany. After working as head of development of driveline systems and torsion vibration dampers at LuK GmbH & Co. KG, he moved on to the University of Karlsruhe – today's KIT – in 1996. His research focuses on product development processes as well as the support of product development by methods for computer-aided engineering, innovation and knowledge management in mechanical and automotive engineering.

INTERNATIONAL CONFERENCE ON ENGINEERING DESIGN, ICED11
15 - 18 AUGUST 2011, TECHNICAL UNIVERSITY OF DENMARK

EMPIRICAL CONSIDERATION OF PREDICTING CHAIN FAILURE MODES IN PRODUCT STRUCTURES DURING DESIGN REVIEW PROCESS

Otsuka, Yuichi (1); Takiguchi, Sho (1); Shimizu, Hirokazu (2); Mutoh, Yoshiharu (1) 1:
Nagaoka University of Technology, Japan; 2: Toyota Motors, Japan

ABSTRACT
Failure Modes and Effects Analyses (FMEA) have widely been used in design review processes. In the cases of complex systems, failure modes on a top system are resulted from failure chains among component in the system. Predicting method to reveal failure chains is highly depending on practitioner's experiences and no logical guidelines have been established. This paper describes the modified DRBFM (Design Review Based on Failure Modes) to effectively find the latent failure chains. In first step, normal state of functional and structural model is set. Next, by comparing features of the models with the one in past product, failures are specified, After failures are found, interference changes by failures is discussed in order to reveal newly-generated failure chains. To compare the performances of detecting failure chains, 28 mechanical engineering students conducted the conventional FMEA or the modified DRBFM method by themselves. The result significantly showed that the modified DRBFM method is more effective in detecting failure chains which causes the errors of underestimating.

Keywords: DRBFM, FMEA, Product evaluation, Design Reliability Engineering, Safety Engineering

1 INTRODUCTION
Failure Modes and Effects Analyses (FMEA) have widely been used in design review processes in order to prevent failures in service [1]. Designers who make the FMEA datasheet are usually requested plenty of knowledge and experiences about failures modes. Supporting tools, such as logical guidelines or failure records, to effectively make them are then indispensable because veterans are retiring so rapidly. Suzuki pointed out the following difficulties in predicting failure scenarios [2];
1. A procedure for predicting the specific contents of failure scenario is unclear. It highly depends on a designer's knowledge or experiences.
2. It is very difficult to use past failure records in design review processes to proactively predict future failures because the form of the record itself is not suitable .,

Accident record including its faults have been arranged in some fields [1,2]. However, it is only effective in alarming people to take care of the failure, because these records include various terms even in the case of similar types of accident. Finding latent failures in their product by using past accidents is still the responsibility of designers. Yoshimura in TOYOTA motors argued that the connections among elements in products are main problem in considering failures [3].
Before discussing latest approaches, we have to specify two types of failure modes;
1. Functional failures; abnormal states in functional components (*Fig. 1*), e.g., loss output, excessive input etc. These expressions do not include mechanisms of failure.
2. Physical failures; abnormal conditions in structural model (*Fig. 2*), e.g., deformation, rupture, etc. These expressions directly relate to their mechanisms, e.g., when deformations is considered, deformations can be easily connected its mechanism of Hook's law.

It should be noted that physical failures often lead to functional failures. Therefore, in order to prevent functional failures, the process of considering physical failures, not only functional ones, is much important. However, this process is extremely difficult to be completed only based on past accident data because physical failures are often happened by the deviations in their service environment. The specific procedure for predicting failure scenarios is not established yet.

There are various approaches in order to effectively find latent problems in a design stage. Wright [4] reviewed the process of managing engineering changes in a product. He argued that effective process of visualizing the effect flow by the engineering changes through a design process is very important to

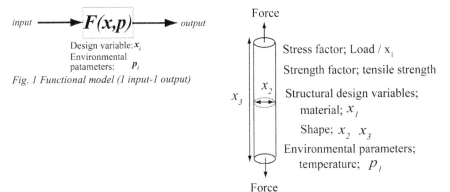

Fig. 1 Functional model (1 input-1 output)

Fig. 2 Structural model

analyze the effect of product's quality. Eckert *et.al.* [5,6] discussed the various methodologies to visualize risks involved in engineering changes; risk matrix, cascade model of effect propagation by changes and component connection network. They pointed out the critical path if risk by the engineering changes should be visualized by using the discussed methods in order to secure the reliability of the considering products. Wirth [7] pointed out the lack of guidelines for practitioners of FMEA. They also suggested the use of input-output functional model to detect functional failures. Deviations in the functional parameters are connected to specific failures, retrieved from failure taxonomy database. Their approach focuses on functional failures, not included physical failures. Tumer and colleagues [8,9,10] also used input-output model to identify functional failures. Hirtz *et.al.* [11] constructed standard function taxonomy. They focused on calculating the probability of system failure and little discussed the process of setting failure scenario. Huang and Jin [12] discussed conceptual stress-strength model to express functional failure. Their approach also showed little in setting the contents of failures. Clarkson *et.al* [13] argued that interferences among element should be considered through FMEA process. Dependency matrix they proposed is beneficial to reveal interferences among element, which highlights potential failure chains. However, in the cases of physical failure, failure chain is not always going along the path of functional interferences. They did not separate functional models and structural models. Physical failures can lead to functional failures. In order to completer FMEA process including the determinations of causes of functional failures, predicting physical failure chains is necessary. However, conventional FMEA process did not distinguish physical failures from functional failures and then should be improved.

This study aims at establishing the process of predicting chain failure modes by using failure modes networks[5] and updating the boundary conditions of structural model of an element interfered by other failures. Section 2 explains the contents of revised DRBFM process that includes the above proposal. Section 3 demonstrates the case study of the revised DRBFM for a motor bicycle. Section 4 shows the experimental result of the performances of the revised DRBFM compared with the conventional FMEA in predicting a failure chain.

2 REVISED DRBFM METHOD

2.1 Stress-Strength model

Stress-Strength model (SSM) is general concept of determining physical failures [1-3]. Failure states in SSM are expressed by the following equation.

$$\text{(Stress factors)} > \text{(Strength factors)} \qquad (1)$$

Stress factors show the degree of energy to break structures, such as forces, chemical attacks or electric voltages etc. Strength factors describe the degree of resistance in materials to be broken. When we consider physical deformation, stress is physical stress (force per internal area) and strength is tensile strength (maximum stress at breaking.) Both factors are affected by service environment, e.g., tensile strength is decreased in corrosive media. Consequently, we have to discuss the balance of stress and strength under a certain environment.

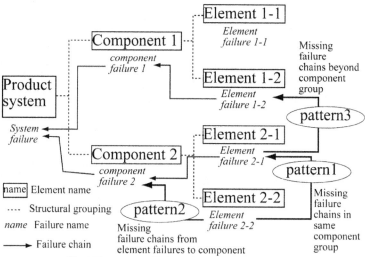

Fig. 3 Three types of failure chains in product structure

2.2 About DRBFM

Shimizu proposed Design Review Based on Failure Modes (DRBFM) in order to support designers to predict failure modes with ease [14]. The DRBFM process contains as the follow steps,
1. Setting past reliable conditions of parts in service
2. Setting design conditions of newly developed parts
3. Determining the changes in design conditions (intentional changes in shapes, materials, manufacturing method etc and environmental changes in service conditions) by comparing both the design condition
4. By using the detected changes (sources of failures) and failure modes list, designers predict failure modes from the changes.
5. Considering the main factor of failure modes by using FTA
6. Thinking countermeasures for the root of main factors.

2.3 Underestimation errors of DRBFM by missing hidden failure chains

Though the DRBFM process is effective in predicting latent failure modes [1], contains three types errors that suffers its performances.
- Unnoticed errors; A problem that designers did not consider. No measures were taken.
- Underestimation errors; A problem that designers noticed but took no measures to prevent it because the designers estimated only partial damage and missed failure chains from a considering failures to entire system failure.
- Misunderstanding; A problem that the designer noticed and took insufficient measures, because of his lack in knowledge or experience.

Unnoticed errors can be reduced by using structural part diagrams [15]. Misunderstanding errors can also be found by setting suitable roles of participants and discussion processes [16]. However, solutions to reduce underestimation errors are not specified because it relate to complex connections among parts in product hierarchical structures. Real accident tends to happen by the complex causes that owe to failure chains. The solutions are then necessary to prevent accidents effectively.

Fig. 3 shows the type should be subdivided as the follows, each of which needs distinct preventive measures.
Pattern 1 Missing failure chains in same component group
Pattern 2 Missing failure chains from element failures to component
Pattern 3 Missing failure chains beyond component group. Links in structural groupings do not always much the failure chains that is resulted from physical connections.

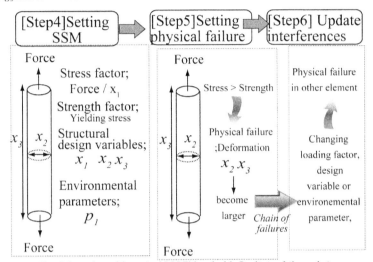

Fig. 4 Procedure of the revised DRBFM method1; Predicting failures in one elements

Fig. 5 Procedure of the revised DRBFM method 2; Predicting failures chains

In the cases of considering pattern 1 and pattern 2 errors, failure links normally match the links among structural grouping; the conventional DRBFM process is applicable. However, in the case of considering pattern 3 errors, the failure chain is not along with the structural grouping. In this case, designers have to prevail interferences of failures emerged from the changes of conditions in functional connections among parts, physical contact or environmental situations etc. To specify predicting process of pattern 3 errors, we revised the DRBFM process by combining the concept of failure networks proposed by Eckert [13] and robust modelling process [17] as the follows,

1. Making structural dependency matrices in the same level of product structure, shown in Table 1.
2. Setting failure modes in one element by comparison(Fig. 4).
3. After setting a failure in one part, designers will consider the changes of "normal conditions" of neighbour elements whether new deviations can be emerged. If the new deviation is defined, new interferences between two elements will be drawn in product structure.
4. Designers consider whether the new interferences can affect stress-strength state of others. If the interfered deviations can result in failures of other parts, it becomes failure chain(Fig. 5).

2.4 Procedure of the revised DRBFM method (Figures 4 and 5)

2.2.1 STEP 1 Setting normal model

1. Specifying design conditions (service conditions, normal user's configurations or environmental conditions)
2. Deposing systems into a set of sub-systems, parts and elements
3. Definitions of functions for all the sub-systems, components and elements.

2.2.2 STEP 2 Setting design changes by comparison
1. Selecting base target for comparison which has same function for a considering elements or was the similar element.
2. Comparing design conditions between base target and the considering part in order to specify differences to be named intentional or incidental changes. Intentional changes mean that changes are made by designer's decision and incidental changes mean that changes are resulted from changes in service environment etc.

2.2.3 STEP 3 Predict physical failures by deviations
1. Considering specific failures of an element caused by predicted changes. By combining the predicted failures and functionally abnormal conditions, failure mode is determined.
2. Checking whether all the engineering changes were considered to lead to some failure modes to prevent unnoticed errors.

2.2.4 Setting structural dependency matrices (Table1)
1. Setting existing interferences , such as functionally connections or physical fixing, among elements.
2. Fulfilling the interference conditions among parts in part interference matrices.

2.2.5 STEP 4 Setting stress-strength model of element.
1. Setting loading factor and strength factor of an element.
2. Normal states of structural model is also determined from the result of step 1

2.2.6 STEP 5 Setting physical failures
1. Considering the cases of excessive loading or decreasing of strength.
2. Each cases are connected to specific names of physical failures, normally failure taxonomy is prepared by knowledge data base.

2.2.6 STEP 6 Update interferences to consider failure chains.
(A) Predicting physical failures corresponding to Pattern 2 error
1. Considering the stress-strength model of a component including excessive loading factors or weakened strength factors.
2. Checking the contents of physical failures of interconnected part.
3. Checking whether the physical failure of the interconnected element can affect excessive loading factors or weakened strength factors, which can generate new deviations.
4. Failure mode of the components from the failure chains is named by the combination of the newly detected physical failure and functional failures.

(B) Predicting physical failures corresponding to Pattern 3 error
1. Considering the stress-strength model of a component, including excessive loading factors or weakened strength factors.
1. Checking the contents of failure modes of interconnected element.
2. Checking whether physical failures of the interconnected element can change the normal conditions of the neighbour element. If changes. The new interferences are drawn in Table 1.
3. Determining new physical failures by new deviations generated by failure chains.

After all processes are completed, result is filled in DRBFM worksheet.

3 CASE STUDY OF THE REVISED DRBFM METHOD

3.1 Procedure of the case study

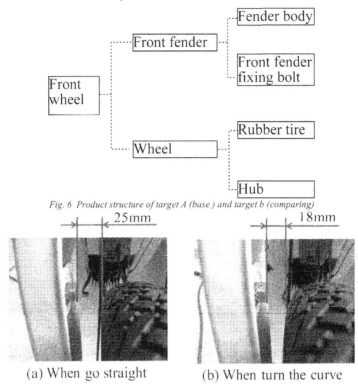

Fig. 6 Product structure of target A (base) and target b (comparing)

Fig. 7 Normal conditions Target A; no interferences found between front fender fixing bolt(left) with rubber tire(right)

(a) When go straight (b) When turn the curve

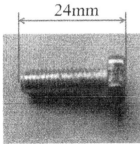

Fig. 8 View of Front Fender's Fix Bolt. If bolt is completely loosened, it cannot contact tire,

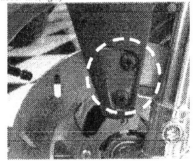

Fig. 9 Structure of fender (right) and rubber tire(center) in target B. bolt head is embedded and has no possibility of protrusions

We chose the parts of front wheel of two types of motorbikes. *Fig. 6* shows the product structure of Target A and Target B.

3.1.1 STEP 1 Setting normal model

Table 2 Normal states of functions and structural model.

		Target A(base)	Target B (comparison)
Function model	Function of front wheel	Running on roads	
Structural model	Number of riders	One	one
	Normal loading (Rider's weight + goods)	Max 80 +15 kg	Max 80 +30 kg
Environmental factors	Weather	Sunny Cloudy , Rainy	
	Road conditoins	asphalt-paved	

Table 3 Finding deviations (incidental changes) by comparing detailed structure of both targets.

Component	Target A(base)	Target B (comparison)
Front fender	Front Fender's Fix Bolt exist between fender body and rubber tire	No protrusions are found.

Table 2 shows the normal states of functions and structural model. There are little differences in both states. Figure 7 and 8 shows the detailed normal states of element. These figures reveal no interferences between them is existed.

3.1.2 STEP 2 Setting design changes by comparison, STEP 3 Predict physical failures by deviations
Table 3 shows the differences of the structures between target A and target B. Figure Fix bolt of front fender locates the side near a tire in the case of target A as shown in Fig. 7(a). There is no such a bolt in the inside of the front fender of target B as shown in Fig. 9.
Table 4 shows the example of DRBFM worksheet for rubber tire's main body. In this case, a normal condition is an equivalent condition between the value of load from grounds and internal pressures. An abnormal condition is a hole on the body which leads to the loss of internal pressures to result in functional loss condition of the rubber body. Failure mode is name by "Impossible to absorb shocks or load from ground by the hole on the body".

3.1.3 Definition of interferences
In the normal condition, there is no physical contact observed among tire, fix bolt of front fender and a hub. However, Fig.7(b) shows the narrowed space between tire and the head of fix bolt when the front wheel is inclined in curving, which emerges the concern of friction (types of interferences)

3.1.4 STEP 4 Setting SSM to STEP 6 considering the failure chains
As shown in Figure 7, space between the tire and the fix bolt is 25mm in the normal condition (a) and 18mm in the inclined conditions (b).Fig. 10 shows the entire shape of the fix bolt of front fender whose total length is 24mm. That means in the normal conditions no physical contact are occurred between the tire and the fix bolt, as shown in Fig. 10(b).
Next, the fix bolt, is considered including its failure conditions. Fig. 10(b) shows its process visually. When considering the inclined condition of the hub (in curving or failure condition), inclined angle increases compared with that in normal condition. In addition, typical failure modes, loosening of bolt are easily predicted that narrows the distances to tire from the head of bolt. By combining these failures, new interferences, frictions on the tire surface by the head of the fix bolt is then predicted. This is new physical failures by new interferences. This failure is called "failure A" in latter sections.

3.1.5 Drawing failure chains
Fig. 11 shows the summary of failure influences. The failure A is emerged by the interferences beyond the different structural groups. Failing to predict the failure A is corresponding to the pattern3 errors of underestimations.

Table 4 Example DRBFM work sheet for failure chains at one component-element structure

Ele-ment	Function	Failure Mode	Mecha-nism	Cause	Solution In design	Solution In evaluation	Solution In production	Attention
Rubber Tire's Main Body	The tire absorbs a shock and grounds	The hole opening ↓ Impossible to absorb	Wear	Front Fender's Fix Bolt loosening + Hub tilting in turn ↓ The bolt contacting and friction is generated	To do durability test and deciding period of use		Covers the bolt to prevent loosening	Please check user's manual

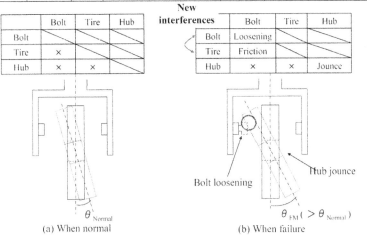

(a) When normal   (b) When failure

Fig. 10 Illustrated process of determining failure A by newly-generated failure chain

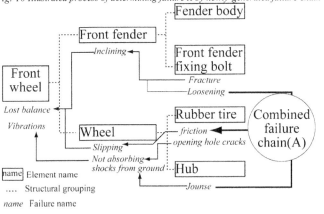

Fig. 11 Failure chains including failure A, generated by combined failure chains

4 EXPERIMENT FOR THE PEFORMANCE OF THE REVISED DRBFM METHOD IN FINDING FAILURE CHAINS

4.1 Subject and the detailed process of practicing.

We collected 28 mechanical engineering students. They were divided in to two groups; FEMA group and revised DRBFM group. The gender and the age of each group are shown in table 5.

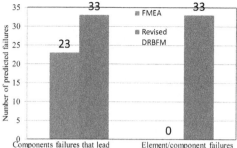

Fig. 12 Results of predicting failure chains

Table 5 Face sheet of participated students

Group	The number of people	Sex		Averag age
		Man	Woman	
FMEA	14	14	0	21.6
DRBFM	14	14	0	21.2

Table6 Statistical analysis result of the ratio of persons who could find failure A

Group	Ratio of persons who could find hidden failure chain A
FMEA	4/14 =0.286
Revised DRBFM	11/14=0.786

T-statistics for t-test ; T=2.65 (significance level p=0.004)

Table7 Statistical analysis result of the number of predicted failure chains by both methods

Group	Participant	Mean	Standard deviation	MAX	MIN
FMEA	14	1.36	0.71	3	0
Revised DRBFM	14	2.57	2.26	6	0

The authors gave all the students following document to conduct either method by him/her self.
- Practice guide of the conventional FMEA method or the revised DRBFM method.
- Failure influence flowchart (only part names are written)
- Pictures for both targets (Figs 6 to 9)
- Agreement document for academically use of their results

Target of analysis is limited the front wheel structure shown in Fig. 6. The structure has two parts; Front fender (elements; front fender body, front fender fix bolt) and Wheel (element; rubber tire, hub). The student is requested to predict as many as failure modes they could by using either method. All the data sent to the students by e-mail and answers are also sent by e-mail up to 2weeks. To emphasize the student's motivation, rewards according to the top-3 obtained number of failure modes were prepared to the students. They are told these awards before practicing. Time of practice is not limited because rewards are not increased by increase of thinking time.

4.2 Statistical analyses
Each result was checked by the authors to determine unique failure mode to accurately count its number. Failure modes that place on the root of failure chains to another parts or higher components were also counted. Subsequently, the difference of the ratio of finding failure A in each group and the difference of the average number of the failure modes to lead another failure modes in each group were statistically analysed by using t-test and Welch's test[18].
Table 6 and 7 show the result of finding ratio and the number of inducement failure mode. Both difference are significant at p=0.05. Therefore, the revised DRBFM can effectively support users to predict complicated failure chains more than using the conventional FMEA. This result demonstrates the use of the revised DRBFM method can improve the quality of design review by reducing the error of underestimation.

4.3 About the performances of preventing the errors of underestimations.
Figure 12 shows the predicted result of failure modes by using conventional FMEA method or the revised DRBFM method by same person. Using the revised DRBFM method enabled the user to predict more failure chains. During the revised DRBFM procedure, some deviations from previous

reliable conditions are emphasized by comparisons to help users to focus the changed point as the candidates of concerns for failure modes [3]. In addition, introducing the considering process of failure chains combined by normal functional conditions, deviations from the normal conditions and updating interferences by the deviations can support users to see explicitly an complicated failure chains such as Failure A in Fig.11. Missing such failure chains leads to the error of underestimation and then the revised DRBFM is effective in reducing the error of underestimation by providing users to chase failure chains with clearer and easier logic.

5 CONCLUSIONS

The revised DRBFM method is arranged in order to reduce the error of underestimation which is yielded by failing the complicated brunches of failure chains. Users of the revised DRBFM method could significantly find more failures which lead to other failures. The revised DRBFM can effectively support users to predict complicated failure chains more than using the conventional FMEA. This result demonstrates the use of the revised DRBFM method can improve the quality of design review by reducing the error of underestimation.

ACKNOWLEDGEMENT

Part of this study was supported by Top Runner Incubation system through Academia- Industry Fusion Training in "Promotion of Independent Research Environment for Young researchers" by Ministry of Education, Culture, Sports, Science and Technology, "*MEXT*" JAPAN. The authors would like to thank Associate Prof. Yukio Miyashita for his critical comments to this study.

REFERENCES
[1] M.G.Stewart and R.E.Melchers, Sakai,S.(dir.),Probabilistic risk assessment of engineering systems 2003(Morikita Publishing INC.), p190-227.
[2] Suzuki,K., Principle of preventive solutions and its system,2004(Nikkagiren Publishing), p89-96.
[3] Yoshimura,T.,Toyota-style preventive solutions GD3, 2000(NikkagirenPublishing), p52-91.
[4] Wright, I.C., Design Studies, 2001, Vol. 18, No. 1, pp.33–39.
[5] Eckert, C.M., Keller, R., Earl, C. and Clarkson, P.J., Reliability Engineering and System Safety, 2006.Vol. 91, No. 12, pp.1521–1534.
[6] Jarratt,T. A. W., Eckert,C. M., Caldwell,N. H. M., Clarkson,P. J., Research in Engineering Design,2011, pp.1-22.
[7] Wirth,R., Berthold,B., Kramer,A., Peter,G., Engineering Applications of Artificial Intelligence, 1996, Vol.9,No.3, pp. 219-229.
[8] Arunajadai,S. G., Uder,S. J., Stone,R. B., Tumer,I. Y., Quality and Reliability Engineering International, 2004,Vol.20,No.5, pp.511-526,.
[9] Kurtoglu,T. , Tumer,I. Y., Jensen,D. C. Research in Engineering Design,2010, Vol.21,No.4, pp. 209-234.
[10] Tumer,I. Y. Stone,R. B. Research in Engineering Design,2003,Vol.14,No.1, pp. 25-33.
[11] Hirtz,J. , Stone,R. B. , McAdams,D. A. , Szykman,S. , Wood,K. L. Research in Engineering Design, 2002, Vol.13,No.2, pp. 65-82.
[12] Huang,Z. ,Jin,Y. Journal of Mechanical Design, Transactions Of the ASME, 2009,Vol.131,No.7 pp. 0710011-07100111.
[13] Clarkson,P. J., Simons,C., Eckert,C., Journal of Mechanical Design, Transactions Of the ASME,2004, Vol.126, pp.788-797.
[14] Shimizu,H.,Yoshimura,T., Trans. Japan Soc. Mech. Eng.Series C, 2004, Vol.71,No.706, pp.230-237.
[15] Shimizu,H.,Otsuka,Y.,Noguchi,H., International Journal of Vehicle Design, 2009,Vol.53, No.3, p.149-165,.
[16] Otsuka,Y.,Shimizu,H., Noguchi,H., International Journal of Reliability, Quality and Safety Engineering, 2009, 16(3), p. 281-302..
[17] Otsuka,Y.,Noguchi,H., Trans. Japan Soc. Mech. Eng. Series C, 2010,Vol. 75, No.761, p.207-216.
[18] Kunisawa,H.,Clear practice Mathematic statistics (in Japanese), 1986(Kyoritsu Publishing9, p129-130.

INTERNATIONAL CONFERENCE ON ENGINEERING DESIGN, ICED11
15-18 AUGUST2011,TECHNICAL UNIVERSITY OF DENMARK

A DESIGN METHODOLOGY FOR HAPTIC DEVICES

Suleman Khan and Kjell Andersson
KTH - Royal Institute of Technology, Sweden

ABSTRACT
This paper presents a design methodology for optimal design of haptic devices, considering aspects from all involved engineering domains. The design methodology is based on parametric modeling with an iterative and integrated design approach that leads to easier design space exploration for global optimal design and initial verification in the conceptual design phase. For design optimization, performance indices such as; workspace volume, isotropy, stiffness, inertia and control of the device, from all involved engineering domains were considered. To handle this complex and non-linear optimization problem, a multi-objective algorithm together with a new proposed optimization function was used, to obtain an optimum solution. A case study, where the methodology has been applied to develop a parallel haptic device is presented in detail in this paper. The simulation and experimental results obtained from this test case show significant improvements in the performances of the device.

Keywords: Design methodology, haptic devices, parallel mechanism, optimization and performances.

1 INTRODUCTION
A haptic device is a robot-like mechanism that provides an extra sense of touch; force/torque feedback capability to an operator based on what he/she discovers and interacts within a virtual world or remote environment. Application of these devices is emerging in various fields such as medicine, telerobotics, engineering design, and entertainment [1, 2]. The work presented in this paper is related to the design methodology for design and development of these devices. Basically, haptic devices present a difficult mechatronic design problem, as they are required to be backdrivable and light (low inertia and friction), as well as being able to provide enough stiffness, feedback forces and torques when reflecting forces from stiff contacts. It is also desired that motion, forces and stiffness provided by the device are isotropic (same in all direction). Furthermore, structural transparency and stability is required so that the operator feels free space motion as free, while during interaction with virtual objects feels the dynamics of the manipulated objects, not of the structure of the haptic device.

The design of the haptic devices is an iterative process, and an efficient design requires a lot of computational efforts and capabilities for mapping design parameters into design criteria, hence turning out to be a multi-objective design optimization problem. Thus it presents a high level of computational complexity for finding an optimal design solution. The main focus of this research is to *develop a methodology for design and optimization of haptic devices*. The methodology will be based on parametric, iterative and integrated modeling design approach that leads to easier design space exploration and early verification during product development.

In traditional mechatronic design methodologies, the mechanical system is developed independently of the electronic and control system, and at a later stage they are integrated with each other [3]. For example the sequential design approach as shown in Figure 1a [4] has the advantage of dividing a large and complex design problem into several smaller design problems. Here the mechanism, actuation and control design are designed independently, which reduce the computational complexity of the problem. However, neglecting to include aspects from dynamics and control point of view into the design of mechanical system, may result in a system with non-optimal dynamic performance. This may, in the worst case, require major redesigns of the electromechanical system late in the design process, e.g. as reported in [3, 4, 5, 6, 7].

Fathy et al [6] identify four different design approaches for integrated optimization of mechanical and control system design: sequential, iterative nested and simultaneous (Figure 1b). The first two approaches have the potential of finding designs that are optimal within each domain, but sub-optimal on the system level. The forth one "simultaneous" consider the whole system at a time for optimization, it can provide the global optimal solution, but at a high computational cost for complex systems.

Roos [4] proposed a new integrated design methodology for design of electro mechanical servo systems. This approach is based on two types of models; static and dynamic models. Static model include parameters related to the physical model, and dynamic model include the dynamic parameters (required for find an optimal solution. This design methodology works efficiently for simple design problems, but its performance becomes worse for complex design problems due to the increased level of computational complexity. A similar approach, based on design decision variables from all involved disciplines for optimal design of product has been proposed by Bart at al [7].

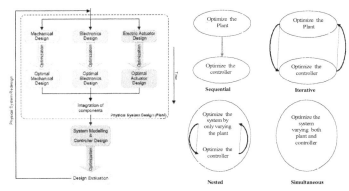

Figure 1.a) Traditional mechatronic design approach [2] & b) Other design methodologies [4].

The approach taken in this paper for a methodology can be categorized to the "Nested approach" by Fathy et al. [6] and [7]. The motivation for developing a methodology for development of haptic devices, specifically using parallel kinematic structures, is that this type of device has complex structures which give many structural advantages like high stiffness and low inertia, but also give complex optimization problem and a complicated control system. The remaining part of the paper is organized in sections. Section 2 explains the design methodology, Section 3 presents the case study, and section 4 presents results and discussion respectively.

2 DESIGN METHODOLOGY

In this work, a methodology has been developed for design and optimization of haptic devices. This methodology provides a model based parametric, iterative design approach that leads to an easier design space exploration and initial verification during process development as shown in Figure 2.

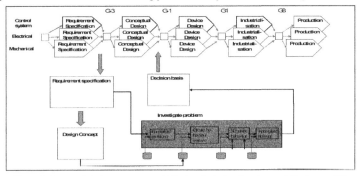

Figure 2. A general design process model for haptic device development (Andersson [4]).

The first stage of the methodology is to define the more direct device requirements and market. These requirements include on an abstract level: Degree Of Freedom (DOF), workspace, force/torque capability, stiffness and control dynamics. The second stage of the methodology is conceptual design; here the methodology should include preliminary analysis of the number of DOF, workspace, actuator requirements and singularity points (which shouldn't exist within the workspace). In parallel, a rough layout of the mechanical structure with preliminary material properties should be made as well as an investigation of possible control strategies and components to use. Next is device design which includes design of the mechanical structure, actuation, transmission, and also analysis of workspace, stiffness, inertia, force/torque capabilities and backdrivability. In parallel with designing the mechanics and actuation, the models necessary for control design are derived. For the control design, sensors and control strategies are selected and designed. Before the device is finally built and the control implemented, thorough work should be made for optimal design using simulation and rapid prototyping to verify performance and if necessary iterate within the design process.

Apparently there is a large number of design parameters that needs to be fixed before a final design is achieved. In addition to the direct specifications it is important to consider other design criteria towards an overall optimal design. Such criteria can include: (1) minimum footprint/size to workspace ratio; (2) uniform motions, forces and stiffness capabilities over the workspace (kinematic isotropy); and (3) minimum inertia of structure, transmission and actuation capabilities (dynamic and control characteristics). All these design parameters are almost mutually dependent, thus leads to a large complex design problem with high computational complexity. To cope with this problem, a global optimal solution is determined using a multi-objective optimization criteria based on efficient computational tool such as multi-objective genetic algorithm. The different phases of the design methodology are discussed in detail in following sections.

2.1 Requirement Specification

Design starts with a need, when satisfied, results in a product that fits into existing market or creates market for its own [7]. In the first step, a literature review and market analysis should be performed in order to identify the potential users and their requirements. From the statements of needs a requirement specification is formulated.

As a minimum for haptic devices, this should include requirements for size (footprint), workspace, Degrees Of Freedom (DOF), force/torque capability, and stiffness.

2.2 Conceptual design

In the conceptual design phase the development of structure, mechanical device and control system should be performed in parallel since the performance of a haptic device is highly dependent on the interaction between all of these systems.

Some activities (and suggested tools to use) that should be performed during this phase are listed below;

- Selection of alternative structures to examine for further development. This information is given by the literature review and market analysis that have been performed earlier, when stating the requirement specification. This should be complemented with a more detailed study of possible structures for the intended application for the device.
- Modeling and analysis to determine numbers of DOF, preliminary actuator requirements and preliminary dimensions for the wanted workspace for selected structures. These are some of the basic requirements for a haptic device to achieve capabilities for feedback in the required DOFs and workspace. For these types of analysis MBS modeling and analysis software, e.g. Adams View® [8], is recommended.
- Investigation and preliminary selection of motors based on the calculated actuator requirements. In addition evaluation and preliminary selection of encoder and transmission should be made.
- Inverse and direct kinematic modeling of the selected structures. Development of inverse and direct kinematic models is a pre-condition for performing kinematic optimization and is also needed for development of the control system. For this type of modeling and analysis, Matlab [9] is recommended.
- Optimization of the kinematic structure. This is a crucial task for haptic devices that are based on parallel kinematic structures. The optimization turns out to be a multi-criteria optimization problem. For these types of problem the use of a genetic algorithm has been proved to be successful in finding a global optimum solution. The goal function should include indices for workspace, isotropy, torque/force and stiffness requirement and inertia of the device. Suggested software to use here is Matlab [9] and MOGA (Multi Objective Genetic Algorithm) toolbox [10].
- Rough layout of the mechanical design based on the MBS analysis and optimization results. This is a traditional engineering design task to make a preliminary assembly layout of the device based on optimization results and MBS analysis. Tasks to consider in this phase are selection of motor, transmission and search for standard components to use for e.g. joints, as well as basic design and preliminary material of support structure. For these tasks any CAD 3D modeling tool is feasible.
- Alternative control strategies for the haptic device. The requirement on the device is to get a frictionless feeling when moving the device in free space and to achieve force/torque feedback when entering contact with an object. This means that the control system have to compensate for the inertia

and friction that always occur in real systems. The task here is to investigate optimal control strategies and different approaches to compensate for these effects.
After selecting the candidate structures to consider for the device in hand, above steps can be done in parallel assuming that a parametric modeling approach is used for all these activities.

2.3 Device design

The outlined activities during the conceptual design phase all follow the verification process described in Figure 2 which has the purpose to produce a decision basis to decide how to proceed to the next design phase. This results in selection of one (or maybe two) candidate structures for further development and final design. The following design phase is the device design phase. Some activities (and suggested tools to use) that should be performed during the device design phase are listed below.

- Mechanical design to make the detail design of the device based on the optimization results. This includes careful selection of standard components, if possible (e.g. joints, electric motors), detail design, material selection and manufacturing documents of components to be manufactured. For these tasks any CAD 3D modeling tool is feasible.
- Prototype creation. Once the mechanical design is determined a physical prototype should be built. This includes the manufacturing of some components and ordering of standard components.
- Control design. As soon as we have a physical prototype we can start testing different control strategies being investigated during the previous design phase. For the initial tasks dSpace [11] can be used but for the final implementation a suitable micro controller should be selected as well as a software development tool for implementing the control system in the micro controller.
- Testing of the prototype. After the prototype being built we should start with the testing of the device. Initially mechanical stiffness and clearance can be tested using a CMM (Coordinate Measuring Machine). After that, testing of the complete device should be made in a controlled and restricted environment. First, simple tests of contact conditions and free space motion should be made and thereafter more complicated contact conditions, requiring many DOF's feedback as a result of a contact, should be investigated.

3 APPLICATION EXAMPLE: DESIGN OF A 6 DOF HAPTIC DEVICE

The proposed design methodology in section 2 has been applied to the development of a parallel 6-DOF haptic device. The intended application of the device is a milling simulator that will be used in curriculum for surgical training of vertebral operations [1]. In this scenario a haptic device is used to achieve manipulation capabilities and force/torque feedback in 6-DOF during simulation of vertebral operations to achieve a user interaction that gives a realistic impression due to the milling process of a virtual modeled bone tissue. Such procedures involve removing bone by drilling or milling, including processing of channels and cavities, hence requiring 5-6 degrees of freedom and stiff contact feedback to the user.

3.1 Requirement specification

In this first step, a literature review and market analysis has been performed in order to identify the potential users and their requirements. From the statements of needs a requirement specification is formulated. The preliminary specifications given here have been obtained in dialogue with a tentative user, in this case a surgeon. The application domain is completely new and unique, thus it is difficult to obtain specific requirements. The initial requirements for the haptic device are as follows [12].

- The device should have 6 actuated degrees of freedom.
- The whole device should fit within the space of 250x250x300 [mm].
- The translational workspace should be a minimum of 50x50x50 [mm].
- The stiffness of the device including actuation and control should be a minimum of 50 [N/mm].
- The TCP peak force/torque performance should be at least 50 [N] and 1 [Nm] in all directions.
- It should be possible to place it on a table in front of the operator, easy to access for the user.

The outcome from this stage is a requirement specification on an abstract level, based on identified users of the device.

3.2 Conceptual design

From the literature review in the first stage of methodology, haptic devices that are currently available in the market or at a prototype stage, both serial and parallel structures are being used [2, 13-23]. However, since parallel structures have some significant advantages as compared to serial ones, e.g. high stiffness, high accuracy and low inertia, we have chosen two concepts based on parallel kinematic structures. In the next step, these concepts were investigated for structural analysis such as numbers of DOF's, workspace and force/torque requirements. For structural analysis, these concepts were modeled using Adams View® MBS software [8] as a main tool.

The first concept is a modified Stewart Gough mechanism [21, 23, 24], which consists of a fixed base, a moving platform, and six identical legs connecting the platform to the base shown in Figure 3a. Each leg consists of an active linear actuator fixed to the base, a spherical joint, a constant length proximal link, and a universal joint. This 6-PSU (active Prismatic, Spherical and Universal) joint configuration was used to get 6 DOF. For parametric design of this structure, six design parameters were considered: range of actuators motion (L_{min}, L_{max}), length of proximal link c_i, radius of base r_b, radius of platform r_p, angle between the base pair of joints 2β and angle between the platform pair of joints 2α, see Figure 3b. The attachment point pairs are symmetrically separated 120° and lie on a circle, both on the base and the platform. The platform attachment points are rotated 60° clockwise from the base attachment points.

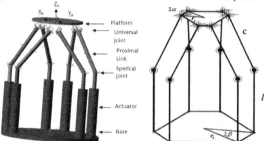

Figure 3. a) Conceptual model of concept 1 in Adams View b) Kinematic structure of the concept 1.

The second mechanism is based on a hybrid parallel kinematic structure called TAU, shown in Figure 4a. This concept consists of fixed I-column, a moving platform and three parallel chains (1, 2 and 3) which connect the base frame to the moving platform. In this structure chain 1 and chain 2 are symmetrical while chain 3 is unsymmetrical as shown in Figure 4. Each symmetrical chain has two active rotational actuators, one attached to the I-column while another one is mounted on the upper link U_1, U_2. Furthermore chain 1 and 2 have extra two proximal links connecting the platform to upper links U_1 and U_2 to increase the structural stiffness. The third chain, chain 3, has also two active rotational actuators, one attached to the I-column and the other mounted at the top of the device.

For parametric design of this structure, five design parameters were considered: position of each parallel chain with respect to the base coordinate system {N} is at 1.5d, 3d and 4.5d, which is function of parameter d, length L_1 of the upper arm, length L_2 of proximal links in each chain, radius of platform R_p, elevation angle θ_{32nom} (nominal angle for θ_{32}) of the upper arm U3 of chain 3 with orientation of the base frame as given in Figure 4b.

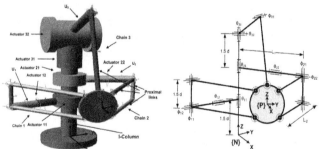

Figure 4. a) Conceptual model of concept 2 in Adams View b) Kinematic structure of the concept 2.

In the next step of conceptual design phase, we investigated the basic performance of these concepts, utilizing the verification process proposed by Andersson et al [12]. First we assign initial dimensions to the device that fulfills device size requirements. Thereafter, we focus on investigating three main properties; No's of DOF, device workspace, and actuator performance giving wanted force/torque performance around TCP.

The first concept (1) provides 6-DOF motion at TCP. The translation workspace provided by the concept is ± [50, 50, 50] mm in X, Y and Z direction as shown in Figure 5a. The maximum range of rotation measured at the center and at each corner of the selected cube within translational workspace was ± 40° around X, Y, Z direction, while in combination it ranges from ±35° around all directions.

The second concept (2) also provides 6-DOF motion at the TCP. The translation workspace provided by this concept is ± [85, 85, 100] mm in X, Y and Z direction, shown in Figure 5b (right). The results from the rotation analysis show that the rotation angles for X and Y axis are ±52° in all eight corners, when rotating one axis at a time. While in combination the range of rotation is decreased to ± 30° in all the corners. Around the Z-axis the structure can provide rotation up to ± 40°.

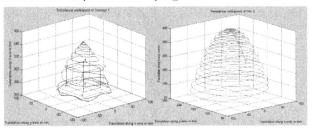

Figure 5. a) Workspace for concept 1(left) and b) for concept 2 (right) in 3D space.

To measure the force and torque capability, a constant force of 50 N was applied on TCP, then TCP was moved on a specified circular path within workspace and reaction forces was measured on each actuator. The force/torque analysis of concept 1 shows that the measured reaction forces at active linear joints increased as the TCP moves along the specified path to the outer circle see Figure 6a. The torque analysis of the second concept shows that higher torque is required on actuator 32, see Figure 6b with a few high peaks. These peaks occur as a result of an incorrect modeling of the load when moving in all directions (xyz) at the same time and should be disregarded.

The outcome of these preliminary analyses in the conceptual design phase is used as a decision basis to select the mechanism that we will consider further for design optimization. Based on the torque requirements and low inertia due to the fixed motors, concept 1 was selected. Next, it is important to consider other design criteria towards device design optimization. Such criteria can include: (1) minimum footprint/size to workspace ratio (workspace); (2) uniform motions, forces and stiffness capabilities over the workspace (kinematic optimization); and (3) minimum inertia of structure, transmission and actuation capabilities (control design). The kinematic and control optimization were performed in parallel based on the defined performance indices in the following section.

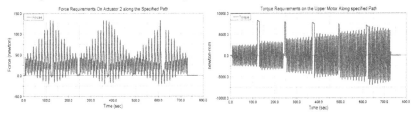

Figure 6 Force requirements (left) concept 1 & (right) concept 2 - 50N force on TCP within workspace.

1. Workspace Index

Workspace is the working space that the haptic device can operate within. It is defined as a three dimensional space that can be reached by TCP. The boundaries of this space were determined using inverse kinematics. A Cartesian workspace within a range of ±75 mm along all three axes was scanned using an evenly spaced grid. Finally, the volume of the workspace can be calculated as $v = \int_v dv$. Where dv is the volume of a grid element. The optimization criterion is to maximize the workspace volume while keeping the footprint (size) of the device as a constraint.

2. Isotropy Index

The kinematic isotropy index (II) indicates how evenly the device produces motions (velocities) in all directions in the workspace. A haptic device is called "isotropic" if at least in one point of the workspace some of its kinematic properties are homogenous with respect to all directions. The isotropy index is defined as the ratio of minimum singular (σ_{min}) to maximum singular (σ_{max}) values of the Jacobian matrix (J) [24], according to

$$II = \frac{\sigma_{min}(J,w)}{\sigma_{max}(J,w)}, 0 \leq II \leq 1, \tag{1}$$

where w is the pose of TCP in workspace. If, at a certain point, the isotropy index approaches unity, the haptic device can produce a more uniform motion in all directions. While on the other hand if the isotropy index approaches zero, it indicates operation close to singular points in the workspace, which needed to be excluded from workspace ($II \geq 0.005$). To represent the average of the device isotropy index over the whole workspace, a global isotropy index is defined as

$$GII = \frac{\int_v II \cdot dv}{v}. \tag{2}$$

A higher value of GII represents a mechanism with a better isotropy characteristic within its workspace, and thus the criterion is to maximize this index.

3. Force requirement Index

The force requirement index (FI) is defined as the maximum magnitude of an actuator force required for a unit applied load on the tool center point (TCP). As the applied load on the TCP is related by the Jacobian matrix to the forces required on the actuators, the force requirements index is defined as the maximum singular value of the Jacobian matrix as $FI = \sigma_{max}(J,w)$. A global force requirement index which represents the average of the device force/torque performance over the selected workspace is defined as

$$GFI = \frac{\int_v (FI) dv}{v}. \tag{3}$$

A smaller value of the force requirement index implies that less capacity of the actuators is required i.e. this index should be minimized.

4. Stiffness Index

From mechanics point of view, stiffness is the measured ability of a body or structure to resist deformation due to the external forces. For the selected mechanism, the stiffness at a given point in the workspace can be characterized by its stiffness matrix [25]. This matrix relates the forces and torques applied at TCP to the corresponding linear and angular Cartesian displacement. If F represents the external applied forces on TCP, then the corresponding linear and angular Cartesian displacement can be determined from ellipsoid sphere with the lengths of horizontal axis and vertical axis being the maximum value and minimum value of the deflection, respectively. The direction with largest deflection of the moving platform has the lowest stiffness. Thus, the maximum value of deflection of the moving platform can be regarded as the evaluating index of stiffness when a unit force F acts on the moving platform. The maximum and minimum deformations can be obtained from the eigenvalues of the stiffness matrix $(K^{-1})^T K^{-1}$ as $\|p_{max}\| = \sqrt{max(\lambda_p)}$ and $\|p_{min}\| = \sqrt{min(\lambda_p)}$. The global stiffness index representing the average stiffness within the workspace is defined as

$$GSI = \frac{\int_v \|p_{max}\| dv}{\int_v dv} \tag{4}$$

Here the criterion is to minimize the global stiffness index and so maximize the stiffness of the structure.

5. Inertial Index

An inertial index is based on the mass matrix of the device that represents the dynamic characteristic of the device. The mass matrix is obtained by computing the masses and inertia of all the moving components (platform, actuators including motor inertia and proximal link) in the task space [26]. In the case of a haptic device it is needed to minimize inertial effects (minimizing the maximum singular value of the mass matrix). Thus the inertial mass index can be defined, using the maximum singular value of the mass matrix (M) as

$$IMI = \frac{1}{1 + \sigma_{max}(M,w)}. \tag{5}$$

The criterion here is to maximize the inertial index (minimize max. singular value), to obtain lower dynamic effects in the workspace.

6. Multi-objective optimization

As in our case all the actuators are identical to each other, thus they have the same stiffness and thus the stiffness matrix K reduce to a diagonal matrix, which simplify the criteria as $K=kJ^TJ$ in task space. Thus the condition number or singular value of the matrix J^TJ *need to be optimize instead of* kJ^TJ [27]. Also in the case of isotropy index we minimize the maximum singular value of the Jacobian matrix, the same criteria as for force and stiffness indices (dependent on Jacobian matrix), thus we effectively reduce this MOO problem to three main indices see equation (6). Furthermore, the selected indices are normalized such that all indices contribute equally in the optimization process. In this normalization each index is divided by a numerical value, calculated from the mid values of the given design parameters space according to equation (8) and their design parameters input space. Finally, a multi-criteria design objective function is defined based on these indices as

$$GDI = \min\left[\frac{VI}{VI_m}, \frac{GII}{GII_m}, \frac{IMI}{IMI_m}\right], \quad (6)$$

where subscript m indicates mid values of the parameter space. The main advantage of this new approach as compared to the traditional objective function presented in [28-30], is to assure that all design indices are equally active in the optimization process. For optimization we also need to define the constraints and allowed range for the design parameters (DP) as per the specification of the device for all sub levels.
Finally the optimization problem can formulated as

$$\begin{array}{l}\text{maximize } GDI \\ \text{subject to } J(X), M(X) \succ 0 \quad X \in v \\ L_i\_\min \leq L_i \leq L_i\_\max \\ Dp\_\min \leq Dp \leq Dp\_\max,\end{array} \quad (7)$$

where (L_i) represents the stroke of actuator. To solve the above described nonlinear and non convex MOO problem, we applied three different approaches/algorithms; Weighted sum, MOGA-II [31] and NSGA-II [32] to find the Pareto optimal solution [33]. These approaches were implemented in Matlab and run with 100 as initial population size and maximum number of generations as 100.

7. Result from Optimization and pareto fronts

The Pareto front resulted from the above described optimization approaches is shown in Figure 7a, and b. The pareto optimal solution obtain from MOGA-II is shown as dense points in Figure 7b, where the performance of all the indices are best and can't be improved more, unless it deteriorate the other one. The solution obtained from these three approaches is approximately the same.

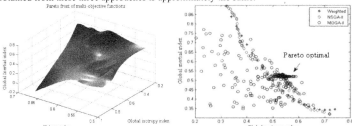

Figure 7. a) Pareto front of the volume, global isotropy and inertial indices (left) and b) pareto optimal solution obtained from the applied approaches (right).

The results from the design optimization process using MOGA-II, with design parameters are presented in Table 1.

Table 1. Design parameters bounds and optimal values

Parameters	Min	Max	Optimal
l [mm]	120	150	129.4159
c [mm]	120	150	125.4555
R_b [mm]	100	125	118.1799
R_p [mm]	40	60	54.9920
β [deg]	10	30	18.1519
α [deg]	10	30	10.5485
Volume index, VI	-	-	0.9790
Global isotropy index, GII	-	-	0.255
Inertial index IMI	-	-	0.8522

Furthermore the set of optimal design parameter values, obtained from genetic algorithm was used to evaluate the performance of the device. In order to visualize the variation of isotropy and force requirements indices in the optimized workspace, the TCP is moved in a circular path in the x-y plane with small incremental changes in radius. When the radius reaches the maximum, the TCP is shifted to the next x-y plane with a small increment in the z-direction. At each small grid isotropy and force requirements indices are measured. Figure 8a shows that the device has good "isotropic" behavior around the central position of the workspace. The force requirements is small for unit applied force around the center of workspace while it increases as the TCP moves away from central point (see Figure 8b). This characteristic is also quite obvious from the isotropy definition of the device. From the index values corresponding to the optimal parameter set and by analysis made in Adams View®, it is concluded that workspace and isotropy requirements as represented in section 3.1 are fulfilled. The variation of stiffness K within the workspace (see Figure 8c), which shows the structure is stiffer when the actuators are at lower limits, and less stiff when the actuator reaches its maximum position.

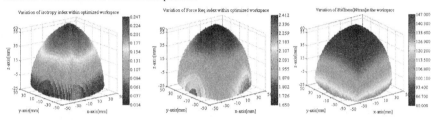

Figure 8. a) Variation of isotropy(left), b) variation of force requirements(middle) and c) variation of stiffness(right) within in the workspace.

In conceptual design phase, we also perform in parallel control design optimization, to obtain a structure optimized both from kinematic, dynamic and actuation point of view. Here the main performances that needed to be considered are transparency and stability of the device. The requirement on transparency means that motion in free space should feel free while motion in contact with a virtual or remote object should result in feedback forces and torques as close as possible to those appearing in the remote or virtual world.

In free space motion, transparency is affected by the dynamics (moving inertia, friction) of the device and dynamics of the operator. Keeping the device inertia as low as possible as well as compensating for it in control design will increase the transparency of device. The task here is to investigate optimal control strategies and different approaches to compensate for these effects. The modeled optimal design control strategy is shown in Figure 9. The control design is based on optimal load from the optimal kinematic structure of the device (complete integrated system).

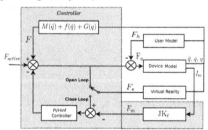

Figure 9. Optimal control structure of the 6-DOF haptic device.

As shown in the Figure 9, the control design is based on computed torques of the device dynamics and current feedback. We measure the current I_m in each motor and thus indirectly torque and forces produced by the haptic device (using motor torque constant K_t and Jacobian matrix J). A force/torque error feedback control is obtained using a PI controller with low pass filter. Input to the PI control is the error between reference force from virtual world F_e and filtered measured force F_m. Then a compensation for the dynamic influence F of the device is added to the control signal as a feed-forward term. The aim of this feed-forward term is to increase the transparency of the device, i.e. the user should not feel the inertia and friction of the device itself, only of the tool.

Outcome from the conceptual design phase is the complete optimal design of the 6-DOF haptic device.

3.3 Device Design

The CAD model for the prototype was developed based on the final set of design parameters from GA, pareto diagram, control strategy model and sensitivity plots [24, 26]. The developed model is shown in Figure 10 below. The size of the model is 250x250x300 mm. Six Dc motors model GR 53x58, 60W were fixed at base and a cable transmission mechanism with pulley was used to convert the angular motion to the linear actuator motion. The cable transmission makes the system backdrivable. The developed 6-DOF haptic device is connected to a personal computer using a dSpace 1103 board as shown in Figure 10. The proposed control structure is implemented in Simulink on that PC and the target controller code is executed on the dSpace board with 1kHz sampling rate. The haptic collision detection and force torque feedback program is implemented on the same computer. The position measurement resolution in each actuator leg is 0.01mm and the update rate of the controller is 1kHz.

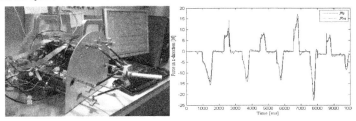

Figure 10. a) Prototype b) comparison of reference force and measured response from the device.

Figure 10b presents the response of the system (measured forces) and the reference forces from the virtual environment both in free space and while interaction with virtual objects. It has been observed that the optimal controller and optimal structure improves the performance of the 6-DOF device, as desired and thus its transparency, as shown in Figure 10b and Table 2.

3.4 Design validation

In the final stage, experiments for workspace, forces and torque capabilities, stiffness capability of the prototype were performed. The experimental results are shown in Table 2.

Table 2.

Characteristics	Values
No. of DOF	6
Dimension	250x250x300 mm
Workspace	Translation:75x75x100 mm Rotation: Pitch=Yaw=± 45° Roll=±40°
Maximum and continuous forces	52N and 20 N
Maximum and continuous torques	1.2Nm and 0.85 Nm
Stiffness	54 N/mm
Resolutions	Linear 0.01mm and Angular 0.01deg
Time step	1 ms

4 CONCLUDING REMARKS

The design process of the haptic devices particularly based on parallel mechanism, presents a complex design, due to multi-disciplinary mechatronic product design. It was concluded from this research work that following a systematic design methodology, one can develop an optimal haptic device, from the prospect of all involved engineering aspects. The proposed design methodology is based on parametric modeling, iterative and integrated design approach that leads to simple design space exploration of a pareto optimal design solution and initial verification in the conceptual phase of the product development. The methodology has been applied on a test case, i.e. the design of a parallel 6-DOF haptic device for a milling simulator for surgical training of vertebral operations. It has been concluded both from simulation and experimental results that the performance of the optimally designed device has been improved and satisfies the user requirements. This indicates that the methodology can support development of an optimal haptic device. However, more test cases are needed to verify this methodology.

REFERENCES

[1] Eriksson M. G., Haptic and Visual simulation of a Material Cutting Process. Licentiate thesis in Machine Design, KTH, Royal Institute of Technology, Stockholm Sweden 2006.
[2] Feng L. L., Analysis and design optimization of in parallel haptic devices PhD thesis, Department of Mechanical and Aerospace Engineering State University of New York at Buffalo Buffalo, New York 14260, 2011.
[3] Li Q. Zhang W.J., Chen L., Design for control-A concurrent engineering Approach for mechatronics systems design. IEEE/ASME Transactions on Mechatronics, vol. 6, no. 2, 2001.
[4] Roos F., Towards a methodology for integrated design of mechatronic servo systems. PhD thesis, Machine Design KTH-Royal Institute of Technology, Stockholm Sweden 2007.
[5] Van Amerongen J., Breedveld P., Modeling of physical systems for the design and control of mechatronics systems. Annual review in control vol. 27 pp. 87-117,2003
[6] Hosam K. F, Julie A. R, Panos Y. P. and A. Galip U, On the Coupling between the Plant and Controller Optimization Problems, Proceedings of the American Control Conference, Arlington, Virginia, USA, June, 2001.
[7] Frischknecht, B., Gonzalez, R., Papalambros, P. and Reid, T., A Design Science Approach to Analytic Product Design, International Conference on Engineering Design(ICED 2009), Stanford, CA, Paper No. 148, 2009.
[8] ADAMS, www.mscsoftware.com, www.adams.com
[9] Matlab, www.mathworks.com,
[10] Murata T, Ishibuchi H. MOGA: multi-objective genetic algorithms. IEEE international conference on evolutionary computation, Perth, WA, Australia, December 1995.
[11] Control disk, www.dSpace.com
[12] Andersson, K., Khan, S., Investigation of parallel kinematic mechanism structures for haptic devices, presented at 2nd Nordic Conference NordPLM'09, Gothenburg January 2009.
[13] Hayward V. Haptic interfaces and devices in sensor review, Vol.24, number 1 pp.16-29, 2004.
[14] Sensible Tech. Phantom 6-DoF haptic device, (1-04-2011) www.sensable.com/haptic-phantom-premium-6dof.htm
[15] Massie T.H. and Salisbury J.K., The PHANToM haptic interfaces: A device for probing virtual objects, Proc. Of the 1994 ASME int. Mechanical Engineering Exposition and congress, Chicago, Illinois, 1994, pp. 295-302.
[16] Haption Tech. Virtous 6D35-45 6-DOF haptic interface http://www.haption.com/site/eng/html/materiel.php?item=1
[17] Freedom 6S. 6-DoF haptic interface from MPB Technology, website visited last time on 1-04-2011 http://www.mpb-technologies.ca/mpbt/mpbt_web_2009/_en/6dof/index.html.
[18] Khan S. A literature review of haptic interfaces. Technical report, KTH Machine Design 2010.
[19] Gosselin C. Kinematic analysis optimization and programming of parallel robotic manipulators. Ph.D. Thesis, McGill University, Montreal, June, 15, 1988.
[20] Gosselin F. and Martins J. P. Design of a new Parallel Haptic Device for Desktop Applications Proceeding of the first Eurohaptics Conference and Symposium on haptic interfaces for virtual environment IEEE, 0-7695-2310-2/05, 2005
[21] Faulring E. L. and Colgate J. E., Peshkin M. A. A high performance 6-Dof haptic cobot. Proc. of the 2004 IEEE Int. conf. on Robotics & Automation New Orleans. LA.
[22] Merlet J-P. and Daney D. Dimensional synthesis of parallel robots with a guaranteed given accuracy over a specific workspace. In IEEE Int. Conf. on Robotics and Automation, Barcelona, April, 19-22, 2005
[23] Hao F. and .Merlet J.P. Multi-criteria optimal design of parallel manipulators based on interval analysis Mechanism and machine theory 40 (2005) 157-171, sept 2004.
[24] Khan S., Andersson, K., Wikander, J. A Design Approach for a new 6-DoF Haptic Device Based on Parallel Kinematics, presented at IEEE International Conference, ICM 2009 Malaga, Spain.
[25] Ahmad A. and Andesson K, "A Novel Approach for the stiffness analysis of 6 DoF haptic device" submitted to ASME IDETC/CIE 2011 August 23-31, 2011, Washington, DC, USA.
[26] Khan S., Andersson, K., Wikander, J. Dynamic based control strategy of the hyptic device, Accepted at IEEE/worldhaptics International Conference, 2011 Turkey.
[27] Legnani G., Tosi D., Fassi I., Giberti H. and Cinquemani S. The point isotropy and other properties of serial and parallel manipulators. Journal of Mechanism and Machine theory 45(2010) 1407-1423.
[28] Lee SU., Shin H., and Kim S., Design of a new haptic device using a parallel Gimbal Mechanism, ICCAS 2005.
[29] Lee J.H., Eom K.S., Suh I.H., Design of a new-6DoF Parallel haptic device, Proc. IEEE ICRA Seoul, Korea, 2001.
[30] Stan S.D., Maties V., Balan R., Rusu C., Besoiu S., Optimal link design of a six degree of freedom micro parallel robot based on workspace analysis, 10th IEEE International workshop on Advanced Motion Control AMC'08 Trento.
[31] Konakla A., Coitb D. W., Smith A. E. Multi-objective optimization using genetic algorithms: A tutorial, Journal of Reliability Engineering and System Safety 91 (2006) 992–1007
[32] Deb K., Agrawal S., Pratap A., and Meyarivan T. A fast and elitist multi-objective genetic algorithm: NSGA-II. IEEE Trans. Evolutionary Computation, 6(2):182–197, 2002.
[33] Tomonori H., Francesco C. and Maria C. Y. Achieving pareto optimality in a decentralized design environment, International Conference on Engineering Design, ICED'09. USA.

Contact: Suleman Khan, KTH Royal Institute of Technology, School of Industrial Technology and Management Department of Machine Design 100 44 Stockholm, Sweden. Tel: 004687907897 Email: Sulemank@kth.se

Suleman Khan is a PhD student at the Dept. of Machine Design at the Royal Institute of Technology.
Kjell Andersson is an Associate professor at the Dept. of Machine Design at the Royal Institute of Technology.

INTERNATIONAL CONFERENCE ON ENGINEERING DESIGN, ICED11
15 - 18 AUGUST 2011, TECHNICAL UNIVERSITY OF DENMARK

A METHODICAL APPROACH FOR DEVELOPING MODULAR PRODUCT FAMILIES

Dieter Krause and Sandra Eilmus
Hamburg University of Technology

ABSTRACT
To offer individualised products at globally marketable prices, Institute PKT's integrated approach for developing modular product families aims to generate maximum external product variety using the lowest possible internal process and component variety. Methodical units of design for variety and life phases modularization support the creation of modular product structures on the level of conceptual design. During embodiment design modular attributes are enhanced through module and interface design according to corporate needs integrating further requirements on product properties. The methodical approach is explained in example of a product family of herbicide spraying systems

Keywords: Design for Variety, Modularization, Methodical Product Development, Product Family, Product Architecture

1 INTRODUCTION
For new products, the extent to which a product meets the challenges of modern market situations is determined during product development. It is important to address contradictory and competing factors and developments. Globally intense pricing competition as well as the megatrend of individualisation is reflected in the conflicting customer requirements of low prices and personalised products. These two scenarios result in two product development strategies. The aim is to develop standard mass-market products to offer competitive prices - the focus being on the advantage of large quantities of the same products. On the other hand, to be able to make a profit, a high number of individualised products can be a successful way to meet individual customer requirements. Both strategies involve chance and risk. In product development, the strategy for developing modular product families is ideal for combining the advantages, such as individual customer demands, with low costs to be well prepared in the future.

The aim of developing a modular product structure for a product family is to maintain the external variety required by the market and reduce internal variety within the company to handle, reduce or avoid the associated complexity of corporate processes in product development. A major advantage of this strategy is the larger quantity of standard modules derived that contribute to cost reduction, for example, with better utilisation of economies of scale and learning curve results, especially in procurement, production and assembly. Modular structures provide the opportunity to parallelise any processes, e.g. to develop different modules in parallel or to test or produce them separately.

This paper presents an overview of the integrated approach for developing modular product families developed during several research projects at the Institute PKT, starting with an introduction of the basics of modular products and strategies for controlling external variety before describing the approach itself using a case study on herbicide spraying systems. Continuing and future research projects within the development of modular product families are then presented, motivated and integrated.

2 THE FIVE ATTRIBUTES OF MODULAR PRODUCTS AND THEIR EFFECTS
The literature defines modularity and modularly structured products in various ways. A comprehensive definition permits the description of common attributes of modular products [1]:

- *Commonality of modules*: Components or modules are used at various positions within a product family.
- *Combinability of modules*: Products can be configured by combining components or modules.
- *Function binding*: There is a fixed allocation between functions and modules.
- *Interface standardisation*: The interfaces between the modules are standardised.

- *Loose coupling of components*: The interactions between the components within a module are significantly higher than the interactions between components of various modules.

Figure 1 is a summary of the five attributes of modular product structures. These attributes of the modularity are characteristics of a product in various forms and degrees. Just as these attributes are gradual, modularity is a gradual characteristic of a product as well. Consequently, the aim of modularization is not the development of a modular product but the realisation of a suitable degree of modularity adapted to the corporate strategy.

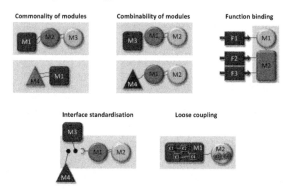

Figure 1. Attributes of modular products (F: Function, C: Component, M: Module) cf. [1]

The modular structure of products and product families can have advantages in every life phase of a product (Figure 2). Yet the potential and limitations of modular product structures have to be considered. The modular structure of a product may inhibit the optimisation of the overall function of each individual product variant. This results in risks in the modularization, such as over-dimensioning, additional interfaces and a lack of product differentiation for the customer.

Analysing the potential and limitations of modular product structures shows that during development of modular products the degree of modularity chosen has to take full advantage of the potential of modular product structures while fulfilling company-specific goals and avoiding negative effects.

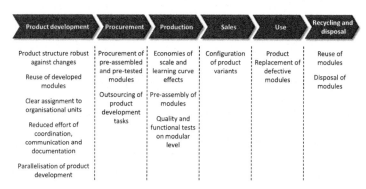

Figure 2. Advantages of modular structures for the product life phases cf. [2]

3 STRATEGIES FOR CONTROLLING EXTERNAL VARIETY

To control the variety demanded by a broad spectrum of customers, companies can follow product-based and process-based strategies, or combinations thereof. As well as the strategy of providing a modular product family, product-oriented strategies also include the *platform strategy*, which is an expansion where a platform, as a basic module applied to a product family, is defined as standard.

A modular product structure adapted to corporate goals allows orientation towards complexity-reducing process strategies, as they are closely related to the product structure. *Process commonality* describes the strategy of using the same processes for different products to counteract the variance of a product family by unifying the processes. A *postponement* strategy enables the greatest possible part of the production process to be independent of variants. Postponement is the delay of processes that are variant. This means that the variant-specific process steps are at the end of the process, if possible.

4 EXISTING APPROACHES IN THE DEVELOPMENT OF MODULAR PRODUCT FAMILIES

To reduce internal variety of product families, approaches in design for variety provide the possibility to develop a product structure that is optimized for effective variant derivation through communality. This results in component designs and product structures that reduce internal variety but do not meet the requirements of other stages within the product life phases (such as sourcing or manufacturing) [3].

Modular product structures can on one hand enhance combinability and on the other hand reduce the negative impacts of internal variety on a company by exploiting the benefits of modular product structures over all phases within the product life (Figure 2). There are three main steps in existing modularization methods [4]:
1. Decomposition of the product up to the level of the components.
2. Analysis and documentation of the components and their couplings.
3. Analysis of the possibility of reintegrating the components.

These existing, often highly matrix-oriented, approaches, such as the Modular Function Deployment [5] and the Design Structure Matrix [6], were developed and will be further developed at a number of institutes, for example, Structural Complexity Management [7]. An approach based on a product's function structure is presented in [8]. These approaches mostly led to a regrouping of components.

In [3], existing methods in design for variety as well as in modularization are outlined to find an integrated approach that reduces variety in modular product families comprehensively, taking effective variant derivation as well into account as reducing the negative impacts of internal variety across all product life phases.

A broad literature review of product family design is given in [9], including configurational product family design as well as production and supply chain issues. The need to align appropriate variety with reduced complexity over the product life phases and appropriate commonality is explored in [10], which provides basic definitions and relationships.

5 PKT'S INTEGRATED APPROACH FOR DEVELOPING MODULAR PRODUCT FAMILIES

PKT's integrated approach extends the idea of designing for variety while meeting requirements from other life phases. All life phases need to be considered to exploit the advantages of modular product structures over the whole product life. The benefits of modular product families can be enhanced by including new technical solutions instead of just regrouping existing components.

The aim of the approach is to reduce internal variety, integrating these comprehensive aspects without cutting the customer-required external variety through different methodical units (Figure 3). It includes the unit of design for variety, which means the redesign of components in terms of variance reduction, and allows the integration of new requirements or functions. This step is followed by the actual modularization, which considers all specific requirements defined by the relevant product life phases and is therefore called life phases modularization. During embodiment design, the modular attributes are enhanced. These units are described in detail in the following sections. Ongoing research is done on a more strategically focussed unit, product program planning, to find ways to reduce internal variety at an earlier level within the redesign/design of a product family.

To carry out a process-based evaluation of alternatives for modular product structures for a product family, a further important methodical unit is the integration and coordination between the product development processes for commonality, the postponement strategy and the product architecture.

The methodical units build PKT's integrated approach for developing modular product families, integrating comprehensive ways of reducing internal variety in one framework of matched tools and methods.

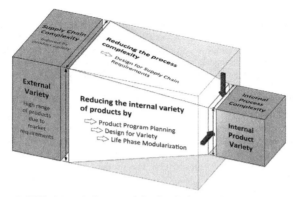

Figure 3. PKT's integrated approach for developing modular product families

6 CASE STUDY OF A MODULAR PRODUCT FAMILIY

A detailed case study using PKT's integrated approach for developing modular product families was conducted on a family of spraying systems for herbicides that were the subject of the AIF-founded research project AUXESIA. The MANKAR-Roll family by Mantis ULV consists of Ultra Low Volume (ULV) Spraying Systems for herbicides. The existing product families consist of 12 actively advertised variants as well as 24 additional variants provided on special customer request (Figure 4). These variants adjust the spraying systems to the individual application conditions of the customers working within professional in-row cultivations or public places via different spray widths or sizes of wheels, for example.

Figure 4. Product variants of the product family of herbicide spraying system MANKAR-Roll

7 DESIGN FOR VARIETY

Design for variety is a methodical unit developed within a research project at PKT [3]. It brings the product families closer to an ideal, allowing a description to be made. This ideal is defined by four characteristics:
1. Differentiation between standard and variant components.
2. Reduction of the variant components to the carrier of a differentiating attribute.
3. One-to-one mapping between differentiating attributes and variant components.
4. Complete decoupling of variant components.

In the first step of the method, the external, market-based and the internal company variety of the product family are analysed. A tree for differentiating attributes aids analysis of the external variety (Figure 5). This tree visualizes the selection process of the customer. Internal variety is analysed at the levels of functions, working principles and components. The variety of functions is shown in an enhanced function structure that makes representation of variant and optional functions possible. The variety of working principles is determined from sketches, where the necessary variance of the functional elements is marked in colour. The specially developed Module Interface Graph (MIG$^©$) is used to analyse the variety of components [2]. The MIG provides a schematic representation of the rough shape and arrangement of the components and their variance, as well as the structural connections and the power, material and information flows. This enables their inclusion when defining modules and reducing variant components.

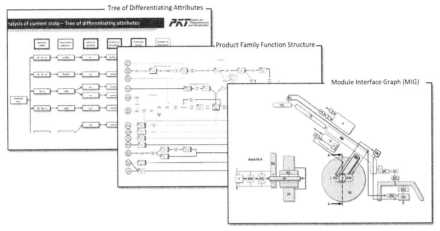

Figure 5: Tools for the analysis of product variety [3]

All relevant information required to carry out design for variety when preparing constructive proposals is visualised in the Variety Allocation Model (VAM^{C}) [3]. The connections between the levels demonstrate the allocations between differentiating attributes, functions, working principles and components (Figure 6). In this way, VAM allows analysis of the degree of fulfilment of the four ideal characteristics. For variant conformity, any weak points in the design can be identified at all levels of abstraction. Thus, VAM is the basis for solution finding and selection of solutions in the methodical unit of design for variety.

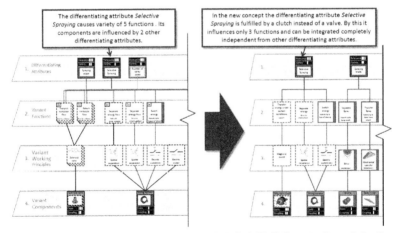

Figure 6. Applying the Variety Allocation Model (VAM) cf. [3], [11] as a tool to optimize the product family of herbicide spraying systems

The result of this methodical unit is a newly designed set of components with an increased number of standard parts. Multiplication effects of the variance are avoided, with the result that each component is required in a small number of variants. The simplified allocation structure between components and differentiating attributes simplifies the variant configuration. These benefits were achieved by using the VAM as a tool to optimize product structure following a product's differentiating attributes, functions and working principles. By considering differentiating attributes as well as functions and working principles, the methodical unit enriches the field of existing approaches with a method that aligns a market-oriented view with a function-oriented one.

8 LIFE PHASES MODULARIZATION

Life phases modularization is a second methodical unit in the development of modular product families developed within a dissertation at PKT [2] in order to use the results of the product design for variety for each individual relevant product life phase, as well as to check their consistency and adjustment to a continual module structure. Product structure requirements can be better met by considering different product structures for individual phases. The procedure is divided into the following steps:
1. Development of a technical-functional modularization as the modularization of product development life phase
2. Development of modularizations for all relevant other product life phases
3. Combination of modularizations
4. Derivation of the modular product structure

The starting point is the technical-functional modularization of the product development phase. Modules are provided that are largely decoupled to reduce the complexity of the development task and allow parallel development of modules. Technical functional approaches, as for example described by Stone [8], can be applied at this step. The development of modularization perspectives of all relevant product life phases is made by module drivers associated with individual life phases. For instance, the production phase is mapped by the module driver 'Separate Testing' (Figure 7).

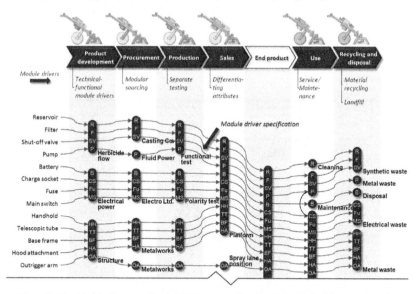

Figure 7. The Module Process Chart (MPC) as a tool to allocate module drivers and module driver specifications to modules cf. [2]

The module drivers are a known concept from [5] but have been supplemented with concrete specifications for the development of modules. In the module driver 'Separate Testing', the tests to be carried out demonstrate the product-specific specifications. In network diagrams, these specifications are linked to the components of the product. The preparation of modules is made by grouping the components that relate to a common module driver specification into one module. Subsequent to the development of modular product structures for the individual life phases, the modularizations are visualised in a MIG to allow consistency checks between the different life phases and demonstrate any conflicts. It was found that it is not sufficient to develop the same module structure for all life phases that cannot be realised because of the different and contradictory criterions. Rather, it is important that the module structures of the individual phases are adapted and continuous but not 100 percent congruent. For assembly, it may be advantageous to install a module that is as large as possible. For purchase, it may be necessary to buy this module in the form of smaller modules from different suppliers which, in case of a well-adapted structure, must not be contradictory. The *Module Process Chart (MPC)* transparently combines the various perspectives of different life phases and makes the coordination process more clear (Figure 7). Finally, the product structure can be derived.

9 MODULARITY IN EMBODIMENT DESIGN

After the modular structure of the product family is defined during conceptual design, the modules and their interfaces are further specified through embodiment design. Figure 8 shows how the steps of PKT's integrated approach for developing modular product families are related to the product design process for single products, following VDI Guideline 2221. In PKT's integrated approach, conceptual design concludes with the definition of a modular product structure. Other approaches place modularization within embodiment design [12] as they do not include redesign in aspects of variety in conceptual design.

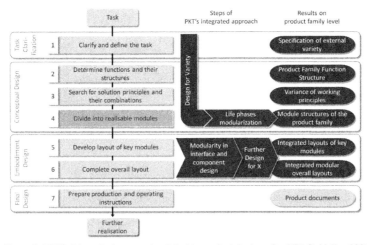

Figure 8. PKT's integrated approach related to product design after VDI Guideline 2221

At the beginning of the embodiment design phase, the level of operation is within modules and on interfaces between modules. This change of level enables clear allocation of the design phases. Yet the development of product families is an iterative process.

Regarding the attributes of modular products we see that even the design of single modules and interfaces affects attributes such as loose coupling or interface standardisation (Figure 1). During embodiment design, these attributes can be given to the product family and so influence how the modularity of the product family meets corporate needs. The challenge during embodiment design is thus to realize all the benefits of the modular product structure defined during conceptual design and to further enhance the modularity and its corporate benefits with the design of single modules and interfaces.

Further requirements of the product family that are not directly related to variety and modularity need to be considered during embodiment design as well. Examples of these requirements are ergonomics and aesthetics, and others commonly summarised under the term Design for X [13], or lightweight design or eco design. Addressing these requirements often influences or is influenced by the modular product structure. Thus, a second challenge within embodiment design of modular product families is meeting different product requirements, e.g. fulfilling specific Design for X guidelines while realizing optimal modularity.

Initial studies of how to meet these challenges during embodiment design of modular product structures were carried out on herbicide spraying systems. Industrial design was one of the requirements addressed after defining the modular product structure. Industrial design concerns the product family in particular regarding the corporate styling over the whole product program and ergonomic design. Applying the attribute of modular products 'loose coupling' components of the product platform are integrated in one body housing. The body housing itself can now integrate aspects of corporate styling and ergonomics without compromising the benefits of modular product structure. For instance, the generic product module is centred close to the wheel to optimize the product's balance point and, by this, its ergonomics. Lines, colours and shapes of the product platform can be freely designed according to corporate design without being influenced by variant components. The resulting concept is shown in Figure 9.

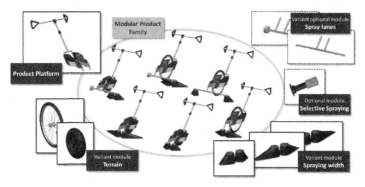

Figure 9. Modular sales concept of the MANKAR-Roll family with integrated aspect of ergonomics and corporate styling

10 RESULTS OF THE CASE STUDY

During the case study described above, a product concept for a modular product family of spraying systems was developed and detailed through embodiment design. The new product family enables the configuration of 64 product variants, including the required existing variants, where 32 are supposed to be actively advertised while 32 are more seldom needed on customer request. Through design for variety, internal variety was reduced from 46 components to 32 components (Figure 10). Through life phases modularization, the product structure was additionally adapted to the specific needs of product life.

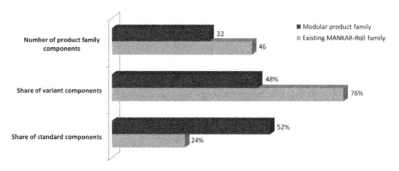

Figure 10. Results of the case study on spraying systems, derived by applying VAM

The new modular product structure enables a postponement strategy, which allows distribution partners to purchase a low number of product platforms and variant modules while offering a high number of variants. This new flexibility enables effective pre-assembly before order inflow and reduces capital and material commitment through stock [14] as every ordered variant can be derived from the same platform and specific variant and optional modules. A high share of standard components (52%) within the product platform, includes the higher part of total component costs (Figure 10).

11 CONCLUSION AND PERSPECTIVES

PKT's integrated approach for developing modular product families enables the reduction of internal variety while maintaining external customer variety. This was verified by the case study as well as in further industrial case studies carried out in the last two years. The approach contains the methodical unit of design for variety. This unit includes steps within conceptual design to optimise the structure of a product family at the levels of functions, working principles and components by partial redesign. The conceptual design ends with the modularization of this redesigned product structure considering specific module drivers of generic product life phases to adapt the optimised product family structure to the needs of the whole product life. By this, the integrated approach aims to optimize the product family structure through conceptual design while other approaches allocate modularization within embodiment design as it is classically seen as a grouping of technical solutions without adapting the solutions themselves to the needs of a variant modular product family.

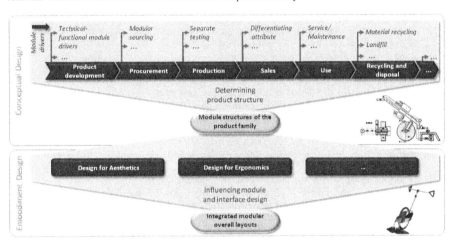

Figure 11. Influence of requirements on product life and product characteristics in conceptual and embodiment design of modular product families

Within embodiment design, the product structure itself is set. The design of interfaces and modules is still to be accomplished with the aim of realising the modular attributes of interface standardisation and loose coupling according to corporate needs. This has to go along with embodiment design according to specific Design for X guidelines. Figure 11 shows how requirements of generic product life phases and product properties are integrated into the development of modular product families. Within the case study, requirements of generic life phases are defined as module drivers to determine the modular product structure. Requirements of specific product properties influence the embodiment design of interfaces and modules, for example, through Design for X guidelines. In Figure 11, the set of requirements and their influence on the case study are shown. The next step for research is using the generic life phases as a basis for defining module drivers that need to be individualised to specific corporate priorities. The requirements that have priority in determining modular product structure as a module driver can then be chosen. Requirements that are subordinate to those considered as module drivers or that can be met within separate modules are considered during embodiment design.

How the specific requirements of lightweight design and design for assembly can be integrated along the process of developing a modular product family is considered in the current research projects of Module Lightweight Design and Modularization for Assembly.

A further field of research is to extend PKT's integrated approach to the redesign of product programs to enhance the share of carry-over parts through modular product structures within the product program.

REFERENCES

[1] Salvador F. Toward a Product System Modularity Construct: Literature Review and Reconceptualization. *IEEE Transactions on Engineering Management 54*, 2007, pp. 219-240.
[2] Blees C. *Eine Methode zur Entwicklung modularer Produktfamilien insbeondere Modularisierung*, 2011 (TuTech Verlag, Hamburg).
[3] Kipp T. and Blees C. and Krause D. Development of Modular Product Families: Integration of Design for Variety and Modularization. In *Proceedings of norddesign 2010*, Gothenburg, August 2010, pp.159-168.
[4] Blees C. and Jonas H. and Krause D. Perspective-based Development of Modular Product Architectures. In *Proceedings of the 17th International Conference on Engineering Design (ICED)*, Vol. 4, Stanford, August 2009, pp.95-106.
[5] Erixon G. *Modular Function Deployment: A Method for Product Modularisation*, 1998 (Royal Institute of Technology, Stockholm).
[6] Pimmler T. and Eppinger S. Integration Analysis of Product Decompositions. In *Proceedings of the 6th Design Theory and Methodology Conference*, New York, 1994, pp.343-351
[7] Lindemann U. and Maurer M. and Braun T. *Structural complexity management: an approach for the field of product design*, 2009 (Springer, Berlin).
[8] Stone R. B. *Towards a theory of modular design*. 1997 (University of Texas, Austin).
[9] Jiao J. et al. Product family design and platform-based product development: a state-of-the-art review. In *Journal of Intelligent Manufacturing*, Vol. 18, 2007, pp. 5-29.
[10] Andreasen M.M. et al. Multi-product Development: new models and concepts In *Design for X - Beiträge zu 21. DfX-Symposium*, Neukirchen, October 2004, pp.75-85.
[11] Kipp T. and Blees C. and Krause D. Anwendung einer integrierten Methode zur Entwicklung modularer Produktfamilien. In *Design for X - Beiträge zu 21. DfX-Symposium*, Hamburg, September 2010, pp.157-168.
[12] Blackenfeldt M. *Managing complexity by product modularisation*, 2001 (Royal Institute of Technology, Stockholm).
[13] Pahl G. and Beitz W. *Engineering Design*, 2007 (Springer, Berlin).
[14] Feitzinger E. and Lee H.L. Mass Customization at Hewlett-Packard: The Power of Postponement. *Harvard Business Review*, 1, 1997, pp.115-121.

Contact: Sandra Eilmus
Hamburg University of Technology
Product Development and Mechanical Engineering Design (PKT)
Denickestr. 17 (L)
D-21073 Hamburg
Germany
Tel: +49 40 42878 2149
Fax: +49 40 42878 2296
Email: Sandra.eilmus@tuhh.de

Prof. Dieter Krause is head of the Institute Product Development at Hamburg University of Technology. He is a member of the Berliner Kreis and the Design Society. The main topics of his research are new design methods for product variety and modularization, as well as lightweight design for aircraft interiors.
Sandra graduated from Mechanical Engineering at TU Berlin and worked as a consultant in the automotive industry before becoming a research assistant at PKT (Institute for Product Development and Mechanical Engineering Design, TUHH). Her research focuses on the development of methodical support that aims to increase product communality within existing product programs.

INTERNATIONAL CONFERENCE ON ENGINEERING DESIGN, ICED11
15 - 18 AUGUST 2011, TECHNICAL UNIVERSITY OF DENMARK

PRODUCT MODEL OF THE AUTOGENETIC DESIGN THEORY

Konstantin Kittel (1), Peter Hehenberger (2), Sándor Vajna (1), Klaus Zeman (2)
(1) Otto-von-Guericke University Magdeburg, Germany, (2) Johannes Kepler University Linz, Austria

ABSTRACT
Product development plays the key role in defining all product properties and benefits. There is a need for appropriate supporting methods that are able to serve and to satisfy multi-criteria and multi-domain requirements. These requirements usually go beyond function fulfilment. Developing mechatronical systems raises the requirements for the applied product development methods, because of the high number of linked elements from different technical und physical domains. Integrating and combining solutions from different domains does not only increase the number of possible solutions, but also the complexity of such systems. Product development methods need to be able to handle these requirements appropriately in order to support the designer within the product development process.

Keywords: Product Development, Evolution, Mechatronics

1. INTRODUCTION
The Autogenetic Design Theory (ADT) uses analogies between natural evolution and product development to ensure that the best possible solution can be found within given requirements, conditions, and boundaries [1][2][3]. These requirements, conditions, and boundaries can also contradict each other and can change over time, i.e. "best possible" has always to be interpreted in relation to the actual situation. The ADT describes the development of products as a continuous improvement and optimisation process. The ADT is called "autogenetic" due to its similarity to processes of autogenesis of the natural evolution.
The ADT is not another variety of Bionics (where results of an evolution, e.g. the structure of trees, are transferred to technical artefacts). Rather, the ADT transfers procedures from biological evolution to accomplish both a description and broad support of product development with its processes, requirements, boundary conditions, and objects (including their properties).

2. THE AUTOGENETIC DESIGN THEORY
The Autogenetic Design Theory (ADT) applies analogies between biological evolution and product development [4] by transferring the methods of biological evolution (and their advantageous characteristics) to the field of product development. Such characteristics are for example the ability to react appropriately to changing environments (requirements and boundary conditions), so that new individuals are in general better adapted to the actual environment as their ancestors.
The main thesis of the ADT is that the procedures, methods, and processes of developing and adapting products can be described and designed as analogies to the procedures, methods, and processes of biological evolution to create or to adapt individuals. Main characteristics of biological evolution (with the underlying principle of trial and error) are continuous development and permanent adaptation of individuals to dynamically changing targets, which in general have to be accomplished in each case at the lowest energy level and with the minimal use of resources, i.e. the evolution process runs optimised in terms of energy consumption and resource employment. The targets can change over time because of (unpredictable) changing requirements, resources, conditions, boundaries, and constraints, and they can contradict each other at any time.
The result of a biological evolution is always a set of unique solutions being of equal value but not being of similar type. Consequentially, the result of the ADT is for the very most part a set of equivalent, but not similar unique solutions that all fulfil the actual state of requirements and conditions best.
Furthermore, biological evolution doesn't have prejudices. This means that new individuals (described by their chromosomes) will not be discarded because they are different. Each individual has to prove

itself in its natural environment. If it turns out that an individual with a new chromosome set is superior to already existing individuals, then this individual gets a better chance for reproducing. By transferring this behaviour of impartiality to product development, new concepts would not be discarded because they were totally different than former concepts but only if their properties were proven not to be superior.

This suggests that both evolution and product development can be described as a continuous but not straightforward improvement process or as a kind of multi-criteria and continuous optimisation [2][4].

One may argue that a weakness of the ADT is the processing (creation, evaluation) of a high number of individuals for reaching a certain product progress due to the evolutionary based approach. Compared to other methods, the number of solutions, which need to be evaluated, is in fact much higher. But one has to keep in mind that, by exploring this high number of possible solutions (individuals) within a solution area, the chance of finding the really best set of solutions to a set of requirements, conditions, boundaries, and constraints is much higher than with traditional approaches that continuously delimit the solution area and thus result only in a single "next best" solution [5].

The analysis of product development from an evolutionary perspective leads to the following insights [1]:

- In every phase of the product development process, various alternatives are developed and compared. These alternatives are in competition with each other, because only the actually best were selected for further processing.
- The processes of searching, evaluating, selecting, and combining are also typical approaches of biological evolution.
- Regardless of the phase of product development or of the complexity level of the emerging product, always similar patterns of activities can be identified that modifies existing or that generates new solutions. This patterns can be compared to the TOTE-Scheme [6][7]. Self-similarity can be found at all levels of complexity of product development as well as in all stages of the emerging product [2].
- According to chaos theory small changes or disruptions in the system can cause unpredictable system behaviour [8]. The fact that the result of the development of a product usually can't be predicted definitely because of the influence of the creativity of the product developer leads to the assumption that the product development process also contains elements of a chaotic system or at least shows a chaotic behaviour in some aspects.

At the present state, three major components of the ADT have been researched. First, a process model describing how the ADT works and what the steps are, which the product developer has to perform (see section 2.1). Secondly, the solution space model, which shows how the space, in which product development takes place, is structured (see section 2.2). Thirdly, the underlying product model that holds the description of how product information is structured and used (see section 3).

The development of the ADT is content of a research project running right now. This means that the concepts presented in this paper are not in a final state. Especially the practical appliance will be in the focus of further work.

2.1 Process Model

There are numerous ideas and concepts aiming on describing the complex and often chaotic process of developing. Caused by the complexity of most products, the focus within the development process isn't mostly on the complete product, but rather on a specific part of the product. This means that each development step is focused on improving or modifying only one specific product property or a specific set of properties by varying certain design parameters.

A challenge in the development of the process model is to define a model, which is not only able to describe the development process in the later stages of the development process but also in the early stages. FAN, SEO, ROSENBERG, HU and GOODMAN describe an approach, wich provides an automated system-level design for mixed domain systems [17]. But as most other methods [9][10] this approach uses a set of predefined elements which can be used by the algorithm to create new solutions. Depending on the available predefined elements the achievable solution quality is more or less limited (see section 2.2).

The ADT process model currently under research aims on providing a holistic development process model that is also able to describe the processes of partial improvements/modifications, which normally do not follow a predefined pattern. The ADT process model describes the development process on two levels.

Level one
Level one provides an overall look on product development within the ADT. It starts with the definition and description of the target function[1] and the solution space based on requirements, starting conditions, boundary conditions, conditions of the environment of the solution space, and (internal and external) constraints. Within the solution space, possible solution patterns are searched, combined, and optimised in a random order. To evaluate the actual state of development, the particular fitness[2] is determined. Because requirements, processes, and both internal and external influence factors are all dynamic due to unforeseeable changes, it is clear that it is only possible to describe (rather small) process patterns, which can be used randomly, instead of specifying a sequence of steps or any predefined "way" the designer has to follow.

To nevertheless support the process of partial improvements or modifications (and allow the designer to address only a limited set of properties) without loosing both consistency and the overall picture, different views can be applied (see section "Product Model"). These views act like filters that ensure that only a specific set of product properties are considered. Thus, the development process becomes a set of activities, each containing a product improvement or modification under a specific view.

Level two
Level two describes under a certain view the activities of improving or modifying a set of properties that are determined by design parameters. The ADT uses the steps creation, evaluation, and updating to modify and to improve a product.

Creation

In analogy to biological evolution, the creation step consists of the four sub-steps selection – recombination – duplication – mutation.
- **Selection** The parent solutions for the next generation are designated.
- **Recombination** The design parameters of two already existing solutions are combined to create (in most cases) a more evolved solution. This is the usual way to create new solutions.
- **Duplication** Creation of an identical copy (duplicate, clone). Creating clones is recommended if solutions with superior properties exist, which should be inherited to the next generation.
- **Mutation** A random change of the design parameters of a solution that was created by recombination. Mutation is necessary to ensure dynamics in the evolution process.

Evaluation

In this step each new solution is evaluated to determine the fulfilment of the optimisation criteria described in the target function. Based on this information the actual fitness (representing the quality of the actual solution) is calculated.

Updating

The update of both solution space and target criteria is necessary to take dynamic requirements into account. Often requirements change within the development process. Such a change influences the solution space (also with the result that a specific solution is not allowed any further) as well as the target criteria (with the result that a specific target criterion gets less important or that another criterion should have a bigger influence on the product fitness).

2.2 Solution Space
In general, the term "solution space" is understood to be a set of all feasible solution elements, which can be used within product development. This includes all elements that a product developer may use for the evolution of a solution or several solutions, on the basis of requirements, inner and outer

---

[1] The target function calculates the fitness of a solution based on the optimisation criteria of this specific solution.
[2] The fitness represents the actual degree of development of a specific solution. The fitness is used to compare two or more solutions.

conditions, ecological/environmental conditions and others. This definition of a solution space can be compared with the mathematical term "domain".

Every product development method has its solution space, which usually is spanned by the requirements and limited by both starting and boundary conditions. The inner structure is influenced by constraints. Some approaches, e.g. TRIZ [9] and Gene Engineering [10], use a solution space with a particularly structured dataset. This dataset contains the solution elements for the emerging solutions. It is common to all such solution spaces that the product developer is offered only a limited amount of possible solution elements. However, the more limited the quantity and possible configurations and combinations are the lower is the achievable solution diversity and quality.

Thus, to improve both solution diversity and quality, it is necessary to not artificially limit the quantity of solution elements, but rather to include permissible elements for all concrete tasks. Thus follows, however, the task of holding on to all permissible elements in the solution space description. As the diversity of existing solution elements (materials, manufacturing methods, operating principles, etc) is immense, a complete solution space description at reasonable time and costs is in most cases impossible. Since the optimal configuration and combination of solutions elements are not available under these circumstances, the maximal possible solution quality can't be achieved.

In order not to limit the product developer and to permit the maximal possible number of allowable solution elements, the ADT applies an inverted solution space. The only limitations of such an inverted solution space are the laws of natural science, i.e. the space is virtually infinite. The inverted solution space contains prohibited areas (taboo zones). Taboo zones are formed by those solution elements of which the use for possible solutions is explicitly forbidden. This inverted solution space is referred to in the following as Prohibition Space.

Another advantage of the Prohibition Space is evident in the early phases of the development process. Due to the lack of knowledge about the relationships between requirements and solution elements, it isn't often possible to determine the forbidden criteria, based on the forbidden requirements. The Prohibition Space therefore contains too few taboo zones at the beginning of the product development process. At this time, the product developer is able to use product criteria, for example, that should actually be forbidden. If, for example, impermissible solution elements are used, it will be noticed upon evaluation that the resulting solution possesses impermissible product criteria. This newly obtained information can be used to refine the Prohibition Space.

A traditional solution space would in the same case be incomplete as well (due to the lack of knowledge about the coherences). Here, however, "incomplete" means that permissible elements are missing, thereby restricting the product developer's solution possibilities. Since a traditional solution space consists by definition of only permissible solution elements, only solutions with permissible criteria will be generated (exceptions could be certain combinations of permissible elements). This means that the existence of an incomplete solution space (some permissible solution elements are missing) is difficult to detect. The definition of the Prohibition Space is based on the requirements and the various starting and boundary conditions, constraints, and the environment, which can arise from different sources. Through the inversion, requirements and conditions turn into appropriate bans that can be (quite easily) formulated and the solution elements that are forbidden can be derived.

The Prohibition Space dynamically changes whenever an external event (for example a requirement modification, a change of a condition, etc.) occurs during the evolution, because, as a result of this external event, changed possibilities for the evolution can arise or existing ones have to be omitted. To reflect these, taboo zones within the Prohibition Space have to be redesigned, which may result in changing taboo zones, in omitting existing, or in adding new zones (figure 1).

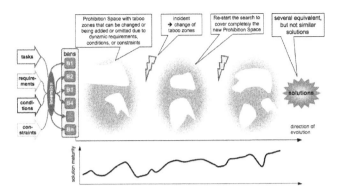

Figure 1: Inversion of requirements and changes of the Prohibition Space due to external events

3. PRODUCT MODEL

The aim of a product model is to provide a framework to capture all product data, which are necessary to describe the product and its life cycle in a structured way. This framework shall be stable for the whole development process, so that data can be completed step by step. In "traditional" product development, different data structures arise along the sequential phases of the product development process.

The design of the product model shall regard the requirements of the modeling of multidomain systems. Mechatronic systems differ from conventional technical systems in the higher number of coupled elements from different technical and physical domains. By combining and integrating solutions from various domains the number of possible solutions increases together with the complexity of such systems. This raises the need for a structured approach for designing and modelling.

An incontestable precondition for successful product development is the interdisciplinary definition, description and presentation of product information. The following issues and requirements for the model representation need to be realised with the product model.

3.1 General Requirements

- Description of product information from all phases of the product life cycle: Taking into account the lifespan of a product is necessary when storing the product relevant information within a single structure.
- Coupled analysis of different physical product properties: Caused by the increasing integration of different applications and functionality in technical systems, the demand arose to find a uniform description of the various product properties, to define, describe, dimension, calculate the product.
- Taking into account the view on an application domain: Different development phases and mechatronic disciplines require different views on the object to be analysed. It follows for a general description of the requirement that model parameters should be provided in various combinations and granularity for a further (more detailed) processing. The influence of the system should be possible on the need for the individual points of view characteristics.

3.2 Mechatronical Requirements [12][13][14]

- Description of divergent aspects, which are related to the complex characters of and dependencies between the involved domains.
- Support product information from the design phase.
- The product model has to represent the physical behaviour of the technical system, which depends on the properties of the subsystems and their interactions. A sub-model can be a model of a single mechanical component or of a complex system e.g., a model of an integrated mechatronic system including an embedded control system.

- The product model has to represent the interface model between sub-models form different disciplines.

3.3 Types of properties
For the description of the ADT the product properties are divided into required properties, resulting properties and definable properties. The required properties can be retrieved from the requirements on the product (customer requirements, boundaries, conditions). The required properties describe the goal state of the product. The actual state of the product is described by the resulting properties. The resulting properties are set by the definable properties (according to physical relations). The definable properties can directly be determined by the product developer. At the end of the development process, the definable properties need to define a product, whose resulting properties match the required properties. A complete matching of resulting properties and required properties at the beginning of product development process need not be given. Along the product development process, the properties are expanded and detailed.

3.4 Concept
The structure of the ADT product model aims on storing the product information from the entire product lifecycle. This requires a structure, which can already be used in the early phases and which does not change along the product development process. Normally, product information is structured according to the geometry of the product or according to the assembly structure of the product. This results in problems in the early phases of the product life cycle, because the geometry or assembly structure is a result of the development process and neither exists in the early phases nor is it stable during the development process.

The "object" which is most stable during product development is the set of resulting properties, which defines the product. As stated before, requirements can change over time, but when being precise, changed requirements mean to development a new product. Trying to find a product information structure which is suitable for every set of requirements needs to result in a structure which is identical for every product development process.

The ADT product model at present is inspired on the extended feature model of the FEMEX. The FEMEX model provides a unified structure to store all product data and information from the whole life cycle [15]. The ADT product model uses the basic matrix structure of the FEMEX model and the method of classifying the product information according to product properties. Contrary to the FEMEX model the ADT product model does not use the product life cycle phases, but multiple views[3] as a second way to classify the product information. In the actual ADT product model a product is described by a certain number of definable properties (represented by the different coloured boxes in Figure 2). To organise all the definable properties, a structure is derived from the existing resulting properties on the one hand and from the different views on the other side.

---

[3] Views are like filters on the product. They include a subset of definable properties as well as a subset of resulting properties.

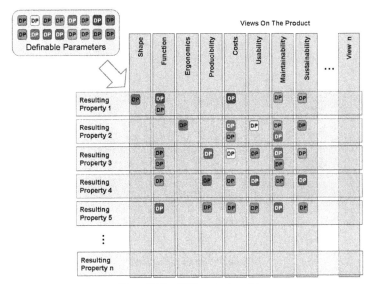

Figure 2: The ADT product model

It has to be mentioned that a definable property is not equivalent to a geometric parameter. Basically, a definable property can be any property the designer defines directly (e.g. materials, manufacturing parameters, geometry, …). The totality of all definable properties defines the complete product with all its properties and its behaviour.

The resulting properties are used to structure the generally high number of definable properties. This is done, by assigning each definable property to the resulting properties, which were influenced by this definable property. As mentioned before, it is possible that a definable property influences more than one resulting property (for example: the definable property "material" influences the resulting properties "maximum weight" and "maximum stress"). A second level of structuring is achieved by assigning the definable properties to the different views. Each definable property can appear in a single view or in multiple views (for example: the definable property "material" appears in the view "producibility" and "costs"). In order to classify the definable properties in the matrix, a meta information (tag) can be assigned to each of them. This meta information can be the view or the product property influenced by the definable property.

Another advantage of this representation form is that the influence of definable properties can be quickly determined. It is obvious, which definable property influences which resulting property. Resulting properties that depend only on a single definable property can be determined at the very beginning of the development process without taking into account dependencies with other resulting properties. This dependence can be checked very simple by analysing the tags of the definable properties.

The ADT product model uses chromosomes to cluster the product information: A chromosome contains a resulting property and all the definable properties on which it depends on (marked with the purple line in Figure 3).

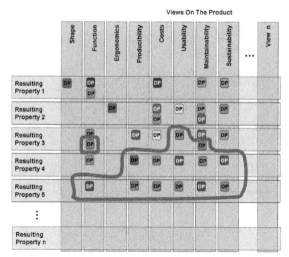

Figure 3: Chromosome model

The values of the included definable properties of the chromosome can change along the development process. Such a change can be triggered by different events. Possible types of events are:
1. Optimisation of a definable property. This is the most common case. The product developer adjusts a definable property to achieve an improvement of a resulting property.
2. Changes due to dependencies. When a definable property changes, it is possible that a thereof dependent definable properties must be adjusted, e.g. if a definable property dictates a material and another definable property dictates the wall thickness. In this constellation it can occur that the definable property wall thickness has to change when the definable property material changes, because not all former values were permitted any longer.
3. External event. By a change in the requirements new or adjusted taboo zones arise. This new circumstance can create the need that definable properties need to be adjusted, because certain values are not permitted any longer.

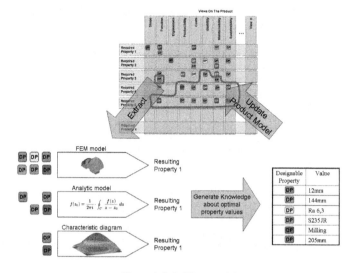

Figure 4: Subsidiary models

Subsidiary models

When working with the product model, the product developer will never work on all definable properties parallel. In most cases, he will focus on the properties clustered by one chromosome (a resulting property and all definable properties it depends on) or a few chromosomes. Subsidiary models are used to describe the behaviour of the product reduced on the interrelation between definable properties and resulting properties within the chromosome (see Figure 4). Normally there is more than one subsidiary model available to describe this interrelation. The subsidiary models differ in the model aim and the method which describes the interrelation (FEM, analytic, diagram). The model aim is defined by the resulting properties, which shall be described and the effects, which were considered or were explicitly not considered (e.g friction, inhomogeneity).

Different subsidiary models use the information from product model. These models differ in their complexity, which depends on:
- Considered definable properties
- Simplifications (considered effects)
- Method to describe the interrelation (numeric, analytic, graphical)

In most cases more than one subsidiary model can be used for the description of a specific resulting property, whereas the less complex subsidiary models were used in the early phases of the product development process, when there is only little information about the product. During the product development process more and more complex subsidiary models can be used, to describe the interrelation between definable properties and resulting properties.

The knowledge about the interrelation between definable properties and resulting properties, gathered together from the subsidiary models can be used to set the values of the definable properties, such as geometric dimensions, material type, properties of the production processes. For example the product designer uses a fem model to determine the occurring stresses. This subsidiary model than provides him information about the relation between definable properties (geometric parameters and material parameters) and resulting properties (maximum stress, maximum deformation). Due to the information from this subsidiary model, the product developer is than able to choose the right values for the definable properties in order to match resulting properties and required properties.

Property dependencies

Increasing the number of definable properties is going along with an increased number of dependencies between these properties. In order to ensure the consistency of the product model the dependencies between the definable properties need to be managed. In [16] a meta model is used to describe the dependencies between the different domains of the product. A similar approach shall be used to ensure the consistency within the ADT product model.

The goal of this system of rules is to support the designer in generating permissible solutions, by providing information about depended definable properties. The system of rules displays possible consistency problems, which can arise by changing a specific definable property.

The dependencies are stored as meta information for every definable property. At the moment the system of rules is designed to provide the following information:
- Which other definable properties depend on the regarded definable property (e.g. a material forces a specific manufacturing process).
- Which values were forced for the dependent property, if the regarded property takes a specific value (e.g. if definable property "material crossbar" takes the value "aluminum", a value of ">0,5mm" is forced for the definable property "wall thickness crossbar").
- Which values were forbidden for the dependent property, if the regarded property takes a specific value (e.g. if definable property "surface roughness" takes the value "<Ra 0,5", the value "milling" is forbidden for the definable property "manufacturing process").

4. CONCLUSION & OUTLOOK

The research on the ADT has shown that both complexity and disciplinarity of products increase towards incorporating domains beyond mechanical engineering. This overview over the different components of the ADT shows the concepts that have been developed so far.

The ADT product model offers an approach to describe a product detached from its actual physical structure. This allows the application of the product model even in the early phases of the development process. The system of rules supports the product developer by providing important information about definable properties.

Future research work on the ADT will deal with process model of the ADT to describe what the product development process looks like and what steps the product developer has to take to develop product with the ADT.

ACKNOWLEDGMENTS

We gratefully acknowledge that this work has been supported by the German Research Foundation (DFG) and has been supported in part by the Austrian Center of Competence in Mechatronics (ACCM), a K2-Center of the COMET/K2 program (which is aided by funds of the Austrian Republic and the Provincial Government of Upper Austria)

REFERENCES

[1] Vajna, S., Clement, St., Jordan, A., Bercsey, T., "The Autogenetic Design Theory: an evolutionary view of the design process". Journal of Engineering Design 16(2005)4 pp. 423 – 444
[2] Wegner, B., "Autogenetische Konstruktionstheorie – ein Beitrag für eine erweiterte Konstruktionstheorie auf der Basis Evolutionärer Algorithmen", Dissertation Universität Magdeburg 1999
[3] Vajna, S., Clement, St., Jordan, A., Bercsey, T., "The Autogenetic Design Theory: an evolutionary view of the design process". Journal of Engineering Design 16(2005)4 pp. 423 – 444
[4] Bercsey, T., Vajna, S., "Ein Autogenetischer Ansatz für die Konstruktionstheorie". CAD-CAM Report 13(1994)2, pp. 66-71 & 14(1994)3, pp. 98-105
[5] Clement, S., " Erweiterung und Verifikation der Autogenetische Konstruktionstheorie mit Hilfe einer evolutionsbasierten und systematisch-opportunistischen Vorgehensweise", Dissertation Universität Magdeburg 2005
[6] Miller, G.A., Galanter, E., Pribram, K.H.: Strategien des Handelns. Pläne und Strukturen des Verhaltens (2. Auflage). Klett-Cotta Stuttgart 1991
[7] Ehrlenspiel, K.:, "Integrierte Produktentwicklung". Carl Hanser Verlag München 2007
[8] Briggs, J., Peat, F. D., "Die Entdeckung des Chaos". Carl Hanser Verlag München 1990
[9] Altshuller, G., "40 Principles – TRIZ Keys to Technical Innovation" (translated and edited by L. Shulyak, S. Rodman). Technical Innovation Center Worcester MA (USA) 2003
[11] Chen, K.Z., Feng, X.A., "A Framework of the Genetic-Engineering-Based Design Theory and Methodology for Product Innovation", in: Norell, M. (editor): Proceedings of the 14th International Conference on Engineering Design ICED03, presentation 1745, Design Society 31, Stockholm 2003, Abstract p. 671
[12] Avgoustinov, N. Modelling in Mechanical Engineering and Mechatronics, Springer Publishing Group, UK, 2007
[13] Bishop, R.H. Mechatronic Fundamentals and Modeling (The Mechatronics Handbook), CRC Press Inc, New York, 2007
[14] Alvarez Cabrera AA, Foeken MJ, Tekin OA, Woestenenk K, Erden MS, De Schutter B, van Tooren MJL, Babuska R, van Houten FJAM, Tomiyama T. Towards automation of control software: A review of challenges in mechatronic design, Mechatronics ISSN 0957-4158. http://dx.doi.org/10.1016/j.mechatronics.2010.05.003
[15] Ovtcharova, J., Weber, C., Vajna, S., Müller, U., "Neue Perspektiven für die Feature-basierte Modellierung" VDI-Z 140(1997)3, pp. 34-37
[16] Hehenberger P., Egyed A., Zeman K.: Hierarchische Designmodelle im Systementwurf mechatronischer Produkte, VDI - Tagung Mechatronik 2009, Komplexität beherrschen, Methoden und Lösungen aus der Praxis für die Praxis, 12.- 13. Mai 2009, Wiesloch bei Heidelberg, Deutschland
[17] Z. Fan, K. Seo, R. Rosenberg, J. Hu, E. Goodman, "A Novel Evolutionary Engineering Design Approach for Mixed-Domain Systems". Engineering Optimization, Vol. 36, No.2, April 2004, pp127-147

PRODUCT DEVELOPMENT SUPPORT FOR COMPLEX MECHATRONIC SYSTEM ENGINEERING– CASE FUSION REACTOR MAINTENANCE

Simo-Pekka Leino[1], Harri Mäkinen[1], Olli Uuttu[2], Jorma Järvenpää[1]
(1) VTT Technical Research Centre of Finland, (2) Eurostep Oy, Finland

ABSTRACT
Development of a multidisciplinary mechatronic system, like a remote operated maintenance system of ITER fusion reactor, requires system engineering approach. System engineering is a leadership approach for designing totally new concepts and technology. On the other hand, system engineering needs support for managing all related processes and information. Product lifecycle management (PLM) can be seen as IT-aided enabler of such processes and information management desires. Divertor Test Platform 2 (DTP2) is a full scale mock-up and test facility for developing, testing and demonstrating remote operated maintenance equipment as well as planning and training future maintenance operations. Characteristic for DTP2 is that its development and operational lifecycle will be several decades long. History of the system has to be traceable and all data must be available during the whole lifecycle. This work in progress paper aims to introduce the first results of the ongoing project, which defines and implements PLM support for DTP2 system engineering. The preliminary results include requirements specification for the PLM platforms, and a concept for mechatronic product model and data model.

Keywords: System engineering, mechatronic system, design support, PLM

INTRODUCTION
ITER is a large-scale scientific experiment intended to prove the viability of fusion as an energy source, and to collect the data necessary for the design and subsequent operation of the first electricity-producing fusion power plant [1]. DTP2 (Divertor Test Platform 2) facility (Figure 1) is supporting the ITER project. *DTP2 is a full scale physical test facility* intended for testing, demonstrating and refining the remote handling (RH) equipment designs with prototypes. The facility will also be used for training future ITER RH operators. Effective and efficient remote replacement of the ITER divertor is central to the successful execution of the ITER project. The aim of the DTP2 is to ensure that the cassette movers supplied to ITER during its construction are based on well-matured designs which have benefited from the experience from the building and operation of a first generation of prototypes. [2], [3]

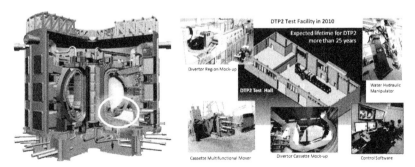

Figure 1. ITER fusion reactor and the DTP2 mock-up facility

System Engineering and Product Lifecycle Management Approach

System engineering is an interdisciplinary approach for designing totally new concepts and technology. There exist many definitions of the system engineering. According to Kossiakoff [4] the characteristics of a system whose development require the practice of system engineering are that the system: is a engineered product which satisfies a specified need, consists of varied components that have complicated relationships, is multi-disciplinary and relatively complex, uses advanced technology, involves development risk and relatively high cost, it's development requires several years, many interrelated tasks and several organizations to complete. The role of system engineering is to seek the best balance of the critical system attributes and engineering disciplines. All of these characteristics do not exist always, but in case of fusion reactor and divertor development they are very relevant. The system engineering method can be thought of as the systematic application of the scientific method to the engineering of a complex system, though it does not correspond to the traditional academic engineering disciplines [4].

A system (or product) lifecycle means evolution stages from requirements through concept development, design, production, operation and maintenance to disposal. In system engineering these stages can be referred as 1) concept development, which is the first stage of system concept and functional definition in order to satisfy a specified need, 2) engineering development, which translates the system concept and functions into validated physical system design, and 3) post-development including production, deployment, operation and support throughout the system life. The post-development is also very important stage, since system engineering is often needed for instance in problem solving during the operation and maintenance of the system. This is likely one way to utilize DTP2 facility in the future. Therefore, traceability of the system development and documentation is essential.

System engineering is integral part of a system development project as well. A large project involves often hundreds of people and several organizations. One of the most important functions of system engineering is to guarantee communication between disciplines and parties through documentation and other communications. Besides conventional text and drawing engineering documents, these include nowadays also e.g. simulation models and videos. These documents are revised periodically. Therefore, it is essential that documents and other information are well managed. System engineering method (Figure 2) can be considered as consisting of four basic activities: 1) Requirements analysis (problem definition), 2) Functional definition (functional analysis and allocation), 3) Physical definition (synthesis, physical analysis, and allocation), 4) Design validation (verification, evaluation) [4].

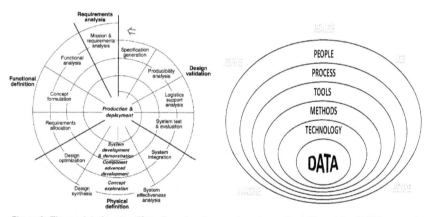

Figure 2. The model of system lifecycle and system engineering method [4], and the PLM framework

Development of a system including totally new concepts and technology faces often unknowns regarding performance of the implemented system. Experiments and simulation are powerful means for verification and evaluation of system performance, as well as for decision making and problem solving throughout the system lifecycle. Simulations and experiments can be utilized both for system functional and material behavior assessment, and they offer complementary information [4].

Product lifecycle management (PLM) is an integrated, holistic information-driven business approach comprised of people, processes/practices, and technology to all aspects of product's life, from its design through manufacture, deployment and maintenance. Components of PLM include the products themselves, organizational structure, working methods, processes, people, information systems and product data. [5],[6]

The right side of Figure 2 illustrates the components of product lifecycle management. Product definition data is the core of PLM. It includes specification of the product itself, but also production, maintenance, disposal and all other specifications during its lifecycle. The data is nowadays often in digital format, but it includes physical documentation as well. PLM technology includes data management and other systems, like PDM (Product Data Management) and ERP (Enterprise Resource Planning). PLM tools and methods include systems and procedures for producing, analyzing and utilizing product data, e.g. CAD (Computer Aided Design), VR (Virtual Reality), AR (Augmented Reality), DFX (design for lifecycle) and CAE (Computer Aided Engineering). There are many kind of processes involved during product lifecycle, from concept design to detailed design, production planning, usage and maintenance. People involved in product lifecycle management are organized in a complex network including different internal departments of a manufacturing company and external partners and suppliers. In present implementations, PLM is usually utilized in detail engineering and design phase, not in the fuzzy concept or research phases.

System engineering and PLM approaches have lot of similarities. System engineering is strong in multidisciplinary engineering and systematic chain from end-user requirements to functional and systematic breakdowns and finally to physical breakdowns, technical solutions and system validations. PLM should support these activities for instance with good requirements and engineering change management capabilities. It is currently a popular research topic how to better implement System engineering in PLM. System engineering is quite an old theoretical approach which seems to have now revival because of development of PLM capabilities. Simulations and virtual prototyping of mechatronic systems are enablers of System engineering as well. On the other hand, PLM is strong both in product type and product individual data management. Those capabilities could be better utilized in System engineering. There are also design theories and many applicable product development methods defined for instance by Andreasen & Hein [7] and Ulrich & Eppinger [8]. Their strengths can be exploited in development and engineering of a mechatronic product. Those product development methods should be also supported by PLM [9]. We see System engineering and design methodology as complementary approaches. System engineering is more like a higher level framework for managing development projects flows from requirements to verifications and validations of a mechatronic multidisciplinary system, when design methodology supports delivering more detailed technical solution alternatives. Obviously there is overlapping between System engineering, design methodology and product lifecycle management.

Motivation and Objective

DTP2 is a complex high-tech mechatronic system consisting of many engineering disciplines and domains like mechanics, electronics, hydraulics, software development, virtual prototyping and virtual reality, and special new technologies like mobile robotics, water hydraulics, remote operation, and machine vision. In current phase of the project, management of the fuzzy front end of product development, including research activities, concept design, simulations, and virtual prototyping, is challenging. Technical and functional requirements for the DTP2 are high, since water-hydraulics driven remote operated system has to be capable for moving 9 ton divertor cassette with few millimeter accuracy and compensate mechanical and hydraulic flexibilities. Virtual engineering (VR, simulations, etc.) are utilized in several lifecycle stages: concept development, engineering design, remote operation, problem solving, visualization, etc. Virtual simulation models versions and revisions must be managed well. Projects around DTP2 involve a large network of organizations from universities to research institutions and companies. This kind of novel system development needs *system engineering approach*.

PLM is a framework to gain benefits in this kind of system development environment. With PLM it can be secured that right product information (version, status, change data etc.) is available at right time for different stakeholders during development, engineering, manufacturing, in-test and in-use. DTP2 has a very long life-cycle of several decades. From PLM perspective DTP2 is an environment which requires full product lifecycle support, from early development concepts until the in-test, and finally in-use period of the fusion reactor itself. PLM has to support the unique long in-test and in-use timeline of approximately 50 years. Besides the "traditional" PLM aspects (e.g. design configuration and individual configuration) of managing the definitions of product types and individuals, there is an important requirement to manage the information required for planning of execution of the divertor itself (e.g. maintenance programs and maintenance execution). Different product structure configurations and components, as well as simulations and test operations must be traceable during the whole life-cycle of the system. DTP2 will also be utilized in testing, risk analyses, problem solving, and training during ITER commissioning and operation. This means that even if the used IT –solutions of PLM most likely will change, the product data itself should be accessible for several decades. History of developing concepts and specifications of DTP2, and testing the DTP2 individual system must be traceable later during the whole life-cycle of the system, since cumulated information will be utilized in the actual ITER reactor construction, operation, maintenance, and problem solving. This requirement is in high importance level simply because of the fact that when the final fusion reactor operates no human can access directly the platform. There are extremely high safety critical requirements to fulfill. Systematic and accurate requirement management and change management processes are essential. Presently there is lack of IT-support for PLM of DTP2 development. From PLM approach point of view, there are also challenges in implementation because conventional PLM installations do not support system engineering of a mechatronic system in best possible way [10].

Because of increasing involvement of software development in a mechatronic system, the role of Software Configuration Management (SCM) systems have become important. Due to the lack of functionalities in the PLM systems, they cannot support the software development process. The software lifecycle including the development process is usually supported by the Software Configuration Management systems [11].

Virtual engineering, i.e. digital mock-ups and virtual prototyping, have been extensively used in DTP2. The digital mock-ups are used for planning and verifying task procedures of future remote handling test trials. During DTP2 operations, digital mock-ups will still be employed for improving, and validating test procedures. The final goal is to produce a complete digital mock-up of the divertor region and its RH equipment. There exist also challenges in virtual engineering of DTP2. The amount of virtual engineering related data is relatively large and increasing continually. Because DTP2 facility is subject to revisions, updates and upgrades, engineering changes together with virtual models' versions and revision should be managed well. Data and specifications originate from different locations, therefore synchronization and reconciliation of data between parties is essential. These challenges need to be tackled with PLM features in future [3].

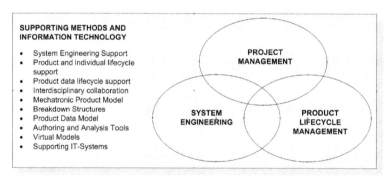

Figure 3. The concept of supporting system engineering approach and project management with PLM

Objective of the paper is to introduce a concept and the first preliminary implementation results of applying system engineering approach to DTP2 system development case supported by PLM and novel virtual engineering tools (Figure 3). The focus of the paper is in needs and requirements and conceptual solutions for the PLM.

METHODS AND MATERIAL

ITER related rules and procedures

As examples, two concept design tasks of the DTP2 are presented in this chapter. These are: 1) concept design of the Cassette Toroidal Mover, and 2) concept design of additional end-effectors for Cassette Multifunctional Mover (CMM). Concept design that is done at DTP2 shall follow ITER rules. Design tool for mechanical design is Catia V5. Modelling work is done according to ITER CAD-manual. CAD-manual includes modelling rules and naming rules for 3D-models. Some attributes for 3D-models need to be added also because ITER will store all CAD information under their PLM system. First phase in the concept design process (Figure 3) is the gathering of the design input data. Design input data includes requirements for the design task. Source for these requirements is ITER System requirement documents (SRD). Typically SRD describes system basic configuration and boundaries, design, safety and quality requirements. Applicable codes and standards are also presented in the SRD.

After the collection phase requirements need to be analysed to evaluate their maturity and completeness. Priority of the requirements is defined and classified. Next phase in the system design approach is system concept definition. In the CTM concept design process this mean that CTM was divided to subsystems and comparison of the alternatives was done. For example the divertor cassette lifting devices electrical and hydraulic actuators options were compared. When the CTM was divided in subsystems, mechanical interfaces between CTM, target plant (divertor cassette) and environment were defined. Motion trajectories for divertor cassette handling were defined and loading conditions were estimated in the design specification phase. After that iterative concept design process was started. After the first draft of the CTM concept was modeled, divertor cassette handling was simulated and checked that no collisions with environment occur. Preliminary Finite Element Analysis (FEA) for the structure was done. After the results next design cycle was done. This was continued until concept design fulfilled requirements. [12]
In the Figure 4, which has been taken from the DTP2 quality plans, this previous described process is presented for the additional end-effectors design development process.

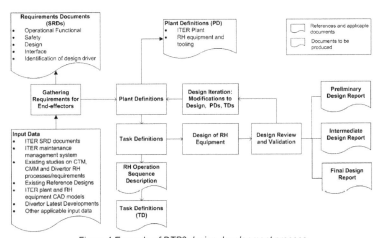

Figure 4. Example of DTP2 design development process

Present DTP2 facility and PLM Architecture

Presently the DTP2 facility consists of Divertor Region Mock-up, Divertor Cassette Mock-up, Cassette Multifunctional Mover (CMM), Water-Hydraulic Manipulator, Control Software, Control Room and DTP2 Test Hall. Current main PLM architecture of DTP2 includes authoring and analysis tools (Dassault CatiaV5, Dassault SolidWorks), software development management system (Subversion SVN), electrical engineering design tool (Eplan), virtual reality software (Dassault Virtools), simulation and analysis software (Dassault Delmia), common office programs, a project management system and a document and quality management system.

Interviews and Workshops

PLM platform requirements specification was created collaboratively with the PLM end users during several interviews and workshops. The interviewees and workshop participants consisted of stakeholders from project management, system engineering, quality and information management departments, and representatives from different engineering disciplines and domains, i.e. mechanical engineers, software developers, electrics engineers, virtual reality and virtual prototyping experts, remote operation personnel, and testing engineers. The interviews and workshops were led by a PLM expert. The requirements specification document was used as basis for PLM supported system engineering platforms definition. Based on the requirements specification, a mechatronic product data model and needed PLM processes were defined and modeled using UML-language. The PLM –platform maps together PLM related requirements from relevant stakeholders and defines a concept how to harmonize product data and processes. It also forms a base for requirements related to a PLM –system architecture.

Product Life Cycle Support Standard

PLCS standard (ISO STEP AP 239) supports the product lifecycle management approach (Figure 5). There are also capabilities in PLCS that support system engineering approach. So called breakdown structures and relation between them support the system engineering method by utilizing functional breakdowns, system breakdowns, physical breakdowns, and actual product structure. [13]

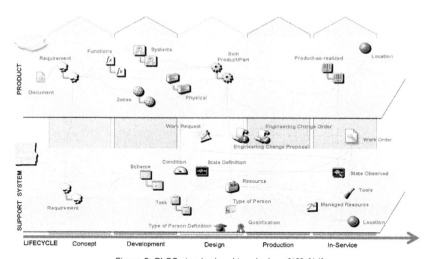

Figure 5. PLCS standard and terminology [13], [14]

RESULTS

System Engineering Platform Requirements

In Table 1 the main requirements for DTP2 system engineering support are listed. They are divided into general and DTP2 special requirements, as well as engineering domains based requirements. All requirements were gathered in DTP2 PLM platform interviews and workshops.

Table 1. Main system engineering support requirements

General PLM requirements	Item management, document management, structure and relationships Engineering change process management, change history, product information history management Product information related to different lifecycle phases is harmonized and centralized Harmonized management of product information Easy to access right and up-to-date product information Increased product information quality
DTP2 special requirements	Intelligent collaboration between different DTP2 engineering disciplines and stakeholders System engineering support Requirements management Virtual Models management: Virtual prototyping, VR, remote operation, product structures Manage the mechatronic product and the individual configuration of DTP2 and its maintenance process.
Mechanical and electrical design requirements	Intelligent management of 3D-CAD models, CAD –document revision and version management Support for E-Plan ECAD Support for ITER mechanical design metadata and formats
Software design and control room requirements	Subversion SVN integration support, relations to SRD and other domains Hardware and software configuration SW design authoring tools interface
System testing requirements	Test planning, system test sequences Relations to requirements and SRD documentation data formats
Project management and administration requirements	Project management and project documentation management support Project task breakdown relation to product structure model, project workflows support Relationships with F4E and ITER codes and naming policy, workflow Automated and smoothen the usage of reporting and deliveries

Mechatronic Product and Data Model

Product data model defines concepts needed to describe the DTP2 system. An item and item structure is used to represent one configuration of DTP2 or a part of a DTP2 configuration. In the PLM –system there exist a number of DTP2 design configurations needed for different purposes, e.g. conceptual design, new design or simulation design. Those designs have always a relationship to product individuals. The product individual is used to represent physical on-floor configuration of the DTP2. Both item and product individual configurations include up-to-date configuration as well as configuration history. An item i.e. design configuration can fulfill one or more required functionalities and those are defined with the functional breakdown –object. These object together with the document- and requirement –objects are mainly used to represent the configuration of DTP2 and relevant design information. Some of the object and definitions of required product data model are based on the ISO Product Lifecycle Support (PLCS) – standard (Figure 5).

Figure 6 presents one example of DTP2 Mechatronical Product Model. The Mechatronical Product Model includes the structure of DTP2 divided to Platform, Divertor Casette, WHMAN Manipulator, Remote Operation and CMM Robot. All these main modules of the DTP2 configuration are classified as Systems and are presented with an item and item structure. Both Main Systems and Systems are used to manage the whole Mechatronical Product Model of DTP2, so in other words they cross-over the boundaries of different engineering disciplines.

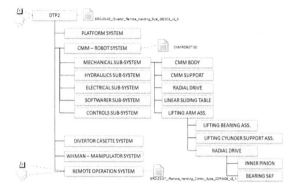

Figure 6. Conceptual example of the main DTP2 mechatronic product model

Supported System Engineering Processes

In order to meet the DTP2 system engineering requirements, PLM processes of Table 2 have to be supported:

Table 2. Supported System Engineering Processes

CORE PROCESSES	BENEFITS
Core Product Lifecycle Management	
• Mechatronics Product Process: Process to design and maintain the configuration of DTP2 Main System in collaboration with engineering disciplines in question. • Product Individual Process: Process to maintain the configuration of DTP2 Main System Product Individual in collaboration with engineering disciplines in question	• Increase the visibility and understanding of DTP2 Main System configurations between different stakeholders • Manage product data in a cost efficient way, • Boost the usage of existing design competencies, • Reduce costs by providing easy access to up-to-date product information
Smooth and Effective PLM Process Support	
• Design Process: Implement a more accurate, effective and automated Design Process • Engineering Change process: Implement a more accurate, effective and automated Engineering Change Process • Product Individual Maintenance Process: Manage the information and process to store information about maintenance and simulation tasks carried out.	• Better quality of product data • Product In-use maintenance history • Optimized collaboration processes with different stakeholders • Shorten development time • Cost efficient engineering change management • More accurate, effective and automated Engineering Design Process
Project Management and Delivery Support	
• Project Management process: Implement processes and tools in PLM –solution to manage project. • Publication and Reporting Process • Delivery Process	• Automate the project management process • Shorten project delivery and reporting times
System Engineering Process Support	
• System Engineering process: Implement processes and tools in PLM –solution to the system engineering process and related information.	• Automated and accurate system engineering process • Shorten project delivery and reporting times • Reduced cost by providing easy acess to up-to-date whole product lifecycle information

The system engineering processes will be supported by following core PLM system functions: Item and document management, requirement management, engineering change management, project and delivery management, project workflow management, product individual management, product in-use management. Some of the defined processes are based on the PLCS standard (Figure 5)

DISCUSSION AND CONCLUSIONS

DTP2 is a full scale mock-up and test facility for developing, testing and demonstrating remote operated maintenance equipment as well as planning and training future maintenance operations of ITER. It is a complex mechatronic system consisting of many engineering disciplines and domains, and special new technologies. Characteristic for DTP2 is that its development and operational lifecycle will be several decades long, and history of the system has to be traceable and all data must be available during the whole lifecycle. System engineering approach is applied in DTP2 development. It should be supported by systematic process and data management system, i.e. PLM.

System engineering is strong in multidisciplinary engineering and systematic chain from end-user requirements to functional requirements and finally technical solutions and validations. Design methodology supports System engineering by delivering more detailed technical solution alternatives. PLM should support these activities for instance with good requirements and engineering change management capabilities. On the other hand, PLM is strong both in product type and product individual data management. Those capabilities could be better utilized in system engineering. Obviously there is overlapping between System engineering, design methodology and product lifecycle management.

Bergsjö [9] and Abramovici [10] and have stated that currently PLM systems do not support multi-disciplinary development of a mechatronic system very well. On the other hand, Bergsjö has concluded that it is feasible to create an integrated information model for mechatronic development and implement it in a PDM/PLM system, although such PLM implementations for inter-disciplinary engineering are not yet reality in companies. The companies have though strategies moving in that direction. Anyway, Abramovici concluded that progress of future PLM models and implementations development will include better support for the whole product type and individual lifecycle, management of engineering support processes, multi-disciplinary engineering integration, and visualization of the product and process.

Esque [3] and Muhammad [11] have introduced special requirements for DTP2 development from software development and virtual engineering viewpoint. They have also proposed concepts and architectures for SCM-PLM integrations and digital mock-up model management in their research papers. Complete integrations of PLM and SCM are currently unavailable, but a suitable solution could be a loose integration between the two systems by using the APIs already built in these systems. They are good basis for further development of DTP2 system engineering support.

Purpose of this paper is to introduce the first results of the ongoing project which aims to define and implement PLM support for DTP2 system engineering. The preliminary results include requirements specification for the PLM platforms, and a concept for mechatronic product model and data model. The requirement specifications and the PLM platform were defined co-operatively with the end users (i.e. system engineers, project management and different engineering domains) and PLM experts. So far, it can be concluded that this liaison between engineering domains and a common understanding of system engineering requirements and PLM approach is essential for success. In the upcoming work, the PLM specifications and mechatronic data model will be further defined and implemented in IT-systems. Alternative PLM architectures will be studied and evaluated. DTP2 is good setup to study and evaluate the feasibility of PLM support for system engineering.

REFERENCES

[1] *ITER web-site,* http://www.iter.org/
[2] Palmer, J., Irving, M., Järvenpää, J., Mäkinen, H., Saarinen, H., Siuko, M., Timperi, A., Verho, S. *The design and development of divertor remote handling equipment for ITER.* Fusion Engineering and Design 82, 2007, 1977-1982
[3] Esque, S., Mattila, J., Siuko, M., Vilenius, M., Järvenpää, J., Semeraro, L., Irving, M., Damiani, C. *The use of digital mock-ups on the development of the Divertor Test Platform 2.* Fusion Engineering and Design 84, 2009. 752–756
[4] Kossiakoff, A., Sweet, W. N. *Systems Engineering Principles and Practice.* 2003. John Wiley and Sons, Inc
[5] Stark., J. *Product Lifecycle Management: 21st Century Paradigm for Product Realisation.* 2004. Springer
[6] Grieves, M. *Product Lifecycle Management – Driving the next generation of lean thinking.* 2006. The McCraw-Hill Companies. New York
[7] Andreasen, M.M. Hein, L. *Integrated Product Development.* 1987. IFS (Publications) Ltd./Springer Verlag, London
[8] Ulrich, K. & Eppinger, S. Product Design and Development, 3rd ed. 2004. Boston: McGraw Hill Irwin.
[9] Bergsjö, D. *Product Lifecycle Management – Architectural and Organisational Perspectives.* Thesis for the degree of doctor of philosophy. Chalmers university of technology, Department of Product and Production Development, Division of Product Development. Göteborg, Sweden, 2009
[10] Abramovici, M. *Future Trends in Product Lifecycle Management (PLM).* The Future of Product Development. 2007, Part 12, 665-674
[11] Muhammad, A., Esque, S., Aha, L., Mattila, J., Siuko, M., Vilenius, M., Järvenpää, J., Irving, M., Damiani, C., Semeraro, L. *Combined application of Product Lifecycle and Software Configuration Management systems for ITER remote handling.* Fusion Engineering and Design 84, 2009. 1367–1371
[12] Mäkinen, H., Järvenpää, J., Valkama, P., Väyrynen, J., Amjad, F., Siuko, M., Mattila, J., Semeraro, L., Esque, S. *Concept Design of the Cassette Toroidal Mover.* SOFT 26th Symposium on Fusion technology, Porto, Portugal, Sep 27 - Oct 1, 2010
[13] *ISO 10303 AP 239 (PLCS) standard,* http://www.oasis-open.org
[14] *Eurostep,* http://www.eurostep.com/

Contact: Simo-Pekka Leino
VTT Technical Research Cetre of Finland
Systems Engineering
Tekniikankatu 1
33101, Tampere
Finland
Tel: +358 40 7377184
Fax: +358 20 722 3365
Email: Simo-Pekka.Leino@vtt.fi
URL: http://www.vtt.fi/?lang=en

Simo-Pekka is research scientist in Systems Engineering knowledge centre of VTT Technical Research Centre of Finland. His current research topic is fusion of engineering design with system engineering and PLM approach, in particular integration of virtual engineering tools, methods and data management into product processes.

INTERNATIONAL CONFERENCE ON ENGINEERING DESIGN, ICED11
15 - 18 AUGUST 2011, TECHNICAL UNIVERSITY OF DENMARK

ANALYZING THE DYNAMIC BEHAVIOR OF MECHATRONIC SYSTEMS WITHIN THE CONCEPTUAL DESIGN

Frank Bauer[1], Harald Anacker[1], Tobias Gaukstern[1], Jürgen Gausemeier[1], Viktor Just[2]
[1] Product Engineering, Heinz Nixdorf Institute, University of Paderborn
[2] Control Engineering and Mechatronics, Heinz Nixdorf Institute, University of Paderborn

ABSTRACT
The increasing penetration of mechanical engineering by information technology enables considerable benefits. This is expressed by the term mechatronics, which means the close interaction of mechanics, electric/electronics, control and software engineering to improve the behavior of a technical system. The development of such systems is a complex and interdisciplinary task. Consequently a domain-spanning specification is required, which describes the system in total and builds the basis for all further communication and cooperation between the experts from the involved domains in the concretization. In order to validate the system, different tests are accomplished during the concretization phase. In this contribution we present how dynamics analysis may be integrated already during the conceptual design phase. For this purpose a simulation tool is used for the validation of the dynamic behavior of the system already on the basis of the principle solution. The refinements effected during the simulation are transferred back into the principle solution. This improves the provided information for the following domain-specific concretization.

Keywords: mechatronic, conceptual design, principle solution, solution pattern, dynamic analysis

1 INTRODUTION: DOMAIN-SPANNING CONCEPTUAL DESIGN AND DYNAMICS ANALYSIS OF MECHATRONIC SYSTEMS

The products of mechanical engineering and related industrial areas are often based on the synergetic interaction of mechanics, electrics/electronics control and software engineering. The term mechatronics expresses this interaction. The aim of mechatronics is to optimize the system's behavior by using sensors, actuators and information processing. Sensors obtain information about the systems environment and the system itself. The processing of this data and the controlling of the actuators enable the system to adapt to the current situation. The design of such systems is a challenge. Existing methodologies for the development of mechatronic systems reached a high level; but they focus on specific tasks of the involved domains. There is too less consideration to the domain-spanning interactions. There is still missing a domain-spanning design method and supporting software tools integrating mechanical-, electrical-, control- and software engineering.

In accordance with existing methodologies, mainly the VDI guideline 2206 [1] the development process of mechatronic systems can be divided into two main phases: the domain-spanning conceptual design and the domain-specific concretization. Within the conceptual design, the basic structure and the operation mode of the system are defined. All results of the conceptual design are specified in the principle solution. The description of the principle solution has to include all necessary information for the following concretization, which takes place for the involved domains. These processes are carried out in parallel and aim a complete description of the system. The domain-specific models are therefore integrated into the final solution. Throughout the conceptual design the involved domains have to cooperate. For the specification of mechatronic systems, a lot of different modeling and specification languages are available, e.g. SysML [2]. These specification and modeling languages do not fully meet the requirements for a domain-spanning description of the system. For this reason, a new specification technique for the description of the principle solution has been developed [3]. The following aspects need to be taken into account: environment, application scenarios, requirements, functions, active structure, shape, behavior and system of objectives (Figure 1). The latter is only required for self-optimizing systems. The behavior is considered as group of partial models because there are various types of behavior, e.g. dynamic behavior of a multibody system, electromagnetic compatibility or heat transfer.

The mentioned aspects are mapped on computer by partial models. A software tool which can used to describe mechatronic systems using the specification technique is the Mechatronic Modeller. The Mechatronic Modeller offers a separate editor for each partial model and is based upon a metamodel. The principle solution is computer-internally represented as a data model, which is the instance of this metamodel [4]. At this design stage first analysis are possible. This includes computer-aided analysis of robustness, reliability, product-structuring or manufacturing costs. Further tests and analysis were conducted within the concretization phase, where the involved domains use specific methods and tools. As a consequence the validation of the specified principle solution is deferred to later point of the design process. In order to verify the information provided by the principle solution, a validation should be done as early as possible during the design process.

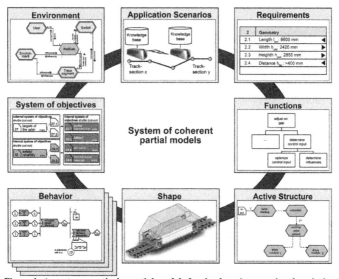

Figure 1: Aspects respectively partial models for the domain-spanning description of the principle solution of mechatronic systems [3]

A simulation-based design process including a dynamics simulation and analysis of a mechatronic system is completed in two abstraction steps. We illustrate these using the example of an active suspension module of a fully x-by-wire vehicle. First, a simple idealized model of the active suspension is built up. This model is used for dimensioning the passive suspension components and for the control design. The idealized model of the vertical dynamics consists of the two masses vehicle and wheel (Figure 2, left). A spring-damper element is used for the active suspension and the connection between wheel and ground. Each mass has only one vertical degree of freedom each. Due to a reduction of the vertical dynamics to only one wheel (quarter-vehicle), the axle kinematics being neglected, and an ideal actuator being used for an active adjustment of the spring pre-load c_A between the vehicle body and the wheel. The scope of the model is limited.

The dynamics control for the lifting motion is designed on the basis of this idealized quarter-vehicle model, with the controller design being based on the skyhook strategy [5]. This inserts a virtual spring and a virtual damper between a virtual coupling point in the "sky" and the vehicle body; along with an additional active spring-damper force between the body and the wheel. Figure 2 displays the simulation results of the passive vehicle behavior and of the closed-loop-controlled with an ideal suspension actuator. The red graph marks the desired behavior of the simulation.

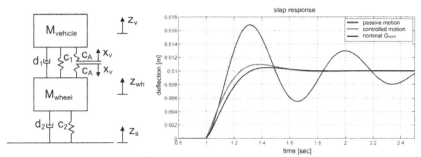

Figure 2: First analysis model and simulation results

In a second step a more detailed multi-body system model is built that takes into account the axle kinematics and the actuator system. This information is used for designing and fine-tuning the control strategy. Based on the requirements of the lifting range of the wheel suspension and on the stiffness of the passive vehicle-body spring, an idealized axle kinematics is designed. The necessary active force F_A is induced via a torsional moment M_A. To take into account the axle kinematics, a multi-body system model is built up whose torque M_A is commuted by means of an electric motor. In order to apply the active torque M_A on the suspension, the motor is coupled to the traverse link via a torsion spring. Thus the active torque can be preset via the motor-angle control which is realized by a cascade control and laid out on the motor model. Figure 3 shows the resulting simulation model which can be used in order to create first control strategies. Also the multi-body system defines the necessary kinematics for the suspension module [6].

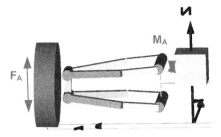

Figure 3: Multi-body system of the active suspension module

2 DYNAMICS ANALYSIS WITHIN THE CONCEPTUAL DESIGN PROCESS

In the following we present an approach which includes a dynamic analysis into the conceptual design process. The aim is to derive all required information for the dynamic simulation from the aspects of the principle solution. Furthermore the results of the analysis are transferred back and refine different aspects. An early validation of the systems dynamics supports the following domain-specific concretization. The domains control- and software-engineering benefit from a first dynamic model of the system, in order to start with domain-specific refinement. Also the dynamic affects the required kinematic and shape of the system which is elaborated within the mechanical-engineering. The electrical-engineering is targeted by the two other domains.

2.1 Design Process

The design of the principle solution (Figure 4) covers all partial models of the presented specification technique except from the system of objectives. In the following we consider advanced mechatronic systems which are not self-optimizing. The design process is divided into three steps; specify aim, synthesis and analysis. The first step starts with the clarification and definition of the task. This includes the aspects environment, application scenarios and requirements. Based on these aspects the systems functionality is defined in terms of a function hierarchy. The synthesis begins with the choice of possible solution pattern to fulfill the required functionality. We will describe the structure of solution pattern later in more detail. Each solution pattern can affect the structure as well as the behavior of the system. The structure of the system is modeled with the active structure. The initial behavior is modeled with states and activities. The partial model behavior – states defines the states of the system and the state transitions. The state transitions describe the reactive behavior of the system towards incoming events. The partial model behavior – activities describes the logical sequence of activities in the system. Especially, parallel executed activities and their synchronization can be described this way. The behavior marks the target for the later dynamics simulation. An initial geometry model is required for a dynamics analysis. Modeling the shape is the last step within the design of the principle solution as presented in Figure 1.

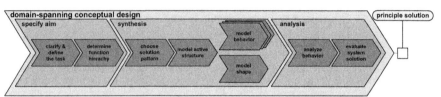

Figure 4: Design Process including an early validation

Due to the early design phase, the domain-spanning concept does not describe all aspects of a system in detail. Especially the number of precise physical parameters is limited. A simulation of the system at this development stage has to regard the provided level of information. Therefore a first analysis has to be done with models that consider this restriction. We refer to these types of models as idealized models, which only contain the necessary level of detail for a first validation. The analysis starts with the derivation of an idealized simulation model based on the active structure and the shape model. The system is simulated and the results will be analyzed with respect to other partial models. The simulation results are compared to the initially defined states and activities. It is checked if the dynamic model shows the predefined behavior. The parameters set within the simulation are compared to the requirements. In case that the simulation does not match the requirements the simulation model is adjusted, simulated and analyzed again. The adaptations within the analysis process are transferred back to the partial models of the principle solution. The active structure is enhanced when a new element is included into the simulation in order to achieve the desired behavior. This also causes an adjustment of the function hierarchy. Parameters set within the simulation model are added to the requirements list which is improved step by step. The behavior activities and states are refined by the simulation results. The geometric information assigned with the simulation is considered in a modified shape model. This process is repeated until the dynamic behavior fulfills the required dynamics necessary for the development task. This indicates the end of the conceptual design phase. In the following subsections we illustrate the design process for the already introduced active suspension module.

2.2 Design of the principle solution

Active suspension modules are used in vehicles to improve the driving characteristics in general. This includes for example a higher driving comfort for the passengers or a higher driving safety. The environment and an application scenario are described by a driving vehicle; where the ground caused disturbance on the wheel. Several requirements can be deflected for active suspension to achieve the outlined defined behavior. The overall function of the module is to keep the chassis in a rest position. A sub-function is the absorption of disturbance. A first simplified active structure, requirements and functions are presented in Figure 5. The whole module consists of a chassis, an active suspension and a wheel. Relations between system elements are represented by flows. Three flows are feasible: material, energy and information flow. In the example only energy relations between the system elements are modeled. Related to the active structure the behavior of the system elements can be characterized. Only two states are considered at this design step. The chassis can be in the rest position or in vibrations, which are caused by the disturbance. In the later no active suspension is applied. The activities are therefore, to transmit the disturbance to the active suspension, absorb them and transmit a lower disturbance to the chassis.

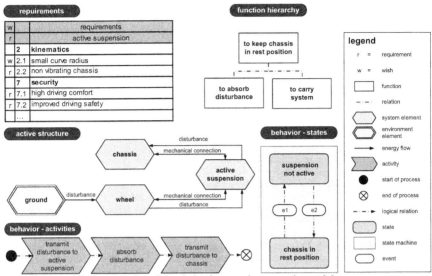

Figure 5: Principle solution for an active suspension module

A fundamental aspect of the design of mechatronic systems is the reuse of once successfully proven solutions in form of solution patterns. A pattern describes a context-specific problem and the core of the solution [9]. Furthermore a pattern is an established instrument to externalize and store the knowledge of experts. Generally the involved experts use their own terminologies during the development process. In order to deal with this challenge we use an adapted uniform specification of solution patterns, which was developed by DUMITRESCU ET. AL [10]. In the following we describe the specification of a solution pattern, which is structured in seven aspects (Figure 6).

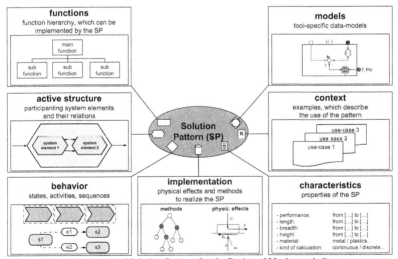

Figure 6: Aspects of Solution Pattern for the Design of Mechatronic Systems

In order to come up with the definition by ALEXANDER, the developed specification of solution patterns for mechatronic systems describes the problem and all necessary aspects of the solution and its implementation. Within the described design process the solution patterns are used for two tasks: Firstly for modeling the principle solution. Secondly for deriving the idealized simulation model. The aspect functions express the problem description, which may be solved by the solution patterns. The description of the solution is separated into the aspects active structure (necessary system elements and their interrelation) and behavior (states, activities and sequences). In order to integrate the solutions into the design process, it is essential to clarify physical effects and methods for information processing. The aspect characteristics describes the significant different properties of the solution patterns. Once successfully applied solutions are generally attached to a specific context, so this aspect contains different use-cases. In addition to the specification of pattern by DUMITRESCU ET AL. the aspect models is implemented. A model is a tool-specific mapping of the pattern for a specific problem or development task; in the presented design process it is a Modelica model. The simulation model of the whole system is build up from the single idealized models of the elements. Predefined models are selected to create a multi-body system which can be simulated.

2.3 Adjustment of the principle solution models

The result of the simulation is a new model, which describes the dynamic behavior of the system. This model is added to the behavior group of the principle solution. Moreover the other partial models have to be extended. The active structure shown in Figure 5 just includes a system element "active suspension". During the analysis this system element has been further refined; new system elements have to be added to the active structure. The multi-body system determines the necessary kinematics for the active suspension. The mechanical connections are mapped as energy flows between the system elements (Figure 7). The chosen combination of a electric motor, the torsion spring and the traverse link affects the behavior of the system. Further the activities are more detailed, compared to the initial model. The outlined sequence is valid for the different control strategies which can be applied. The state model consist now of three states. When the module is not in progress no active suspension is available. The two other states characterize different control strategies executed by the control unit. Although this example is quite simple, it shows the benefits of an early validation already within the conceptual design phase. The enhanced partial models build a better foundation for the following concretization phase. Especially for the control and software engineering a detailed description of the activities, states and events is useful. The three-dimensional simulation model can be used as a first shape model, which will be concretized within the domain mechanical engineering. Also specific solution elements will be selected during the concretization. The analysis of the required torques and motor dynamics provides information that is essential for selecting the appropriate actuators. These specific parameters will be added to the requirements list.

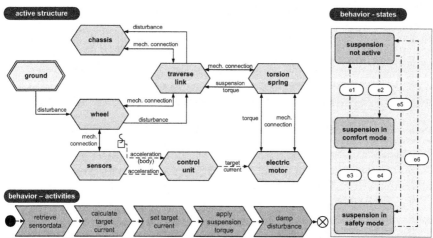

Figure 7: Adjusted active structure and behavior models

3 TOOL SUPPORT

The described design process is supported with two major software tools: The Mechatronic Modeller for modeling the required partial models of the principle solution and Dymola for the dynamics analysis. Dymola is an editor for the object-oriented physical description-language Modelica [7]. These tools are used during the conceptual design. Further a knowledge-based system is integrated to support the developer during the design process. The knowledge-based system contains a database of solution patterns and an ontology with domain specific and procedural knowledge. Originally "Ontology" is a philosophy discipline, used to structure entities and their relationships in certain area. This concept has been adapted to computer science in which an ontology refers to an intelligent data model. Today ontologies are a basic technology for semantic applications used to formalize and exchange knowledge. The Web Ontology Language (OWL) is the most used modeling language for ontologies. The basic elements from OWL are classes for collection of objects, individuals for specific objects and properties to model the different types of relations [8]. We use an ontology to model the relation between aspects of the specification technique, the solution patterns and the Dymola simulation model.

3.1 Modeling the principle solution

The design of the principle solution is supported by the Mechatronic Modeller. First the requirements list and the function hierarchy of the system is specified. The knowledge base assists the engineer during this conceptual task. The data base contains functions for the description of a mechatronic system. Within the ontology the dependencies between functions are modeled. For example the function "accelerate material" requires the function "convert energy" to produce the necessary acceleration energy. The stored information allows a computer-aided modeling of the function hierarchy. The mentioned dependencies support the engineer in adjusting or completing the function hierarchy. This formalization is required for the search and selection of possible solutions.

Similar to the functions, the solution patterns are stored in the database and structured by the ontology. The search is realized by a relation between functions and patterns. Based on the function hierarchy of the designed system the possible solution patterns are selected. For this task the requirements list is considered. A requirements list is normally just a text-based document written by the developer. With respect to a computer-understandable interpretation the requirements needed to be formalized. A first step is a template for the Mechatronic Modeller which includes basic requirements with necessary parameters valid for the mechatronic system. For example the maximum electric power supply. These parameters are compared with the characteristics of the patterns. Thus, patterns are selected which fulfill the required functionality and requirements. The active structure and behavior of the selected patterns are combined in the active structure and the initial behavior of the principle solution.

3.2 Analyzing the principle solution

Based on the partial models, that specify the principle solution, a semi-automated generation of a Dymola simulation model is achieved. This means that some parts of the simulation model can be generated automatically. Nevertheless the final assembly is up to the developer. Within Dymola a system is represented by components. A component could be a single element, like a sensor or a complex system which includes other components. Each solution pattern is attached to a Dymola component; an idealized model. Therefore each element within the active structure can be directly transferred into a Dymola simulation. The connections between components can be derived from the connections inside the active structure. Dymola offers different ports in order to model the connections between components. An energy port can be distinct into a mechanical or electrical port, which can again split up into one-phase and three-phase. Also the ports are subdivided by incoming and outgoing ones. The material, energy and information flows within the active structure also consider the direction. For a dynamics simulation information about the geometry has to be added. This information is provided by the solution patterns and a rough shape model of the specified principle. The developer uses this for the final assembly of the Dymola model and run the analysis.

4 CONCLUDING REMARKS

The presented approach combines the domain-spanning conceptual design of mechatronic systems and an early validation of the desired behavior. The result is a validated principal solution of the system which marks the starting point for the following concretization phase. The validation is realized by a dynamics simulation. During the design process the developer uses the Mechatronic Modeller for the domain-spanning description of the mechatronic system. Partial models specify functionality, structure, behavior and shape of the system. The behavior is modeled with activities and states. Based on this information an idealized simulation model is derived. The analysis phase is done within the tool Dymola. The simulation results are compared with the required system behavior and the simulation model is adjusted in order to achieve it. The adjustments are transferred back to the partial models of the principle solution to keep consistency.

Our future work covers the formalization of the aspects of the principle solution. For example, the required geometric aspects for a physical simulation have to be considered. In addition, we will also focus on software engineering, achieving a transformation from the evaluated principle solution to domain-specific models of the domain software engineering. Finally the mentioned tools need further concretization. The algorithms for the selection and combination of solution patterns have to be concretized.

ACKNOWLEDGEMENT

This work was developed in the project "ENTIME: Entwurfstechnik Intelligente Mechatronik" (Design Methods for Intelligent Mechatronic Systems). The project ENTIME is funded by the state of North Rhine-Westphalia (NRW), Germany and the EUROPEAN UNION, European Regional Development Fund, "Investing in your future".

REFERENCES

[1] Verein Deutscher Ingenieure (VDI): *VDI-Guideline 2206 – Design Methodology for Mechatronic Systems*, Beuth Verlag, Berlin, 2004
[2] Weilkiens, T.: *Systems Engineering mit SysML/UML*. dpunkt.verlag, Heidelberg, 2. Auflage, 2008
[3] Gausemeier, J.; Frank, U.; Donoth, J.; Kahl, S.: *Specification Technique for the Description of Self-Optimizing Mechatronic Systems*. Research in Engineering Design, Vol. 20, No. 4, Springer-Verlag, London, 2009
[4] Gausemeier, J.; Dorociak, R.; Pook, S.; Nyßen, A.; Terfloth, A.: *Computer-aided cross-domain modeling of mechatronic systems*. In: Proceedings of the 11th International Design Conference (DESIGN 2010), Dubrovnik, 2010
[5] Gevatter, H.-J.; Grünhaupt, U.: *Handbuch der Mess- und Automatisierungstechnik im Automobil*. Springer- Verlag, Berlin, 2005
[6] Nachtigal, V.; Jäker, K.-P.; Trächtler, A.: *Development and Control of a Quarter-Vehicle Testbed for a Fully Active X-by-Wire Demonstrator*. In: Proceedings of 9th International Symposium on Advanced Vehicle Control (AVEC 2008), Kobe, Japan, 2008
[7] Tiller, M.: *Introduction to Physical Modeling with Modelica*. Kluwer Academic Publishers, Nowell, 2001
[8] Hitzler, P.; Krötzsch, M.; Rudolph, S.; Sure, Y.: *Semantic Web – Grundlagen*. Springer- Verlag, Berlin, 1. Auflage, 2007
[9] Alexander, C.; Ishikawa, S.; Solverstein, M.; Jacobson, M.; Fikshdahl-King, I.; Angel, A.: *A Pattern Language*. Oxford University Press, New York, 1977
[10] Dumitrescu, R.; Anacker, H.; Gausemeier, J.: *Specification of Solution Patterns for the Conceptual Design of Advanced Mechatronic Systems*. In: Proceedings of International Conference on Advances in Mechanical Engineering (ICAME 2010), Selangor, Malaysia, 2010

Contact: Prof. Dr.-Ing. Jürgen Gausemeier
Heinz Nixdorf Institute, University of Paderborn
Product Engineering
Fuerstenallee 11
D-33102 Paderborn
Germany
Phone: (0049) 5251 606267
Fax: (0049) 5251 606268
E-mail: Juergen.Gausemeier@hni.uni-paderborn.de
www.hni.uni-paderborn.de/en/pe/

INTERNATIONAL CONFERENCE ON ENGINEERING DESIGN, ICED11
15 - 18 AUGUST 2011, TECHNICAL UNIVERSITY OF DENMARK

LINEAR FLOW-SPLIT LINEAR GUIDES: INFLATING CHAMBERS TO GENERATE BREAKING FORCE

Nils Lommatzsch, Sebastian Gramlich, Herbert Birkhofer, Andrea Bohn
Technical University of Darmstadt, Germany

ABSTRACT
The linear flow-splitting technology developed within the Collaborative Research Center (CRC 666) "Integral Sheet Metal Design with Higher Order Bifurcations" offers new options to manufacture innovative products. Especially using the technology to continuously produce linear guides is focused in this research. With linear flow-splitting and linear bend-splitting, chambered steel profiles provide possibilities to integrate functions into linear guides.
In this contribution, an approach to develop functions for linear flow-split linear guides is presented. Basing on calculation models and property networks, optimized solutions can be created while design modifications can be derived from the property networks. These property networks are very well suited to present an easy overview over the so called "set screws" with which the fulfillment of the requirements can be influenced. The approach also includes the validation of the calculation models and the functionality with finite element models and experiments. The approach is explained on the example of the function "clamping".

Keywords: linear motion guides, product development, property network

1. INTRODUCTION
Within the CRC 666, new massive forming processes for sheet metal are developed and researched. These new technologies enable the production of novel chambered steels profiles that show special properties. These properties predestine the components to be used in linear guides that can be produced in integral style.
In this contribution, the development of a clamping function for linear flow-split components is shown. A systematic approach is chosen to achieve optimized solutions for the functions and modifications. Therefore, the property networks developed in the CRC are generated to describe the effect used to fulfill the clamping function. Property networks show independent and dependent properties with their corresponding relations in a manageable manner. In the property network, the function's "set screws" become visible which allows the designer to derive design modifications to simplify or improve the functionality.
With the chosen approach in this study, an optimized solution for the desired function is found while the approach is usable in the process chain of the CRC.

2. BASICS OF THE NEW PRODUCTION TECHNOLOGIES
In the following section, the newly developed core technologies researched in the CRC concerned with "Integral Sheet Metal Design with Higher Order Bifurcations" are presented. In specific, linear flow-splitting and linear bend-splitting are explained. Both of these production technologies are massive forming processes for steel at ambient temperature that can be used for continuous production with potentially up to 100 meters per minute [1]. A schematic of both processes, linear flow-splitting to the left and linear bend-splitting to the right, is displayed in Figure 1.

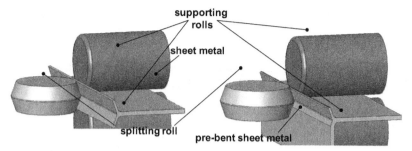

Figure 1: Schematic of the linear flow-split and linear bend-split processes

In the beginning of the process, the sheet metal is rolled off of a coil. During the linear flow-splitting process, the sheet metal band is fixated and guided by two supporting rolls above and below the sheet metal while the splitting roll splits up the coil's edge as seen in the left of Figure 1. This process occurs simultaneously on both sides of the band's edges. Thereby, the characteristic Y-shape of the two created flanges develops [2]. The depth of the splitting and thereby the length of the flanges is incrementally increased over several linear flow-splitting stands. Today, flanges with a total length of 20mm can be created without damaging the material [3].

Linear bend-splitting is relatively similar compared to linear flow-splitting. However, two major differences between the processes exist. First, linear bend-splitting requires pre-bent sheet metal as input. The bending edge is the contact point for the splitting roll as displayed on the right of Figure 2. Secondly, in contrast to linear flow-splitting, the process is only conducted on the one bending edge of the material while the other side (not shown in Figure 2) needs to fixated towards the force. A further difference is that linear bend-splitting can be used to create flanges anywhere in the material as long as the material can be bent in that place. Linear flow-splitting can only create flanges on the material's edges.

If combined with roll-forming, the new technologies enable the production of bifurcated steel structures as shown in Figure 2.

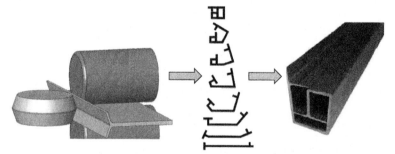

Figure 2: Bifurcated structures with the linear flow-splitting, flower diagram according to [4]

The geometry of the three-chambered profile as shown in Figure 2 is achieved by using linear flow-split material to roll-form the final geometry. Thereby, the whole product is produced in integral style. In the above scenario, the linear flow-split component's characteristic Y-shaped flanges have been modified to a 180° and 90° opening angle as shown in the middle of Figure 2. Additionally, adjacent processes as laser-welding and milling can be used prior to, between or after the linear flow-splitting process. In the three-chambered profile above, the chambers could be welded shut or holes, for e.g. connections, could be induced into the material. The inclusion of these processes into the process chain to be used integrally is also researched within the CRC.

The flanges of linear flow-split and linear bend-split components have similar technologically-induced properties resulting from the massive forming processes. First of all, the microstructure at the flanges has an ultra fine grained continuum [5]. This results in an increased hardness and lower surface roughness in comparison to the base material [6]. Thereby, the hardness is falling gradually towards the backside of the flanges. As preliminary tests revealed, the flanges additionally have an increased rolling contact and sliding contact fatigue-life [7]. Evidently, these properties predestine the linear flow-split flanges to be used as rolling contact surfaces.

These technologically induced properties make the new components very eligible to be used in linear guides. Moreover, there are further advantages resulting from the technology. For once, the created bifurcated structures have a high area moment of inertia and therefore a high stiffness while being rather light compared to full material which common rail guides are made of. This provides stability while also offering potential for light weight design. The continuous production of linear flow-split components and their eligibility to be used as rolling surfaces, straight from the integral production without further effort, provide a potential cost advantage over traditionally produced linear rail guides. Another advantage is the chambered structures of the linear flow-split profiles, since they can be used as "vessels" for additional functions. This could give linear flow-split linear guides a functional benefit which, combined with the production in integral style, could present a unique selling point.

Besides the obvious technological benefits of linear flow-split components for linear guides, the wide areas of application including various shapes and different degrees of technology for linear guides as well as their wide price span offers a large market and therefore also a large variety for possible problem solutions that the new linear guides could target. The ultimate goal for innovative linear guides from linear flow-split components should be to provide high quality functional guides for comparably low costs. For the conducted research, the main orientation has been ball rail guides.

In the following section, the idea for function integration into the chambers of linear flow-split profiles used in linear guides is further described. Due to the functional benefit of an innovative, integrated function, an attractive product can be created. Mainly produced in integral style, a "clamping" function for the newly developed linear guides is created. Possible fields of application for such a function could be an emergency stop or positioning tasks.

3. INTEGRATION OF THE CLAMPING FUNCTION

The integration of the clamping function is the goal of this function integration. The application of a breaking force can be solved in different ways that could be produced in integral style with the new technology. The chosen solution for the analysis is to create the breaking force by clamping a part of the sled between to inflated chamber walls of the rail, thereby creating friction that decelerates and eventually stops the sled. The inflation of the chamber wall is performed with pneumatic pressure. Therefore, the chamber of the linear flow-split rail needs to be sealed by welding.

3.1 Classification of properties

The multitude of design options for the embodiment of the function requires a systematic approach for design to find an optimized solution. The approach is based on the modified property classification developed in the CRC. The basics of the property classification are presented in Figure 3.

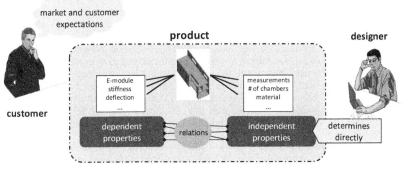

Figure 3: property systematic in the CRC 666

The above shown property classification assumes that the customer has certain expectations towards a product that he perceives as a whole. Opposite to the customer is the designer who can determine the product with a certain sets of properties while other properties can only be influenced over the connections with the influenceable ones. In the following, the properties that the designer can influence are called independent properties. Independent properties are e.g. the number of chambers or the material. Further, the properties that can only be influenced over independent properties are called dependent properties. Dependent properties are for instance the E-module or the deflection. [8]

If the dependent and independent properties as well as their relations are known, they can be displayed in a property network. From the property network, a designer can receive an overview over the influenceable and non-influenceable properties. The independent properties represent the so called "set screws" that the designer can turn to fulfill the customer's requirements according to the constraints regarding the functionality.

3.2 Approach to design functions for linear flow-split components

The general course of action of the here presented approach to design functions for linear flow-split components is shown in Figure 4.

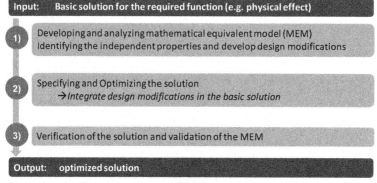

Figure 4: Approach to generate an optimized function for linear flow-split components

The required input for the approach is a feasible solution for the function as e.g. the physical effect.
Based on the input, the first step of the approach is to formulate a mathematical equivalent model (MEM) that displays the relevant dependencies between properties, sometimes partially simplified. Normally, the MEM is based on physical models that are often used in engineering. Furthermore, the independent and dependent properties and their corresponding relations are compiled and the according property network is spanned. With the knowledge of the independent properties for the fulfillment of the function, the feasible solution that was used as input is varied and modified so that a simplified or improved functionality is generated. The goals from this first step are the preparation of an MEM for the optimization as well as the development of design modifications with regard to the function fulfillment.

In the second step of the approach, the solution is specified and optimized. With the help of the MEM, a mathematically possible solution for the function is determined and then optimized with regard to the constraints. Modifications generated in the first step should be implemented into the MEM if otherwise the validity would be put in question. If the MEM should no longer be usable due to the design modifications, it needs to be adapted accordingly or the occurring error should be estimated with the help of the measurements from the third step. The results of the second step are a solution based on the MEM as well as the implementation of design modifications either by reformulating the MEM or by fault estimation.

In the third step, the function and the MEM are validated with the help of finite elements simulations and test rigs. Thereby, it is determined how well the simplified MEM is representing reality and how good the developed solution fulfills the desired function.

The steps of the approach can be used iteratively until no better solution can be found. Due to the variety of possible design measures and the determination of the physical principle as input, it cannot be assumed that a global optimum is found. However, the developed result represents an optimized solution.

3.3 Analysis of the chamber inflation and determination of the breaking force

As described above, the application of the breaking force is achieved by inflating two chamber walls in the rail, thereby clamping a part of the sled due to the created pressure. The active principle that is used to inflate the chamber walls is pneumatic pressure. A schematic of the function is displayed in Figure 5. The geometry of the rail and the sled is abstracted to only show the relevant areas for the function.

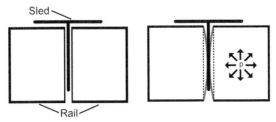

Figure 5: Schematic of the function principle

As shown in Figure 5, a part of the sled called "flag" is clamped between two expanding chambers of the rail, thereby creating the breaking force. According to the approach, a MEM is formulated to describe the effect in a simplified way. The MEM should include all the properties with high relevance to the fulfillment of function. As a point of origin, the simple chamber geometry shown in Figure 6 is used.

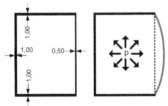

Figure 6: Regarded cross section of the chamber to formulate a MEM for the inflation process

Figure 6 shows that the regarded cross section of the chamber is a simple rectangular geometry with one wall thinned out to half the thickness of the base material. The biggest movement is to be expected at the thinnest chamber wall. To calculate the deflection of this wall, beam theory is used. Effectively, the deflection of a beam under an area load is used to approximate the movement due to inflation. In doing so, only the thinnest chamber wall is regarded. The basis to develop the MEM from beam theory is shown in Figure 7.

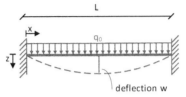

Figure 7: Abstraction of the chamber wall to apply beam theory

As displayed in Figure 7, a beam model with fixed support on both ends was chosen to represent the chamber wall. The biggest deflection is in the middle of the beam at *L/2*. The assumption that the inflation of the chamber can be approximated with the deflection of the beam is the basis for the calculation of the breaking force. To calculate the breaking force, first, the maximum deflection of the beam under the area load resulting from the applied pneumatic pressure is determined. Then, the necessary area load for the chamber wall to tangentially touch the sled's flag is determined. With the difference of those area loads, the length of the flag and the contact surface of flag and wall under maximum pressure are used to determine the breaking force. These calculations contain simplifications which are not further specified at this point. Nonetheless, it needs to be mentioned that the calculation error increases towards the ends of the beam. However, the following results constitute that the essential influence factors have been considered. Considering the aspired pressure range from one to six bar with further restrictions for the chamber wall length with up to 50mm, the MEM provides adequate results.

Results from the MEM regarding breaking force and deformation are shown in Table 1.

Table 1: Results for breaking force and deformation from the MEM

pressure (bar)	force (N)	deformation (mm)
1,5	59	0,15
2	95	0,2
3	190	0,27
4	275	0,39
5	490	0,49
5,5	530	0,54

This procedure complies with a part of the first step of the developed approach.

3.4 Development of the property network

In the following, the property network derived from the before formulated MEM is presented. For simplification, the property network displayed below only shows the properties that are also represented in the MEM. The property network is displayed in Figure 8.

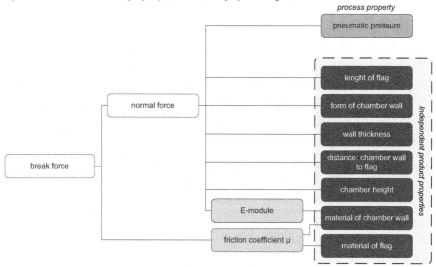

Figure 8: Property network based on the MEM

On the right side of Figure 8, the independent properties are displayed while the dependent properties develop to the left eventually resulting in the breaking force that is required. In contrast to the other independent properties, the independent property on top on the right side is not a product property but a process property. In this case, process properties cannot be used to derive design modifications. On the other hand, all other independent properties can be modified to ease or improve the functionality. Especially geometric modifications regarding the form of the chamber wall will be present later.

The benefit of this approach is the clear identification of the "set screws" for the designer which allow to systematically developing design modifications. These enable the designer to overview and widen the solution area for the desired function. With the development of the property network, the first step of the above presented procedure is finished.

The design modifications that were derived from the property network are presented in a later passage. The specification and optimization is not presented in detail since it is essentially depending on the restrictions that are derived from the application that the function was chosen for. Based on these restrictions, the specification and optimization of the function can be achieved with the equivalent model.

3.5 Validation of the MEM with experiments

To validate the results of the chosen MEM based on beam theory, a test rig has been developed. It allows measuring the occurring forces and deformations due to the inflation process under a given pressure. The test setup is displayed in Figure 9.

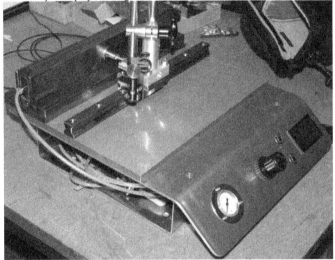

Figure 9: Test setup to measure force and deflection during inflation

Averaged measurements of the occurring forces and deformations during inflation of the chamber are listed in Table 2.

Table 2: Force and deflection of a chambered steel profile depending on the pressure

pressure (bar)	force (N)	deformation (mm)
1,5	65	0,17
2	100	0,22
3	208	0,3
4	301	0,41
5	513	0,52
5,5	581	0,57

Comparing the results of the MEM with the experimental data, a deviation of less than ten percent is perceived. However, the deviation is relatively small in the preferred pressure range for the function of one to six bar.

The data documented in Table 2 was measured on a rectangular profile where one chamber wall has been thinned out to 0,5mm, half the thickness of the base material. The material was removed from the outside of the chamber. However, linear flow-splitting would allow thinning-out the inside by first removing the material and then forming the geometry.

As shown in Figure 9, the deformation of the profile is constrained on two sides. The measurements were taken on the longer side of the rectangular cross section. It is reasonable to assume that the deformation on the shorter side of the profile is significantly smaller.

It is necessary to underline again that the MEM can only be used for this or similar cross section geometries. Design modifications to optimize the functionality were not validated and would probably only represent the real behavior insufficiently.

Regarding the comparison of the empiric and theoretic data, deviation also results from the neglect of the profile length in the direction of the translatory movement of the linear guide. This is based on the assumption that the inflated chamber wall regarded in the calculation is significantly larger than the shorter side of the cross section but simultaneously significantly smaller than the neglected length. The length of the real sample only partially reflects that assumption. Therefore, smaller experimental results for force and deformation can be assumed which corresponds to the collected data.

3.3 Validation of the MEM with finite elements

The validation of the MEM with finite elements simulations is important to check the compliance with the elastic yield of the material. Therefore, the geometry of the test sample was modeled in CAD and then transferred to a FE model. The analysis reveals critical areas in the inflation process which gives hints for an appropriate optimization in this direction. Since plastic deformation of the material is not desired, the elastic yield cannot be exceeded.

The results from the FE simulation with regard to the deformation of the geometry are found in Table 3.

Table 3: Results of the FE-analysis

pressure (bar)	deformation (mm)
1,5	0,187
2	0,25
3	0,375
4	0,5
5	0,625
5,5	0,68

Since the geometry of the FE model is ideal compared to the real sample, the results for the deformation in the simulation can be expected to be bigger which corresponds to the collected data. The same effect results from the chosen constraints of the simulation. Nonetheless, the result of the validation is that the deformation of the profile is adequately captured in the MEM and well within the elastic range. The MEM can therefore be used for the optimization within the desired pressure range and geometric restrictions.

With the validation of the equivalent model, the first iteration of the proposed approach is finished. When design modifications are considered, further iteration could provide a better solution to the task.

3.4 Recapitulation

The validation of the MEM shows that the real behavior of the chamber wall under the given constraints is adequately displayed. Those constraints especially determine the cross section geometry of the chamber. Additionally, the application scenario also constrains the pressure range. Accordingly, the MEM is not universal but only valid under the given restrictions. However, the testing revealed that the function can be integrated with regard to achievable deflection and forces without exceeding the material's elastic yield. Moreover, the property network enables the designer to get a good overview of the properties influencing the functionality.

3.5 Design options derived from the property network

Based on the property network, the set screws of the clamping function were determined. In particular, these independent properties are the wall height, material thickness as well as the shape or the cross section of the chamber. The influences of the chamber height and the material thickness are known from the MEM. However, the variation of the cross section geometry of the chamber allows a variety of modifications that cannot be reproduced directly in the MEM's current form. To reproduce the following modifications in the MEM, it needs to be adapted appropriately. Only one of the derived modifications is presented here to demonstrate the potential of the proposed approach. This modification that is presented here targets the form of the chamber wall. It is shown in Figure 10.

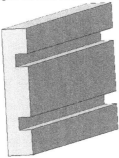

Figure 10: Options for design modifications

In Figure 10, there are only two grooves milled into the chamber wall instead of the whole wall being milled off. This reduces the milling effort and is more material efficient. Additionally, the deflection curve develops a more leveled contact area for the sled's flag. Another option could be a geometry analog to a bellows with wich bigger displacements could be achieved. If a solution with a reversed active principle is aspired, a bellows-like geometry with grooves instead of forming areas could provide a solution.

As both of the exemplary described modifications massively change the geometry of the chamber, the mathematical equivalent model as determined above does no longer comply and needs to be modified to provide equivalently accurate solutions.

4. CONCLUSION AND OUTLOOK

The presented contribution clearly underlines the strengths of the linear flow-split technology which also underlines the potentials of the technology for practical applications.

The outlined approach for the optimization of functions by providing a MEM for mathematical optimization and developing a corresponding property network to derive design modifications has proven to be efficient. The property networks show the relations of separate influencing factors and provide an insight into the subject even with more complex connection matrices. Especially for designers, the "set screws" to create a well working function become obvious. The weakness of the approach lies in the predetermination of the active principle. However, intelligent design principles based on the technological possibilities and strengths can help to chose proper working principles from the start.

With regard to the development and optimization of the function and geometry it becomes evident, that the new production technologies provide plenty of options. The possibilities range from modified semi-finished parts as tailored blanks to process extensions as flexible linear flow-splitting. The optimization of functions has to be conducted not only in experiments but also with the help of FE-simulations. Of great importance is a further form optimization of the chamber where the embodiment of the transitions between thinned-out and base material as well as critical areas in the edges has to be researched.

Future research focus will lie on the further integration of new and innovative functions into linear flow-split linear guides in integral style.

ACKNOWLEDGMENTS
Thanks to the German Research Association (DFG) for funding this work (Research Grant CRC 666).

REFERENCES
[1] Vucic D. And Groche P. Erweitern der Verfahrensgrenzen beim Spaltprofilieren. *Tagungsband 1. Zwischenkolloquium Sonderforschungsbereich 666*, TU Darmstadt, 2007, pp. 67-72
[2] Groche, P. and Vucic, D. Multi-chambered profiles made from high strength sheets. *Product Engineering, Annals of the WGP, Vol. XIII/1*, Hannover Germany, 2006
[3] Jöckel, M. Grundlagen des Spaltprofilierens von Blechplatinen, Shaker Aachen, 2005
[4] Groche, P., Ludwig, C., Schmitt, W. and Vucic, D., Herstellung multifunktionaler Blechprofile. *wt Werkstatttechnik online 10-2009*, 2009, pp. 712–720
[5] Bohn T., Bruder E. and Müller C. Gefüge und mechanische Eigenschaften von Spaltprofilen aus ZStE500. *Tagungsband 1. Zwischenkolloquium Sonderforschungsbereich 666*, TU Darmstadt, 2007, pp. 97 – 102
[6] Groche P., Vucic D. and Jöckel M. Basics of Linear Flow Splitting, *Journal of Materials Processing Technology 183*, 2007, pp. 249–255
[7] Lommatzsch N., Gramlich S. and Birkhofer H. Linear Guides of Linear Flow-Split Components. *International Conference on Research into Design – ICoRD'11*, January 2011, Bangalore, India, pp. 439-446
[8] Birkhofer, H. am Wäldele, M. Properties and Characteristics and Attributes and…- an Approach on structuring the Description of Technical Systems, *Applied Engineering Design Science - AEDS 2008*, 2008, Workshop in Pilsen, Czech Republic, pp. 19–34

Contact: Nils Lommatzsch
Technical University of Darmstadt
Department of Product Development and Machine Elements
Darmstadt, 64287
Germany
Tel: Int +49 6151 163982
Fax: Int +49 6151 163355
Email: lommatzsch@pmd.tu-darmstadt.de
URL: www.pmd.tu-darmstadt.de

INTERNATIONAL CONFERENCE ON ENGINEERING DESIGN, ICED11
15 - 18 AUGUST 2011, TECHNICAL UNIVERSITY OF DENMARK

A KNOWLEDGE-BASED MASTER MODELING APPROACH TO SYSTEM ANALYSIS AND DESIGN

Marcus Sandberg[1], Ilya Tyapin[1], Michael Kokkolaras[1], Ola Isaksson[2]
(1) Luleå University of Technology, Sweden, (2) Volvo Aero Corporation, Sweden

ABSTRACT

The jet engine industry relies on product models for early design predictions of attributes such as structural behavior, mass and cost. When the required analysis models are not linked to the governing product model, effective coordination of design changes is a challenge, making design space exploration time-consuming. Master modeling (MM) approaches can help alleviate such analysis overhead; the MM concept has its origins in the computer-aided design (CAD) community, and mandates that manual changes in one model automatically propagate to assembly, computer-aided manufacturing (CAM) and computer-aided engineering (CAE) models within the CAD platform. Knowledge-based master models can also be used to communicate changes in the product definition to models that are external to the CAD platform. This paper presents details of the knowledge-based master modeling approach as applied to mechanical jet engine analysis and design, where different fidelity models and analysis tools are supported in the early design stages.

Keywords: Knowledge-based engineering, master model, multidisciplinary analysis and design optimization, jet engine

1 INTRODUCTION

The decisions made in the early phases of product development (PD) have a decisive impact on the product and its use throughout the life cycle. As a consequence, effective design requires modeling approaches that can predict the properties and behavior of forthcoming products in early PD stages. The Systems Engineering discipline provides overarching methods for how to approach such design problems systematically [1], but the quality of data available for evaluation is a bottleneck. A better and more precise description of the product and its environment is thus required. Using virtual product modeling techniques, such details can be generated by creating conceptual design solutions with a greater resolution than ever before.

Virtual product models are commonly used to predict life-cycle effects of alternative designs, with the aircraft industry being a driving force of development of computer aided design (CAD) and computer aided engineering (CAE) techniques. CAD and CAE technologies have enabled the creation of product models for digital mock ups, weight estimation, and rotordynamics, stiffness, fluid dynamics and performance analysis. The need of a number of analysis models for each discipline creates a coordination challenge for product changes since there is seldom a single product definition during early design stages. The synthesis of analysis results often leads to re-modeling and design iterations necessary to re-assess the behavior of products after the separate disciplinary design and simulation activities have been conducted. Many analysis models and product representations have to be created several times, and merely the co-ordination of these modeling activities tends to be a costly and time-consuming exercise (see Figure 1).

A master model (MM) approach means having one managing model to control other models. Once the master model is changed, then the associated models are updated accordingly. One of the first master model approaches was reported by Newell and Evans [2]. Within the CAD field MM technology is more or less taken for granted but within the CAE field MM technology is less established. There exists a lot of work on multidisciplinary analysis, but not much of it focuses on how to manage product definition changes that occur when domain-specific models are used concurrently for different analyses. This makes it challenging to conduct design optimization. Commercial software environments such as iSIGHT, Optimus, ModeFRONTIER or ModelCenter provide techniques to link one product definition to different models. To provide more flexibility in geometry change and analysis model linking compared to traditional parametric CAD, knowledge-based engineering can be employed to control the MM. Previous work has demonstrated that it is possible to use KBE to create MMs [3, 4]. However, there is a need to further detail the actual use of the KBE technology to create the MM. This paper provides details about this process by means of a whole jet engine model.

The paper is structured as followed: In section 2 a research background is given to present research within the fields of knowledge-based engineering, CAD-CAE integration and knowledge-based master-models. Section 3 explains the knowledge-based MM approach and section 4 presents an application of this approach. Section 5 wraps up the paper with concluding remarks.

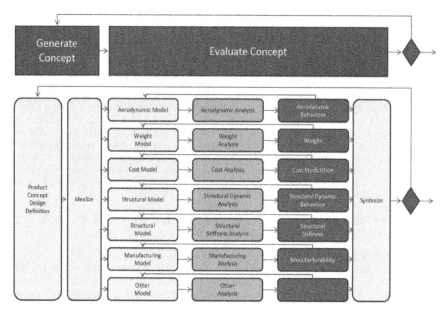

Figure 1. Conceptual design and analysis using unlinked product models

2 BACKGROUND

Automating chains of engineering tasks has been the approach of experienced engineers since the dawn of the computer. Knowledge-based engineering stems from knowledge-based systems, [5], and is claimed to have been coined at the release of the CAD software iCAD [6]. Stokes defines KBE as *"the use of advanced software techniques to capture and re-use product and process knowledge in an integrated way"* [7]. The core is about creating a generative model that can generate product development items such as geometry, reports, BOMs, or finite element models [8]. By using rules, geometry objects can be modeled in a way beyond traditional parametric models. Radical topological changes, e.g. changing a cylinder into a rectangular prism, are possible. For routine engineering tasks KBE applications were found useful [9, 10]. During the last decade, the major CAD/PLM vendors have adopted KBE modeling capabilities in, for example, Siemens NX and Dassault Systemes CATIA.

There exist numerous approaches where the challenge of integrating CAD models with CAE models is targeted. Lee presented a CAD-CAE (computer-aided engineering) integration strategy for feature-based design [11]. The strategy is based on a MM that creates the required CAD and CAE models. CAD model creation is done interactively with the user. The abstraction and dimensional function is semi-automatic. Since the Lee framework is not fully automatic, further work is needed to use it in an optimization loop. Hong-Seok and Phuong integrated CAD and CAE using scripts, programming languages, application programming interfaces and meta-modeling to perform structural optimization [12]. Their approach is limited to traditional parametric capabilities; more radical geometry changes, permitted by KBE, are absent.

In the field of master modeling techniques Hoffman and Joan-Arinyo suggested a master model architecture centered around a server and a repository to which different clients can connect to [13]. These clients can be CAD systems, geometrical dimensioning and tolerancing agents, manufacturing process planners or other downstream clients. Each client receives their view of the design. Each design change made by one of the clients causes changes to other clients' views according to a change protocol and permissions. The architecture is semi-automated and user interaction is needed. La Rocca and van Tooren presented a framework to enable MDO supported by KBE [3]. The core unit of the system consists of a multi-model generator (MMG) that can generate numerous aircraft component (exemplified with an aircraft wing) configurations based on a high-level primitives concept. The MMG can extract data and information from the product definition to specific analyses. Design (product definition) changes are propagated in an automated fashion to all analysis models. A toolbox checks the analysis convergence and compares results with the design specification. If failing to satisfy the specification, the toolbox can trigger new design iterations.

Despite the relative success of KBE approaches proven, the design methodology of using a governing master model is not well established. Best practice that maximize the use of software functionality offered by vendors implies a risk over time where methodology and rule dependencies may become obsolete as new software tools and versions are launched. A system-independent, yet system-implementable, design logic is needed. Detailing the actual constituents of the knowledge-based master model (KBMM) is of primary interest. It is also of interest to develop KBMMs for jet engine structures, as exemplified in this paper, in order to complement the application presented in [3]. The investigation of implementation issues in KBE software environments other than ICAD, which was used in [3], is also of interest.

3 THE KNOWLEDGE-BASED MASTER MODEL APPROACH

This section explains the knowledge-based master model (KBMM) approach in terms of the general idea and its constituents. The general idea of the KBMM approach is to use a CAD system and its KBE software to link all analysis models to one centralized product definition so that early product development can be made more effective. When new ideas need to be tested the product definition can be changed and these changes are automatically propagated to the linked models. An additional goal is to completely automate the design and analysis activities, so that the managing unit can be used to handle the optimization process. By using the capabilities of KBE software within a CAD system it is possible to further enhance the master-model ideas since KBE can enable more flexible geometry configuration compared to traditional parametric CAD. The rules within the KBE classes are also suitable to be used to link the analysis models to the governing product definition.

The KBMM contains of a managing unit, KBE classes and API (Application Programming Interface) calls and Macros as shown in Figure 2.

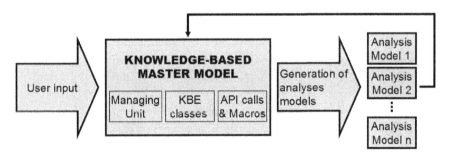

Figure 2. Overview of the knowledge-based master model approach

3.1 The managing unit

The managing unit coordinates the design and analysis loop by taking the user input and initiating required models and analysis activities. Since KBE software is often coupled to the CAD software the managing unit is needed to either operate through the API or the CAD software or in another way to be able to fire the instantiation of KBE classes, API calls and Macros. Optimization functions are included into the managing unit.

3.2 The KBE classes

The KBE classes can generate geometry, finite element objects (e.g. mesh, boundary conditions) and geometry analyses (e.g. weight calculation). Object-oriented KBE software (such as Knowledge Fusion in Siemens NX) usually has predefined classes for fundamental geometry objects (e.g. block, cylinder, ellipse, datum-plane) and predefined methods or functions for parameter handling (e.g. max, min, floor, sin) and CAE operations (e.g. meshing, boundary conditions, loads). The rules that govern the analysis model generation reside within the user defined geometry classes in the KBE module of the CAD software. These classes use predefined classes to create specific geometry. A number of parameters (e.g. dimensional) are used in the object definitions inside each of these classes. These parameters are used when generating the analysis models.

Figure 3 shows an example of how the KBE classes can be organized. The geometry classes are ideally organized in a hierarchy and contain rules needed to generate all geometrical objects of the system. The system contains of a number of subsystems that in turn contains a number of components. Each component is built up by a number of instantiations of part classes but also geometrical objects that are instantiated exclusively for each class. The part classes usually instantiate geometry that is used by several components.

The analysis classes can be of several types: 1) a complete analysis – i.e., a class that uses the geometry generated by the geometry class and analyze one or several properties, 2) pre-processing – using the CAD functionality to perform e.g. meshing, boundary conditions, or 3) creating an API call or Macro for later use during the analysis cycle. Some analysis classes can have children but these children still operate on system level. Some analysis classes start from the component level and use all component level results to create the system level results while other classes read parameters and other geometrical object data from the system level geometry class (the system level connection is indicated by the dashed lines).

The reason of not arranging all analysis classes under one main analysis class is that the API calls and Macros are used to instantiate the classes instead of having one main class that instantiate all analysis classes.

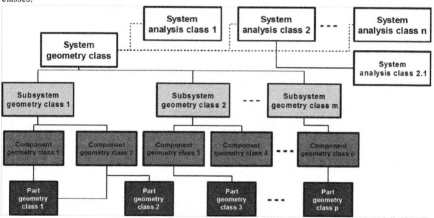

Figure 3. Example MM KBE class scheme

3.3 The API calls and Macros

The API and Macros are a part of the CAD software and aid the creation of the design and analysis loop where the KBE module of the CAD software needs to be complemented in functionality. Therefore, API calls and Macros can be used to instantiate KBE classes that help to automate the design and analysis loop. Macros can both be recordings of design and analysis activities as well as calls of CAD/CAE functions through the API. KBE classes can be used to name geometrical objects (e.g., faces, bodies) that are used by API calls later in the design and analysis loop. KBE classes can also be used to actually code (in e.g. Visual Basic, C++) some API calls by writing geometrical parameters (e.g., dimensions) into the code.

3.4 Generation of analysis model
Based on user input the KBMM approach can be used to generate or link a number of analysis models. The analysis model generation can be done in two ways: CAD-based and rule-based. The CAD-based method uses the actual CAD model generated by the KBE classes to create other models. The rule-based way uses the rules in the KBE classes to generate models. One example of the rule-based method is when an input file to a solver is written using geometrical parameters (or rules) that are defined within the KBE class. The file writing can be done using KBE functions to create, open and write files. An additional option to accomplish this is by external programs (e.g., MATLAB) called by a function within the KBE software.

Some models need input from other models. When all analysis activities are done the managing unit compares the results with the defined design objectives and constraints. If needed, a new iteration is initiated until an optimal design is found or a maximum number of iterations is reached.

4 JET ENGINE EXAMPLE
In the research project METOPIA (http://www.ltu.se/tfm/fpd/research/projects/METOPIA?l=en), a simplified, yet illustrative, turbo fan jet engine example demonstrates how the KBMM approach can be applied. Figure 4 presents an overview of the design and analysis loop that was used to find an optimal jet engine structure. It contains of six major activities; (1) *User interaction* (2) *Optimization*, (3) *Automated geometry generation, mass and manufacturing cost analysis*, (4) *Automated finite element model generation*, (5) *Automated rotordynamics analysis*, (6) *Automated displacement due to rotor-dynamics loads analysis*.

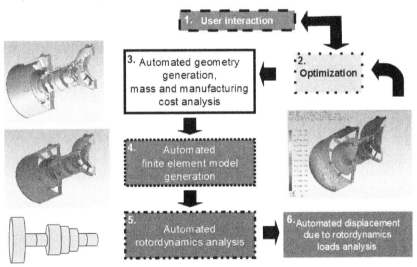

Figure 4. Design and analysis loop

4.1 Automated analysis activities
Further details of the automated loop are shown in Figure 5, where sub-activities and their corresponding software are presented. Activity 1 starts with the user choosing optimization variables,

objectives and constraints and then starts the second activity: optimization. The second activity writes the input file that is used by the geometry generating KBE classes. The CAD-software NX is started where a Macro is run in the Gateway application to initiate a chain of journals (visual basic-based code that through the API executes NX functions), METOPIA actions, which are started from the user-defined menu and marks the start of Activity 3. In the NX Modeling application, all geometry is generated (i.e. fan case, low pressure compressor, intermediate compressor case, high pressure compressor, combustion chamber, high pressure turbine, turbine mid frame, low pressure turbine and turbine rear frame) and .dat-files containing mass and manufacturing cost analysis results are written. The file containing the mesh and material properties for the rotordynamics analysis is also generated (part of Activity 5). In Activity 4 the geometry is united to four bodies, each being one subsystem: 1) Fan case, 2) low pressure compressor and intermediate compressor case, 3) high pressure compressor, combustion chamber, high pressure turbine, turbine mid frame, low pressure turbine and 4) turbine rear frame. Note that subsystem 1 and 4 have only one component. A journal needed for the later coming add material activity is written. In the next stage the geometry faces and bodies are named to be used in e.g., mesh-mating and when material is added. NX Advanced Simulation is started and boundary conditions are added to the finite-element model and mesh-mating conditions are defined for the body interfaces. Nodes are generated for each bearing that will interact with the structure and these nodes are connected with 1D elements to the structure creating a so-called spider-mesh. Each body is 3D meshed with Tet4s, mesh size is adjusted automatically and separately for each body according to body size and mesh-mating conditions. Material is assigned for each body, then NX is switched to the .sim-file where requested output (e.g. displacements) from the finite-element analysis is assigned and input file (.dat) for NX Nastran is written, continuing Activity 5. This input file is edited by adding two lines of a code to punch out the stiffness matrix for the bearing position nodes. The earlier generated input file is then used to run the rotordynamics analysis in MATLAB. Activity 6: The resulting forces are added back to the finite-element model and NX Nastran solves for displacements. All NX files are closed as well as NX and results are analyzed. Based on the results a new input file is written for the next iteration.

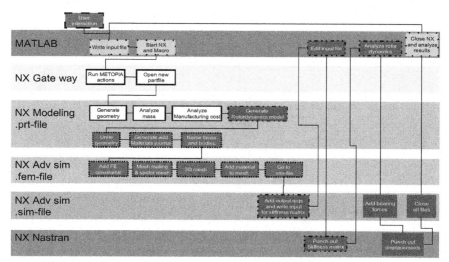

Figure 5. Design and analysis loop activity details

4.2 Analysis models

The analysis models used in the example include geometry, weight and manufacturing cost models as well as a finite element model and rotor model. Siemens NX 7.5 and Mathworks MATLAB 2009B were used to implement the models. All geometry and models are generated particularly for each iteration and no models are reused in the later operations.

The geometry is generated based on 18 user defined KBE classes and each jet engine component has its own class. The rest of the classes represent common geometry and are instantiated in several component classes. Common geometry includes: cases, flanges, struts and mount lugs. All geometry can be configured in different ways, e.g. changing lengths, radii, number of struts, thicknesses, cone angles (for intermediate compressor case, turbine rear frame).

As volume is easy to compute in CAD-models the weight can also be found. The following equation was used to compute the manufacturing cost ($Cost_{manufacturing}$):

$$Cost_{manufacturing} = m\ Cost_{material}\ k_{manufacturing} \quad (1)$$

where m (kg) is the mass, $Cost_{material}$ ($/kg) is defined by material choice (Titanium 6Al 4V, Titanium 6Al 2Mo 4V, Inconel 718, aluminium, steel) and its price per kilo, manufacturing method (cast, forged or fabricated) is also included into the manufacturing cost. For cast 15% more material is added and for forged and fabricated 10% and 5% were added respectively. The coefficient $k_{manufacturing}$ is used to tune the model to higher fidelity models where the manufacturing cost calculation is more elaborate.

The finite element model consists of Tet4 elements and has approximately 33000 nodes depending on the geometry configuration. The bodies with different material and element properties are connected together using mesh mating conditions. The rotor model is connected to the structure using 1D rigid elements. The finite-element model is regenerated at each iteration.

The rotor model consists of 9 cylindrical beam elements; each length and diameter is governed by the product definition and is an example of a rule-based analysis model generation. The stiffness matrix is created for the bearing nodes and used in the rotordynamics analysis governed by MATLAB.

4.3 Design study

The ultimate goal of developing analysis and simulation models is to have the ability to conduct design studies and evaluate what-if design scenarios. In this paper, a relatively simple design scenario was used to demonstrate the usefulness of the presented modeling approach. In particular, it was investigated how relatively limited design changes in component level impact system behavior. This seemingly limited design optimization study is in fact quite significant from a tier-1 supplier point of view: being able to evaluate such design scenarios rapidly increases the supplier's advantage against competitors and its ability to negotiate system-level design with the original equipment manufacturer. Specifically, it was investigated how the number of intermediate compressor case struts impacts mass (and thus cost, as the latter is a proportionate function of the former) and structural integrity when considering a fan-blade-off loading condition.

An optimization approach was used to investigate this design scenario. An optimization problem was formulated to minimize the mass of the jet engine structure subject to a maximum displacement constraint caused by the load generated by the imbalanced mass due to the lack of one fan blade (the actual impact of the blade on the engine is not considered). The design optimization variable was the number of struts for the intermediate compressor case. The initial guess was 12 struts while any number from 5 to 20 struts was considered. The displacements constraint was set to $0.5*10^{-6}$ m.

Even though the design optimization variable was discrete, a gradient-based optimization algorithm was used to solve the optimization problem. Specifically, the MATLAB implementation of the sequential quadratic programming (SQP) algorithm was utilized. A gradient-based algorithm was possible to use because when considering a single integer optimization variable, the gradient computation can be manipulated to evaluate the neighbors of the optimization variable value (e.g., the values 7 and 9 of an incumbent iterate equal to 8), and thus guide the gradient-based algorithm to find a solution to a uni-variate discrete problem. Given the displacement constraint value mentioned above, the optimization converged to the lower bound of 5 struts. The displacement analysis results for this design are shown in Figure 6.

The next step of our research effort will be to formulate and conduct full-blown optimization studies. Obviously, the size and extent of such studies will depend on practical issues such as wall-clock time for one function evaluation (currently about 5 min) and the presence or not of mixed variable types (continuous and discrete). In the latter case, advanced derivative-free optimization algorithms based on mesh-adaptive direct search (MADS), such as NOMAD [14] will be used. Such algorithms may require hundreds of function evaluations for moderate-size problems (tens of variables), but are remarkably effective.

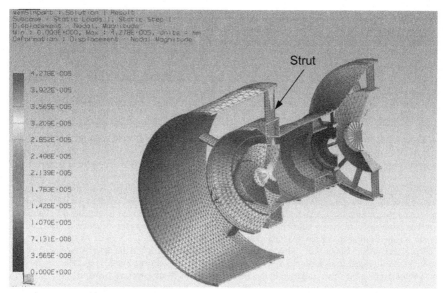

Figure 6. Displacement for 5 intermediate compressor case struts.

5 CONCLUDING REMARKS

A knowledge-based master model approach has been presented in this paper. In comparison to conventional master model approaches, e.g., [13], this approach is argued to be more flexible since the KBE classes can generate more radical geometry topology changes compared to traditional parametric models. In addition, the KBE classes can be reused since they are created in an object-oriented environment that promotes easy instantiation of the classes. In comparison to [3] this paper elaborates on the details of the constituents of the KBMM approach and presents new design optimization capabilities.

By using Macros and API calls to instantiate the KBE classes, the same Macros and API calls can be used even if the KBE classes are slightly updated. It was found to be beneficial to create smaller NX Journals that are NX session independent to maximize re-usability instead of grouping many functions into one Journal.

By having a top class that contains all optimization variables only one file needs to be edited by the managing unit at each iteration. The geometry classes was during this research effort updated for a new jet engine configuration, based on data from the EU FP7 integrated project: CRESCENDO (http://www.ltu.se/tfm/fpd/research/projects/crescendo?l=en), and most of the rules could be reused for the new configuration. Knowledge acquisition for the KBE model creation is not the focus of this paper; readers interested in such techniques are referred to [7].

When KBE software is part of CAD software the transition to detailed design is easier compared to dealing with discipline-specific models since detailed design is often based on CAD models.

Compared to using commercial software environments e.g. iSight, ModeFRONTIER, the KBMM approach is argued to be implementable in any CAD-software that has an API from where all CAD functions can be reached. Using MATLAB as the managing unit is argued to be beneficial since many organizations have MATLAB licenses. The geometry data export capability of the CAD platform determines which analysis models can be linked. Nevertheless, KBE software can be used to write input files to CAE software external to the CAD platform even though definition of mesh coordinates may need extensive coding of the KBMM.

The optimization will be developed further for larger and more elaborate design problems, and other optimization algorithms (e.g., effective and efficient derivative-free instead of gradient-based) will also be considered. Currently one iteration takes around 5 minutes using a 4GB RAM, dual core (2.8Ghz) processor computer. The example in Section 4 is relative simple, but demonstrate the concept for the KBMM approach and shows how whole jet engine design and analysis can be conducted, which is one of the objectives in CRESCENDO.

As the KBMM links analysis models to one product definition and enables optimization it is argued to be beneficial for early design and analysis. Automated model reconstruction implies potential time savings but also quality assurance since company-approved work practices can be used. As KBE is found most useful for routine design tasks the KBMM approach is believed to mitigate problems for standard engineering changes in designing complex systems such as jet engines.

ACKNOWLEDGMENTS

The financial support of VINNOVA NFFP5 (National Aeronautical Research Program, phase 5) is gratefully acknowledged. The financial and technical support of Volvo Aero Corporation is also gratefully acknowledged. Company-specific data of the jet engine model have been altered due to proprietary reasons.

REFERECES

[1] Kossiakoff, A. and Sweet, W.N., *Systems Engineering Principles and Practice*, (Wiley Series in Systems Engineering and Management, 2003).
[2] Newell, M.E. and Evans, D.C., Modeling by computer. in IFIP Working Conference on Computer-Aided Design Systems. Austin, Texas, USA, 1976.
[3] La Rocca, G. and van Tooren, M.J.L., Enabling distributed multi-disciplinary design of complex products: a knowledge based engineering approach. *Journal of Design Research*, 2007, 5(3), pp.333-352.
[4] Sandberg, M., Kokkolaras, M., Aidanpää, J.-O., Isaksson, O. and Larsson, T., Whole jet engine analysis and design optimization: a master modeling approach, in *8th World Congress of Structural Multidisciplinary Optimization.* Lisbon, Portugal, June 2009.
[5] Dixon, J.R., Knowledge-Based Systems for Design, *Journal of Mechanical Design*, 1995, 117, pp11-16.
[6] LaCourse, D.E., *Handbook of Solid Modeling*, (McGraw-Hill, Inc, New York, 1995).
[7] Stokes, M., "*Managing Engineering Knowledge - MOKA: Methodology for Knowledge Based Engineering*", (ASME Press, 2001).
[8] Chapman, C.B. and Pinfold, M., Design engineering - a need to rethink the solution using knowledge based engineering, *Knowledge-Based Systems*, 1999, 12, pp.257-267.
[9] Isaksson, O., A generative modeling approach to engineering design, In *International Conference on Engineering Design,* Stockholm, Sweden, 2003.
[10] Sandberg, M., Boart, P. and Larsson, T., Functional product life-cycle simulation model for cost estimation in conceptual design of jet engine components, *Concurrent Engineering: Research and Applications*, 2005, 13(4), pp.331-342.
[11] Lee, S.H., A CAD-CAE integration approach using feature-based multi-resolution and multi-abstraction modelling techniques, *Computer Aided Design*, 2005, 37, pp.941-955.
[12] Hong-Seok, P. and Xuan-Phuong, D., Structural optimization based on CAD-CAE integration and metamodeling techniques, *Computer Aided Design*, 2010, 42, pp.889-902.
[13] Hoffmann, M.C. and Joan-Arinyo, R., CAD and the product master model, *Computer-Aided Design*, 1998, 30(11), pp.905-918.
[14] Audet, C., and Dennis, Jr., J. E., "Mesh adaptive direct search algorithms for constrained optimization," SIAM Journal on Optimization, 2006, 17(2), pp.188–217.

Contact: Marcus Sandberg
Luleå University of Technology
Department of Business Administration, Technology and Social Sciences
SE-971 87 Luleå
Sweden
Tel: Int +46 920 493072
Fax: Int +46 920 491399
Email: marsan@ltu.se
URL: http://www.ltu.se/?l=en

Dr. Sandberg is currently an Assistant Senior Lecturer at the Department of Business Administration, Technology and Social Sciences, Luleå University of Technology, Sweden. His research interests are in the areas of knowledge-based engineering, systems engineering and simulation-driven design.

Dr. Tyapin is a Post Doc at the Department of Business Administration, Technology and Social Sciences, Luleå University of Technology, Sweden. He has among others done research considering optimization of industrial robots such as Parallel Kinematic Machines in collaboration with ABB Robotics, Sweden. In 2009 Ilya has completed his PhD in Mechatronics and Computer Science at University of Queensland, Australia.

Dr. Kokkolaras is a Visiting Professor at the Department of Business Administration, Technology and Social Sciences, Luleå University of Technology, Sweden. His primary appointment is Associate Research Scientist at the Department of Mechanical Engineering at the University of Michigan, in Ann Arbor, Michigan, USA. His research interests include multidisciplinary design optimization of complex engineering systems, uncertainty quantification, platform-based design of product families.

Prof. Isaksson holds a position as senior company specialist in Engineering Design at Volvo Aero in Trollhättan, Sweden. He has an MSc in Mechanical Engineering and a PhD in mechanical engineering, both from Luleå University of Technology. He joined Volvo Aero in 1994 and is currently responsible for technology and methods development in the area of engineering and product development.

SOCIAL SYSTEMS ENGINEERING – AN APPROACH FOR EFFICIENT SYSTEMS DEVELOPMENT

Thomas Naumann[1], Ingo Tuttass[1], Oliver Kallenborn[1], Simon Frederick Königs[1]
(1) Daimler AG

ABSTRACT
Our objective is to establish an understanding of product development as a sociotechnical system and as a foundation for new methods to manage social and technical complexity. Existing approaches experience a gap in explaining today's phenomena and in managing rising complexity. One reason is a decoupled view on either technical or social systems. To bridge this gap, we describe product development as a dynamic sociotechnical system where interrelated functions link social and technical systems. The sociotechnical system is shown in a schematic meta-model. System functions are represented by interaction (human/machine) and communication (human/human). Based on this approach, complexity, with its major impact on the systems self-governance, can be identified and related to social and technical subsystems. This view helps to develop new approaches to measure and manage complexity. The meta-model is the foundation for development of engineering services and tools to improve effective system modelling and efficient system development. As an example, we'll show a derived UML meta-model currently applied to improve collaboration between different social systems in automotive development.

Keywords: social systems engineering, meta-model, complexity, network analysis, communication analysis, system modelling, system development, UML

1. INTRODUCTION - CURRENT SITUATION IN AUTOMOTIVE INDUSTRY
The development of a sociotechnical systems approach is a consequence of recent dynamics and changes in the automotive industry and a current lack of a comprehensive theoretical meta-model, which combines different areas of science, domains and aspects of efficient as well as effective product development. The automotive industry faces immanent challenges and hence changes (i.e. differentiation, integration of new technologies, process optimization) leading to a continuous learning and exploitation of knowledge. Recently the situation has significantly changed especially caused by CO_2 restrictions, market drifts and the economy crisis. As a consequence, completely new technical, economical, and organizational requirements have to be fulfilled: Due to the demand for fuel efficient green vehicles, product development has to leave existing paths of technology to new drive systems like hybrid and e-drive as well as light weight construction materials. Economically, budgets for individual projects are far more restricted to maintain flexibility in uncertain and still dynamic times. Therefore, intensifying cooperation, joint venture and network structure in manufacturing and product development seems to be the organizational solution to meet these requirements.

Summing up the empirical evidence, product development has to handle the drastically increased types of complexity and is simultaneously forced to operate more effective and efficient. Changes in organizational structures, processes, methods and tools like CAx- and PLM technologies to foster innovation are additional challenges. Furthermore, intensified, distributed collaboration in intercultural design teams exacerbates the governance and coordination of projects. Established management approaches and design methods are in place to solve these problems, but reach their limits because they are unable to explain, model and systemize today's phenomena. Literature provides different theories applicable for a subset of these phenomena. These theories often decouple technical and social aspects in product development to highlight science specific questions. To describe products/artefacts theories of technical systems are available. Design Theory and Methods take into account the process to develop technical systems. In Management Science and Organizational Theory different aspects of social behaviour and its governance play a role. As a consequence of decoupled theories, we have developed a meta-model based on Social and Technical Systems Theory combining different aspects of science and domains to handle increasing technical and social complexity and to understand their relationship. Especially the occurrence of digital communication in engineering context is a key to understand how social systems cope with complexity and how self-organization works. The schematic meta-model of sociotechnical systems is a comprehensive approach, which allows the measurement of complexity in social and technical structures and functions. With the aid of statistical indicators, the system performance will be assessable and allows the derivation of improvements. Furthermore, the meta-model is the basis to collect, structure and provide information within a context in form of documents and data, so that IT services and tools can be developed and evaluated. Finally, the meta-model should help to improve coordination issues related to certain product configurations under consideration of social behaviour, design methodologies and IT services.

2. SCHEMATIC META-MODEL OF SOCIOTECHNICAL SYSTEMS

The schematic meta-model of sociotechnical systems should be able to describe the development and utilization of technical systems by a social system, to make complexity determinable, to reveal and intensify the ability for self-organization as well as to provide a basis for the development of IT-services. The formal structure of the meta-model regards the three system concepts from Ropohl: structural, functional and hierarchical concept [1]. The meta-model also respects the functional-genetic System Theory of social systems from Willke [2]. It regards the distinction of a system to its environment via evolutionary characteristics of internal complexity in structures and processes, the nesting of subsystems, the interaction with the system environment as well as the control of complex systems through symbolic methods, strategies and tools. Every complex system reproduces itself continuously by generating new structures and elements according to self-organization patterns. The used mechanisms are functional differentiation and symbolic-generalised governing mechanisms, which create a space of possible variations, where potential structures, processes and system status come to exist [2].

2.1 Structure

A sociotechnical system consists of one or more social or technical subsystems (see figure 1). The boundary of a sociotechnical system excludes the natural surroundings – that is: everything not included in the sociotechnical system. Every subsystem consists of elements or subsystems and is able to develop specific structures and properties. The specific system element of social systems is the human operation system, the human being. Under certain condition, at least two human operation systems are able to create a social system. All systems realize system functions by transforming input into output parameters throughout all events. In this situation, parameters can originate from the system environment and the structures and functions of sociotechnical systems themselves. Through the system functions, the particular states of subsystems and the whole system change. In the same way, system attributes and properties can be changed by system functions.

The essential system principle of sociotechnical systems is the coupling of social systems and technical systems via the system functions of technical genesis. Technical genesis contains all sub functions for the development and utilization of technical systems through one or more human operation systems. A technical system can also be a tool. For instance, an engineer can be considered as a sociotechnical system at her workplace while using a CAD-system (human operation system plus technical system used as a tool). Similarly, a car driver and the vehicle display a sociotechnical system as well. In simple words: we always form a sociotechnical system when in contact with a technical system.

2.2 Human operation system

Detailing the meta-model approach starts with the human operation system. Its rough structure can be separated into the psychological and organic system. The psychological system contains rationale mental and intellectual structures as well as functions. The organic system consists of all vegetative controlled structures and functions. According to Ropohl, the corresponding sub structure is expressed via the objective, information and active system [1]. Simply spoken, the active system carries out work by transforming material and energy whereas the information system gathers information via receptors from the active system and system environment and interacts with other external systems. Within the objective system, all objectives for activities are constituted and processed in relation to system, state and environment.

One fundamental function within the human operation system can be found in all target-oriented and reasoned activities. This scheme starts with defining objectives e.g. travelling somewhere with a fast and efficient vehicle or (if this vehicle does not exist) to develop it. The operation transforms the plan and carries out a check for objectives' achievement. If not achieved, a repeated flow through the scheme of operation takes place. If achieved, the operation scheme or the function results in the designated output. All functions of social and sociotechnical system, especially functions of the technical genesis are considered target-oriented and composed of the scheme of operation. After all, couplings of the sociotechnical system result from interactions of the human operation system with technical systems. In addition to that, couplings between more than one individual, say a group with common objectives, are a social system.

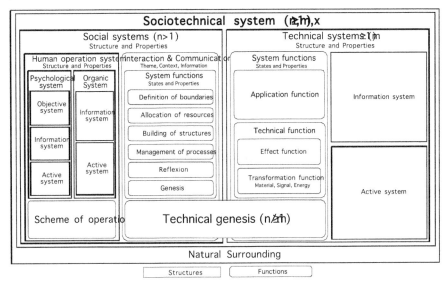

Figure 1: Schematic meta-model of sociotechnical systems

2.3. Social systems

Workgroups, teams, departments, business units, a corporate entity, society (customers) – these are typical social systems related to a formal, hierarchical line organization as known from product development. Depending on organizational structure, structures of project organization and forms of informal organization can be added to the list [3]. As an engineer, I belong to a team, a department, a business unit, a corporation. I work in one or more projects. I regularly meet colleagues from different domains and gather information about topics yet beyond the contexts mentioned above. All these things are couplings within a superordinate system – the respective corporation. Beyond this system, I am coupled with my family and friends, to an association and several institutions, to my household and my country. Each coupling usually has a meaning or a purpose. Via coupling, I become aligned to the system and if I decouple from the system, it still exists. So, how does coupling come about and how does a coupling of elements lead to a social and cooperative system?

2.4. Interaction and communication

As linguistic findings and new anthropological studies show, evolved gestics and language are fundamental abilities for social cooperation. Human gestic and language differ from animal's communication by a far more sophisticated ability for mutual exchange of cognition and intention. Therefore, it enables collective intention for conjoint action, which leads to cooperation [4]. Consequently, each social system can only exist because of communication. At the same time, cooperation determines and supports the development of communication and interaction. Interaction means interdependent acts between a sociotechnical system's human and technical systems. For our approach, we are interested in structurally evaluable communication. For example, observing spoken language shows that information is exchanged and processed via communication events regarding a specific topic. Topics may be predetermined by an agenda (i.e. a report about a test run) or may emerge spontaneously. They estab-

lish a semantic framework for information and a reference framework for communication events (information, inquiry, discussion, decision), which form the communication process. The central aspect is information, which is to be transferred (i.e. test run was successful). Additionally, each communication can be codetermined by a specific context (i.e. the test run was conducted within the scope of a specific project). Along with the exchange of information, communication has to fulfil functions to enable conjoint actions, respectively cooperation. Namely, these functions are availability of information, maturation of knowledge, objectives-instrument-identification, commitment and regulation of affect. To avoid dysfunctions in cooperation and to maintain connectivity, the following should be assured:
- All available information is accessible for everyone → availability of information.
- All exchanged information is consistent and comprehensible → low-loss knowledge integration.
- Objectives and instruments must fit → suitable objectives-instrument-identification.
- Tasks and follow up appointments between involved parties should be mandatorily arranged and should be documented → commitment for follow up activities and communication.
- Discussions should be moderated and reflected → sufficient regulation of affect.

As long as these basic functions are assured by interaction and communication, a social system can develop via the system functions proposed by Willke [2].

2.5. System functions of social systems
The individual formation of social system functions enables systems to process external and internal complexity. This is necessary for its self-preservation. Those functions include 'definition of boundaries', 'allocation of resources', 'building of structures', 'management of processes', 'reflection' and 'genesis'. To give an example, the system functions could be processed as follows: the product development receives an order from the company to develop a more fuel efficient vehicle. A project for the development of a hybrid drive is initiated and a project manager is appointed. She receives objectives (boundaries), assembles a team (resources) and appoints roles, responsibilities and tasks (structures). She monitors task processing and target achievement and intervenes for corrective actions when necessary (process). The team meets on a regular basis and rechecks that all requirements have been considered (reflection). Finally, additional requirements cause continuous expansion of the team (genesis). In this example, a coupling is established between the newly assembled team and the new hybrid drive which has to be developed. Technical genesis is the binding function for realising the development of the hybrid drive and the coupling between social system and technical system. Prior to examining the technical genesis function, we will go into detail about structure and functions of technical systems.

2.6. Structure and functions of technical systems
In parallel to the social and the human operation system, technical systems can also be described by the structural system concept by Ropohl. That leads to a differentiation of information and execution system. Until today, no objective systems exist within technical systems. Structure and characteristics of information and execution systems can be exceedingly diverse. However, system functions of technical systems can be sub classified into two types of sub functions. The first type is the application function which could be travelling from A to B. Ropohl's principle of sociotechnical differentiation of labour, which has to be applied to the human operation system and the technical system, states that technical systems take over human action i.e. performing physical work or processing information. The other type of sub function is the technical function which has no distinct contribution to the utilization of the technical system but enables the technical systems to perform its required behaviour. In order to do that, technical functions transform, transport and store energy, material and signals and realize a change or a maintaining of system states. Application and technical function of technical systems have to be specified within the technical genesis.

2.7. Technical genesis
The technical genesis represents the coupling between the social and the technical system, figure 2. It comprises both, the utilization and the development of technical artefacts.
The flow diagram of the technical genesis from Ropohl describes the utilization of a technical system by a human operation system with reference to the earlier mentioned scheme of operation. We would like to set the focus to the function of product development. The function comprises a number of sub functions, known from common design methods [6].

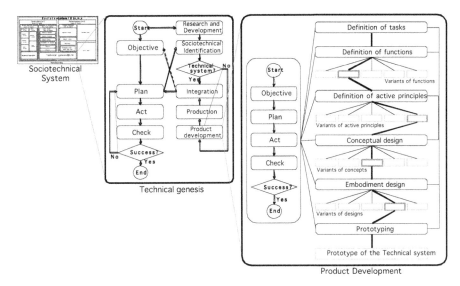

Figure 2: Technical genesis

The execution of the different sub functions leads to the generation of several partial models that represent the technical system. Variants of one model will be selected and provided to the next sub function. Finally, a complete product model has been generated that can be forwarded to production. All sub functions follow the scheme of operation which means they have an objective, are usually planned, and realized. They are completed when the final examination states the accomplishment of the objective. The sub functions are performed step by step iteratively until the synthesis of the product is completed. The social system of product development has to be capable of performing all those functions. The equivalence principle can be applied between the social systems' complexity and the complexity of the developed product. The social system has to adapt its own structures and functions in order to cope with those of the technical system. The principles to achieve are self-organization and self-governance. The more efficient a social system is in adapting to new conditions and requirements by applying the two principles, the more efficient it will be in developing technical systems.

3. COMPLEXITY OF TECHNICAL SYSTEMS

This section addresses the question on which principles technical complexity becomes obtainable, measurable and objective assessable. The implementation of complexity indicators will be the basis for the development of further methodic- and governing-oriented approaches. A definition of complexity in respect to the most important characteristics in different areas of science is illustrated in figure 3. Taking into account the conclusion from literature research and the ideas from the sociotechnical meta-model, the implementation of a dualism by an objective and subjective perspective towards complexity is worthwhile [5]. The observer has to be taken into account when assessing the complexity of a situation, because the evaluation depends on subjective experiences and cognitive abilities. Another important characteristic of the dualism lies in the fact that the governance of objective complexity can only be done by the human active system. The governance of complex situation is not only based on objective descriptions, but also on human perspectives and their limitations.

3.1. Objective complexity

Objective complexity can be differentiated into structural and functional complexity. Structure represents the static shape of elements and relations including hierarchical perspective, whereas functions stand for a description of transformation processes. Using this description there is an increase of complexity caused by an increase of quantity and variety of relations and elements. Moreover complexity increases by the time dependent changes in structure and function, which is known as dynamic phenomena. Based on these characteristics, the complexity of technical systems as interacting elements will be discussed below. For the completeness of figure 3, common properties of complexity are mentioned. But they were not capable for the development of complexity indicators in our approach.

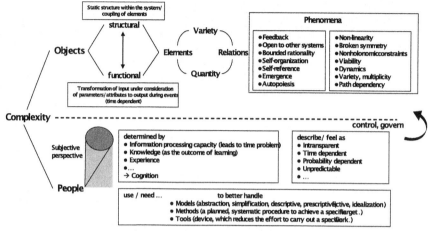

Figure 3: Understanding of complexity

3.2. Subjective complexity

Subjective complexity originates in situations where individuals try to capture and control objective complexity by means of existing instruments. Product development is already identified as a complex problem solving process [6]. Therefore a psychological insight of people dealing with complexity is useful. First of all, humans are physiologically limited in processing information, which leads to a specific duration to assess complexity. These time and processing boundaries of individuals can be extended by experience and knowledge about the issue in question and strategies to cope with those situations [1, 7]. Another aspect of subjective complexity lies in perception. How do people observe and describe complex situations? In this context properties like networked, dynamic, non transparent, probability-dependent etc. are mentioned, which lead to typical mistakes in coping with complex situations such as over steering or ignorance of side effects [8].

To increase efficiency and effectiveness in processing complexity, people develop and utilize models, methods and tools for obtaining and governing complex situations. The purpose of models is to abstract, simplify, describe or idealize real situations to an appropriate dimension. However methods specify procedures to use models and tools for objective achievement. Via the usage of tools, functions of the human active system are substituted or complemented by technical systems, which points to the relevance of the sociotechnical systems approach.

3.3. Complexity in technical systems

In literature, several approaches to measure complexity of technical systems are discussed. The field of computational complexity focuses on the computability of algorithms with the inherent space and capacity problem [9]. In mechanical engineering, structural and functional complexity is mathematically described in a formal scheme, which is based on approaches from Computational Complexity, Information Theory [10] and Axiomatic Design [11]. In this section, we focus on describing technical system complexity on the basis of elements and relations along with their quantity and variety. We concentrate on relativizing a measurement, which reveals interdependencies in the meta-model and enables us to fully understand the sociotechnical system.

3.4. Structural complexity in technical systems

Structural complexity can be found in the information and active system of technical systems. The basis of the structure is constituted by elements and relations. Elements can be separated into peripheral and internal subsystems, depending if they are coupled to other systems in an input output relation or coupled internally to realize transformations. In addition, the hierarchical concept of Systems Theory distinguishes in classifying element clustering. Couplings can be separated from each other by their energetic, material, informational, time and space attributes. Another differentiation characteristic represents the coupling type, e.g. serial and parallel connections as well as feedback. Based on these characteristics, the structural complexity indicators can be defined as followed:

- Quantity and variety of relations of artefacts defined via attributes
- Quantity and variety of internal / external couplings
- Quantity and variety in coupling types
- Quantity and variety of elements / hierarchies
- Quantity and variety of peripheral and internal subsystems

3.5. Functional complexity of technical systems

Functional complexity can be separated into two categories: Application functions focus on how an artefact is used by a human system and technical functions express physical principles, which highlight the description and correlation between input and output attributes. The technical functions are partitioned into effect functions (if the output attributes change) and transformation functions (if there is a change in state). Attributes can be quantified or described via physical laws. According to this explanation, indicators for functional complexity are:

- Quantity and variety of technical functions defined via effect and transformation functions
- Velocity of function completion
- Quantity and variety of input and output attributes
- Quantity of system and environment states defined via attributes
- Variety of system and environment states

After sketching complexity in technical artefacts, it is notable that structure and function are different perspectives on an equivalent relational entity [1]. The difference appears in the relative stability of structure and in most cases the time dependency of functions. Finally application functions describe the utilization of a technical system in relation to the social system and its environment.

3.6. Existing methods to evaluate complexity in technical systems

Design structure matrices (DSM) facilitate the detection of dependencies between items. Primarily activities and their relations were considered by means of square matrices to optimize via sequencing [12]. Moreover this method was used to analyze and improve product architecture, information flow and organization structure [13, 14, 15]. Although this method identifies system internal dependencies, Danilovic developed the domain mapping matrices (DMM) to identify intra system dependencies [16]. On this basis, further research in analyzing dependencies between product architecture and the organization as well as product architecture and requirements were carried out [17]. Recently another approach to combine multiple domains (MDM) has been developed [18]. Matrices can be visualized by using networks in which items are represented via nodes and their relation via edges. Based on Graph Theory as the formal language and on findings from social network research, the network can be further analyzed through social network metrics (SNA) [19, 20]. Furthermore some findings in Network Theory about large networks are accessible and complement existing analyzing techniques. Research in applying these methods in design context increased during the last decade by focusing on connections between different domains like process, organization, product and communication [21, 22, 23].

Today these methods are only partially introduced into industrial practice, because of a high effort to obtain appropriate data, the variety of metrics which exist and the lack of standardized recommendation for action. Whereas the analysis of static networks is covered by methods and software applications, the analysis of dynamic networks has been explored less. Trier investigates an interesting approach in the context of knowledge management and developed initial methods to analyse the dynamic of networks [24].

4. COMPLEXITY OF SOCIAL SYSTEMS

The following section is about the social system in general, its inherent kinds of complexity and its position within the meta-model. As shown, the social system operates via communication. Social systems start relatively small. For the most part, a group of people may initialize a social system. With sufficient growth, the system no longer depends on any single individual. Corporate entities for example, may exist without their initial founders. It is the same for justice systems and states. At any time, specific individuals greatly influenced its history. But at any point, things had the potential to develop in completely different ways. Technically speaking, each operation influenced the systems further development. If you put a timeline on this, you could mark every single operation as an event. The sum of these events effectively developed the state the system is in today. The fact that each event could have played out in a vast variety of ways is also true for future events.

For each event, a contingency space of possible operations exists. The number of possible operations, as well as the finally chosen operation, depends on the respective state of the system at the time of the event. This marks a degree of complexity as well as the systems mode of operation for dealing with this complexity. However, the contingency space is not bound to a single individual. As pointed out, individuals are temporarily bound to the system rather than being part of the system. Otherwise, one could not be bound to several different systems at a time (working environment, family etc.). The contingency space is not even limited to the inner environment of the system, for it may be irritated from the outside, which might lead to completely different outcomes. Therefore, the contingency space – and the probability of a single operation to be chosen – seems quite volatile. The contingency space is usually assumed as infinite and the probability for the occurrence of an event is usually treated as identical for each event and asymptotically zero. As my former argumentation points out, I do not agree with this static view of a contingency space. Even though, the result is the same: An evaluation of a social systems contingency space – and therefore an event based view of a social system's complexity – is, at this point, not possible. Therefore, a different approach on complexity is necessary.

Willke proposed different kinds of complexity connected to different functions of a social system. To utilize this concept, each event is retrospectively linked to one of the systems functions and therefore with the respective kind of complexity [2]. The social System needs to handle each kind of complexity and each solution leads to a new kind of complexity. In other words, the system evolves with each step of processing complexity. According to Willke [2], a constituted system, performing the function of setting system limits, competes with other systems for resources and is therefore faced with resource complexity. Assimilating resources causes a new situation, where the coupled persons need to be kept involved in close interactive relations [2]. In this situation, social complexity arises, making it necessary to clarify accountabilities via role structures, causing functional structures to develop. Additionally, with this higher degree of connections between internal operations [2], process rules are necessary to handle the resulting temporal complexity. At this level, the systems operations have a higher capacity for independence from the environment. Up to this point, the system basically reacted to its environment. Now, the operations become contingent, marking the need for reflexion of the systems operations. This finally causes cognitive complexity - various opportunities for own operations, limited by the internal and external environment. Now, objective complexity is the sum of these kinds of complexity and is represented as a function of the complete communication process of the social system.

As a practical matter, the complete communication process can be recorded and conceived in a non-experimental environment, but extensive reduction of data content "on-the-fly" is necessary. For example, Bales developed the interaction process analysis (IPA) in 1950 and, in 1982, "SYMLOG" as a multiple level observation method for group behaviour. A comprehensive coverage of a communication process requires a degree of observation only possible in experimental environments. Even a complete record of each event still needs a high degree of post-processing. Current attempts to automatically analyse communication include the development of Social Badges [25], which record interpersonal communication and analyse it based on behavioural patterns. However, in these examples, the actual context of each operation is not taken into consideration. Relating actual events to their respective kinds of complexity can be seen as a compromise between close regard of content and rather descriptive approaches. In this context, an experimental study was conducted with focus on system functions. It has shown that a better fulfilment of each system function (each represented by specific hypotheses, i.e. a higher degree of integration of the person with the lowest share of communication represents a better fulfilment of the function "internal integration") correlated with better team results (measured via level of progress in the respective assignment), indicating the importance of the described system functions. Each event was categorized manually in a time consuming manner. However, a real-life project will create massive amounts of data, which indicates the necessity of automatic ways for post processing the incoming data. A study of a real-life project confirmed this necessity. Additionally, all mentioned approaches interfere to some degree with normal work life. Therefore, it is required to find a balance between observation of communication and interference with the observed system. We propose to predominantly use sources for communication data that are already used by the observed system. This includes protocols and email data. However, local laws and further restrictions must be taken into consideration. Our current approach includes pseudonymization of Email address data as well as black lists to anonymize the actual content of Mails. While evaluation of the social systems internal structure can be realized via social network analysis and steps toward Email analysis are already taken (i.e. [26]), the evaluation of interdependencies of single events (or sequences), their content and the outcome of the process, is a more difficult matter. It has been proposed, that "creating a robust coding scheme" and the "need for a more complete capture of data [...] where analyses can be done in real time" [27], are important implications for future group research in development processes. We assume this is also mandatory for the evaluation of said interdependencies. It is proposed, that the presented aspects of the sociotechnical system need to be taken into consideration when coding schemes for real time analysis of cooperation and development processes are designed.

5. EFFECTIVE MODELLING AND EFFICIENT DEVELOPMENT OF SYSTEMS

The following section briefly introduces a systems engineering approach that respects aspects of the sociotechnical meta-model. It establishes a framework to support effective modelling, efficient development as well as complexity determination of technical systems. A fundamental basis of contemporary engineering is the generation and use of specialized, partial models to facilitate the different tasks that occur during the development process of modern products. Those tasks and their corresponding models range from the specification of a product (specification models), its functional and detailed design (functional and geometrical models) to the validation of its components (simulation models). The rising complexity within technical systems and within the technical genesis forms the need to establish new methods that help to display and design systemic dependencies between technical and social system entities. In order to reach effective modelling of technical systems the approach is to develop a comprehensive, detailed UML meta-model considering the sociotechnical meta-model. It comprises all partial models, including their relations, and embeds those models into the process of their technical genesis. This UML meta-model will serve as a scheme for PDM systems, enables engineering services for the social system and hereby supports an efficient development of technical systems.

5.1. Detailed UML meta-model

The developed UML scheme describes the coherence between the most important types of entities within the technical genesis of automotive systems. A detailed description of the model would go beyond the scope of this paper. However, this paragraph briefly introduces an extract from the scheme. The model of a technical system represents an aggregation of different partial models that can correspond to function, structure and hierarchy of its original when put into relation.

The UML meta-model distinguishes macro systems (*product models* e.g. the Mercedes-Benz E-Class), meso systems (*technical subsystems* e.g. the powertrain, the lighting system etc.) and micro systems (*elements* e.g. screws, gears, wires etc.). A micro system represents the smallest technical system. Its internal structure and hierarchy are not considered by the OEM. An analysis has led to the definition of eight major dependencies that are used to compose the meta-model (see Figure 4). 'Inheritance' indicates a child / parent relation and can be read as "A is a type of B". Children inherit relations from their parents (e.g. *requirements* inherit the ability to have *parameters* from *partial model*). 'Derived from relation' indicate an existence due to the existence of another entity. 'Specifies relation' can be read as A specifies B or B fulfils A. 'Defines relation' express that existence and attributes of B are defined by A. 'Coupling' refers to the link between in- and outputs e.g. between *technical functions*. 'Composition' and 'aggregation' follow the UML specification.

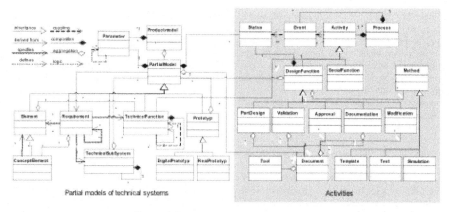

Figure 4: Extract from the UML meta-model

The figure also shows some of the classes that are necessary to integrate the technical system into its sociotechnical development process. Three different types of *activities* within the sociotechnical system are distinguished. The *technical function* which is at the same time *partial model* and *activity*, describes the relation between in- and output attributes (see technical function in the sociotechnical meta-model). *Social functions* represent the functions of the human operation system and the social system. These functions are not considered in detail as the focus of the scheme lies on the development process (see social functions in the sociotechnical meta-model). The *design function* represents an activity within the technical genesis. Although there are a great number of different design functions, only few are worthwhile considering in the scheme. Five of them are exemplarily displayed in the extract. *Activities* and *events* represent a *process* whereas an *event* is defined by a preceding *activity*. *Events* finally define *statuses* that belong to *partial models* or to the relations between them. The UML meta-model is currently being implemented for two examples in the automotive industry. The mechatronical rear view system focuses on the challenges of systems engineering at the interface between mechanical and electrical development. The development of a vehicle's side panel is looked at in respect to the collaboration between engineering design and styling.

5.2. Engineering services to improve Social Systems Engineering

The developed UML meta-model was translated into a class diagram and is currently implemented in a software prototype. The software represents a framework that supports different graphical representations like SysML diagrams, realizes interfaces to specialized modelling tools like CAD or requirements management software and provides access to the data-backbone that is implemented as a SQL database. The framework embeds engineering services which support efficient development, e. g.:

- The visualization of context related views enables the providence of relevant information to specific engineering tasks and supports a domain comprehensive understanding of the system. Back and forward tracking between the different entities allows engineers to browse, understand and analyze the system. Impacts of modifications can be determined and rationales be retraced which helps engineers to cope with the technical complexity.
- Synchronization of engineering data supports the process of coordination between different social systems. That can e.g. be applied to validation processes where the provision of consistent, up-to-date input data, post processing of results and propagation of status between different departments are supported.
- The system models enable a fast creation and evaluation of systemic alternatives and support the decision-making process. The models can be used as documentation of the technical genesis including the performed changes, chosen and rejected concepts and their corresponding responsibilities.
- The provision of system indicators like technical and social complexity enable an analysis of product development's efficiency and enable derivation of actions for system governance.

6. CONCLUSION

The presented approach enables a combined description of social and technical systems at the same time. This facilitates the determination of system parameters, impacting systems self governance, and represents the basis for the definition of system models that support the sociotechnical process of technical genesis. Both aspects present the basis for the analysis of organizational efficiency for an entire self-organization of social systems as well as the efficient development of technical systems = Social System Engineering (SSE). However, there is still a need to improve methods for real-time interaction and communication analysis, improve dynamic network analysis as well as to enable an effortless generation of technical system models in order to enable a real-time capture and evaluation of the system parameters and to support efficient system development.

REFERENCES

[1] Ropohl G. *Allgemeine Technologie,* 2009 (Universitätsverlag Karlsruhe, Karlsruhe).
[2] Willke H. *Systemtheorie I: Grundlagen,* 2000 (UTB, Stuttgart).
[3] Schreyögg G. *Organisation, Grundlagen moderner Organisationsgestaltung,* 2008 (Gabler Verlag, Wiesbaden).
[4] Tomasello M. *Die Ursprünge der menschlichen Kommunikation,* 2009 (Suhrkamp Verlag, Berlin).
[5] Flood R. L. Complexity: A definition by construction of a conceptual framework. *Systems Research,* 1987, 4(3), 177-185.
[6] Ehrlenspiel K. *Integrierte Produktentwicklung, Denkabläufe, Methodeneinsatz, Zusammenarbeit,* 2003 (Carl Hanser Verlag, München).
[7] Dörner D. *Die Logik des Misslingens,* 1989 (Rowohlt Verlag, Reinbeck bei Hamburg).
[8] Reither F. *Komplexitätsmanagement,* 1997 (Gerling Akademie Verlag, München).
[9] Papadimitriou C. H. *Computational Complexity,* 1994 (Addison-Wesley).
[10] Shannon C. E. and Weaver W. A mathematical theory of communication, *Bell Systems Technical Journal,* 1948, (27), 379-423.
[11] Suh N. P. *The Principles of Design,* 1990 (Oxford University Press, New York).
[12] Steward D.V. The design structure system: a method for managing the design of complex systems. *IEEE Transactions on Engineering Management,* 1981 28(3), 71-74.
[13] Eppinger S. D., Whitney D. E., Smith R. P. and Gebala D. A. A model-based method for organizing tasks in product development. *Research in Engineering Design,* 1994, 6(1), 1-13.
[14] Morelli M. D., Eppinger S. D. and Gulati R. K. Predicting technical communication in product development organizations. *IEEE Transactions on Engineering Management,* 1995, 42(3), 215-222.
[15] Browning T. R. Exploring integrative mechanisms with a view toward design for integration, advances in concurrent engineering. *In Fourth International Conference on Concurrent Engineering: Research and Applications,* CE´97, August 1997, pp.83-90 (Technomic Publishing).
[16] Danilovic M. and Börjesson H. Managing the multi-project environment. *In Proceedings of the 3rd international design structure matrix workshop,* Cambridge, October 2001.
[17] Maurer M., Pulm U. and Lindeman U. Tendencies towards more and more flexibility. *Proceedings of the 5th dependence structure matrix (DSM) international workshop,* Cambridge, October 2003.
[18] Maurer M. *Structural Awareness in Complex Product Design,* Dissertation, 2007 (Technische Universität München, München).
[19] Wassermann S. and Faust K. *Social Network Analysis: Methods and Application,* 1994 (Cambridge University Press, Cambridge).
[20] Borgatti S. P., Everett M. G. and Freeman, L. C. UCINET for Windows: Software for Social Network Analysis, *In Harvard: Analytic Technologies,* Cambridge, 2002.
[21] Batallas D. A. and Yassine A. A. Information Leaders in Product Development Organizational Networks: Social Network Analysis of the Design Structure Matrix. *Engineering Management, IEEE Transactions on Engineering Management,* 2006, 53(4), 570-582.
[22] Collins S. T., Yassine, A. A. and Borgatti, S. P. Evaluating product development systems using network analysis. *Systems Engineering,* 2009, 12(1), 55-68.
[23] Gokpinar B., Hopp W. J. and Iravani S. M. R. The Impact of Misalignment of Organizational Structure and Product Architecture on Quality in Complex Product Development. *Management Science,* 2010, 56(3), 468-484.

[24] Trier M. *Virtual Knowledge Communities*, 2007 (VDM Verlag, Saarbrücken).
[25] Fischbach K., Gloor P. A., Putzke J. and Oster D. *Analyse der Dynamik sozialer Netzwerke mit Social Badges*, In Stegbauer, C. (Publ.), Netzwerkanalyse und Netzwerktheorie, 2008 (VS Verlag für Sozialwissenschaften | GWV Fachverlag GmbH, Wiesbaden).
[26] Uflacker M., Skogstad P., Zeier A. and Leifer L. Analysis of virtual design collaboration with team communication networks. In International Conference on Engineering Design, ICED'09, Vol. 08, Stanford, August 2009, pp.275-286.
[27] Törlind P., Sonalkar N., Bergström M., Blanco E., Hicks B. and McAlpine H. Lessons learned and future challenges for design observatory research. In International Conference on Engineering Design, ICED'09, Vol. 08, Stanford, August 2009, pp.371-382.

Contact: Dr.-Ing. Thomas Naumann
Daimler AG
Research & Development
Wilhelm-Runge-Str. 11
89081 Ulm
Germany
Phone: +49 731 5052174
Fax: +49 711 3052176083
E-mail: thomas.t.naumann@daimler.com

Contact: Ingo Tuttass
Daimler AG
Information Technology Management
Epplestraße 225
70567 Stuttgart
Germany
Phone: +49 711 17 91412
Fax: +49 711 1779052607
E-mail: ingo.tuttass@daimler.com

Contact: Oliver Kallenborn
Daimler AG
Information Technology Management
Epplestraße 225
70567 Stuttgart
Germany
Phone: +49 711 1794327
Fax: +49 711 1779052657
E-mail: oliver.kallenborn@daimler.com

Contact: Simon Frederick Königs
Daimler AG
Research & Development
Wilhelm-Runge-Str. 11
89081 Ulm
Germany
Phone: +49 731 5054843
Fax: +49 711 3052119935
E-mail: simon.koenigs@daimler.com

IMPROVING DATA QUALITY IN DSM MODELLING: A STRUCTURAL COMPARISON APPROACH

Steffen F- Schmitz[1], David C. Wynn[2], Wieland Biedermann[1], P. John Clarkson[2] and Udo Lindemann[1]
[1]Technische Universität München, [2]University of Cambridge

ABSTRACT
The Dependency Structure Matrix (DSM) has proved to be a useful tool for system structure elicitation and analysis. However, as with any modelling approach, the insights gained from analysis are limited by the quality and correctness of input information. This paper explores how the quality of data in a DSM can be enhanced by elicitation methods which include comparison of information acquired from different perspectives and levels of abstraction. The approach is based on comparison of dependencies according to their structural importance. It is illustrated through two case studies: creation of a DSM showing the spatial connections between elements in a product, and a DSM capturing information flows in an organisation. We conclude that considering structural criteria can lead to improved data quality in DSM models, although further research is required to fully explore the benefits and limitations of our proposed approach.

Keywords: Design Structure Matrix, Knowledge elicitation, Structural similarity

1 INTRODUCTION
The Dependency/Design Structure Matrix (DSM) is a modelling approach which can help to visualise and manage the structure of dependencies in a complex system such as a product, process or organisation [1]. It is a useful tool for eliciting the structure of dependencies, as each cell in the matrix can be systematically considered to determine whether a dependency exists and what its nature is. Many analysis approaches based on DSM modelling have been proposed to assist the design, optimisation and maintenance of complex systems. To illustrate, Table 1 summarises some methods which have been used to analyse system structures modelled as a DSM.

 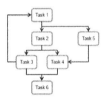

Figure 1: The Design Structure Matrix represents the dependencies in a complex system

The DSM has proved to be a useful tool for eliciting and analysing system structures. However, as with any modelling approach, the insights gained from analysis are limited by the quality and correctness of input information [1,2]. Relatively few studies have provided methods to create better quality DSMs or to evaluate the quality of DSMs. Furthermore, in a recent survey of DSM modellers, "methods for data elicitation" was identified as one of the most pressing opportunities for improving the methodology [3]. In this paper, we discuss methods which allow DSM modellers to raise the quality of their models by cross-checking, helping to pick up accidental mistakes during the elicitation process. In turn this should increase the accuracy, objectivity, and confidence in DSM-based analyses. We limit the analysis in this paper to consider acquiring only the existence, or not, of binary dependencies between a predefined list of elements in a system. We do not consider procedures for eliciting the 'strength' of connections or for eliciting the elements that comprise a system.

Table 1: Some approaches to analysing a system structure represented as a DSM

Analysis	Description, application
Clustering	Identifies clusters, where elements within a cluster possess many dependencies with each other but few with elements in other clusters.
Distance matrix	The distances between the system elements. It can be realigned to identify groups of indirect dependencies.
Matrix of indirect dependency	The number of paths between any pair of elements.
Partitioning, sequencing	The reordering of the rows and columns of a DSM in order to place all elements a) on one side of the diagonal or b) at least close to the diagonal has several names such as partitioning or sequencing.
Banding	Indicates mutually-independent groups of consecutive elements in a given sequence.
Change propagation	Identify how change initiated in one subsystem can propagate through dependencies to ultimately require rework in many others
Process simulation	Identify how rework generated by interdependent tasks adds to project lead time

2 BACKGROUND

It has been said that all models are created for a purpose; and that while all models are wrong, some are more useful than others. This section discusses some of the aspects of data acquisition which influence the utility of a model to support a given purpose, and reviews some prior research regarding data acquisition for the Dependency Structure Matrix (DSM).

2.1 Different sources of information

A number of acquisition-related activities are required to create any DSM. Some of the key activities are shown in Table 2 (in reality these steps may be disordered and iterative).

Table 2: Some key activities in acquiring information for a DSM model

Activity in acquisition process	Example (for vacuum cleaner)
1. Identify breadth of modelling	Entire product (but not user or use context)
2. Identify depth of modelling	Major parts (all mouldings but not screws etc.)
3. Identify types (and sub-types) of element	Modules (e.g., cyclone) and parts (e.g., mouldings)
4. Identify types (and sub-types) of dependency	Physical connections only – spatial mechanical, spatial static
5. Create (possibly hierarchical) list of elements	Handle, cyclone rear moulding, cyclone top moulding, etc.
6. Elicit dependencies between elements	Via dismantling workshop
7. Check data	Via cross-checking

Considering the elicitation of dependency information, which is the focus of this paper, several methods may be used. These include:

- **Direct extraction of dependencies from databases or other 'hard data'** (e.g., a PLM system or links on the world-wide web)
- **Analysis and superposition of existing documentation** (e.g., process maps or org. charts, each covering some part of the system to be modelled)
- **Questionnaires/Surveys in which dependencies are elicited directly** (e.g., asking: who do you talk to in an organisation?)
- **Workshops with domain experts** (e.g., working systematically across rows and columns and considering each possible dependency in turn)

2.2 Different perspectives of dependencies in a system

The more complex a system is, or the more stakeholders interact with it, typically the more perspectives are available from which information can be elicited and from which a model of dependencies can be built. A complex system such as an organisation or a process, which to a large extent exists only in the minds of its stakeholders, can be considered to contain several sorts of information that can be viewed from different angles and perspectives. For instance, some people are closely involved in a particular design process and therefore have detailed knowledge, whereas others may only have a broad overview of that process, on a more abstract level. Some would consider work in a process to be divided into business functions whereas other would organise their view according to lifecycle phases. Some participants would see the pertinent border of a system to be wide, whereas others would set the modelling scope more narrowly. Furthermore, not all participants have equal knowledge of a system – which should be considered when acquiring data. Depending on the type of a system, knowledge about its dependencies may take several forms. These can be considered along different axes, including: *Objective vs. Subjective dependencies* (e.g., connections in a product vs. communication flows in an organisation; *Tacit vs. Explicit knowledge*; and Internal vs. *External information* (i.e., information which must be elicited directly from process participants vs. that which can be identified from documents).

2.3 Prior research on data acquisition for the DSM

One of the main issues associated with data acquisition for the DSM is overcoming problems of scaleability. While a benefit of the DSM method is its suitability for systematic consideration of possible connections, this benefit is quickly eroded as the number of elements increases, because the number of potential dependencies is $n^2 - n$, where n is the number of elements. Thus, eliciting the connections between 10 elements requires consideration of up to 90 cells, while a matrix with 100 elements requires consideration of 9,900 cells. Complex engineering systems, such as aircraft or design organisations, may easily contain thousands of elements which could be modelled.

In general, moderately large DSM models may be difficult to elicit for a number of reasons, including:

- **High effort.** A lot of effort is required to consider all possible dependencies, as explained above.
- **Distributed knowledge.** People may not know about the whole system; it may be difficult to identify who should be asked about what, and it may be necessary to negotiate regarding the existence (or not) of particular dependencies.
- **Disorientation.** In a large matrix, dependencies may be easily placed in the wrong cells – due to alignment difficulties or disorientation when faced with a large grid.
- **Ambiguity.** System elements may be misinterpreted if they are similar or given similar names, so that dependencies may be mistakenly identified.
- **Interface limitations.** It is difficult to visualise large matrices, using computer software or paper methods – "like viewing a map through a letterbox".
- **Fatigue.** Modellers may become fatigued; typically the cells in the first few rows of a large matrix are more carefully considered than the last few when identifying dependencies.

A number of authors have discussed approaches to mitigate such concerns when acquiring data for the DSM, aiming to reduce acquisition effort, to improve modelling quality, or both. Many of these authors have quantified the effort reductions of their approaches, in terms of the number of elicitation operations required for different schemes. Some of the existing approaches are summarised in Table 3.

Table 3: Examples of approaches to support DSM data acquisition in the literature

Approach	Explanation	Effort implications	Quality implications
Element hierarchical-sequential (effort reduction)	Elicit one matrix connecting dependencies between subsystems, to rule out possible dependencies at lower level (e.g. [1])	Reduce effort by focusing elicitation (not all cells require consideration)	Possibly leads to missed dependencies between weakly connected subsystems
Element hierarchical-sequential (quality improvement)	Elicit one high-level matrix and one low-level matrix. Look for discrepancies (e.g. [1])	Increase effort (subsystem dependencies also required)	Improves quality by hierarchical comparison
Element hierarchical-concurrent	Divide the system into subsystems. Have groups of experts elicit dependencies within a subsystem, and interface dependencies (e.g. [4])	Reduce effort by focusing elicitation (although interfaces elicited multiple times)	Improve quality (of interface descriptions only)
Dependency hierarchical-concurrent	Break down dependency type into sub-types, elicit matrices, then combine. (e.g. [5])	Increases effort	Improves quality by deeper consideration of the existence (or not) of a dependency

2.4 Summary

The ultimate objective of methods to support data acquisition for the DSM is to reduce effort and/or improve quality. In one sense, improving quality can be considered as minimising errors in the model. However, for many systems which are too complex or subjective to access directly, it is difficult to

quantify 'error' – even in principle, because the model itself provides the only baseline for comparison. Therefore, in this paper we aim to develop methods to help identify and minimise disagreement between views of a system, which we view as helping to 'triangulate' between perspectives and thus reach a better understanding. We consider that such disagreements may include:
- **Disagreements in perspective/interpretation** – i.e., different models take a different scope or focus, or disagree on the existence (or not) of a dependency.
- **Disagreements in data** – i.e., data does not reflect the system as represented by other data.
- **Disagreements in transcription** – i.e., the modeller may make mistakes while creating a model, which as a result unintentionally differs from their understanding of it.

3 APPROACH OVERVIEW

The premise of this paper is that disagreements in models can be highlighted by comparisons between different points of view, allowing their reconsideration and ultimately improving data quality. We consider comparisons between a given model and:
- A prior understanding of the system that model represents;
- Another model elicited by another person or method;
- Another model elicited from a different perspective or different level of abstraction.

In particular, comparisons are made using the following guiding principles:
- **Network structure constraints can highlight disagreements between a model and an understanding of the system type.** Knowledge of the type of system represented in a DSM can provide information to the elicitation process which results in better-quality DSM models.
- **Structural comparison between models can highlight disagreements between models.** Multiple views of the dependencies in a system, elicited from different perspectives, can be compared from a structural point of view, highlighting potential weaknesses in the data and thus assisting the modeller in raising its quality.

Each of these principles is discussed in greater detail below.

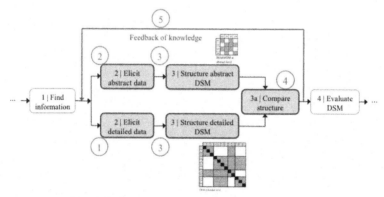

Figure 2: Comparison of data on different levels of abstraction

3.1 Network structure constraints

If prior knowledge about a system is taken into account, it is possible to place very basic network structure constraints on the DSM. For instance, a matrix of structural interactions between physical parts, or of communication frequency between individuals in an organisation should be symmetric. Likewise, a reporting network in an organisation, or certain workflows, should contain no cycles. Considering these constraints, it is possible to identify marks which are not bidirectional or cycles in the network, and highlight them for consideration by the modeller. Each item thus identified suggests that a mistake has been made, and helps pinpoint its possible location.

3.2 Structural comparison between models

A system can be viewed from many different perspectives. However, all these views should have some aspects in common. Thus it should be possible to spot some inconsistencies in a system model by comparing two or more different perspectives. This strategy is used by several of the approaches listed in Table 3; but is expanded here as it is central to our proposed approach.

The most conceptually straightforward comparison approach is to elicit the dependencies between a set of system elements twice, by different methods or from different stakeholders, and compare directly. However, this is effort-intensive.

A second approach, summarised in Figure 2 is to compare a detailed DSM with a more abstract DSM, where multiple elements in the detailed DSM can be mapped to one in the abstract DSM. (For instance, a set of parts in the detailed DSM of a product might be mapped to a single module in the abstract DSM). Apart from reducing effort in comparison to the first approach, a secondary benefit of this method is that the dependencies in the abstract DSM can potentially be elicited by people that are broadly familiar with the system concerned but do not necessarily have all the detailed knowledge. The two matrices can then be compared directly, by assuming that a mark between two elements in the abstract DSM should correspond to one or more dependencies between their sub-elements in the detailed DSM. The approach taken in this paper aims to extend this idea by comparing matrices according to their structural characteristics, as explained below.

From a structural analysis standpoint, all discrepancies between two supposedly-congruent views of a system are not equally important. Where multiple discrepancies exist, it should be possible to focus on those which are likely to have most significance to a structural analysis of the system. Thus, we propose that dependencies in two perspectives of a system can be ranked according to their structural importance, according to multiple criteria. If the two models do not agree, but the structural significance of the disagreement is low according to some criteria, it might be said that comparison of the perspectives indicates high-quality data. On the other hand, if the comparison indicates that the matrices have high structural disagreement, it might be said that further elicitation work is necessary to refine the information. Furthermore, it may be possible to pinpoint the important disagreements, thus helping to focus efforts to improve the data.

4 CRITERIA FOR STRUCTURAL COMPARISON BETWEEN DSM MODELS

In this section, we discuss metrics which were considered as the basis for structural comparisons between DSM models in this paper. The comparison approach is then illustrated through two case studies of dependency structure elicitation.

4.1 Network comparison metrics

Numeric comparison metrics convert each matrix into a single number. The numbers can then be compared directly to see if the matrices agree according to the criterion of interest.

Degree of Connectivity

All existing edges (EE) between the elements (n) are put in relation to the quantity of all possible edges (PE) for both DSMs being compared. The ratio (R) between EE and PE is known as the degree of connectivity. Mathematically this can be described as follows:

$EE = count\ of\ all\ existing\ edges$	Equation 1
$PE = n \times (n - 1)$	Equation 2
$R = EE / PE$	Equation 3

This ratio results in the degree of connectivity (DoC) which can assist plausibility checks in similar systems or models of the same system [6]. To enable comparisons, the DoC for a detailed matrix should be less than or equal to that for a abstract matrix of the same system (because one connection in a cell in the abstract matrix implies at least one in the equivalent cells of the detailed matrix).

4.2 Dependency comparison metrics

Dependency comparisons convert each matrix into a matrix of numbers, where each cell in the result indicates the importance of the corresponding dependency in the original matrix according to a particular criterion. These metrics can then be compared for different views of the same system to highlight large discrepancies. To compare metrics from matrices of different sizes, it is necessary to map the score for each dependency in the abstract DSM onto the equivalent dependencies in the detailed DSM. The means by which this is achieved depends on the metric under consideration, as discussed below.

Minimum distance

The minimum distance metric describes the shortest possible distance between two given elements by traversing dependencies in the structure. A distance of one, for example, means that a direct dependency exists between the two elements. A distance of two indicates that the shortest path between the two elements is of length 2. A matrix containing binary direct dependencies (DD) is equivalent to a distance matrix of length one (ID1), and can be transformed to a distance matrix with indirect dependencies (ID2) of length two by squaring – and so on for higher powers. Mathematically:

$ID1 = DD$ *Equation 4*

$ID2 = ID1 \times DD = DD^2 = ID1^2$ *Equation 5*

$Dn = ID(n-1) \times DD$ *Equation 6*

The minimum distance metric between any pair of elements may thus be calculated by finding the lowest value of n for which the dependency first appears in Dn.

For the structural comparison, it is necessary to compare minimum distances for matrices of different size. This requires consideration of the number of detailed sub-elements which are wrapped up into a single abstract element. For instance, if two elements in the detailed matrix are wrapped into one in the abstract matrix, the four distance values are added, divided by four and divided by its square value. This considers the fact that that elements which are connected through a short chain of dependencies generally have stronger influence on each other than pairs that are connected only via long chains.

$RV\ (detailed) = \sum 1/(4 * SEV2)$ *Equation 7*

A further step is required to take the higher number of subelements on the detailed level into account. For each of the two matrices, cell values that are above the average value of all cells in the given matrix are set to 1; otherwise, the cell value is set to 0. The matrices are then combined by setting the value of a given cell to 1 if the two matrices are in agreement for that cell (both 0 or both 1); otherwise the value is set to 0. This single matrix is then unwrapped to yield an x-y graph as shown in Figure 4. To illustrate application of this metric, the top graph shows comparison of a detailed DSM to an abstract DSM which exactly summarises it. The lower graph shows the same matrix with one incorrect dependency introduced. This highlights the impact of the incorrect dependency on the minimum distance between several other pairs of elements. The 'amount' of discrepancy as revealed by this metric depends both on the structure of the data and the location of the error. It can be used to highlight and compare the potential importance of discrepancies.

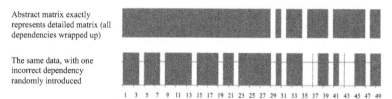

Figure 4: Example comparison using minimum distance criterion

Number of indirect dependencies of length N

Whereas the distance matrix is built from the lowest possible length of indirect dependency, the number of indirect dependencies metric, concerns the number of indirect dependencies between a given pair of elements for a given path length.

The value (RV) of a given cell in the abstract matrix is generated as follows. The value (V) shown in the matrix of indirect dependencies of a specific length is multiplied by the amount of relations (AR) in the detailed DSM. This takes into account that the relations between subelements do not all need to exist if the corresponding relation exists on the abstract level.

$$RV\ (abstract) = V \times AR \qquad\qquad Equation\ 8$$

The value (RV) of the same cell for the detailed matrix is calculated as follows. The degree of filling of subelements is calculated dividing the number of existing relations (EE) by the number of possible relations (PR) for the given matrix. This value is multiplied by the amount of indirect dependencies per sub-element of length four and scaled by an appropriate value to allow direct comparison.

$$RV\ (detailed) = 0.125 \times (EE\ /\ PR) \times \sum SEV \qquad\qquad Equation\ 9$$

These values are calculated for each cell in the matrix and plotted on a 2D chart using similar technique to that described above. Using the same example as above, this is plotted in Figure 5 to illustrate. In each chart, the RV for each cell in the detailed matrix is shown in grey; the corresponding RV for the abstract matrix is shown in blue. This metric clearly highlights the location of discrepancies, but also the magnitude of their impact in terms of the number-of-indirect-dependencies metric.

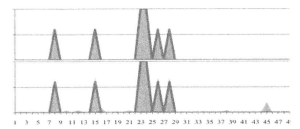

Figure 5: Example comparison using number of indirect dependencies criterion

4.3 Summary

The three metrics discussed above are not an exhaustive selection of the comparison criteria that could be used. They were chosen for illustrative purposes and are used in the case studies described below. Any network structure metric could be used for identifying important discrepancies; for instance, Kreimeyer [7] identifies over 100 metrics that could be adapted to serve this purpose. Ultimately, the particular set of metrics which should be used would depend upon the purpose for creating the matrix. For instance, the sensitivity of a process simulation or change propagation analysis to differences in certain metrics might be more pronounced than others. In such cases, the more critical metrics should be identified and used as the basis to support elicitation. On the other hand, if the objective is to elicit an overall system structure without a particular analysis in mind (for instance, to gain overview of a product architecture), or to create a model for multiple purposes, it might be appropriate to consider a broader set of metrics when comparing multiple views of a system to consider their quality.

5 CASE STUDIES

To illustrate the proposed approach, case studies were undertaken in which multiple models were elicited of 1) the structure of components in a product; and 2) the communication flows in an organisation. These DSMs were created using the Cambridge Advanced Modeller software, and a structural comparison was performed to highlight discrepancies in the DSMs for each case.

5.1 Product DSM of a vacuum cleaner

System overview

The vacuum cleaner fulfilled the criteria of being simple enough to understand and comprising a manageable amount of elements, while remaining complex enough to justify modelling and being able to sensibly cluster the list of parts into modules to obtain an abstract DSM for structural comparison. Figure 6 shows the device which was used, the *Argos Value VC9730S-6*.

Data acquisition
Acquiring lists of elements. The vacuum cleaner was taken apart and 30 parts identified, focusing on 'major' parts such as mouldings and not including connectors, screws, clips etc. On the abstract level two different modularisations were identified – namely, a 5 x 5 and 7 x 7 DSM.
Acquiring dependencies for detailed DSM. To identify all spatial connections between the parts, disassembly workshops were held with each of eight participants. Each participant was given the vacuum cleaner with fasteners removed to ease disassembly. They were also given the empty 30x30 matrix, with instructions to capture spatial connections between the listed parts. The time taken to elicit the detailed level DSM (435 possible relations) was about an hour for each participant.
Acquiring dependencies for abstract DSMs. All three abstract level DSMs (37 possible relations) were filled during similar workshops. Five people did this independently. These people were not involved in the detailed elicitation workshops, thus had not encountered the detailed description of the vacuum cleaner. On average, this took about 10 minutes for a participant to elicit all three abstract level DSMs, for a total of 20 possible relations.

Overview of acquired data
In the abstract DSMs shown in Figure 6, each relation between clusters is shown if at least three of the participants see a relation between clusters. In the detailed DSM, a mark is shown if at least four of the eight participants had identified a relation between those elements.

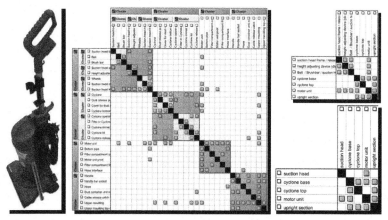

Figure 6: Different views of the vacuum cleaner and connections within it.

Analysis of acquired data
Comparing the 30x30 and 5x5 matrices according to the hierarchy criterion results in a 90% match. Both DSMs are entirely symmetric, which does not highlight any possible errors on the symmetry criterion and suggests a good-quality model. As expected, the degree of connectivity for the detailed DSM is lower than for the abstract DSM. Finally, Figure 7 shows the comparison between the detailed (30 x 30) and the abstract level DSM (5 x 5) using the two dependency comparison metrics.

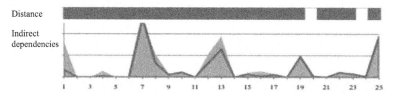

Figure 7: Cell-wise comparison of the dependency comparison metrics shows few disagreements between models

The metrics thus indicate that the two DSMs are strongly in agreement with regards to the basic structural criteria discussed. This suggests that the model is of high quality, which was expected because the system being modelled was largely objective, not too complex, and the two models each represented an agreement between multiple modellers working independently.

5.2 Organisation DSM of a research group

System overview
The second case study is the elicitation of an organisational DSM based on the communication flows in the Engineering Design Centre (EDC) in the Engineering Department at the University of Cambridge. This differs from the first case in that the existence (or not) of a dependency is potentially far more subjective, and because knowledge of communication flows is distributed among many people. Thus, more disagreements would be expected than for the vacuum cleaner model.

Data acquisition
Acquiring lists of elements. The lists of elements were acquired directly from the EDC website. For the detailed DSM, 47 researchers working in the EDC were identified (not including academic staff). For the abstract DSM, the seven research themes within the EDC were listed.

Acquiring dependencies for detailed DSM. An online survey was constructed and distributed to capture the interaction between individuals. Each member of the EDC identified on the list of elements was asked to rate the frequency and intensity of communication with every other member (i.e., to work down the list of 47 and select either none, low, medium or high for each of frequency and intensity). Individuals were also asked to identify which of the 6 research themes they work in, where each person may work in more than one theme. 45 of 47 members completed the survey.

Because only binary dependencies are considered in the structural comparison metrics, the responses were filtered to show a dependency between two people if a medium or high level of communication was described for both frequency and intensity. This filtered out the weaker dependencies within what would otherwise be a very strongly-connected model, and allowed its treatment as a binary matrix.

Acquiring dependencies for abstract DSMs. The interaction between research themes was extracted in two ways, resulting in two abstract matrices. Firstly, each EDC member was asked, while filling out the survey described above, to also indicate the levels of communication (frequency and intensity) that they had with each of the 7 themes as a whole. Thus, if they spoke to any person within a given theme on a daily basis, they would select 'high' for frequency of communication with that theme. These responses were compiled into a single 7x7 abstract DSM using the filtering procedure outlined above. Secondly, five Senior Research Associates were asked to separately fill a 7x7 DSM indicating the frequency and intensity of communication they believed occurred between the seven research themes. These 5 DSMs were filtered individually, then accumulated into a single abstract DSM by including only those dependencies which at least 4 of the 5 participants had identified.

Overview of acquired data
The three DSMs which were acquired are shown in Figure 8.

Analysis of acquired data
The comparison between the detailed matrix and each of the two abstract matrices, according to the structural metrics, is shown in Figure 9 and discussed below.

Comparison of detailed DSM to abstract DSM obtained through survey
The hierarchy constraint shows well over 50% correlation between the detailed organisation DSM and the abstract DSM obtained through the survey. This shows that relationships between members of the EDC and between research themes match fairly well. In cases where no relations between research themes exist, no or few relations between their members exist, and vice versa. Considering the symmetry constraint, both levels of abstraction were expected to be completely symmetric, yet this was not entirely the case. Closer examination of the underlying data set suggested that certain respondents responded across all interactions that the level of communication to their colleagues is far higher than their colleagues perceive it. This systematic bias, perhaps due to imprecise wording in the survey, accounted for the missing symmetry and could be corrected.

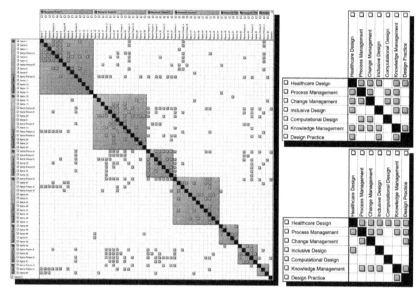

Figure 8: Three views of the communication flows in the EDC, between individuals elicited from survey (left), between themes elicited from survey (right, top) and elicited directly from senior researchers (right, bottom)

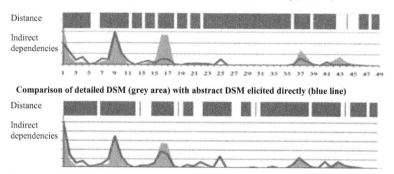

Figure 9: Comparison of detailed DSM to abstract-survey DSM (top) and SRA-DSM (below)

Comparison of the two matrices using distance metrics also highlights quite a number of mismatches. This suggests that the data on the two levels do not perfectly fit together. This suggests that there are still discrepancies that need revisiting if the two matrices are to agree with regards to this metric.

Considering the indirect dependencies, the graph suggests the two matrices match reasonably well. There are several issues that might be worth looking at, specifically cell numbers 1, 16, 17, 25, 37 and 43 as the difference is quite high in these.

In summary, the match between the data elicited through the EDC online survey at different levels of abstraction does not completely match, even though the same people have elicited it. Some aspects are quite good and suggest a strong basic match. Nevertheless, there are some inconsistencies especially with the minimum distance and the amount of indirect dependencies. A possible suggestion could thus be: certain relations need to be reconsidered on both levels of abstraction. To localise these, the metric *high amount of indirect dependencies* might need to be broken down into several areas of the whole detailed DSM to spot where exactly mismatches might be.

Comparison of detailed DSM to directly-elicited abstract DSM
The abstract DSM filled directly by the SRAs is not symmetric either; just as the detailed one. The reason is the same as with the detailed DSM and has been described earlier in this Subsection. The degree of connectivity seems of equal quality just as the distance matrix; the according graph shows less but larger inconsistencies. The matrix of indirect dependencies, however, shows a better matching abstract DSM. Especially cells number 1, 16, 17, 37 and 43 have improved. Cell number 25 turns out to be a worse match.

5.3 Summary of case studies
Product DSM. A product DSM is less complicated and more straightforward than an organisational DSM as less subjectivity and personal perception play a role. In this paper, the product DSM illustrates how the structural comparison metrics can differentiate between high- and low-quality data.
Organisation DSM: The metrics clearly show that, while the models have major consistencies, they are not totally alike. This is clearly shown by the presentation of the comparison along a single axis, which facilitates interpretation of the differences between values for particular cells. The data itself is highly subjective, as the symmetry constraint showed through indicating that the different participants had different views on the meaning of a strong or weak dependency. During the processing of survey data to create the DSMs, the data turned out to be affected by transcription errors. Some relations were missed, others which should not have existed were marked. Running the metrics resulted in an unusual looking set of graphs. Especially the metric *matrix of indirect dependencies* was able to locate where errors had been made; each error was considered to determine whether it could be easily explained and corrected. The case study seems to illustrate how comparison of models elicited using different means and from different perspectives, using structural metrics, can add value to the modeller and can supply her/him with additional information and insights. These insights can be used to highlight subjectivity as well as potential mistakes in transcription and other sources of error.

6 DISCUSSION
The results suggest that it is possible to raise the quality of data during the elicitation process by taking different views and perspectives of the same system. The comparison of consistency between the two or more levels, using structural metrics, allows insights about the data quality that has been elicited. Interestingly, it seems possible to gain insights that can help improve data quality, even without considering the real-world implications of the metrics that are discussed.

The approach outlined in this paper is only a starting point which aims to highlight the potential for using structural comparisons to assist in data acquisition for the DSM. Clearly, there are many opportunities to improve the analysis which has been outlined here, and systematise it as a paper-based method or even as a process embedded in a DSM software tool. In terms of the theory, a key aspect that needs attention is the multitude of different structural aspects that could be considered. Even if one structural aspect suggests an inconsistency between the matrices, the elicitation could still be correct because data is lost in comparing abstract and detailed models. Likewise, different metrics would most likely suggest different disagreements or different levels of importance for particular disagreements. The approach discussed in this paper only considers binary DSMs. Many of the methods could be adapted relatively easily for binary MDMs; however additional issues arise when considering DSMs or MDMs containing information about the dependencies, such as their strength.

Finally, it is important to highlight that the approach can only pick up inconsistencies between different perspectives. In the event of the same error occurring on both levels of abstraction the metrics will not work. The approach also cannot help distinguish which is the 'correct' value, when multiple models are in disagreement. However, by pointing out the potential discrepancy, we propose that structural comparisons may help modellers focus their efforts and result in better-quality models.

7 CONCLUSION
The DSM is a useful technique to gain insights into a complex system, such as a product, process or organisation. However, the quality of data in the DSM can significantly affect the quality and believability of insights gained through study of that model. Various methods have been proposed to assist with DSM knowledge elicitation, aiming to reduce effort of acquisition or improving quality of

models by considering and comparing data acquired from different levels of abstraction. This paper has proposed and illustrated an extended approach in which the structural importance of disagreements between perspectives of a system is considered. We argue that highlighting the structural importance of disagreements can help focus the modeller's attention on those potential errors which may have most impact on the structurally-oriented analyses for which DSMs are often used – such as modularisation and simulation. Initial application of the ideas to two realistic DSM-modelling case studies seem promising, but much further work is required to systematise the proposed method.

REFERENCES

[1] Lindemann, U.; Maurer, M.; Braun, T.: Structural Complexity Management – An Approach for the Field of Product Design. Berlin: Springer 2009.
[2] Biedermann, W.; Strelkow, B.; Karl, F.; Lindemann, U.; Zaeh, M.: Reducing data acquisition effort by hierarchical modelling. In: Proceedings of the 12th International Dependency and Structure Modelling Conference, Cambridge, UK, July 2010. Munich: Hanser 2010.
[3] Wynn, D.; Kreimeyer, M.; Eben, K.; Maurer, M.; Clarkson, J.; Lindemann, U.: Proceedings of the 12th International Dependency and Structure Modelling Conference, Cambridge, UK, July 2010. Munich: Hanser 2010.
[4] Ariyo.: 'Change propagation in complex design: predicting detailed change cases with multi-levelled product models', PhD-thesis, Cambridge University Engineering Department, 2007.
[5] Jarrett, T.: 'A model-based approach to support the management of engineering change', PhD-thesis, Cambridge University Engineering Department. 2004.
[6] Eichinger, M.; Maurer, M.; Pulm, U.; Lindemann, U.: Extending Design Structure Matrices and Domain Mapping Matrices by Multiple Design Structure Matrices. In: Proceedings of the 8th Biennial Conference on Engineering Systems Design and Analysis (ASME-ESDA06), Torino. Torino, Italy: ASME 2006.
[7] Kreimeyer, M.: A Structural Measurement System for Engineering Design Processes. Dissertation, Technische Universität München, 2010. München: Dr-Hut 2010.

Contact:
Dr. David C. Wynn
University of Cambridge
Engineering Design Centre
Trumpington Street
Cambridge, CB2 1PZ, UK
Phone +44-1223-748565
Fax +44-1223-332662
dcw24@cam.ac.uk
http://www-edc.eng.cam.ac.uk/people/dcw24.html

Steffen Schmitz was a student at the Technische Universität München, Germany, and graduated in 2010. He focused his studies on product development and structural complexity management. He is now working as a management consultant for companies with complex technical products.
David C. Wynn is a Senior Research Associate at the Cambridge University Engineering Department. His research focuses on computer-based modelling of complex collaborative processes, such as engineering design.
Wieland Biedermann is a scientific assistant at the Technische Universität München, Germany, and has been working at the Institute of Product Development since 2007. He has published several papers in the area of structural complexity management.
P. John Clarkson received his PhD from Cambridge University and worked at PA Consulting before returning to Cambridge. He was appointed Director of the Engineering Design Centre in 1997 and Professor of Engineering Design in 2004. His interests are in the general area of engineering design.
Udo Lindemann is a full professor at the Technische Universität München, Germany, and has been the head of the Institute of Product Development since 1995, having published several books and papers on engineering design. He is committed in multiple institutions, among others as Vice President of the Design Society and as an active member of the German Academy of Science and Engineering.

INTERNATIONAL CONFERENCE ON ENGINEERING DESIGN, ICED11
15 - 18 AUGUST 2011, TECHNICAL UNIVERSITY OF DENMARK

ENHANCING INTERMODAL FREIGHT TRANSPORT BY MEANS OF AN INNOVATIVE LOADING UNIT

Dipl.-Ing. Max Klingender, Dipl.-Ing. Sebastian Jursch

ABSTRACT

Within the project "TelliBox - Intelligent MegaSwapBoxes for advanced intermodal freight transport" a new intermodal loading unit was developed and built as a prototype by an international consortium. This new 45 feet long loading unit is applicable to transport on road, rail, short sea and inland waterways. It combines the advantages of currently available loading units, e.g. manoeuvrability and safety of containers, loading facilities and dimensions of semitrailers and the effective use of loading area of swap-bodies, in one sustainable transport solution. The efficient and successful usage of this new solution could be also verified on a demonstration trial within Europe.

Keywords: Intermodal freight transport, loading unit, container, swap-body

1 INTRODUCTION

In recent decades intermodal freight transport was driven by changing requirements of global supply chains. The improvement of integration and compatibility between modes provides the necessary scope for a sustainable transport system [1, 2]. The encouragement of intermodality brings about the opportunity to better use rail, inland waterways and short sea shipping, which are seldom used at present because individually they do not allow door-to-door delivery. The operation grade of carriers will increase significantly in the coming decades and the European Commission prognoses a suboptimal load of rail and waterways compared to the undue quantity of road transports [3].
Since the road transport system is nearly overloaded and currently does not offer enough potential for technological enhancement to face the future increase of traffic performance, the balancing of the modes of transport i.e. intermodal transport represents a crucial solution within the scope of European transport policies [4].
In 2003, European Parliament and Council proposed a Directive on the standardization and harmonization of intermodal loading units, with the objective to reduce inefficiencies in intermodal transport resulting from various sizes of containers circulating in Europe. Within the actual 7th Framework Program the encouragement of modal shift and intermodal transport focuses to "improve the efficiency of interfaces between modes", to "maximize cargo capacity" and to optimize "logistics services, transportation flows, terminal and infrastructure capacity within European and global supply chains".
Within the last few years new innovations and developments for intermodal transport were made (e.g., Arcus 100, HighCube Container, MegaSwapBody, craneable Semitrailers etc.), but there are still a number of economical, technical and operational obstacles to overcome to provide a beneficial usage of intermodal systems for various freight transports.
Today, the transport system has to face various challenges in terms of safety, reducing traffic congestion and improving loading processes and interoperability of available transport modes. The creation of a sustainable transportation system for Europe and Russia depends furthermore on the cooperation of operators along the transport chains. Interfaces between modes and national transport chains have to gain quality and flexibility, to make intermodality competitive [5, 6].
Additionally, transport costs may be reduced by following the trend towards high-volume loading units. Concerning the dimensions of the loading units, some transport modes may not meet the requirements of every cargo [7, 8]. The lack of standardization concerning intermodal loading units hinders the connectivity of modes and generates costs e.g. by requiring special transhipment technologies [9].
Elementary developed transport solutions concerning dimensions, like the high-cube containers or jumbo semitrailers, but also regarding facilities of loading, for example by curtain-side swap-bodies or

boxes with liftable tops, were introduced to the market. But single solutions often can only be used in special areas of applications and in fact necessitate special operational technologies [10].

The success of intermodal transport solutions compared to road transport depends significantly on cost efficiency of loading processes, improvement of interoperability and the exploitation of a maximized cargo area.

STATE OF THE ART – LOADING UNITS

Currently, many different intermodal loading units (ILU) are used in European freight transport. The three most dominant concepts ISO-containers, swap-bodies and semitrailers have been further developed for certain purposes and markets. A general tendency in the enhancement of loading units concerns the maximization of the cargo area and the facilitation of loading and transhipment processes [11].

These loading units differ amongst others in aspects like dimensions and stability as well as usability regarding handling, transport and loading processes. The following paragraph describes the chief differences and their applicability. It compares the advantages and disadvantages of containers, semitrailers and swap-bodies – focusing their applicability to intermodal transport.

Semitrailers are predominantly used in road transport. Compared to containers and swap-bodies they prove to be more flexible especially in regard to the use of their cargo area. In road haulage semitrailers are preferred because of their flexibility not only in terms of manoeuvrability but also concerning the coupling and uncoupling process. Within the last few years greater sized semitrailers, called "Mega-Trailer" or "Jumbo", have become more and more commonly used. Because of their length and internal height (about 3m) the maximized cargo area is associated with important economic advantages leading to more lucrative transportation [12]. Compared to containers and swap-bodies however, semitrailers are only partially applicable in intermodal transport chains because they are not stackable and require special equipment for handling. In general, only less than 3% of semitrailers are permitted in intermodal transport [13].

Containers are loading units which can be carried on all transport modes and for that reason they offer many advantages for their use in intermodal transport. During the last 50 years maritime containers complied strictly with the ISO standards and were optimized with the introduction of the 40' standard container. Advantages of standard containers are related to efficient terminal handlings and cost effective intermodal operations. Intermodal transport focuses on the use of containers, as they prove to be functional in intermodal transport chains. The stability and stackability of these carriers ensure an efficient transport and loading process. However, when it comes to road transport, containers represent very little per cent of the total tonnage per freight vehicle. The reasons are that compared to semitrailers and swap-bodies containers offer a smaller cargo area and are not adjusted to standardized pallets. They do not effectively utilize the full dimensions allowed on road transport [14].

Swap-bodies are predominantly used in intra-continental freight transport between the European member states carried on road or rail. Therefore, they are adjusted to road transport restrictions and the transport of standardized euro pallets. Swap-bodies are demountable from the chassis but more lightly constructed than containers, so that most of them cannot be stacked or top-handled. The increasing use of curtain-sided and volume-optimized swap-bodies meets the loaders requirements for wide openings on both sides and a maximized cargo area [15]. For example the swap-body "Arcus 100" from Ewals Cargo Care B.V. offers a cargo volume of 100m^3 and has curtains on both sides. These commonly used curtain-sided swap-bodies, however, only resist low horizontal forces in short sea shipping. Also, curtain-sided bodies cannot be certified for TIR and therefore are not able to be transhipped to Russia. Hence, in general swap-bodies are not suitable for intermodal transport including all modes road, rail and waterborne. While European standards in terms of dimensions and general requirements [16, 17], securing of cargo [18] and the coding, identification and marking of swap-bodies [19] have been published in the last 15 years, there are still notable differences in dimensions, fitting and lifting points and other properties concerning the loading and transhipment operations.

When comparing the above listed intermodal loading units a comparison matrix of technical and technological parameters of ILU can be created. For example swap-bodies exist with an internal height of 3m, but none of them can be used for shipping or are equipped with side doors. In the following table (Table 1) the most important demands on an intermodal loading unit for the envisaged usage are summed-up and conferred to existing transport solutions on the market. At present, there is no intermodal unit on the market which meets all, or at least most, listed demands.

Table 1. Comparison of technical and technological parameters of ILUs

	Container (ISO 1A)	Swap-body (series A)	Semitrailer (Jumbo)
Trimodal	+	-	-
Stackable	+	o	-
Handling from top	+	-	-
Cargo volume (100 m³)	-	+	+
Internal height (3 m)	-	+	+
Loading port from three sides	o	o	+
Safety of cargo (theft/pilferage)	+	o	o
Liftable top	o	o	+
Legend: + standard o purpose-build - none			

INNOVATIVE LOADING UNIT – REQUIREMENTS AND DESIGN

For a new efficient intermodal transport system, which can face the current and future transportation, the concept of swap-bodies has to be improved by taking inspiration from the manoeuvrability and safety of containers and the loading facilities and dimensions of semitrailers. Aim should be to combine the individual advantages of the loading units by eliminating the disadvantages. To achieve these objectives innovative and intelligent constructions have to be found respectively combined.

Against this background within the project "TelliBox – Intelligent MegaSwapBoxes for advanced intermodal freight transport" a prototype of an innovative 45 feet long intermodal loading unit applicable to road, rail, short sea and inland shipping has been developed. This new ILU combines all the advantages of containers, swap-bodies and semitrailers integrated within one loading unit for the first time (Figure 1). Current available solutions have not offered such combination. The engineering design of such innovative loading units is complex and has to balance the needs and requirements of customers and users on the hand and necessary sophisticated technical solutions on the other hand. To face this challenge the work of a large consortium consisting of manufactures, freight forwarder, research institutions and associations is necessary.

Therefore, an European consortium consisting of operators, logistic enterprises, manufacturers and research institutes has worked for three years under an European grant (FP7). The scientific methodology of the project is subdivided into six dedicated phases. They follow the proposed general structure by Pahl & Beitz [20]. In the beginning of the development process an extensive analysis phase is carried out. This phase included the analysis and prioritization of all technical, operational and constructive requirements for the innovative loading unit. It results in the solution space. The solution space has to consider new approaches and technological solutions for components which are on the market or prior their introduction to create variants. In this phase all partners have to be involved.

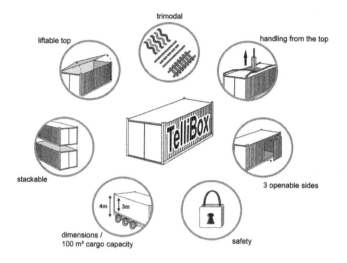

Figure 1. Requirements on the innovative loading unit

The creation of possible solutions within the second phase has been carried out in workshops of the entire consortium. A combination of methods, e.g. scientific speed dating, morphological analysis etc. was used. The result of this work phase consists of approximately three concepts, which appeared to be the most suitable (technically and operationally) for the innovative loading unit.

The elaboration of the chosen variants of the innovative loading unit and the adapted chassis comprised the third phase. Simulation methods like the 'finite element method' were used for the further analysis of the design.

In the fourth phase the solutions were evaluated. Aim of this evaluation phase has been to choose the concept with the best design. A prototype based on this design has been designed in detail. Different calculations of economic efficiency were carried out and in addition the usability of all concepts was evaluated.

The transition to the fifth phase has been the identification of the most economically, operationally and technically suitable design. One prototype of the innovative loading unit has been produced within the construction phase. The last phase, the demonstration phase comprises the testing of the prototype for all transportation modes (road, rail, short sea and inland shipping). Therefore, the handling processes in terminals have been researched. In addition, tests concerning the loading and unloading of commercial freight are carried out.

The result of the test phase has been used for an optimization loop. They also deliver an evaluation of the quality of the prototype that considers both technical and operational aspects. Therefore, an optimization loop is integrated in the scientific methodology. The recommendations out of the demonstration phase are used to improve the design. Therefore, the design process starts again in the third phase. After the needed reengineering, the prototype has been tested and evaluated again.

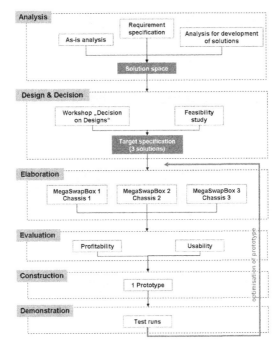

Figure 2. Methodology of design process

The initial main idea of the worked out solution of the new ILU is based on a shoe box. A closed shoe box is more stable, but if one removes the top respectively one or more sides of the shoe box, it becomes unstable. This construction principle has been further developed and applied. So, the loading unit can be stabilized if the side doors are part of the body construction (Figure 3) and in addition linked to the roof with fastening/closing elements. It is therefore important for this construction, that the door hinges and door fastening/closing elements are designed in a way that they can take part of the forces which react on the loading unit when it is being handled.

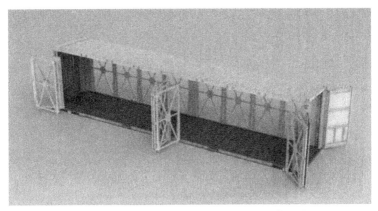

Figure 3. Simulation model of the innovative loading unit

The bottom frame is made up from a simple steel welded construction. To reduce the weight and at the same time maintaining a stable construction, the bottom frame cross members have a trapeze form (Figure 4). The top layer of the bottom frame has a steel and plywood combination as the loading platform. The bottom frame has an integrated goose neck tunnel (Figure 4). It is used for loading the container on the chassis so that the total height of chassis and container does not exceed the maximum allowed height in Europe of 4m (according to [21]).

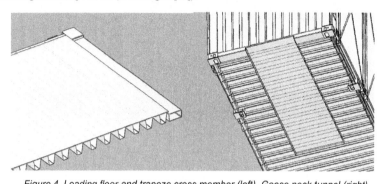

Figure 4. Loading floor and trapeze cross member (left), Goose neck tunnel (right)

The fastening system for the side doors is integrated in the side members of the bottom frame. The bottom frame fastening system is driven using a crank shaft and a gear which are integrated in the bottom frame. The system allows the fastening of one whole side door section using one gear and crank shaft. Because of this system it is possible to close the side doors and fasten them to the bottom frame construction. The front wall is a conventional container front wall of corrugated steel which is currently used also in other container constructions. The side doors are made up of 8 single doors (Figure 3). The door sections are made from a steel construction so that the loading unit is stable. The door hinge fastening systems are integrated in the side doors. The side doors are covered with an aluminium covering. Between the doors, commercial gaskets are mounted that secure the side doors against moisture and water (Figure 5). The rear doors for the new ILU are commercially available container rear doors that guarantee high security. The roof is constructed just like an ISO-Container roof. It is equipped with a rack toothed jack and a gearbox so that it can be lifted at the sides to enable loading and unloading of the system from the sides. Since the new loading unit has a liftable top it needs also a special water drainage system and is therefore included in the roof construction (Figure 5).

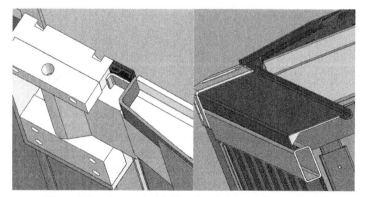

Figure 5. Side door gadget (left), water drainage system roof (right)

TESTING OF THE INNOVATIVE LOADING UNIT

The constructed prototype of the new intermodal loading unit has passed all necessary certifications for commercial use [22]. To verify also the applicability for commercial use, the transport of goods as well as the interoperability of the new intermodal loading unit it was tested and demonstrated on a 5000 km long test run. As demonstration trial a route was chosen where currently automotive goods are transported.

To identify crucial test cases for the test runs fault tree analyses have been worked out in which critical points were defined which need to be thoroughly considered. It has been essential that all project partners – from manufacturers to freight forwarders – have contributed to this process as not only the mechanical properties but also the usability and the applicability of the new loading unit have been subject of the test runs. Based on the definition of undesired events and on an extensive research of all causes with probabilities of affecting the undesired events different fault trees have been worked out (Figure 6). Afterwards the potential causes have been clustered and then used do define the test cases. The work described above was carried out in different guided workshops with members of the TelliBox consortium. From the point of view of the freight forwarders different use cases of the innovative loading unit have been identified and discussed with the manufacturers to specify critical points. For these critical points, e.g. proper door locking mechanism, undesired events have been identified in a following workshop. Based on these findings potential causes of such undesired events have been taken into account to specify test cases for the test runs of the innovative loading unit.

As test cases have been identified:
- above mentioned certifications,
- clearance tests with the tractor/trailer unit combined with the new loading unit,
- function tests of the loading unit under different loading conditions,
- test drives on a test track,
- handling operations at terminals,
- test runs with the entire combination on real roads and on ferries and
- test runs of the loading unit on short-sea ferries and inland-water barges as well as on railway wagons.

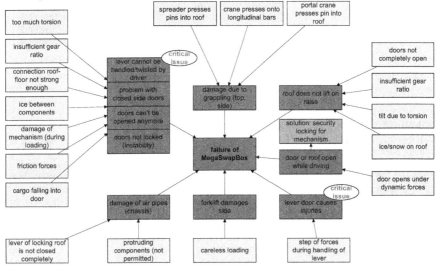

Figure 6. : Example of the analyzed fault trees

All above discussed test cases have been documented by means of photography protocols and standardized test protocols. This extensive documentation forms the basis for an optimization loop of the loading unit.

The practical execution of the test cases has been covered by the following steps (Figure 7): To perform the unloading and loading test, the new loading unit has been transported to a forwarder in Gliwice (Poland) on road. To accomplish the craning from truck onto terminal facilities and onto a railway wagon, the loading unit has been transported by road to the terminal in Katy Wroclwaski (Poland), where the loading unit has been craned onto a Megafret wagon. The subsequent forwarding was carried out by a partner on rail to Hamburg. For the next test case – transport on inland-water barges – the loading unit has been transferred to Mannheim (Germany) on road.

At a terminal in Mannheim, the loading onto the barge have been carried out so that from Mannheim, a test run on inland waters (Rhine) to Rotterdam (Netherlands) have been executed. Having arrived in Rotterdam, the loading unit has been transported on road to Calais (France) to complete the ferry trip to Dover (UK). From Dover (UK), the journey continued on road to Ellesmere Port (UK), where a test loading and unloading has been performed at an automotive company. Subsequently, the loading unit has been transported via Dover and Calais back to Eindhoven.

Figure 7. : Demonstration trial (left), the new ILU in use on the demonstration trial (right)

A mayor result has been that the modes road, rail, short sea and inland shipping could be used successfully within the demonstration. Before using the new intermodal loading unit the goods could not be transported trimodal on this track without exchanging loading units.

CONCLUSION AND OUTLOOK

With the development of the new intermodal loading unit an international consortium consisting of terminal operators, freight forwarders, container manufactures, research facilities, consultants and an association for intermodal transport, succeeded by creating a sustainable solution for the intermodal transport market. Through the orientation on standards and norms the new ILU offers high interoperability and easy application in transport while using the existing handling equipment and facilities. Within the above discussed demonstration trial of the prototype it could be shown that the ILU can be applied successfully commercially in trimodal transport.

For the introduction to the market it would be necessary to modify the new intermodal loading unit according to the requirements of serial production. Above it could be considered to offer the new intermodal loading unit on market in variants for special purposes and customers e.g., only one side openable, with liftable or without liftable roof. This approach would lead to a reduction of the engineering design complexity and would benefit from a modularization of the innovative loading unit introduced in this paper.

REFERENCES

[1] Sondermann, K., *Systemlösungen im kombinierten Verkehr: Rahmenbedingungen, Entwicklungen, Forderungen, Potentiale*, Bochum: RUFIS, 1991
[2] Aberle, G., *Transportwirtschaft - Vierte Auflage,* München: R. Oldenbourg, 2003
[3] European Commission, *Midterm-review of the White Paper on European transport policy*, Brussels, 2006
[4] Commission of the European Communities, *Keep Europe moving – Sustainable mobility for our continent*, Brussels, 2006
[5] Winkler, F., *Schweiz: Kombiverkehr entwickelt sich trotz Ungewißheit*, Internationales Verkehrswesen, (49) 6/97: 334-336, 1997
[6] Polzin, D., *Multimodale Unternehmensnetzwerke im Güterverkehr*, München: Huss, 1999
[7] Bukold, S., *Kombinierter Verkehr Schiene, Straße in Europa : eine vergleichende Studie zur Transformation von Gütertransportsystemen*, Frankfurt a.M.: Peter Lang GmbH, 1996
[8] Kracke, R., *Die Zukunft des kombinierten Verkehrs - was ist zu tun*, 25. - 26. März 1999 in Hannover. München: Huss, 1999
[9] Hoepke, E., *Der LKW im europäischen Straßengüter- und kombinierten Verkehr*, Renningen: Expert, 1997
[10] Seidelmann, C., *Der kombinierte Verkehr – Ein Überblick*, Internationales Verkehrswesen, (49) 6/97: 321-324, 1997
[11] Walter, K., *Intermodale Verkehre: Motor aller Verkehrsträger*, Internationales Verkehrswesen, (57) 10/05, 2005
[12] Buscher, R., *Kombinierter Verkehr Straße Schiene in der Supply-Chain wirtschaftlich tragfähig*, Konradin, 1998
[13] EUROSTAT, *EU Energy and Transport in Figures*, Brussels, 2006
[14] Fonger, M., *Gesamtwirtschaftlicher Effizienzvergleich alternativer Transportketten*, Göttingen: Vandenhoeck & Ruprecht, 1993
[15] Stackelberg, von F., *Kombinierter Verkehr*, Göttingen: Vandenhoeck & Ruprecht, 1998
[16] EN 284 "*Swap bodies – Swap bodies of class C – Dimensions and general requirements*"
[17] EN 452 "*Swap bodies – Swap bodies of class A – Dimensions and general requirements*"
[18] EN 12640 "*Securing of cargo on road vehicles – Lashing points on commercial vehicles for goods transportation – Minimum requirements and testing*"
[19] EN 13044 "*Swap bodies – Coding, identification and marking*"
[20] Pahl, G., Beitz, W., Feldhusen, J., Grote, K.-H.., *Konstruktionslehre*, Berlin: Springer, 2007
[21] Directive 96/53/EC, *Council Directive 96/53/EC, laying down for certain road vehicles circulating within the Community the maximum authorized dimensions in national and international traffic and the maximum authorized weights in international traffic*, 1996
[22] EN 12642 "*Securing of cargo on read vehicles – Body structure of commercial vehicles – Minimum requirements*"

INTERNATIONAL CONFERENCE ON ENGINEERING DESIGN, ICED11
15 - 18 AUGUST 2011, TECHNICAL UNIVERSITY OF DENMARK

REPRESENTING PRODUCT-SERVICE SYSTEMS WITH PRODUCT AND SERVICE ELEMENTS

Yong Se Kim, Sang Won Lee and Dong Chan Koh
Creative Design Institute, Sungkyunkwan University, Korea

ABSTRACT

This paper discusses the new Product-Service Systems (PSS) representation method which configures product and service elements. PSS is composed of a number of product elements and service elements, and they are complicatedly connected to each other to satisfy customer needs. Therefore, it is of much significance to appropriately represent product and service elements and their relations in PSS. In this paper, a new PSS representation scheme to effectively configure the product and service elements is proposed. In the proposed PSS representation scheme, the service elements can be modelled with stakeholders – service provider/receiver, activities and associated product elements. The product elements are included in the service element and serve as media for realizing PSS. To realize the specific function, several service elements can be connected with flows that were identified in PSS functional modelling. Those flows can also be used to connect associated product elements. Finally, case study is conducted to investigate the applicability of the proposed PSS representation method to the real PSS design project.

Keywords: Product-Service Systems (PSS), Product Element, Service Element, PSS Representation

1 INTRODUCTION

Recently, the researches on design and modelling of product-service systems (PSS) have been rapidly growing. PSS, which was firstly introduced by Goedkoop et al. in 1999 to deal with the environmental and economical challenges, have also been of central attention since PSS could satisfy consumers' needs more effectively by providing integrated solutions of products and services [1]. There have been a few meaningful definitions on PSS such as a system of products, services, supporting networks and infrastructure that is designed to satisfy customer needs and have a lower environmental impact than traditional business models and an integrated body of products and services and communication strategies that was conceived, developed and promoted by (a network of) actors to generate values for society [2, 3]. By the definitions given above, PSS usually has a number of diverse aspects such as stakeholders, activities, functions, product elements, service elements, and so forth. Therefore, appropriate representation scheme should be necessary for effective designing.

The research on the PSS modelling and representations could date back to 1980s. Shostack proposed the molecular modelling of service by introducing product elements, service elements, bond and essential evidence [4]. In his molecular modelling, the connection between product and service elements was made by a simple line – bond. He also proposed the service blueprint to identify and visualize the activities of customers and service providers and their relations during service process. In 1995, Congram and Epelman adopted the Structured Analysis and Design Technique (SADT) to represent services [5]. They claimed that SADT focused on activities which could be major building blocks of services and that SADT models could help employees at every level to understand what happens in delivering a service. In addition, SADT could describe who or what performed the activities and what guided or limited the activities.

More recently, Morelli has been among very active for researching systematic PSS design and modelling [6,7]. In his methodological framework for designing PSS, the major functions and requirements for the PSS were extracted, and they were then mapped to the elements of products and services in the case study of an urban tele-centre [5]. Morelli and Tollestrup studied various service representation methods such as the actor network mapping, motivation matrix, IDEF0, system platform and use cases [7].

Shimomura and his colleagues have been another active research group to study PSS design and modelling [8-11]. In their service engineering research, the service model containing several sub-models such as flow model, scope model, view model and scenario model was proposed. They also introduced receiver state parameters (RSPs) to address customers' value and cost. Maussang et al. proposed the functional block diagram to correlate product unit and service unit in the PSS design process [12].

Although there have been some research works on PSS modelling and representation, the systematic and detailed framework to address product and service elements and their relations has not been substantially studied. Therefore, in this paper, a new PSS representation scheme involving product and service elements is proposed and its applicability is examined. The service element can be described with service provider/receiver and their activities. In addition, the product element can be considered as the media for service element to effectively and completely realize PSS. In this new PSS representation scheme, there are two different domains – service element domain and product element domain, and the mapping between service and product elements could be expressed in a 3-dimensional way. Multiple service elements can be connected by flows, which are usually identified in PSS function modelling. Those flows can further be decomposed and they connect several product elements in various ways. Sample case example on the clothes TakeIN PSS is studied to investigate the applicability of the proposed PSS representation method.

2 PRODUCT-SERVICE SYSTEMS DESIGN PROCESS [13]

We have previously proposed the systematic PSS design process, and it is composed of the following 6 phases: (1) Requirement Identification and Value Targeting, (2) Stakeholder Activity Design, (3) PSS Function Modelling, (4) Function-Activity Mapping and PSS Concept Generation, (5) PSS Concept Detailing and (6) PSS Concept Prototyping. Fig. 1 shows the schematic diagram of PSS design process. During PSS design process, alternative PSS concepts can be generated by correlating functions, stakeholders, activities and product/service elements. When generating PSS concepts, the activities and functions are linked in the modified service blueprint, and they should be mapped to appropriate product and service elements. In addition, those product and service elements should be properly connected together to effectively realize intended activities and functions. Therefore, it is necessary to develop a new representation scheme for PSS addressing mapping of product and service elements.

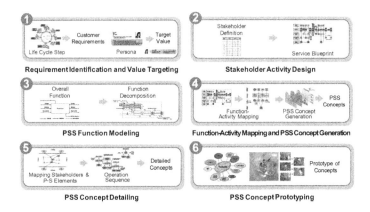

Fig. 1 Product-Service Systems (PSS) design process [13]

3 PSS REPRESENTATION METHOD

Major components composing a PSS concept are functions and related activities, and the functions should be mapped to appropriate product and service elements. When reviewing PSS function modelling in the PSS design process, a PSS function can be expressed as the block diagram which is given in Fig. 2. As can be seen in Fig. 2, the input and output to the function block are represented by three flows – energy, material and information, and they can also be used to logically connect multiple function blocks. In the function block, corresponding service provider and service receiver are expressed as folded nodes in upper left and lower right corners, respectively.

The overall function of PSS is decomposed into a number of sub-functions, and they are connected by flows based on causal and logical relations. The schematic representation of PSS function decomposition is given in Fig. 3. As can be seen in Fig. 3, the flows determine orders and connectivity of function blocks and play a significant role for building up critical functional modules. The service provider and receiver are also decomposed into sub-service providers and sub-service receivers, and they are assigned to each sub-function block.

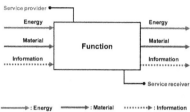

Fig. 2 Schematic of PSS function

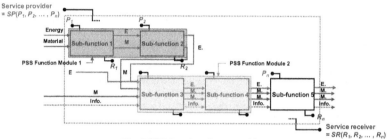

Fig. 3 PSS function decomposition

Each sub-function should be realized as sub-PSS. This sub-PSS is composed of a number of product elements and service elements, and they are connected together in order effectively to realize the corresponding function. Therefore, a proper new representation scheme describing product/service elements and their relationship are needed.

To represent a service element, stakeholders (who), activities (what) and product elements (how) were introduced. In general, service required specific providers and receivers, and their specific activities can also be followed. In addition, proper product elements should be accompanied as media to completely realize service elements. Fig. 4 shows the service and product elements, which are located in two different domains – service element domain and product element domain, respectively.

As can be seen in Fig. 4(a), the service element is represented with the block having service provider (SH_P), service receiver (SH_R) and their activities (ACT_P, ACT_R) in upper left and lower right corners, which is similar to the PSS function block. In addition, those service element blocks can be connected by flows of material and information which were already identified in PSS function modelling. The flow of energy can be used after more detailed attributes of product and service elements are determined. Meanwhile, as can be seen in Fig. 4(b), the associated product elements are also arranged and ordered via the flows of material and information. These product elements will be linked with the appropriate service elements to complete the sub-PSS. In the representation of the service element given in Fig. 4(a), the rectangular nodes have links with the product elements. The service and product elements represented in each domain should be appropriately connected, and 3D view of PSS representation is given in Fig. 5.

A single service element can have multiple product elements, and, on the other hand, a single product element can also be shared by a few service elements. For instance, in Fig. 5, the service element 1 has the product elements of P11, P12 and P13, and the product element of P13 can also be linked with the service element 2. In addition, the flows of material and information can be used to connect associated product elements.

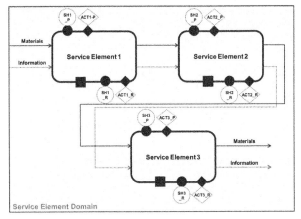

(a) Service element representation

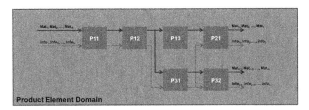

(b) Product element representation

Fig. 4 Representations of service elements and product elements connected with flows in service element domain and product element domain

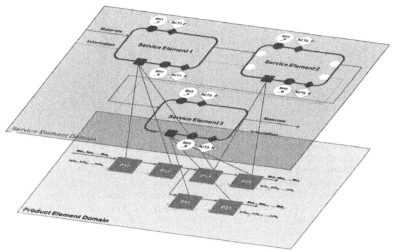

Fig. 5 3D view of PSS representation by mapping service elements and product elements

The patterns of connections among product elements can be different based on the attributes of selected product elements and flows. Besides, different service and product elements can be identified and configured to realize same function, and as a result, several alternative sub-PSSs can be generated and represented. In summary, the following formula can be hold for the proposed PSS representation scheme.

$$\text{Service Element}_i = SE_i (SH_i\_P, SH_i\_R, ACT_i\_P, ACT_i\_R, P_{ik}) \tag{1}$$

In formula (1), SH_i_P and SH_i_R represent stakeholders that serve as service provider and service receiver of the i-th service element SE_i respectively, and ACT_i_P and ACT_i_R mean corresponding activities of service provider and service receiver. In addition, P_{ik} represent product elements of the i-th service element SE_i.

4 CASE EXAMPLE – USED CLOTHES TAKEIN PSS

To demonstrate the applicability of the proposed new PSS representation scheme, the case example of clothes TakeIN PSS was studied. We have been using the name of *TakeIN* to denote PSS series for providing desirable reuse of products, indicating taking them back in use rather than throwing away. In this case example, the new scenario of clothes TakeIN PSS was developed by designing new activities and functions with the change of location context of a current used clothes bin. Fig. 6 shows the scenario of new used clothes TakeIN PSS. As can be seen in Fig. 6, the user (donator) repairs and cleans the used clothes, and he/she packages them at the clothes TakeIN station. The donated clothes can be properly stored in the clothes TakeIN station. Then, the stored clothes are collected by collectors and they are classified and properly kept in the warehouse. Afterwards, appropriate receivers are selected, and the donated clothes are delivered to them.

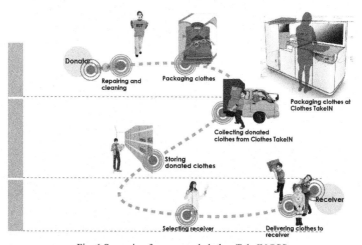

Fig. 6 Scenario of a new used clothes TakeIN PSS

In order realize the new used clothes TakeIN PSS given in Fig. 6, essential function requirements should be listed and properly arranged. Fig. 7 shows the overall function of the new used clothes TakeIN PSS and its corresponding function decomposition diagram. As can be seen in Fig. 7, the overall function of clothes TakeIN PSS was defined as "collect and supply clothes", and total 27 sub-functions were identified. Those sub-functions were connected by flows based on their logical and causal relations. For each sub-function block, corresponding sub-service provider and sub-service receiver were assigned. Total 7 stakeholders were identified for sub-service providers and sub-service receivers for 27 sub-functions, as can be seen in Fig. 7(b).

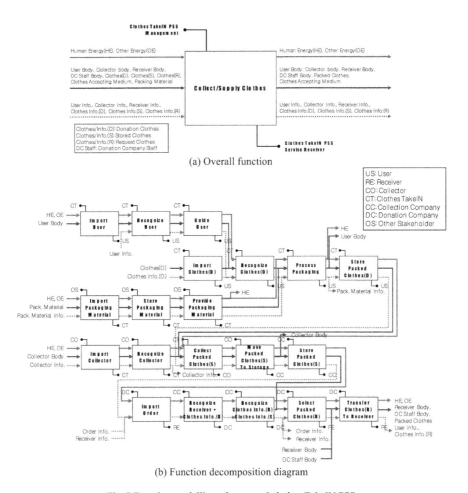

Fig. 7 Function modelling of new used clothes TakeIN PSS

The proposed PSS representation scheme was applied to each function block, and service elements and product elements were defined and linked to each other to completely describe corresponding sub-PSS. Sample sub-PSS representations to a couple of sub-functions are given in Fig. 8. The first three sub-functions of "Import User", "Recognize User" and "Guide User" were considered for case study. As can be seen in Fig. 8(a), the service elements of "welcome service" and "clothes TakeIN introduction service" were selected and associated service provider/receiver and their activities were assigned. In the case of "welcome service", the user (service receiver) entered (activity) the clothes TakeIN PSS space when the CT staff (service provider) welcomed (activity) the user with the product elements of "door" and "welcome sign". Simiarly, in the case of "clothes TakeIN introduction service", the CT staff (service provider) introduced (activity) the clothes TakeIN PSS and the user (service receiver) confirmed (activities) it with the product element of "introduction leaflet". The material of flow of "user body" connected those product elements. Similar descriptions can also be made in the case of the function of "Recognize User", which is shown in Fig. 8(b). The service elements of "user information acquisition service" and "user information confirmation service" were come up with, and appropriate stakeholders and activities were assigned, as given in Fig. 8(b).

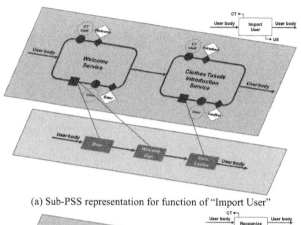

(a) Sub-PSS representation for function of "Import User"

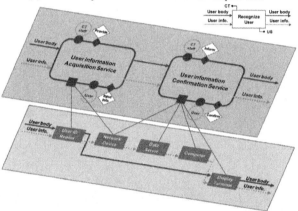

(b) Sub-PSS representation for function of "Recognize User"

(c) Sub-PSS representation for function of "Guide User"

Fig. 8 Sample sub-PSS representations for some sub-functions of used clothes TakeIN PSS

The product elements of "user ID reader" and "network device" were connected to "user information acquisition service", and the product elements of "network device", "data server", "computer" and "display terminal" were linked with "user information confirmation service". As mentioned above, the product element of "network device" was shared by two service elements. There were two input flows of "user body" and "user information", and only "user ID reader" and "display terminal" had the flow of "user body". The flow of "user information" was run through all product elements. Therefore, it can be concluded that the product elements of "user ID reader" and "display terminal" had interactions with user, which could become service touchpoints for user. Fig. 8(c) denotes the sub-PSS representation for the function of "guide user".

Three functions of "Import User", "Recognize User" and "Guide User" can be grouped together to make the functional module. Consequently, associated service elements and product elements should also be linked together to realize the corresponding PSS module. Fig. 9 shows the PSS module representation corresponding to the group of three functions of "Import User", "Recognize User" and "Guide User".

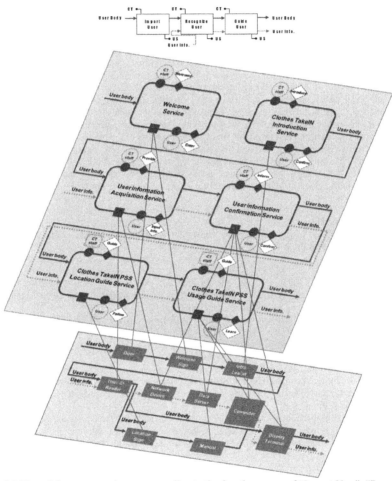

Fig. 9 PSS module representation corresponding to the function group of "Import User", "Recognize User" and "Guide User"

As can be seen in Fig. 9, total six service elements were combined together via the flows of material and information, and associated product elements were arranged subsequently. In particular, the product element of "computer" and "display terminal" could be shared by two different service elements such as "user information confirmation service" and "clothes TakeIN PSS usage guide service".

The proposed PSS representation scheme can help designers to effectively understand the relations among identified service elements and product elements with visual description. It is also possible to effectively display critical flows of material and information running through various product elements and service elements. Therefore, alternative arrangement and configuration of product/service elements can become much simpler and more effective. The assignment of stakeholders and activities to service elements also enables designers to effectively understand and show specific service providers and service receivers and their associated activities.

5 CONCLUDING REMARKS

This paper presented the new PSS representation method by modelling the product element and service element. To effectively represent the PSS, two different domains – service element domain and product element domain – were considered. The service element was modelled with service provider/receiver and their activities, and they were depicted as circular and diamond nodes at upper left and lower right corner of the service element block, respectively. In addition, the product element was included in the service element as a medium, and it could be connected with the service element block at the rectangular node. A single service element could have multiple product elements, and it was also possible for a single product element to be shared by several service elements. Multiple service elements were connected via the flows of material and information, which were usually identified in the PSS functional modelling. Similarly, the connection among the product elements could be made via the flows. These connections could differ when different product and service elements were identified, and as a result, alternative configurations of the product and service elements of the PSS realizing same functions could be generated and compared. Sample case study on the clothes TakeIN PSS was conducted, and the results demonstrated the applicability of the proposed PSS representation method. In the case study, several sub-PSS representations could be concatenated together based on the grouping of their corresponding functions, and the proposed method could also be applicable.

ACKNOWLEDGMENTS

This research was supported by the Korean Ministry of Knowledge Economy under the Strategic Technology Development Program.

REFERENCES

[1] Goedkoop, M. J., van Halen, C. J. G., te Riele, H. R. M., and Rommens, P. J. M. (1999): Product Service Systems: Ecological and Economic Basics, in: *Report for Dutch Ministries of Environment (VROM) and Economic Affairs (EZ)*.
[2] Mont, O. (2004): Product-Service Systems: Panacea or Myth?, *Ph.D. Dissertation*, Lund University.
[3] van Halen, C. J. G., Vezzoli, C., and Wimmer, R. (2005): *Methodology for Product-Service System Innovation*, Royal Van Gorcum, Netherlands.
[4] Shostack, G.L. (1982) "How to Design a Service," *European Journal of Marketing*, 16(1): 49-63.
[5] Congram, C. and Epelman, M. (1995) : How to describe your service: An invitation to the structured analysis and design technique, in: *International Journal of Service Industry Management*, 6(2): 6-23.
[6] Morelli, N. (2003): Product-Service Systems, a Perspective Shift for Designers: A Case Study: the Design of a Telecentre, in: *Design Studies*, Vol. 24, No. 1, pp. 73–99.
[7] Morelli, N. and Tollestrup, C. (2007): New Representation Techniques for Designing in a Systematic Perspective, in: *Proc. Nordic Design Research Conference*, Stockholm.
[8] Tomiyama, T., Shimomura, Y., and Watanabe, K. (2004): *A Note on Service Design Methodology*, in: *Proc. ASME Int'l. Conf. of Design Theory and Methodology*, Salt Lake City.
[9] Sakao, T., Shimomura, Y., Comstock, and M., Sundin, E. (2005): Service Engineering for Value Customization, in: *Proc. 3rd Int'l. World Congress on Mass Customization and Personalization (MCPC)*, Hong Kong.

[10] Sakao, T., and Shimomura, Y. (2007): Service Engineering: a Novel Engineering Discipline for Producers to Increase Value Combining Service and Product, in: *Journal of Cleaner Production*, Vol. 15, pp. 590–604.
[11] Hara, T., Arai, T., and Shimomura, Y. (2009): A Method to Analyze PSS from the Viewpoints of Function, Service Activity, and Product Behavior, in: *Proc. CIRP Industrial Product-Service Systems Conf.*, Cranfield.
[12] Maussang, N., Sakao, T., Zwolinski, P., and Brissaud, D. (2007): A Model For Designing Product-Service Systems Using Functional Analysis and Agent Based Model, in: *Proc. Int'l. Conf. on Engineering Design*, Paris.
[13] Kim, Y. S., Maeng, J. W., and Lee, S. W. (2010): Product-Service Systems Design with Functions and Activities: Methodological Framework and Case Studies, in: *Proc. of Int'l Conf. on Design and Emotion*, Chicago.

Contact: Yong Se Kim
Creative Design Institute
Sungkyunkwan University
Suwon 440-746
Korea
Phone: +82-31-299-6581
Fax: +82-31-299-6582
Email: yskim@skku.edu
URL: http://cdi.skku.edu

Yong Se Kim is Director of Creative Design Institute, and Professor of Mechanical Engr., at Sungkyunkwan Univ., Korea. He received PhD in Mechanical Engr. from Design Division of Stanford in 1990. His research interest is Design Cognition and Informatics, which investigates fundamental processes in design, and provides methodologies and computer-based tools for design and design learning.

Sang Won Lee received B.S. and M.S. at the department of Mechanical Design and Production Engr. in Seoul National University, Korea, in 1995 and 1997. He received Ph.D. at the department of Mechanical Engr., University of Michigan in 2004. Dr. Lee is an assistant professor of School of Mechanical Engr., and is an active member of the Creative Design Institute at Sungkyunkwan University, Korea. His research interest is Design Theory and Methodology and Design Process Modelling.

Dong Chan Koh is an undergraduate research student of the Creative Design Institute at Sungkyunkwan University, Korea. He is majoring in Industrial Engineering, and his research interest is to develop systematic methodologies for Product-Service Systems (PSS) design.

A CLASSIFICATION FRAMEWORK FOR PRODUCT MODULARIZATION METHODS

Charalampos Daniilidis[1], Vincent Enßlin[1], Katharina Eben[1], Udo Lindemann[1]
(1) Technische Universität München, Institute of Product Development, Germany.

ABSTRACT
The modularization of product architectures and the standardization of components and modules across a product family or product portfolio constitute approaches to reduce the internal variety in an enterprise while keeping the range of the external variety. Thus costs and development time can be reduced through scale effects and further transparency within the product portfolio can be enhanced as well. In this context a plethora of methods and approaches to identify modules in product architectures has been introduced. These methods differ in the application area and the procedure and show different benefits and weaknesses. This paper introduces a classification framework for modularization methods and approaches that provides a systematic overview on past and current developments. Therefore an extensive literature survey on modularization was carried out to identify the major methodologies introduced in the last years.

Keywords: Product modularity, product architecture, modularization methods, classification framework.

1 INTRODUCTION
During the last decades the change from a sellers' to a buyer's market has led to an increase of variant products in order to meet individualized customer requirements. Furthermore, a large number of product families has immerged and the product life cycle has been considerably reduced to meet constantly changing and expanding customer needs. On the one hand this has led to more complex products, which incorporate several different functionalities and on the other hand the plethora of different products and variants has led to larger, unclear and complex product portfolios. In this way the internal complexity and variety in an enterprise has increased to satisfy the external variety, which is required in order to operate the market. [1]

Therefore, main objective of a large number of research activities and publications in the last years has been to reduce the internal complexity and variety and at the same time to preserve the variety and the range of a product portfolio, as illustrated in Figure 1. [2]

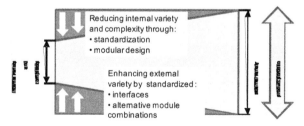

Figure 1. Motivation and scope of research activities.

An approach to achieve the reduction of internal variety and therefore internal complexity is to modularize product architectures and define standardized modules. Whereby, modularization is the process of decomposing product architecture into modules. These modules, as Gershenson et al. [3] defines, can be seen as units, which provide a unique basic function necessary for the product to operate as desired. Further, product architecture is defined by Ulrich [4] as the arrangement of functional elements, the mapping from functional elements to physical components and the specification of the interfaces among interacting physical components.Thereby, modularization of products leads to reduction of design lead time and costs, improves design quality and facilitates redesign and change processes. Furthermore, modularization enables the standardization of components and the definition of standardized physical interfaces and therefore facilitates the development of product variants [3, 5].

Another approach in order to reduce internal complexity is the identification and definition of product platform elements that can be used across the variants and products of a product family. In the literature is mentioned that a product platform can be seen as a common set of subsystems, components, processes, technologies shared by all variants in a product family [6, 7]. Thus can be standardized in order to reach scale effects and reduce design lead time. Furthermore, the identification of modules or platform elements can be carried out at various levels of product aggregation and abstraction, i.e., between systems, subsystems, components or functions [6].

In this work we concentrate on modularization methods and approaches but it is important to note that many of these methodologies can be also used to identify and define platform elements in product architectures.

Although in the last twenty years a lot of research has been done in this field and a plethora of different approaches to modularize single products or whole product families have been published and can be found in the literature (s. Chapter 4) there is a lack of a systematic overview on the existing methodologies. Scope of this work is to provide an efficient synopsis of the major modularization approaches introduced in the last years through the introduction of a classification framework. Thereby, diverse methodologies are categorized according to consistent characteristics, such as the application area of the methodology.

In the next chapter past research works on systematizing and assessing diverse modularization methodologies are presented that are relevant to this paper. In chapter 3, at first a brief introduction to the scope and benefits of modularization is given and then a systematic overview on the major modularization methods and approaches is presented. Subsequent in chapter 4, the classification framework is introduced by elaborating its dimensions and an overview on the methods by means of the framework is presented. Finally, in chapter 5 some critical conclusions and areas of future work are given.

2 RELATED RESEARCH

In the past years several categorization and assessment works for modularization and platform-based design methodologies have been introduced and can be found in the literature [8-11].

Holtta et al. [8] carry out in their work an analysis of three major modularization methods. The Design Structure Matrix [12, 13], the Function Structure Heuristic Method [14] and the Modular Function Deployment Method [15] are being deployed to modularize three different end-user products and a product family. Thereby, Holtta et al. [8] discovered that all three methods, given identical inputs, produce different results. Reason for this is that each method was developed from a different viewpoint and for different application areas.

Fixson [9] provides in his work an extensive literature survey over 15 years on modularization and commonality research and analyzes the included references according to their subject, their effect and the introduced research method.

Jiao et al. [10] provide in their work a comprehensive review of strategies for product family design and platform-based product development. Thereby, they introduce a decision framework, which provides a holistic view on product family design and takes design aspects from the customer to the functional, physical and process perspective into consideration. Modularization of a product's architecture as well as the communalization of physical components constitutes thereby only an aspect of the physical and functional perspective of the framework. In the same sense Gershenson et al. [11] present a review of existing measures of product modularity and methods to modularize product architectures. Each method is briefly described and discussed according to its benefits and boundaries. Finally, Gershenson et al. [11] discuss on which abstraction levels there is consensus and commonalities in modularization methods and measures.

Similarly to Gershenson et al. [11], scope of this work is to provide a systematic overview on existing modularization methods, which facilitates the decision making about which method or approach is more suitable for a specific application area. This includes a clearly visualized classification framework for the different methodologies to modularize product architectures. Whereby, the categorization is carried out according to the application area and the potential of the method.

3 REVIEW ON MODULARIZATION METHODS

As companies strive to rationalize engineering design, manufacturing and support processes in order to reduce and manage internal complexity they are focusing on modularity. Thereby, modularity refers not only to product architectures but also to processes and resources that fulfill various functions and can be considered as distinct building blocks [3, 16].

In this context main scope of modularization is the reduction of product complexity and internal variety in an enterprise. Thereby, standardization of modules or components is enhanced in order to achieve scale effects and to reduce cost and development time. Furthermore, modular designs and the deployment of standardized components improve the reliability and quality of the product and allow an easier product diagnosis, maintenance, repair and disposal [16].

Nevertheless, a large number of influencing factors and constraints has to be considered in order to achieve positive effects by modular design. Thereby, the application area of modularization has to be clearly defined at the beginning of the design or redesign process and must incorporate the product life-cycle, technology and quality issues in order to realize its potential [16]. Furthermore, in accordance to the application area, the abstraction level – physical or functional – to carry out the modularization has to be defined.

Over the past years a plethora of different approaches and methods to achieve modularity in New Product Development and Product Redesign has been introduced. Yet there is a lack of a systematic review on the different methodologies. Figure 2 shows an overview and systematization of important publications and methods on modularization of product architectures and product families. The review begins in the early 90's with the book of Pahl and Beitz [17]. For clarity the initial publications for a modularization method or approach are shown and a small number of subsequent publications of the same research group. Furthermore, the relations between different methodologies are documented.

Design Structure Matrix (DSM) [13], Function Heuristics [14], Modular Function Deployment (MFD) [15] and Design for Variety (DfV) [5, 18] can be seen as the major modularization methods. Whereby, DSM, Function Heuristics and MFD concentrate on modularization, Design for Variety constitutes a more extended approach, which includes the modularization aspect.

In order to reveal and explore alternative product architectures and to improve thereby the quality of the resulting product design Pimmler and Eppinger [13] proposed a component-based Design Structure Matrix to model and analyze a product's architecture. Furthermore, by using the DSM method design coordination demands can be supported, that are required when subsystems interact [19]. Because of the component-based analysis of product architecture the modularization cannot be carried out in the early phases of New Product Development, as in these phases little is known about the exact component structure of a product. In addition, the simultaneous consideration of more than one product or a product family in order to reveal modules applicable across the product family is not possible.

Stone et al. [14] introduced a systematic approach to identify modules of a product on a functional basis. Thereby, modules are identified according to the major flows in the function structure. Stone et al. [14] provide in their work a number of rules (heuristics) to identify modules and a large number of publications can be found which introduce further heuristics to identify modules depending on the specific application. Because of the more abstract analysis and the standardized approach to set product function models this method can be easily optimized to consider a larger variety of products. [20]. Nevertheless, the poor repeatability and the varying results constitute the major disadvantage of the approach [8].

Modular Function Deployment (MFD) [15] considers more strategic and abstract management parameters to develop a modular product structure. MFD is also based on function structures, but modularity drivers other than functionality are considered. Furthermore, MFD is designed to modularize single products and Ericsson et al. [15] provide in their work an amount of twelve modularity drivers to achieve that considering diversity, assembly, life-cycle and re-usability parameters. As by the Function Heuristic approach, the MFD's major disadvantage is the poor repeatability of the method's results [8].

Finally, Design for Variety (DfV) is an approach, which considers customer demands in a product line in order to identify the optimal design for a product while minimizing the life-cycle cost of providing the variety [18]. DfV uses product structure graphs and ratings as model for the product architecture in order to identify the optimal design and is developed for the redesign process of existing products. On this basis a large number of further developments can be found in the literature, which apply not only for single products but also for product family design [21]. Thereby, modularity is an aspect of the redesign process. The initial and the optimized product architecture can be described using metrics in order to assess the result [5].

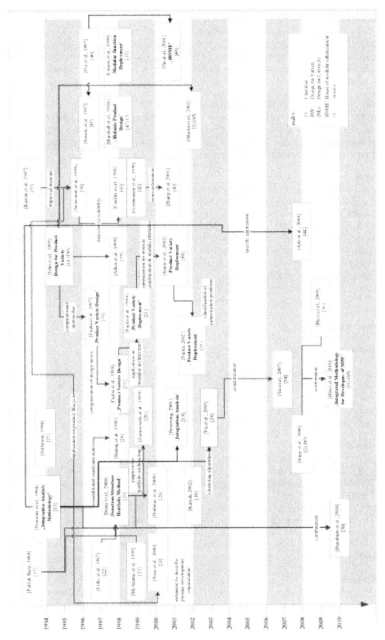

Figure 2. Overview of some important modularization approaches.

Figure 2 shows an overview on the major modularization methods and their further developments and combinations found in the literature. Important to note is that in the years between 2004 and 2006 no major developments could be found.

4 CLASSIFICATION FRAMEWORK

In order to systematize this plethora of methodologies and provide a clear overview this paper introduces a classification framework. Thereby different methodologies to achieve modularity are classified according to parameters that describe the application area and the capabilities of each approach. In the following sub-chapters the dimensions of the framework are elaborated and some examples of classified methods are given to demonstrate the application of the framework.

4.1 Application area of modularization methods

After a thorough literature-survey it was concluded that the parameters, which definitely characterize the application area of each modularization approach are the *product variety*, the *product generation* and the *product life-cycle*. As shown in Figure 3 these parameters are used by means of the classification framework as categorization criteria.

Product variety

Variety is one of the most often mentioned driving factors for modularity (e.g. in [49], [50] and [35]). Due to general market saturation it becomes more and more important to fulfill specific customers' needs. Globalization and therefore globalized competition enhances this impact even further, so that offering specialized products on a wide range is a key factor to success [1]. Providing this variety of products is possible, but only in combination with minimal company-internal variety (s. Chapter 1). This difficulty of offering high external variety on the one hand while minimizing internal-variety on the other hand can be resolved by modularization. Furthermore, product variety can be used to characterize the application area of a modularization approach. Thus we can identify three categories of modularization approaches according to their capability in identifying modules in one or more product architectures. Hence, these categories are as follows:

1. Single product
2. Product family
3. Product portfolio

The difference between product family and product portfolio is that a product family consists of a group of related products derived from a common product platform [51]. Whereas, a product portfolio is defined as the sum of all products offered by a company [1] and usually consists of several product families. Variety within a portfolio therefore stands for creating a set of product families, which serve the customer's demands in the best way [20].

Product generation

The development of a new product or product family is time-, effort-intensive and fraught with risk. Thus it is efficient to take future product generations into consideration when developing a new product. By this way changing market requirements can be easier adapted to the products' specification.

Some methods strive to support New Product Development (NPD), whereas a larger number of methods have been introduced to support product reengineering and redesign. Therefore, product generation constitutes a further characteristic to describe the application area of a modularization method. In this research we distinguish between *product reengineering* and *New Product Development*.

Product life-cycle

A challenging yet very important aspect of creating successful products is the product's life-cycle. The consideration of the product life-cycle during the design phase can have a major impact on manufacturing and assembly cost and a major influence on the customers' satisfaction with the product and the product's image. Thereby, the customer's increasing awareness of the environment brings matters of recycling and reusability more and more into focus [38]. Ishii et al. [52] state that modularity in product design impacts every stage of the product life-cycle and modular design can create benefits for many aspects of a product life-cycle such as design, assembly, services and recycling [53]. Therefore, in the context of the classification framework for modularization methods a standardized product life-cycle [54] constitutes the third parameter to characterize the application area of a method. As three parameters are utilized in order to carry out the classification, the framework can be graphically visualized, as shown in Figure 3. Furthermore, Figure 3 provides an overview of the parameters scaling.

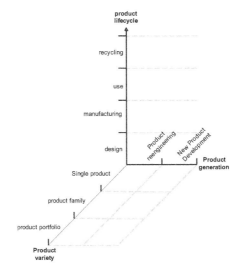

Figure 3. Classification framework for modularization methods.

Finally, further categorization criteria can be used in order to classify modularization methods. For example, the product type – e.g. technical or design product, individual or customized product – might also have an effect on the selection and utilization of a method to modularize a product's architecture. Furthermore, the customer type – e.g. OEM-product or supplier product – can be utilized to classify products and therefore modularization methods. In this research we concentrate on three basic parameters that describe the application area and the goals of a modularization approach. An examination and implementation of further criteria would go beyond the scope of this paper to provide a simple overview. Nevertheless, it would provide a further perspective on the classification and consists an essential future task.

4.2 Categorization of modularization methods

In order to demonstrate the deployment and effectiveness of the classification framework a few modularization methods and approaches are categorized as an example. Important to note is that the information for the classification of the methodologies is acquired from the case studies of the relevant publications. Figure 4 shows the classification of four major modularization methodologies (see Chapter 3) and of a few further developments.

In Figure 4 (a) the application area of *Design for Variety* (DfV) is illustrated. Thereby, Ishii et al. [18] proposed a methodology to capture the broadest customer demands during the development of a product. This methodology is described as *Design for Product Variation* [18]. *Product Variety Deployment* constitutes a further development designed to tackle the simultaneous optimization problem of a product family [21]. In addition, Blees et al. [35] introduce in their work an approach for the utilization of DfV to capture modularization demands from all life-cycle phases of a product.

The *Function Heuristic Method* of Stone et al. [14, 25] is shown in Figure 4 (b) in relation to the dimensions of the classification framework. Furthermore, a development of this methodology is shown that introduces an approach to deploy *Function Heuristics* for portfolio architecture design [20].

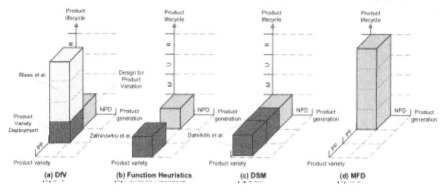

Figure 4. Exemplary categorization of some modularization methods.

Figure 4 (c) illustrates the application area of *Design Structure Matrix* (DSM) as it was introduced by Pimmler et al. [13] to identify modules in a product architecture. Additionally, an approach is shown to utilize DSM for the modularization of product families or for the identification of common modules across a product portfolio [30].

Finally, in Figure 4 (d) *Modular Function Deployment* (MFD) [15] is shown in accordance to the classification framework. Thereby, MFD provides designers and product managers with a holistic modularization potential from a more managerial perspective.

5 CONCLUSION AND FUTURE WORK

This work provides a systematic overview on modularization methods and approaches and on current developments in the field of modularization. A classification framework is introduced in order to enhance transparency and facilitate decision making when a modularization approach is to be deployed for a certain application. Therefore, the application area is utilized as categorization criterion for the classification of different methodologies. Thereby, the application area is explicitly described by three parameters, the *product life-cycle,* the *product generation* and the *product variety,* which constitute the dimensions of the framework. For the purpose of this research an extensive literature survey was carried out in order to identify the major trends on modularization. During this task it became apparent that the three parameters mentioned above constitute consistent characteristics for a homogeneous description of the application area. Each modularization method or approach can be explicitly categorized according these criteria. Thereby, the abstraction level – functional- or component-based methods – on which a modularization method is deployed, is excluded as classification criterion. The reason is that the abstraction level of the method deployment significantly affects the product variety on which a method can be deployed. Hence, one can conclude that functional-based modularization methods possess a wider application area according to the *product variety* dimension of the framework than component-based methods that can be deployed for the optimization of single products. In conclusion the classification of several methods and approaches according to the framework revealed that the three defined dimensions of the framework are adequate to provide a transparent and systematic overview on modularization methods.

Nevertheless, as stated in Chapter 4.2 the information for the classification of the modularization methods was acquired from relevant publications and not from the application of these methods on own product examples. In order to review the current positions of the methods in the framework, the methods must be deployed and utilized using own examples. Thereby, the effectiveness of the methods in relation to the application area needs to be verified. Furthermore, in order to carry out an examination as accurate as possible both end-user and technical products have to be considered and utilized as case studies. Thereby, it is important to note that in most case studies found in the literature only end-user product are used as examples to demonstrate the deployment of a method. In addition, a structured procedure has to be set up in order to facilitate decision making on which method is more adequate for a specific application situation. At this point the examination and realization of further perspectives to the classification using criteria as the product or customer type may be beneficial.

REFERENCES
[1] Renner I. Methodische Unterstützung funktionsorientierter Baukastenentwicklung am Beispiel Automobil, 2007 (Dissertationsverlag Dr. Hut, München).
[2] Kipp T. and Krause D. Design for Variety – Ein Ansatz zur variantengerechten Produktstrukturierung. In *6. Gemeinsames Kolloquium Konstruktionstechnik 2008*, Aachen, 2008, pp. 159-168.
[3] Gershenson K. J. and Prasad J. G. and Zhang Y. Product modularity: definitions and benefits. *Journal of Engineering Design*, 14(3), 2003, pp. 295-313.
[4] Ulrich K. The role of product architecture in the manufacturing firm. *Research Policy*, 24, 1995, pp. 419-440.
[5] Martin V. M. and Ishii K. Design for variety: developing standardized and modularized product platform architectures. *Research in Engineering Design*, 2002, 13(4), pp. 213-235.
[6] Kalligeros K., de Weck O., de Neufville R., Luckins A. Platform identification using Design Structure Matrices *Proceedings of INCOSE '06*, Orlando 2006.
[7] Berglund F., Bergsjö D., Högman U., Khadke K. Platform strategies for supplier in the aircraft engine industry. In *Proceedings of the ASME DETC*, Brooklyn, New York, 2008.
[8] Holtta K. M. M. and Salonen M. P. Comparing three diffrent modularity methods. In *Proceedings of DETC'03*, Chicago, 2003.
[9] Fixson S. Modularity and Commonality research: Past developments and Future Opportunities. *Concurrent Engineering*, 2007, 15(2), 85-111.
[10] Jiao J., Simpson W. T., Siddique Z. Product family design and platform-based product development: a state-of-the-art review. *Journal of Intelligent Manufacturing*, 18, 2007, pp. 5-29.
[11] Gershenson J. K. and Prasad G. J and Zhang Y.. Product modularity: measures and design methods. *Journal of Engineering Design*, 2004, 15(1), pp. 33-51.
[12] Ulrich K. T. and Eppinger S. D. Product Design and Development, McGraw-Hill, New York, 2nd ed, 2000.
[13] Pimmler T.U. and Eppinger S.D. Integration analysis of product decompositions. In *Proceedings of the 1994 ASME Design Engineering Technical Conferences—6th International Conference on Design Theory and Methodology*, Minneapolis, 1994.
[14] Stone R.B., Wood K.L. and Crawford R.H. A heuristic method for identifying modules for product architectures. *Design Studies*, 2000, 21(1)1, pp. 5-31.
[15] Ericsson A. and Erixon G. Controlling design variants: Modular product platforms. *ASME Press.* New York, 1999.
[16] Kusiak A. Integrated product and process design: a modularity perspective. *Journal of Engineering Design*, 13(3), 2002, pp. 223-231.
[17] Pahl G. and Beitz W. Engineering Design: A Systematic Approach, 1984 (Springer-Verlag, Berlin).
[18] Ishii K., Juengel C. and Eubanks C. F. Design for product variety: key to product line structuring. In *Proceedings of the 1995 ASME Design Engineering Technical Conferences—7th International Conference on Design Theory and Methodology*, Boston, 1995.
[19] Browning R. T. Applying the Design Structure Matrix to system decomposition and integration problems: a review and new directions. *IEEE Transactions on Engineering Management*, 48(3), 2001, pp. 292-306.
[20] Zamirowski E.J. and Otto K.N. Identifying product portfolio architecture modularity using function and variety heuristics. In *Proceedings of the 1999 ASME Design Engineering Technical Conferences—11th International Conference on Design Theory and Automation*, Las Vegas, 1999.
[21] Fujita K., Sakaguchi H. and Akagi S. Product Variety Deployment and Its Optimization Under Modular Architecture and Module Commonalization. *In Proceedings of the 1999 ASME Design Engineering Technical Conferences*, 1999.
[22] Little A., Wood K. and McAdams D. Functional analysis: a fundamental empirical study for reverse engineering, benchmarking and redesign. In *Proceedings of the 1997 ASME Design Engineering Technical Conferences—9th International Conference on Design Theory and Methodology*, Sacramento, 1997.

[23] McAdams D., Stone R and Wood K. Functional independence and product similarity based on customer needs. *Research in Engineering Design*, 1999, 11(1), 1–19.

[24] Sosa M.E., Eppinger S.D. and Rowles C.M. Designing modular and integrative systems. In *Proceedings of the 2000 ASME Design Engineering Technical Conferences—12th International Conference on Design Theory and Methodology*, Baltimore, 2000.

[25] Stone R.B., Wood K.L. and Crawford R.H. A heuristic method to identify modules from a functional description of a product. In *Proceedings of the 1999 ASME Design Technical Conferences—11th International Conference on Design Theory and Automation*, Las Vegas, 1998.

[26] Dahmus J.B., Gonzalez-Zugasti J.P. and Otto K.N. Modular product architecture. In *Proceedings of the 2000 ASME Design Engineering Technical Conferences—12th International Conference on Design Theory and Methodology*, Baltimore, 2000.

[27] Hillstrom F. Applying axiomatic design to interface analysis in modular product development. *Advances in Design Automation*, 1994, 67–76.

[28] Huang C.-C. and Kusiak A. Modularity in design of products and systems. *IEEE Transactions on Systems, Man, and Cybernetics*, Part A, 1998, 28(1), 66–77.

[29] Yu T., Yassine A. A., Goldberg E. D. A genetic algorithm for developing modular product architectures. In *Proceedings of the ASME DETC*, Chicago, Illinois, 2003.

[30] Daniilidis C., Eben K., Deubzer F., Lindemann U. Simultaneous Modularization and Platform Identification of Product Family Variants. In *Proceedings of the NordDesign 2010*, Göteborg, August 2010.

[31] Fujita K., Akagi S. and Yoneda T. and Ishikawa M. Simultaneous Optimization of Product Family Sharing System Structure and Configuration. In *Proceedings of the 1998 ASME Design Engineering Technical Conferences*, Atlanta, 1998.

[32] Fujita K. and Ishii K. Task Structuring Toward Computational Approaches to Product Variety Design. In *Proceedings of the 1997 ASME Design Engineering Technical Conferences*, 1997.

[33] Fujita K. Product variety optimization under modular architecture. *Computer-Aided Design*, 2002, 34(12), 953-965.

[34] Yu T., Yassine A. A., Goldberg, E. D. An information theoretic method for developing modular product architectures using genetic algorithms. *Research in Engineering Design*, 18, 2007, pp. 91-109.

[35] Blees C., Kipp T., Beckmann G. and Krause D. Development of Modular Product Families: Integration of Design for Variety and Modularization. In *Proceedings of the NordDesign 2010*, Göteborg, August 2010.

[36] Blees C., Jonas H. and Krause D. Perspective-Based Development of Modular Product Architectures. In *Proceedings of ICED '09*, Stanford, 2009, pp.4-95-4-106.

[37] Kusiak A. and Chow W.S. Efficient solving of the group technology problem. *Journal of Manufacturing Systems*, 1987, 6(2), 117–124.

[38] Newcomb P.J., Bras B. and Rosen D.W. Implications of modularity on product design for the life cycle. In *Proceedings of the 1996 ASME Design Engineering Technical Conferences— 8th International Conference on Design Theory and Methodology*, Irvine, 1996.

[39] Allen K.R. and Carlson-Skalak S. Defining product architecture during conceptual design. In *Proceedings of the 1998 ASME Design Engineering Technical Conference*, Atlanta, 1998.

[40] Fujita K. Product variety optimization: simultaneous optimization of module combination and module attributes. In *Proceedings of the 2001 ASME Design Engineering Technical Conferences and Computers and Information in Engineering Conference,* Pittsburgh, 2001.

[41] Coulter S.L., Bras B., McIntosh M.W. and Rosen D.W. Identification of limiting factors for improving design modularity. In *Proceedings of the 1998 ASME Design Technical Conferences—10th International Conference on Design Theory and Methodology*, Atlanta, 1998.

[42] Gershenson J.K., Prasad G.J. and Allamneni S. Modular product design: a life-cycle view. *Journal of Integrated Design and Process Science*, 1999, 3(4), 13–26.

[43] Zhang Y., Gershenson J.K. and Allamneni S. An initial study of the retirement costs of modular products. In *Proceedings of the 2001 ASME Design Engineering Technical Conferences—13th International Conference on Design Theory and Methodology*, Pittsburgh, 2001.

[44] Arts L., Chmara K. M., Tomiyama T., Modularization method for adaptable products. In *Proceedings of the 2008 ASME Design Engineering Technical Conferences—20th International Conference on Design Theory and Methodology*, Brooklyn, New York, 2008.
[45] Sosale S., Hashemian M. and Gu P. Product modularization for reuse and recycling. *Concurrent Product Design and Environmentally Conscious Manufacturing*, 1997, 195–206.
[46] Gu P., Hashemian M. and Sosale S. An integrated modular design methodology for life-cycle engineering. In *Annals of the CIRP*, 1997, 46(1), 71–74.
[47] Marshall R., Leaney P.G. and Botterell P. Enhanced product realisation through modular design: an example of product/process integration. *Journal of Integrated Design and Process Technology*, 1998, 3(4), 143–150.
[48] Gu P., Watson G. HOME: House of modular enhancement for product redesign for modularization. In *Proceedings of the 2001 ASME Design Engineering Technical Conferences—13th International Conference on Design Theory and Methodology*, Pittsburg, Pennsylvania, 2001.
[49] Catherine da Cunha C, Bruno A. and Kusiak A. Design for Cost: Module-Based Mass Customization. *IEEE Transactions on Automation Science and Engineering*, 2007, 4(3), 350-359.
[50] Gershenson J. K. and Prasad G. J. and Allamneni S. Modular Product Design: A life-cycle View. *Journal of Integrated Design & Process Science*, 1999, 3(4), 13-26.
[51] Simpson T. W. and Maier J. R.A. and Mistree F. A product platform concept exploration method for product family design. In *Proceedings of the ASME DETC'99*, Las Vegas, 1999.
[52] Ishii K. Modularity: A Key Concept in Product Life-cycle Engineering, 1998.
[53] Gu P. and Sosale S. Product modularization for life cycle engineering. *Robotics and Computer Integrated Manufacturing*, 15, 1999, 387-401.
[54] VDI Richtlinie 2221, Methodik zum Entwickeln und Konstruieren technischer Systeme und Produkte. VDI-Verlag, Düsseldorf, 2005.

Contact: Charalampos Daniilidis
Technische Universität München
Institute of Product Development
Boltzmannstrasse 15, 85748, Garching bei München
Germany
Tel: +49 89 289 15154
Fax: Int +49 89 289 15144
Email: daniilidis@pe.mw.tum.de
URL: http://www.pe.mw.tum.de

Charalampos Daniilidis has studied mechanical engineering at the Technical University of Munich. Since 2008 his is scientific assistant at the Institute of Product Development and has worked on several projects on systems engineering and complexity management. His main research fields are methods for the reduction of internal complexity and the establishment of transparency in complex organizations.

INTERNATIONAL CONFERENCE ON ENGINEERING DESIGN, ICED11
15 - 18 AUGUST 2011, TECHNICAL UNIVERSITY OF DENMARK

A META MODEL OF THE INNOVATION PROCESS TO SUPPORT THE DECISION MAKING PROCESS USING STRUCTURAL COMPLEXITY MANAGEMENT

Sebastian Kortler[1], Udo Lindemann[1]
(1) Institute of Product Development, Technische Universität München, Germany

ABSTRACT

The innovation process is characterized by numerous interactions of numerous domains. Cyclic interdependencies intensify the pressure in terms of quality and schedule, causing shortened testing phases, frequent releases of new models, and thus hardly calculable risks. Structural Complexity management is established in order to avoid wrong decisions, instable processes and error-prone solutions. Therefore, Structural Complexity Management evaluates system's characteristics by analyzing system's underlying structures across multiple domains, condensing each single analysis into one big matrix that represents multiple domains at a time.

Identifying suitable perspectives, generating suitable models and using suitable analyze criteria are the challenges in this field.

In order to support the manufacturing of innovative products and thus the evaluation and interpretation of the system's underlying structure this paper proposes a meta model. The created model describes the author's perspective on entities arising during the innovations process and their interactions. The proposed model is used to simplify the decision making processes and to enable the management of cyclic interdependencies during the innovation process.

Keywords: structural complexity management, structural criteria, structural meanings, cycle management, meta model

1. INTRODUCTION-MOTIVATION

Manufacturing technical innovative products implies complex design processes as well as complex product architectures with manifold challenges caused by cyclic interdependencies. Those cyclic interdependencies intensify the pressure in terms of quality and schedule, causing shortened testing phases, frequent releases of new models, and thus hardly calculable risks. The whole innovation process is characterized by numerous interactions of numerous domains. Moreover, manifold artifacts, models and actors are involved. Complexity management is established in order to avoid wrong decisions, instable processes and error-prone solutions. Structural Complexity Management evaluates system's characteristics by analyzing system's underlying structures across multiple domains, condensing each single analysis into one big matrix that represents multiple domains at a time.

However, comparing and evaluating the criteria of a complex structure makes it necessary to interpret different structural criteria and then evaluate their impacts on the system. Identifying suitable perspectives, generating suitable models and using suitable analyze criteria are the challenges in this field.

In order to support the manufacturing of innovative products and thus the evaluation and interpretation of the system's underlying structure this paper proposes the process of deriving a meta model. The created model describes the author's perspective on entities arising during the innovations process and their interactions. The proposed model is used to simplify the decision making processes and to enable the management of cyclic interdependencies during the innovation process. Especially the presented model is used as a systematical basis for Structural Complexity Management.

The paper is structured as follows: After defining relevant terms in section 2, a short review of the current research in structural complexity management is presented in section 3. Section 4 describes which scientific methods were used to derive the meta model. Section 5 presents the meta model. Section 6 demonstrates the use of the model. Finally, the paper proposes a conclusion.

2. DEFINITIONS

2.1 System
A system is created by entities (elements) and their interdependencies (relationships) forming a system's structure. Such a structure possesses individual properties, which contribute to fulfill the system's purpose [1]. Systems are delimited by a system border and connected to their surroundings by inputs and outputs. Changes of system's parts can be characterized by dynamical effects, which lead to a specific system's behavior. However, in this paper variations over time are not considered.

2.2 Domain
Domains represent the classification of elements, which create the system. Examples of domains are "components" or "documents".

2.3 Relationship type
The relationship type describes the meaning of a dependency. Different relationship types can even exist between the same elements and between the same domains [2]. Examples of relationship types are "change impact" or "waiting for".

2.4 Structure
"Structure" is understood as the network formed by dependencies (edges) between a system's entities (nodes). It furthermore relates to the semantics of this network; the structure of a system therefore always contributes – in its constellation – to the purpose of the system. Structures and their subsets can be analyzed by means of computational approaches, primarily provided by the graph theory and related sciences [2].

2.5 Structural criteria
A structural criterion is understood as a particular constellation of nodes and edges, i.e. it is formed by a particular pattern considering nodes and edges [2]. The criterion gains its meaning by the way the pattern is related to the actual system it is part of, i.e. it must serve a special purpose in the context of the overall system [1]. A structural criterion only possesses significance in the context of the system it is describing.

2.6 Structural meaning
Structural meanings relate structural criteria to their respective effects impacting the modeled system. The effects are, amongst other factors, dependent on the modeled domain, the relationship type describing the dependencies between the corresponding entities (see figure 1).

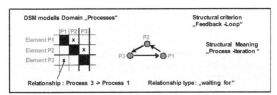

Figure 1. Definition of structural meanings

2.7 Cycle
Cycles are reoccurring patterns of temporal or structural nature that can be subdivided into phases. Their constituting elements are:

- Repetition
- Phases
- Duration
- Triggers
- Effects

3. RELATED WORK

3.1 Structural Complexity Management
To manage a structure efficiently, different methodologies prevail: Most commonly, matrix based methodologies such as the Design Structure Matrix (DSM), Domain Mapping Matrix (DMM), and Multiple Domain Matrices (MDM) are commonly applied, and the underlying theory provides for ample means of analysis. Furthermore, network theory is available, describing how the structure of random systems in nature, which have evolved over time, can be described. Ultimately, graph theory provides for a formal, mathematically founded framework grasping complex interdependencies.

Network and graph theory are closely interconnected. Hence, it is not easy to separate them. Whereas network theory focuses on the global features of any network, graph theory addresses structural features that originate from the interaction of single nodes and edges of a network structure. Graph theory is often traced back to Euler's works (e.g. [3]), while network theory can be dated back to the research of [4].

Research on matrix based complexity management has come a long way. Originating from a process focus with the first published formulation of a DSM [5], a whole community has developed around this research. The DSM is able to model and analyze dependencies of one single type within one single domain. Browning [6] classifies four types of DSMs to model different types of problems: component-, team, activity-, and parameter-based DSMs. However, many other classifications exist (e.g. [2]) nowadays.

There are numerous algorithms to analyze the overall structure of the relationships within a DSM; starting from the original algorithms for tearing, banding and partitioning [7][8] to a still non-exhaustive list provided by [2].

The authors of [9] have extended DSM to DMM, i.e. Domain Mapping Matrices. The goal was to enable matrix methodology to include not just one domain at a time but to allow for the mapping between two domains, as previously postulated e.g. by [10]. [2] has taken this approach further to model whole systems consisting of multiple domains, each having multiple elements, connected by various relationship types. He refers to this approach as Multiple Domain Matrix (MDM). He provides a number of ways to analyze the system's structure across multiple domains, condensing each single analysis into one DSM that represents multiple domains at a time. That way, he is able to apply algorithms for DSM analysis meaningfully across several domains, i.e. across a whole system. As especially the last DSM conferences have shown, matrix-based approaches integrating multiple views "domains" become more and more accepted to manage several perspectives onto a system, especially when it comes to large structures (e.g. >1000 elements per DSM).

3.2 Interpretation of structural criteria
Most of the approaches of structural complexity management look into what criteria qualities can be found in a structure, from the level of a global structure down to the integration of individual nodes. Structural criteria relates to the pattern of nodes and edges. Figure 2 orders the structural criteria, as provided by [2], by the evaluation of the number of edges and nodes that form a structure. In fact, most of the criteria can be traced back to a few basic elements [11][12].

In [2], several structural criteria are identified and interpreted considering change propagations between the elements regarding the modeled domain "components". Therefore, Maurer [2] divided structural criteria depicted in figure 2 into 2 groups: Structural criteria describing the meaning of nodes and edges and structural criteria describing the meaning of subsets. For each of these groups Maurer [2] discovered the structural criteria's meanings considering the development of a race car. The author presented how structural meanings ease structural complexity management by suggesting several interpretations of structural criteria. Until today several structural meanings considering different domains are identified:

Eben [13] analyzed structural meanings of requirements. Elezi [14] identified structural meanings considering processes with the aim of lean thinking. Kortler [15] described structural meanings considering components and their responsible designer. Kortler [16] identified another structural meaning considering the connection of requirements and design artifacts.

But, the structural meanings cannot be transferred from one application to another by implication. Different structural meanings may occur caused by differing data acquisition methods or differing models. In order to ensure a systematically basis a meta model describing which domains interact with each other considering the innovation process can be useful.

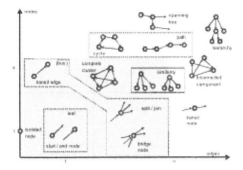

Figure 2. Basic structural criteria [11] [12]

4. SCIENTIFIC APPROACH

The authors are part of a research project which focuses in the area of managing cycles in innovation product development processes. The central topic of the research is the implementation and use of elements of complex solutions, nowadays typically consisting of a combination of product- and service-components, so-called product-service-systems (PSS). The components are subject to development-, manufacturing-, and life-cycles of varying length which are provided by different functional divisions. Availability and maturity of technologies, changes of competences, financial cycles at capital markets or of investments and write-offs as well as changes of customer demands represent external influences on the company. In contrast, the associated business processes underlie different cycles in research and development, manufacturing, logistics, finance, service, and recycling, which are mutually affecting each other as well.

The collaborative research project is performed by an interdisciplinary team from the fields of engineering, social and business sciences at TU München. The project is divided into different subprojects. Each of these subprojects focuses on different perspectives of the innovation process.

In order to derive a meta model describing the interactions between involved entities, the authors performed 14 interviews. Having performed the first interview phase the authors derived about 50 domains with diverse interactions. In order to remove the identified redundancies and gaps of the first version, the authors designed a catalog including gaps and redundancies in the meta model. With the help of this catalog the authors performed a second interview phase focusing the interactions of the modeled domains. Finally, 35 domains were identified. Considering the projects view, the elements of these domains are identified as the most influencing elements during the innovation process. Moreover, the authors derived more than 450 possible relationship types between elements of these domains. The second version of the model was used to implement the subproject's views on cycles during the innovation project. Thereby, the authors derived the most important domains and dependencies included in innovation cycles of the respective subprojects. In conclusion the authors classified the identified domains and derived a group-model in order to understand the essential dependencies of the basic entities during the innovation process.

5. A META MODEL DESCRIBING THE PROJECTS VIEW OF THE INNOVATION PROCESS

Figure 3 depicts the domains whose elements and dependencies are used in the presented meta model (depicted in figure 5b). Table 1 illustrates all of the abbreviations used in the meta model. The meta model depicted in figure 5b represents the project's view of the most important entities and interactions inside the innovation process. The development processes are not described in detail as a meta model describing development processes in detail is provided by Kreimeyer [12]. The proposed model is however not complete considering the innovation process. Instead of that, the first benefit of the model is supporting the author's project by highlighting entities and dependencies which are involved in re-occurring events (so-called cycles) – the dependencies shown with orange color in figure 5b. The presented model collects the most relevant relationship types of the identified domains. Dependencies between elements of the same domain are not included in this model. With the presented model the authors can decide whether changes in one or more elements will lead to change propagations in other elements. Thus, the authors can use the model in order to support the decision making process (taking change propagations into account).

With the help of the presented model and the 'model of four aspects in product development' [2] the authors classified the identified domains. To in order do so, the authors enhanced the product aspect to the Product-Service-System (PSS) aspect. Figure 3 depicts the identified domains ordered to their groups. Moreover, the authors identified for each group its own time reference. Elements of the process domains act in different phases of the innovation process. Elements of the PSS domains can be ordered to one or more PSS life cycle phases. Elements of the group environment or organization-unit have their own lifecycle (machines become obsolete, employee leave the company, laws change over time).

The authors identified super-relationship types describing the basic dependencies between the four groups. Subsequently the authors combined the groups and relationship types to a meta^2 model (depicted in figure 4). This model describes how the elements of the domains of the groups interact with other elements in general. As a second benefit, the model leads the authors to an improved understanding of the whole innovation process.

As a third benefit, this gained knowledge can be enhanced with structural knowledge in order to derive structural meanings describing the impacts of cycles or the possibilities of managing cycles during the innovation process. Thus, the model can be used as a systematic basis combined with structural complexity management. In doing so the authors identified structural criteria and structural meanings regarding special dependencies between elements of different domains.

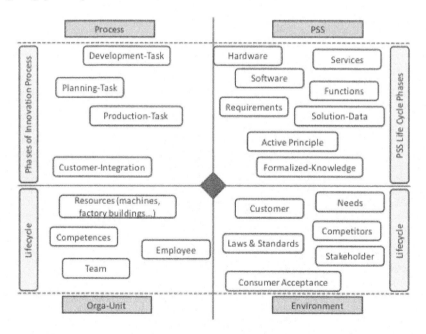

Figure 3. Clustered domains of the meta model

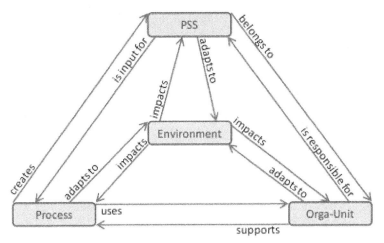

Figure 4. Meta² model

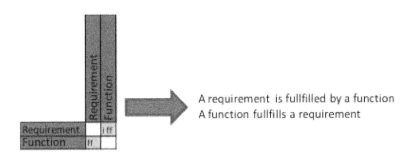

Figure 5a. Small Section of the meta model

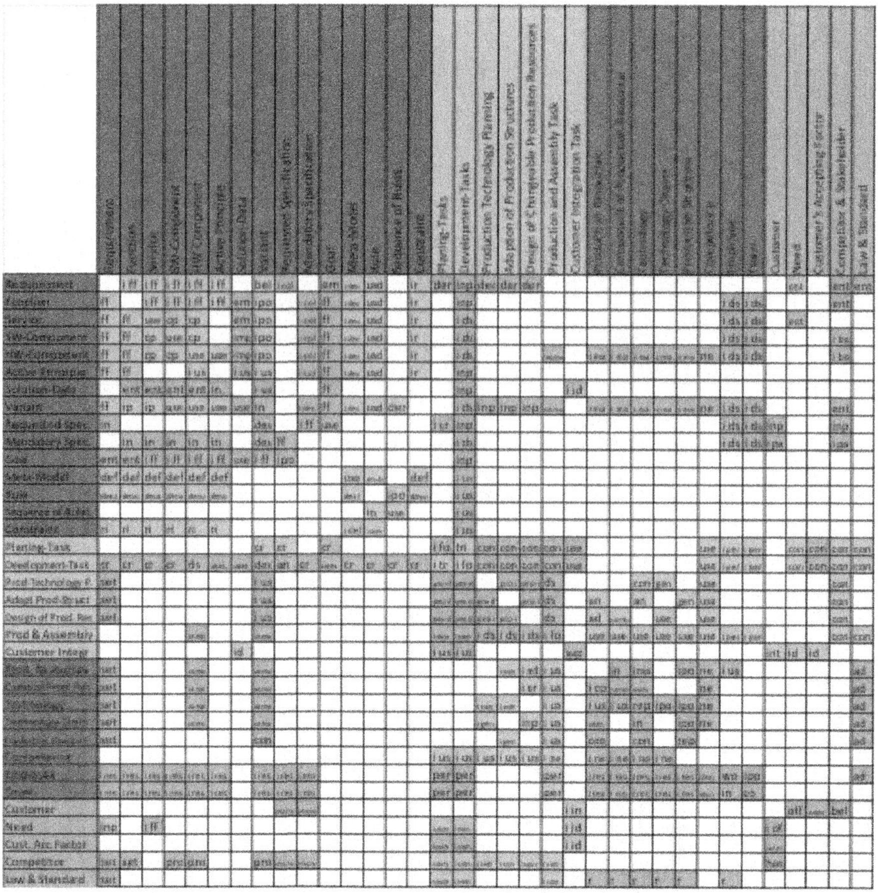

Figure 5b. The created meta model including identified domains and relationship types

Table 1. Table of abbreviations used in the meta model

Abbreviation	Relationship Type	Abbreviation	Relationship Type
ado	adapts	i ma	is manufactured by
an	analyzes	i ne	is needed
asma	assembles / manufactures	i of	is offered by
bel	belong to	i pa	is passed to
c ma	can be manufactured with	i per	is performed by
co	consist of	i res	is responsible for
comb	combines	i tr	is triggered by
comp	is compatible with	i us	is used for/by
con	considers	iasma	is assembled / is manufactured
cp	complements	id	identifies
cr	creates	imp	implements
def	defines	in	includes
der	derived by	inp	is input for
des	describes	int	interviews
des i	describe instantiation	ip	implements
desu	describes usage	ipo	is part of
ds	designs	ir	instantiation is restricted by
em	emerge from	ne	needs
ent	entails	nm	needs to be manufactured with
ff	fulfills	off	offers
gen	generates	per	performs
i ad	is adapted by	pro	provides
i bo	is bought at	pro b	Provides building blocks for
i co	is compatible with	pro d	provides data for
i col	is collected	pro i	provides input for
i con	is considered	r	restricts
i cr	ic created by	re	receives
i dec	is declared by	rep	replaced by
i def	is defined by	ri	restricts instantiation
i des	is described in	set	sets up
i ds	is designed by	sup	supports
i ff	is fulfilled by	tri	triggers
i fo	is followed by	use	uses
i gen	is generated by	usd	usage is described by
i id	is identified by	wo	works with
i in	is interviewed by		

6. USING THE META MODEL

The aim of the meta model is to support the development of innovative products and thus the evaluation and interpretation of the system's underlying structure. Therefore, the meta model provides a systematical basis. The authors' project aims on managing cycles which appear during the innovation process. At this point the authors present a small example using the presented meta model and structural complexity management in order to manage such emerging cycles.

This example demonstrates the applicability throughout the iterative process of refining requirements and concretizing product properties [16]. In this case the meta model prepares the dependencies between requirements and functions (see figure 5b).

The aim of the authors [16] was to control the refinement cycles of requirements and functions. Therefore, they connected stepwise the requirements model to the functional model. Finally they identified the structural criteria (active sum and passive sum) as an instrument to control the refinement cycles. More precisely, they indicated which function and which requirement need to be refined within which iteration. To do so, they mapped the requirements model on the functional model by using inter domain matrices. All functions and requirements as well as their relations are captured within each step of iteration. In order to identify functions and requirements with a high potential of refinement, they used active and passive sum considering the inter domain matrices. In this way the possibility of controlling the refinement cycles in each step of iteration was provided. The presented meta model (depicted in figure 5b) was used as a systematical basis for Structural Complexity Management. The model in this paper highlights domains and relationship types where the interpretation of structural criteria would be useful. Thus, the interpretation of structural criteria can be supported by using the proposed model. Scientists can start identifying further structural meanings.

Moreover, the designed meta model is to generate more transparency in the way of acting for all influencing stakeholders inside the innovation process. With the help of the meta model, all the participants of the innovation process can easily derive whether performed actions will have any influences on other involved partners. Moreover, all involved parties can identify whether the change actions of other partners propagate on their own elements.

7. CONCLUSION AND OUTLOOK

Structural awareness becomes more important considering the development of innovative products. Small changes in structures of the innovation process can cause huge impacts, so all available information about structure should be used in order to avoid wrong decisions, instable processes and error-prone solutions. Structural Complexity Management assumes a systematical basis in order to derive stable and reusable structural interpretations. The created meta model represents such a basis. In future work the authors aim to identify important paths inside the meta model. These paths allow for deriving chains of effects. Furthermore, the authors will include the proposed time references of the four identified groups (depicted in figure 3) into the meta model. Another point is the systematical identification of deliverables being transferred between the groups. In future research the authors will be focused primarily on the interpretation of further structural criteria in order to ease the development of innovative products and to support cycle management.

ACKNOWLEDGEMENTS

We thank the Deutsche Forschungsgesellschaft (DFG) for funding this research as a part of the collaborative research centre "Managing cycles in innovation processes – Integrated development of product service systems based on technical products" (SFB 768).

REFERENCES

[1] Boardman, J. and Sauser, B. System of Systems - the meaning of of. System of Systems Engineering, 2006 IEEE/SMC.
[2] Maurer, M. Structural Awareness in Complex Product Design. Lehrstuhl für Produktentwicklung (Dr.-Hut, München, 2007).
[3] Gross, J.L. and Yellen, J. Graph Theory and its Applications. (Chapman & Hall/CRC, Boca Raton, 2005).
[4] Erdoes, P. and Rényi, A. On Random Graphs I. Publicationes Mathematicae Debrecen, 1959, 6 (290).
[5] Steward, D.V. The design structure system: A method for managing the design of complex systems. IEEE Transactions on Engineering Management, 1981, 28, 71–74.
[6] Browning, T. Applying the Design Structure Matrix to System Decomposition and Integration Problems: A Review and New Directions. IEEE Transactions on Engineering Management, 2001, 48(3), 292-306.
[7] Kusiak, A. Engineering Design: Products, Processes and Systems. (Academic Press, San Diego, 1999).
[8] Steward, D. Partitioning and Tearing Systems of Equations. Journal of the Society for Industrial and Applied Mathematics: Series B, Numerical Analysis, 1965, 2(2), 345-365.
[9] Danilovic, M. and Browning, T.R. Managing complex product development projects with design structure matrices and domain mapping matrices. International Journal of Project Management, 2007, 25(3), 300-314.
[10] Yassine, A., Whitney, D., Daleiden, S., Lavine, J., "Connectivity maps: modeling and analysing relationships in product development processes." Journal of Engineering Design, 2003, 14(3), 377-394.

[11] Kortler, S., Kreimeyer, M. and Lindemann, U., "A Planarity-based Complexity Metric", Proceedings of the International conference on engineering design, ICED 2009.
[12] Kreimeyer, M., "A Structural Measurement System for Engineering Design Processes" Lehrstuhl für Produktentwicklung (Dr.-Hut, München, 2010).
[13] Eben, K., Daniilidis, C. and Lindemann, U. "Interrelating and prioritizing requirements on multiple hierarchy levels", Proceedings of 11th International Design Conference DESIGN 2010 Dubrovnik, Croatia 2010.
[14] Elezi, F. "Reducing waste in product development by use of Multi-Domain Matrix methodology", Proceedings of 11th International Design Conference DESIGN 2010 Dubrovnik, Croatia 2010. Design Paper 2010
[15] Kortler, S., Diepold, KJ. and Lindemann, U., "Structural Complexity Management using domain-spanning structural criteria", Proceedings of 11th International Design Conference DESIGN 2010 Dubrovnik, Croatia 2010.
[16] Kortler, S., Helms, B., Berkovich, M., Lindemann, U., Shea, K., Leimeister, JM. and Krcmar, H., "Using MDM-Methods in order to improve managing of iterations in design processes", Proceedings of the 12th International DSM Conference Cambridge, UK 2010

Contact: Sebastian Kortler
Technische Universitaet Muenchen
Institute of Product Development
Boltzmannstraße 15
D-85748 Garching
Germany
Phone: +49 89 289 151 53
Fax: +49 89 289 151 44
Email: kortler@pe.mw.tum.de
URL: http://www.pe.mw.tum.de

Sebastian Kortler is a scientific assistant at the Institute of Product Development at the Technische Universität München, Germany since 2008. Before, he studied computer science until 2007. Currently, his research is focused on structural complexity management and the development of structural characteristics that govern complex systems.

Udo Lindemann is a full professor at the Technische Universität München, Germany, and has been the head of the Institute of Product Development since 1995, having published several books and papers on engineering design. He is committed in multiple institutions, among others as Vice President of the Design Society and as an active member of the German Academy of Science and Engineering.

INTERNATIONAL CONFERENCE ON ENGINEERING DESIGN, ICED11
15 - 18 AUGUST 2011, TECHNICAL UNIVERSITY OF DENMARK

PARETO BI-CRITERION OPTIMIZATION FOR SYSTEM SIZING : A DETERMINISTIC AND CONSTRAINT BASED APPROACH

Pierre-Alain Yvars[1]
(1)LISMMA, SupMeca, France.

ABSTRACT
In this paper we are studying a deterministic constraint based approach to solve Pareto bi-criterion optimization problems in design. After presenting the use of multi-objective optimization methods in design, the CSP method and the several ways to solve it is introduced. A quick overview of CSP application in product engineering is given too. Moreover, we introduce an optimization point of view for CSP and we propose a deterministic alternative to stochastic methods for solving Pareto bi-objective system sizing problems. An example in mechanical system optimization is given via the case study of the Pareto bi-criterion optimal design of a bolt coupling. The case is modeled as a Constraint Satisfaction Problem on both discrete and real variables. Finally, the numerical results and the Pareto frontier are exposed.

Keywords: Constraint propagation, CSP, Pareto frontier, bi-criteria optimization, system sizing.

1 INTRODUCTION
The context of integrated and collaborative design of mechanical product as became now the usual context of design. Even if some concepts and tools are available to support this process, lacks still exist. In this context optimization of product remains a strong issue. In this paper we are studying a deterministic constraint based approach to solve Pareto bi-criterion optimization problems in design.
A first naïve approach should be used to deal with CSP and multi-objective optimization. It consists on transforming the interval of values of the first criteria into a set of discrete values. For each of them, a mono-objective algorithm is started to minimize the value of the second criteria. Unfortunately, this approach has at least two main drawbacks in design: On the one hand, there are many problems to adjust the set of discrete values and on the other hand, in case of a non convex and discontinuous design problem, we can miss several optimal points. The main purpose of this article is to go over these limitations in case of mixed design problems.
After presenting the use of multi-objective optimization methods in design, the CSP method and the several ways to solve it is introduced. A quick overview of CSP application in product engineering is given too. Moreover, we introduce an optimization point of view for CSP and we propose a deterministic alternative to stochastic methods for solving Pareto bi-objective system sizing problems. An example in mechanical system optimization is given via the case study of the Pareto bi-criterion optimal design of a bolt coupling. The case is modeled as a Constraint Satisfaction Problem on both discrete and real variables. Finally, the numerical results and the Pareto frontier are exposed.

2 MULTI-OBJECTIVE OPTIMIZATION IN DESIGN

2.1 Multi-objective optimization problem
A lot of problems in the design fields are relevant to multi-objective optimization problems. Multi-objective optimization is an answer to the need of satisfying both many conflicting criterion. Because there is rarely a solution better than another at any point, different compromises depending on preferences can be chosen.

A multi-objective problem is defined such that:

$X = (x_1, x_2, ..., x_n)$ called decision vector. In a design problem, the x_i variables are called the design parameters.

$F = (f_1, f_2, \ldots, f_m)$ called performance vector. Of course, in mono-objective optimization problem, F is a scalar value.

$$\exists p \in \mathbb{N}, \forall j \in \{1, \ldots, p\}, \exists g_j : \mathbb{R}^n \to \mathbb{R}, \exists \mathcal{X} \subseteq X, g_j(\mathcal{X}) = 0 \tag{3}$$

$$\exists q \in \mathbb{N}, \forall j \in \{1, \ldots, q\}, \exists h_j : \mathbb{R}^n \to \mathbb{R}, \exists \mathcal{X} \subseteq X, h_j(\mathcal{X}) \leq 0 \tag{4}$$

$$\forall i \in \{1, \ldots, m\} f_i : \mathbb{R}^n \to \mathbb{R}^+ \tag{5}$$

$$Find \quad X / \min F \tag{6}$$

Unfortunately, the solutions to this problem rarely minimizes all the f_i. It is necessary to propose a comparison operator to determine if a performance vector is better than another or if they are equivalents. A possibility is to use the relation of domination according to the definition given by Pareto. Noting $\preccurlyeq$ this relation in its wide sense and $\prec$ in its strict sense, F_i dominates F_j in Pareto sense if and only if:

$$\forall k \in \{1, \ldots, m\}, f_{ik} \preccurlyeq f_{jk} \tag{7}$$

and

$$\exists k \in \{1, \ldots, m\}, f_{ik} \prec f_{jk} \tag{8}$$

2.2 Application in design

The problem consists on determining the non-dominated set of points in the performance space. For any m, the Pareto hyper-surface can theoretically be obtained although its calculation is usually difficult and expensive.

More often, design problems and more precisely sizing ones are characterized by:
- A strong set of requirements.
- Mixed variables: the design parameters and the performance parameters should be both continuous and discrete.
- Many analytical relations: The parameters have to satisfied many algebraic relations.

A lot of works have been done to use stochastic algorithms to solve multi-objective optimization problems in design. A genetic algorithm as NSGA II give the opportunity to draw a good approximation for the Pareto frontier [1]. Unfortunately, those algorithms have several drawbacks: the calculation time is very high and more often they check the satisfaction of the analytical relations of the design problem by using a penalty mechanism.

The constraint based approach, that we will deal with in this article, allows us to express and check the analytical relations posted on the design parameters and the performance parameters. This analytical relations should be all algebraic equations and inequalities.

3 BI-CRITERION OPTIMAL SIZING AS A CSP PROBLEM

3.1 CSP

A CSP is defined by a Triplet (X, D, C) such that [2]:

- $X = \{X_1, X_2, \ldots, X_n\}$ is a finite set of variables called constraint variables with n being the integer number of variables in the problem to be solved.

- $D = \{D_1, D_2, \ldots, D_n\}$ is a finite set of variables value domains of X such that :

$$\forall i \in \{1, \ldots, n\}, x_i \in Di \tag{9}$$

A domain should be a real interval or a set of discrete values.

- $C = \{C_1, C_2, \ldots, C_p\}$ is a finite set of constraints, p being any integer number representing the number of constraints of the problem.

$$\forall i \in \{1, \ldots, n\}, \exists \;_i \subseteq X \;/\; C_i(\;_i) \; x_i \in D_i \tag{10}$$

A constraint is any type of mathematical relation (linear, quadratic, non-linear, Boolean…) covering the values of a set of variables.

More precisely, the constraints can be the following:
- Logical: such that x=1 or y=4; x=3 ⇒ z=5
- Arithmetical: such that x > y; 2x+3y < z

- Non-linear: such that cos(x) < sin(y)
- Explicit: in the form of n-tuples of possible values such that: (x, y) (0, 0), (1, 0), (2, 2)
- Complex: such that: the variable values x, y, z must all be different.

The variable domains can be:
- Discrete: in the form of sets of possible values.
- Continuous: in the form of intervals on real numbers

Solving a CSP boils down to instantiating each of the variables of X while meeting the set of problem constraints C, and at the same time satisfying the set of problem constraints C.

The solving process for a CSP depends on the type of the constraint variables. In fact CSP on integer variables called discrete CSP are different from CSP on real variables also called continuous or numerical CSP.
- On the one hand, for solving discrete CSP, the methods are ones arising from operational research and artificial intelligence. The first work dates back over thirty six years [3]. These discrete CSP methods, of exponential complexity, are based on enumeration and filtering. This filtering, also called constraint propagation, enables the definition domains of variables to be reduced as the resolution process evolves.
- On the other hand, CSPs have been developed with real variables with values in intervals. This interval-based resolution technique is a synthesis between interval-based analysis [4] and CSPs in [5,6]. Several techniques have been developed, one of which is presented as an example in [7].

During the design process, designers used and managed design rules, tables, abacus, relations… All these structures should be modeled as constraints (mathematical relations between variables).
The CSP community has developed work applicable in product and systems design [8-11].

3.2 Numerical CSP solving process
Search algorithms as Branch and Prune start the process by selecting a variable to bisect. The order in which this choice is done is referred the variable ordering. A correct ordering decision can be crucial to perform an effective solving process in case of real-life problems. There exist several heuristics for selecting the variable ordering. After selecting the variable to bisect, the algorithms have to select a subinterval form the variable's domain. This selection is called the value ordering. It can also have an important impact on the duration of the solving process. The Prune subroutine contracts the intervals of D by using interval arithmetic [4] and consistency mechanisms [6]. The goal is to fit the intervals bounds as much as possible without losing any solutions.

```
BP(CSP(X, D, C), {})
begin
D ← Prune(C, D)
if notEmpty(D) then
 if OkPrecise(D) then Insert(D, L)
   else
   (D_1, D_2) ← Split(D, ChooseVariable(X))
   BP(CSP(X, D_1, C), L)
   BP(CSP(X, D_2, C), L)
 endif
return L
end
```

Figure 1. Branch and Prune algorithm[7]

3.3 A deterministic constraint based Pareto bi-criterion optimization algorithm

The mono-objective optimization principle adopted to minimize the value of an objective f is described on Figure 2. Usually, f should be a variable equal to a constraint expression representing the criteria to minimize. The key point is to solve by dichotomy a sequence of CSP where the set of constraints increases from one CSP to the next. At each step, we add a constraint expressing that the next CSP has to be better than the current one according to the minimization of the f variable. The process stops on the CSP which minimize the f variable value when the required precision ε is reached. This is a kind of branch and bound method which works on finite domains as well as intervals.

```
OptimCSP(X, D, C)
begin
  f ∈ [f_min, f_max]
  CSP ← (X, D, C)
  while f_max - f_min > ε
    C ← C ∪ { f < (f_max + f_min)/2 }
    if find a solution for CSP
      f_max ← fval
    else
      C ← C - { f < (f_max + f_min)/2 }
      f_min ← (f_max + f_min)/2
    endIf
  endWhile
  return [f_min, f_max]
end
```

Figure 2: CSP and mono criteria optimization

Usually, the problem of finding the non-dominated frontier is addressed by finding the optimal value in one direction and restart the search with constraints that limit the search space to other Pareto-Optimal solutions. For a bi-criterion problem, a naïve approach consists on solving a mono-objective problem for each value of the second criteria but a lot of drawbacks should appear especially when the Pareto frontier is not convex and/or discontinuous.

Van Wassenhove and Gelders [12] find the non dominated frontier in a bi-criterion (f_1, f_2) scheduling problem. It should be described as follow:

1st step: find the optimal solution (Opt_2) that minimize function f_2; Let $d \leftarrow Op_{t2}$
2nd step: impose that $f_1 < f_1(d)$
3rd step: minimize f_2. If a solution S exists, then it is non dominated; else exit.
4th step: Let $d \leftarrow f_1(S)$. Go to 2nd step.

If we try to use both csp and Van Wassenhove & Gelders principles we obtain what we call the BiPareto algorithm as shown in Figure 3.

```
BiPareto(X, D, C, f₁, f₂)
begin
Opti(X, D, C, f₂)
repeat
    Nondominated ← NonDominated ∪ (X,D,C)
    C ← C ∪ (f₁ < f₁val)
until ! Opti(X, D, C, f₂)
return NonDominated
end
```

Figure 3: CSP and Pareto bi-criterion optimization

4 CASE-STUDY

4.1 Description

We propose to illustrate the contribution of CSPs method with a simple case study of mechanical component. Figure 2 is presenting a bolt coupling that is used to transmit a torque by adherence between two shafts.

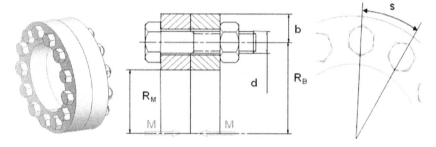

Figure 4: Bolt coupling

We suppose here that, in the design process of the product, technological choices have been made. Once designer have chosen this component, it is possible to identify design parameters {DPs} and functional requirements. In this simple case, functional requirements can be written as explicit relations between design parameters. These relations are equations and inequality relations. The model used here to establish theses relations lies on the expertise of the designer. This model is based on the VDI [13] method to calculate bolt dimension. The relations listed below will give the set of constraints for the CSP method. The mono-objective version of this problem has been previously treated in [14] with a pure genetic algorithm based approach and in [15] with a CSP one.

Table 1. Set of design parameters

		Design parameters
Geometrical parameters	d_s	diameter of stress area (mm)
	d	nominal diameter of the bolt (mm)
	b	radial width of the contact surface
	s	interval between bolt (mm)
	d_2	pitch diameter of thread (mm)
	p	pitch of thread (mm)
	s_m, b_m	size of the tightening tools (mm)
	A_s	area of stress cross section (mm²)
	N	number of bolts
	N_m	minimal number of bolts
	R	radius of the couling (mm)
	R_b	radius of bolts (mm)
	R_m	radius of housing shafts (mm)
Functional parameters	M	torque transmitted by the coupling (N.mm)
	M_T	torque to be transmitted by the coupling (N.mm)
	$F0_{mini}, F0_{maxi}$	minimal and maximal tensile strength in bolts (N).
	C_1	torsion moment in the bolt due to the preload (N.mm)
	σ_{max}	maximal normal stress in the bolt (MPa)
	τ_{max}	maximal tangential stress in the bolt (MPa)
	σeq_{max}	maximal Von Mises stress in the bolt (MPa)
	α_s	accuracy factor of the tightening tool
Material paramete	f_m, f_l	friction coefficient between rim of the coupling and threaded contact surfaces in bolt

4.2 Modeling as a constraint satisfaction problem

An analysis of the bolt coupling gives us the relations below taken as CSP constraints:

$$M = N \times R_b \times f_m \times F0_{mini} \qquad (11)$$
$$F0_{maxi} = \alpha_s \times F0_{mini} \qquad (12)$$
$$\sigma_{max} = \frac{F0_{maxi}}{A_s} \qquad (13)$$
$$\tau_{max} = 16 \times \frac{C_1}{(\pi \times d_s^3)} \qquad (14)$$
$$C_1 = F0_{maxi} \times (0.16 \times p + 0.583 \times d_2 \times f_1) \qquad (15)$$
$$\sigma eq_{max} = \sqrt{\sigma_{max}^2 + 3 \times \tau_{max}^2} \qquad (16)$$
$$s = 2 \times \pi \times \frac{R_b}{N} \qquad (17)$$
$$R_b = R_m + b \qquad (18)$$
$$A_s = \pi \times \frac{d_s^2}{4} \qquad (19)$$
$$R_b \geq R_m + b \qquad (20)$$
$$0.9 \times R_e \geq \sigma eq_{max} \qquad (21)$$
$$M \geq M_T \qquad (22)$$
$$s \geq s_m \qquad (23)$$
$$b \geq b_m \qquad (24)$$
$$N \geq N_m \qquad (25)$$

The screw parameters set of values (Table 2) should be modeled as a constraint table.

A constraint table is a global constraint that represents the possible combination values of a set of constraint variables. By global constraint, we mean a constraint that should be propagated on complex data structures. In our case, each line of a constraint table is a tuple of consistent values. If one or several values of a constraint variable become forbidden during a CSP solving process all the tuple related to this value are removed from the table too.

For example, with Table 2, if we decide that d_2 has to be greater than 10 and p has to be different to 2 then, Lines number 1, 2, 3, 5 and 6 are removed from the table. Only lines number 4, 7 and 8 stay inside the constraint table.

Table 2. Screw parameters set of values table

Num	d	ds	d2	p	bm	sm	dt
1	6	5.062	5.350	1.00	7.50	14.50	6.6
2	8	6.827	7.188	1.25	9.50	18.50	9.0
3	10	8.593	9.026	1.50	12.50	23.50	11.0
4	12	10.358	10.863	1.75	13.50	26.50	13.5
5	14	12.124	12.701	2.00	15.50	29.50	15.5
6	16	14.124	14.701	2.00	17.00	32.00	17.5
7	20	17.655	18.376	2.50	21.00	40.00	22.0
8	24	21.185	22.051	3.00	25.00	48.00	26.0

Our CSP model has been implemented with the IlogCP C++ library [16].

We would like now to illustrate our approach with a set of numerical results. The initial domains given by the expert for the constraint variables are presented on Table 3. After a first propagation (Prune subroutine), the intervals are reduced as in the second column of Table 3. Then after adding the following specifications:

$M_T = 4000000$ (26)
$f_m = 0.15$ (27)
$f_1 = 0.15$ (28)
$\alpha_s = 1.5$ (29)
$R_s = 627$ (30)
$N_m = 8$ (31)
$R_m = 50$ (32)

The next propagation step again reduces the intervals as shown in the fourth column of Table 3.

4.3 Pareto bi-criterion optimization
In this case study we would like to minimize the two following functions:

- The total cost of the bolt coupling in euros :
$$C_{coupling} = K_1 \times d + K_2 \times N \quad (33)$$
$K1 = 0.6$ euros/mm (34)
$K2 = 5$ euros /bolts (35)
- The total mass of the bolt coupling (t_h parameter is the thickness of the coupling) :
$$M_{coupling} = \frac{\pi}{2} \times t_h \times (\rho_j \times (4 \times R_b \times b_m - N \times d_t^2) + \rho_v \times N \times d^2) \quad (36)$$
$\rho_j = 2.710^{-6}$ kg.mm^{-3} (37)
$\rho_v = 7.810^{-6}$ kg.mm^{-3} (38)

Table 3. CSP solving process

Design	Initial domains	1st propagation	After specifications
A_s	[15, 400]	[20.12, 352.49]	[20.12, 352.49]
R_b	[5, 1000]	[13, 1000]	[58, 100]
R_m	[5, 1000]	[5, 992]	50
s	[1, 100]	[14.5, 100]	[14.5, 78.5398]
b	[5, 50]	[8, 50]	[8, 50]
N	[1, 1000]	[4, 433]	[8, 43]
N_m	[4, 1000]	[4, 433]	8
d	[6, 24]	[6, 24]	[6, 24]
p	[1, 3]	[1, 3]	[1, 3]
d_2	[5.35, 22.051]	[5.35, 22.051]	[5.35, 22.051]
d_s	[5.062, 21.185]	[5.062, 21.185]	[5.062, 21.185]
s_m	[14.5, 48]	[14.5, 48]	[14.5, 48]
b_m	[7.5, 25]	[7.5, 25]	[7.5, 25]
dt	[6.6, 26]	[6.6, 26]	[6.6, 26]
M	[1000, 1e+007]	[1000, 1e+007]	[4e+006, 1e+007]
$F0_{mini}$	[10, 100000]	[10, 66666.7]	[6201.55, 66666.7]
$F0_{maxi}$	[10, 100000]	[15, 100000]	[9302.33, 100000]
C_l	[0, 200000]	[2.4, 200000.]	[5840.53, 200000]
σ_{max}	[0, 2000]	[0.0425, 1110.6]	[26.3903, 564.274]
τ_{max}	[0, 2000]	[0.00128, 641.2]	[3.12851, 325.442]
σeq_{max}	[0, 2000]	[0.0426, 1110.6]	[26.9409, 564.3]
α_s	[1.5, 4]	[1.5, 4]	1.5
f_m	[0, 1]	[3.464e-008, 1]	0.15
f_l	[0, 1]	[0, 1]	0.15
R	[10, 1050]	[21, 1050]	[66, 150]

The goal is to minimize both $C_{coupling}$ and $M_{coupling}$. To do that, we implement the BiPareto algorithm and obtain the Pareto frontier of the problem (Figure 5). We notice that in our test case the Pareto frontier is discontinuous and not convex.

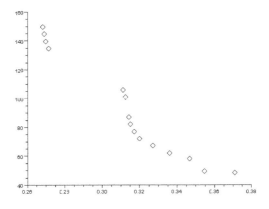

Figure 5: Pareto frontier of the bolt coupling sizing problem

5 CONCLUSION

As a conclusion, we can outline that CSP method are able to support the phase of reducing the size and the complexity of the design space. CSP modelling and solving processes avoid complex analytical manipulation of function requirements. Designer can test several ways of constraining the design by adding or deleting constraints. This incremental process is very useful in this phase of reducing the design space, in an integrated and collaborative design process.

For this simple example, the BiPareto algorithm finds easily the Pareto frontier optimal solution. One of the advantage is that designers don't have to simplify the formulation of the optimization problem. Another one is that our proposal allows to deal with all types of bi-dimensional Pareto frontier even if they are discontinuous or non convex.

Our current research concerns the generalization of this approach to an n dimensional Pareto Surface algorithm based on the CSP mechanisms.

REFERENCES

[1] Gupta S., Tiwari R., Nair S.B., Multi-objective design optimization of rolling bearings using genetic algorithms, *Mechanism and Machine Theory, Volume 42, Issue 10*, October 2007, pp 1418-1443
[2] Montanari U., Networks of constraints: fundamental properties and applications to picture processing, *Information Science* 7, 1974, pp.95-132.
[3] Mackworth A.K., Consistency in networks of relations, *Artificial Intelligence* 8, 1, 1977, pp.99-118.
[4] Moore R.E., *Interval Analysis, Prentice-Hall*, 1966.
[5] Davis E., Constraint propagation with interval labels, Artificial Intelligence, v. 24.3, 1987.
[6] Falting B., Arc consistency for continuous variables, Artificial Intelligence 65(2), 1994.
[7] Benhamou F., Granvilliers L., *Continuous and Interval Constraints. In Handbook of Constraint Programming*, Chapter 16:571-604, 2006.
[8] Vargas C., Saucier A., Albert P., Yvars P.A., Knowledge Modelisation and Constraint Propagation in a Computer Aided Design System, *In Workshop notes Constraint Processing in CAD of the Third International Conference on Artificial Intelligence in Design*, Lausanne, Switzerland, August, 1994.
[9] Yannou B., Harmel G., Use of Constraint Programming for Design, in Advances in Design, El Maraghy W. Editors, Springer, chapter 12, 2005.
[10] Chenouard R., Sebastian P., Granvilliers L., Solving an Air Conditioning Problem in an Embodiment Design Context using Constraint Satisfaction Techniques, CP'2007, 13th International Conference on Principles and Practice of Constraint Programming, 2007.
[11] Yvars, P.A., 2008, Using constraint satisfaction for designing mechanical system, *International Journal on Interactive Design and Manufacturing (IJIDeM)*, Volume 2, Number 3, August 2008.
[12] L. N. V. Wassenhove and L. F. Gelders. Solving a bicriterion scheduling problem. *European Journal of Operational Research*, 4(1):42-48, 1980.
[13] V.D.I.2230, *Systematic Calculation of High Duty Bolted Joints*, 1998.
[14] Giraud, L., Lafon, P., Optimization of mechanical design problems with genetic algorithms, *Proceeding of the 2nd International Conference IDMME'98*, Compiègne, France, pp. 90-98, 1998.
[15] Yvars P.A., LAfon P., Zimmer L., Optimization of mechanical system: Contribution of constraint satisfaction method, *proc of IEEE Computers and Industrial Engineering CIE'39*, Troyes, France, 2009.
[16] Ilog, *IlogCP, Reference Manual, Ilog*, Gentilly, France, 2009.

Contact: P.A. Yvars
Institut Supérieur de Mécanique de Paris (SupMeca)
LISMMA
3 rue Fernand Hainaut
93407, Saint Ouen Cedex
France
Phone : int +33(0)1 49 45 29 25
e-mail : pierre-alain.yvars@supmeca.fr
URL : http://lismma.supmeca.fr

Pierre-Alain Yvars is an Associate Professor at the Institut Supérieur de Mécanique, Paris. His current subjects of research focus on the declarative meta modeling of complex systems, design processes, and production systems, the application of constraint programming techniques and interval propagation to solve and optimize design problems.

INTERNATIONAL CONFERENCE ON ENGINEERING DESIGN, ICED11
15 - 18 AUGUST 2011, TECHNICAL UNIVERSITY OF DENMARK

SICK SYSTEMS: TOWARDS A GENERIC CONCEPTUAL REPRESENTATION OF HEALTHCARE SYSTEMS

Alexander Komashie[1], Thomas Jun[2], Simon Dodds[3], Hugh Rayner[3], Simon Thane[4], Alastair Mitchell-Baker[4], John Clarkson[1]
[1]Egineering Design Centre, University of Cambridge, Cambridge, UK,
[2]Loughborough Design School, Loughborough University, Loughborough, UK
[3]Birmingham Heartlands Hospital, Birmingham, UK
[4]Tricordant Ltd, UK

ABSTRACT
In this paper, we argue that the healthcare systems within which patients are treated are like patients themselves. The systems display symptoms which may give indication of problems in an "organ" of the system. The human system that forms the core of healthcare activities is a complex system and so are healthcare systems. The success of medical diagnosis has been facilitated by a generic concept of the human anatomy and its systems, organs, and corresponding physiology. The lower levels are the building blocks on which the upper levels depend. Disease processes cause failure at the chemical levels and this failure affects organs, systems, and even the whole body. We observe an interesting similarity between the medical diagnosis process and the systems design approach, yet there is no corresponding generic representation of healthcare systems akin to the anatomy and physiology of the human system. Our goal in this paper is not to match the healthcare system to the human system part by part and organ to organ but to discuss how the structured medical diagnosis process can be applicable to healthcare systems if an appropriate conceptual representation of the system can be developed.

Keywords: Conceptual representation, Healthcare, systems, simulation modelling

1. INTRODUCTION
It could be considered impossible to design, model, simulate or analyse a system without a conceptual representation of it. Even when we avoid the use of a formal conceptual representation of the system, in the process, our mental models are always involved (Sterman, 1991; Forrester, 1991). It is therefore surprising to find how little research has gone into the development of a formal and consistent approach to conceptual modelling particularly in healthcare systems. In the field of Operations Research (OR) and simulation modelling for instance, this appears to be a newly discovered area of research (Robinson et al., 2011).
Healthcare systems are undeniably very complex but the human body as a system may be even more complex (Thibodeau & Patton, 2010, p3). If a generic conceptualization of the healthcare system exists as it does for the human system (the human anatomy), it may facilitate the achievement of two goals namely; a diagnostics approach to understanding healthcare system problems and the effective communication of healthcare system issues to practitioners. This paper discusses the conceptual representation approaches in a number of research communities, presents the human system anatomy and physiology analogy and proposes an embryonic generic conceptual representation of healthcare systems for the above stated goals.
For centuries, anatomists, physiologists and pathologists have endeavoured to develop a proper understanding of the structure, functions and mechanisms of disease of the human system. These efforts have led to a well established, fundamental and generic, conceptual representation of the human system that has become the key to the medical diagnosis process (Singer, 1957; Gonzales-Crucci, 2007). For several decades, the bulk of healthcare research has focused on the clinical aspects of care, however, it is now understood that the quality of healthcare must be measured by the quality of the structure within which care is provided, the process by which care is provided and the outcome of care (Donabedian, 1966). All of these together, form the system for providing quality care.

The growing interest in healthcare system research has led to the emergence of several challenges for both researchers and practitioners. Amongst these challenges are the effective application of industrial tools in healthcare (Young et al., 2004; Kopach-Konroad et al. 2007; Young and Mclean, 2009), the understanding of the complex nature of the healthcare system (Palley and Gail, 2010) and the conceptualization of the healthcare system for various purposes including system design, modelling and simulation (Brailsford, 2007). This paper will focus on the challenge of conceptualization.

Healthcare systems research has become a multi-disciplinary endeavour with disciplines ranging from operations research, modelling and simulation, information technology, organisational theory, to name a few. As a result, several conceptual models have been developed that seem to address certain aspects of healthcare that is of particular interest to a discipline. In some disciplines a unified approach does not exist. In the field of operations research (OR) modelling and simulation, conceptual modelling is just emerging as an active field of research (Robinson et al., 2006; Robinson et al., 2011). Some of the commonly used conceptual models are discussed in more detail in section 3.

The generic representation proposed in this paper is novel in a number of ways: first of all it adopts the "Sick Systems" approach which means considering a system with problems as "sick" and requiring diagnosis and treatment as a physician would consider a human being with health problems. Secondly, it is argued that this approach to conceptualising the system would facilitate communication of systems design and modelling concepts and issues to healthcare practitioners and help improve uptake and implementation. This metaphor is intuitive and sometime used informally in some aspects of healthcare management but has not yet become a subject for research. For instance, the renal service managers of an NHS Trust in England have developed an "analogous list of metrics that indicate the health of the local kidney care system", and refer to various elements of the system as organs or body parts. There is also the TRICORDANT consultancy in England that has used the "unwell" systems metaphor for years with clients in healthcare. To underpin these concepts with research should help reap the maximum benefits in application.

As a result, the main motivation for this paper is to make a modest contribution towards a new method of thinking about healthcare systems problems in diagnostics terms akin to healthcare. It is also intended to stimulate debate on the subject within a community that has an interest in the design of healthcare systems.

The next section of the paper presents the study framework, highlighting the ideas that are used to evaluate existing representations and also as the basis for the proposed generic representation. Section 3 discuses the conceptual modelling approaches from a number of fields and theories. Section 4 then presents a summarised evaluation of more forms of conceptual representations. In section 5, we further discuss in detail the proposed generic concept and in section 6, we provide a brief discussion and some direction for future research on the subject.

2. STUDY FRAMEWORK

This paper distinguishes between structure and function; anatomy and physiology. We are not at this stage concerned with the functions of various elements of the system though vitally important. This approach is similar to Henry Mintzberg's proposed generic structure of an organisation which considered the organisation as comprising essentially of the Strategic apex, operating core, middle line, technostructure and support staff (Mintzberg, 1983). However, Mintzberg's representation of the system has a functional connotation. The basic questions in this paper are; "what is a healthcare system essentially made up of and can these be generically represented?" and "is the medical diagnostics process analogy of healthcare systems a feasible one"? In the next section we discuss the five basic elements of a system which we propose as sufficient for representing the essential elements of most healthcare systems.

2.1 Elements of a Healthcare System

We consider that there are five major elements that are common to most systems namely; Resources, Processes, Data/Information, Entities and Environment (Wyllys, 2011). These are essentially what these systems are made up of and should be sufficient for describing the system in a very generic sense. These elements are briefly explained below:

Resources: in the general sense, resources are the elements of the system that use or support processes in transforming entities or delivering results for entities. This would include financial resources, human resources and materials.

Processes: these are the elements of the system that involve designed steps necessary to facilitate the achievement of specific goals for entities.
Data/Information: these are the elements of the system that represent the source of the knowledge necessary to ensure effective interaction between various system elements vertically and horizontally.
Entities: these are the elements of the system that go through the processes using data and information and consuming and often competing for resources.
Environment: this defines the boundary of the system and involves elements outside of the system and/or its elements but with which the system may interact.

Using the examples of a GP practice, an emergency department and a hospital setting, Table 1 shows examples of how these elements may be defined.

Table 1. Examples of system elements in healthcare systems

System elements	GP practices	Emergency care	Hospital care
Resources	**People**: GPs, nurses, practice managers and receptionists	**People**: A&E doctors, nurses, receptionists and consultants	**People**: managers, doctors and nurses
	Equipment: blood pressure gauge and stethoscope	**Equipment**: Blood pressure gauge, Stethoscope and ECG	**Equipment**: various medical devices
	Facilities: reception and consultation rooms	**Facilities**: beds	**Facilities**: wards, beds
Data/Information	Appointment schedule, referral letters, patient records and test results	Referral letters, patient records, test results and discharge letters	Referral letters, patient records, test results and discharge letters
Processes	Appointment booking, diagnosis, treatment, immunisation and referral	Triage, minor treatment, major treatment, resuscitation and referral	Scheduling, diagnosis, treatment and referral
Entities	general patients and patients in special needs	walk-in patient, GP-referred patient and ambulance patient	in-patient and out-patient
Environment	Homes, referral agencies	GP, Ambulance, support services	GP, A&E, Homes,

Whereas these elements are not based on empirical research, we suggest that the role they play in healthcare systems is self-evident and could be reasonably employed at this early stage of this research. These elements may also be identified at various levels of the healthcare system and form systems themselves. The advantage of these elements is that, they are generic and therefore useful for the human body system analogy that is presented in this paper.

2.2 The Human Body System Analogy

In this section, we present a brief discussion on anatomy, physiology and pathology. We focus on the distinction between anatomy and physiology and how the study of these two logically facilitates the study of pathology. The human body system that forms the core of all healthcare activities is a complex system but the structure is fairly well understood. The success of medical diagnosis has been facilitated by a generic concept of the human anatomy and its systems, organs, cellular structure and corresponding physiology. The lower levels are the building blocks on which the upper levels depend. Disease processes cause failure at the chemical or cellular levels and this failure affects organs, systems, and even the body as a whole (Thibodeau & Patton, 2010).
We believe that applying this analogy to the healthcare system itself has considerable benefits. This does not mean matching the healthcare system to the human system part by part and organ to organ but to learn from how the structured medical diagnosis process can be applicable to healthcare systems if an appropriate conceptual representation of the system can be developed.

Anatomy, physiology and Pathology

In medicine, anatomy is fundamentally concerned with the structure of the human body system and the interaction between its grouped systems, split systems and principal organs (see figure 1). Figure 1 (adapted from Thibodeau & Patton, 2010) is a generic conceptual representation of the human body structure (anatomy) as a system of systems. The figure shows the interaction between the various systems and between the entire system and its external environment.

Physiology in medicine is thus concerned with the functions of the body parts or organs. This is therefore closely related to the study of anatomy. Pathology, which is the study of disease, uses the principles of anatomy and physiology to determine the nature of particular diseases (Thibodeau & Patton, 2010). Thibodeau & Patton stress that by knowing the structure and function of a healthy body, physicians are better prepared to understand what can go wrong to cause disease. That is, the formal progression from anatomy to physiology and pathology is a logical one.

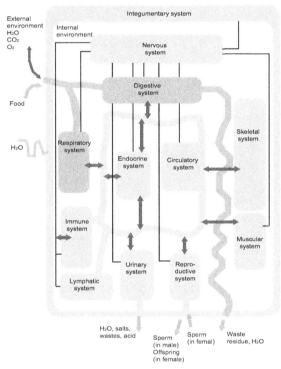

Figure 1. Integration of human body organ systems (adapted from Tibodeau & Patton, 2010)

The medical diagnosis process

Figure 2 shows the typical steps involved in the medical diagnosis process. This was formulated by an experienced vascular surgeon in the NHS. The diagnosis process often starts with the patient recognising that he or she has a problem and needs help resolving it. In medical terms this is known as "adopting the sick role".

The process is not always straight forward as can be seen in figure 2. It involves considerable uncertainty. The computerisation of this process has been the subject of significant research in artificial intelligence and expert systems for decades (Szolovits & Pauker, 1978). Szolovits & Pauker viewed medical decision making as a spectrum with categorical reasoning at one extreme and probabilistic reasoning at the other. They found that experience of the clinician has a considerable impact on the process. In a more recent work, Baerheim, 2001 identify two phases of the diagnostics process as abductive (in which a doctor infers one or more diagnosis from a patient's story) and deductive (where the doctor begins to check his diagnosis with specific tests). Of most relevance here is the emphasis that the deduction from a hypothesis is the process of using logic to check the patient's particulars against a given medical theory. This further highlights the fact of the importance of the underlying knowledge of anatomy, physiology and pathology in the diagnostics process. This is why we believe that the generic concepts of system "anatomy" and "physiology" are important to the development of the subject of "sick systems".

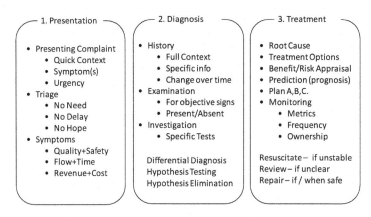

Figure 2. Medical diagnostics process produced by an experienced vascular surgeon

Most healthcare managers know when there is a problem in their system but often may not understand the problem or how to deal with it. For a patient, this is often easier when sickness is dramatic, that is "acute" as it is known in medicine. It is more difficult to adopt the sick role when sickness is degenerative or in medical terms, chronic. A patient, however, may become aware of the creeping disease in a dramatic way as may be the case in a heart attack. Again, an application of this concept to a "degenerative" system problem is not far-fetched.

The above arguments therefore form the framework for the discussions and proposals that follow.

3.0 CONCEPTUAL REPRESENTATION OF SYSTEMS: STATE-OF-THE-ART

Conceptual representations are a part of every field that involves some abstraction of reality to facilitate problem solving. We have reviewed several conceptual models including IDEF0 (Feldmann, 1998), contingency theory, socio-technical system theory (Jackson, 2000), work system design (Carayon et al., 2009) and the Tricord model (Thanes, 2007). All of these models are either, problem or project specific, languages and methods or focused on some aspect of the organisation. Following our review and to the best of our knowledge, no conceptual representation exists that attempts to represent the intrinsic elements of the system in a generic sense as desired in this study. Whereas we have considered the major areas of study that deal with systems and conceptual representations, we acknowledge that this is not a comprehensive review of all the pertinent literature. Brief discussions on five of these models are presented in the following sections due to space restriction. In some areas such as the Operations Research (OR) field, the discussion represents the present state-of-the-art in conceptual representation.

3.1 Operations Research (OR) Community

In the OR community, conceptual modelling is receiving more attention in particular as it applies to the simulation process. Thus conceptual models are vital to the success of a simulation modelling project. Surprisingly, however, very little has been written on the subject. In his review of the subject, Robinson identifies the following issues that require research attention in the OR field (Robinson, 2006):

- Definition of conceptual model(ling)
- Conceptual model requirements
- How to develop a conceptual model
- Conceptual model representation and communication
- Conceptual model validation
- Teaching conceptual modelling
- Other issues in conceptual modelling

This paper seeks to contribute to the conceptual model representation and communication problem. The state-of-the-art in conceptual modelling in the field of OR is such that no unifying definition and unifying approach to conceptual modelling exists (Robinson et al., 2011). The closest to this is described by Onggo, who proposed a unified conceptual model representation by combining a number of different diagramming methods (objective diagram, influence diagram, business process diagram and activity cycle diagram) (Onggo, 2009). The problem with this model is that it is problem focused and does not capture the essential elements of the healthcare system in a generic way.

3.2 Organisation modelling (OM)
According to Morabito et al. OM attempts to unify concepts of organisation structure in Organisation Theory (OT) and information modelling. At the heart of OM is the concept of the organisation molecule. Figure 3 shows a culture molecule of an organisation adapted from (Morabito et al., 1999). OM is a level of modelling that spans all areas of an organisation and it is argued that both biological and organisational systems may be configured as molecules (Morabito et al., 2008).

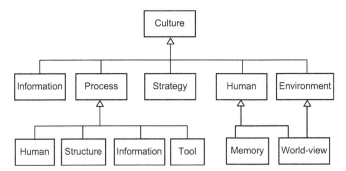

Figure 3. A culture molecule in Organisation Modelling (adapted from Morabito et al, 1999)

The concept of the molecule is used as a generic and useful means of defining aspects of the organisation, for example, process molecules, information molecules or strategy molecules. What is lacking with this approach, however, is that the biological system analogy is limited to the definition of the molecule but that does not fit together as to how the whole organisation works as an organism. As a result this concept is considered a useful complement to the present formulation as discussed in section 5.

3.3 Design Community
In the design community conceptual models or representations are found in two areas. The first area of conceptualisation is a stage in the design process known as "conceptual design" (Pahl et al., 2006). The second area of conceptualisation is that of the capturing and representations of what designers do, also known as the "design process".

There has been extensive research into conceptualising the design process and there are excellent reviews on this (Wynn and Clarkson, 2005; Dubberly, 2005). Of particular relevance to the method proposed in this paper is the chromosome product model proposed by Andreasen (1992).

Andreasen, 2008 presents a useful background to the development of the chromosome model. The discussion here only highlights the mechanism of the model as it is adapted into the current proposed model of the sick system. The model provides a very intuitive way of understanding how to deliver functions and operations to specifications by logically linking processes, parts organs and components together.

Though a product model, the chromosome model seems well suited for a service (or healthcare) environment at the operational level when an operation or process requiring a skill (function) that has to be carried out by a specialty (organ) made up of staff or equipment (parts). This model and that of the organisation molecule are further discussed together with the "sick system" model in section 5.

3.4 Systems engineering (SE) community

Several conceptual representations are used in systems engineering but the most generic representation of a system is shown in figure 4 adapted from the systems engineering handbook (Haskins et al., 2007).

This figure provides a generic structure for representing systems which is provided for addressing specific problem situations. Though this may be modified to apply to other systems, it seems to have been designed with a focus on engineering systems involving parts, components, assemblies and subassemblies.

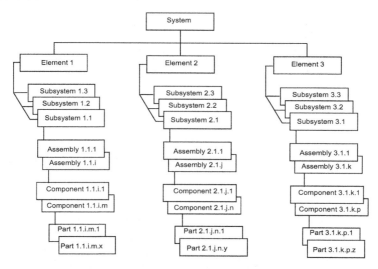

Figure 4. The Systems Engineering Hierarchy within a system model

4.0 EVALUATION OF SELECTED MODELS

In section 3, we discussed four areas where conceptual models are used and their relevance to the present argument. There are several other models which space would not allow for further discussion. Table 2 therefore presents a brief evaluation of some selected areas of conceptual modelling according to which elements of the system they explicitly capture as discussed in section 2 above.

The purpose of this table is to show primarily that various models do not often address all the five elements of a system as discussed in section 2.1. This in part confirms the lack of research focussed on the development of a generic conceptual representation of systems as proposed in this paper.

Table 2. Evaluation of selected conceptual models

System elements	System Conceptualisation Models								
	SC1	SC2	SC3	SC4	SC5	SC6	SC7	SC8	SC9
Resources		♦				♦	♦		♦
Data/Information	♦	♦	♦	♦	♦		♦	♦	
Processes	♦	♦	♦	♦	♦	♦	♦	♦	♦
Entities	♦				♦	♦	♦		♦
Environment								♦	

SC1= Design (chromosome); SC2= OR; SC3 = BPM; SC4 = OM; SC5 = Systems Engineering; SC6 = Contingency Theory; SC7 = Work System Design; SC8 = Socio-Technical Systems; SC9 = Soft Systems Methodology.

5.0 PROPOSED GENERIC CONCEPTUAL REPRESENTATION

This conceptual formulation is based on the view that healthcare systems are Complex Adaptive Systems (CAS) (Begun et al., 2003) but with the assumption that these systems have reducible radical openness and contextuality (Chu et al., 2003). The levels of grouping are arranged according to a modified form of the levels of human body system level after Thibodeau and Patton, 2010. They reduce the entire human body system to three grouped system level namely; Skeletomuscular system, Neuroendochrine system and the Urogenital system. These are further divided into the major body system, split system and then the principal organ level (figure 5a). The organs are also reduced down to the molecular and chemical level.

In figure 5(b) the various levels of system groupings are shown with the system of resources broken further down from the Grouped system level to the operational or "organ" level. The grouped system level is a conceptual level of abstraction and represents the major essential elements of the system. These are made up of systems of resources, processes, information and entities. These concepts have been defined in section 2 above. The emphasis on systems at this level is to indicate that each of these elements is in fact a collection of several interconnected subsystems of the system at the major system level. At the major system level, the elements of the grouped system level become more identifiable but on a bigger scale as indicated in the figure for the system of resources broken down into various types of system groups. The split system level is a further identifiable grouping of the resources of the system. The model also shows the internal and external environments with which the system exchanges dynamic elements generally indicated as in-flows and out-flows. At the operational or "organ" level it should be possible to identify specific resources for specific processes. These processes would be defined as system molecules as explained below.

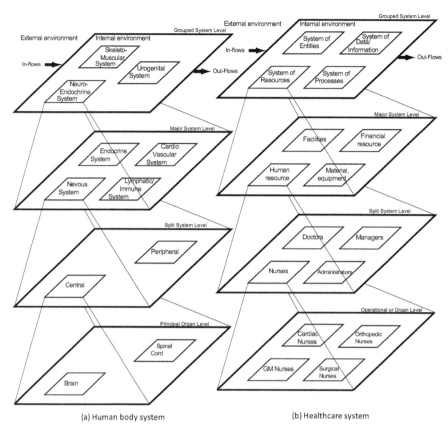

Figure 5. The generic conceptual representation of healthcare systems compared to the human body system

5.1 System molecule

In section 3.2, the concept of the organisation molecule in Organisation Modelling (OM) was presented. At the operational ("organ") level of the above conceptual representation of the system, there is the need to define specific processes or operations and the various other elements that would be involved. We consider the concept of the molecule an appropriate method for achieving this. This has the advantage of flexibility since any type of molecule can be defined. Once molecules are defined, the task of examining the element of the molecule that provides specific functions as achievable by the chromosome model can begin.

5.2 System chromosome

Throughout this paper we have emphasised on our concern for the structure (anatomy) of the system rather than the function (physiology). We briefly discussed the chromosome model by Andreasen, 1992 in section 3.3 above. At this point we adopt the chromosome concept as an appropriate concept for systematically examining the functions of system elements as defined in the molecules. This would be developed as we continue to do further work on this subject.

6.0 DISCUSSIONS AND FUTURE WORK

The application of the human body system metaphor to healthcare systems seems intuitive, however, it has not yet become a research subject. At the same time, researchers particularly in Operations Research modelling and simulation are beginning to focus on new and better ways of representing conceptual models of healthcare systems and by these communicating systems issues effectively to healthcare practitioners.

We have tried to show through the above arguments that existing conceptual models from various fields vary in their focus and none entirely seeks to represent the system entirely in a generic sense as with the human system analogy. We have also stressed the need for a systematic approach to the "anatomy" and "physiology" of healthcare systems as has been the case in medicine showing that this theoretical knowledge significantly underpins the medical diagnostics process. If healthcare systems may be considered "sick" and needing diagnosis and treatment then there is still a lot of work to be done in systematically explaining the disease process in these systems.

In medicine, it is understood that the lower levels are the building blocks on which the upper levels depend and that disease processes cause failure at the chemical or cellular levels and this failure affects organs, systems, and even the body as a whole. It may be obvious that in a healthcare system, if nursing processes on the ground fail consistently, it might lead to the failure of a department and even a whole hospital. We anticipate that if this way of looking at the healthcare system is embraced, it might lead to a paradigm shift.

At this stage, however, we find that the "sick system" view of the system may also raise questions we cannot yet imagine how to address. For instance if the concept of anatomy and physiology helps to explain the physical health of the system, what would be the equivalent of the emotional health of the system? In spite of the potential challenges, however, we find that this way of thinking raises even more exciting questions which may be worth pursuing as outlined next under plans for future work.

FUTURE WORK

This conceptual work presents several opportunities for further research. The following questions are a few that we hope to further explore: What are the symptoms of a sick system? If we had a healthy system how would we tell? That is, what defines a healthy system? What makes a system unwell and can we understand the rate of decay from a healthy to an "unwell" system? What are the pathological processes that can affect a system and how would they manifest? What diagnostics tests would we need? How would we interpret the results and how would they affect management decisions? What determines the "emotional health" of a system? Following diagnosis, what treatment options could be available? What are the risks and benefits of each? As with every new concept, we also hope that this would lead to interesting and important questions that we have not yet imagined.

ACKNOWLEDGEMENT

This research was carried out in association with the NIHR CLAHRC for Cambridgeshire and Peterborough in Cambridge, UK.

REFERENCES.

Andreasen, M. M. and McAloone, T. C. (n.d.). Applications of the Theory of Technical Systems: Experiences from the 'Copenhagen School'. In *Applied Engineering Design Science 2008 Workshop* (pp. 1–18).

Baerheim, A. (2001). The diagnostic process in general practice: has it a two-phase structure? *Family Practice*, *18*(3), 243 -245. doi:10.1093/fampra/18.3.243

Begun, J. W., Zimmerman, B. and Dooley, K. (2003). Health Care Organizations as Complex Adaptive Systems. In S. M. Mick and M. Wyttenbach (Eds.), *Advances in Healthcare Organization Theory* (pp. 253-288). San Francisco: Jossey-Bass. Retrieved from http://www.change-ability.ca/Complex_Adaptive.pdf

Brailsford, S. C. (2008). Tutorial: Advances and challenges in healthcare simulation modeling. In *Simulation Conference, 2007 Winter* (pp. 1436–1448).

Carayon, P., Schoofs Hundt, A., Karsh, B-T., Gurses, A. P., Alvarado, C. J., Smith, M. and, Flatley Brennan, P. (2006). Work system design for patient safety: the SEIPS model. *Quality and Safety in Health Care,* 15(suppl 1), i50-i58.

Chu, D., Strand, R. and Fjelland, R.(2003). Theories of complexity. *Wiley Online Library.* (n.d.). . Retrieved October 25, 2010, from http://onlinelibrary.wiley.com/doi/10.1002/cplx.10059/pdf

Donabedian, A. (2005). Evaluating the quality of medical care. *Milbank Quarterly, 83*(4), 691–729.

Dubberly, H. (2005). [Beta] How do you design? Retrieved January 14, 2011, from http://www.dubberly.com/articles/how-do-you-design.html

Feldmann, C. G. (1998). *The Practical Guide to Business Process Reengineering Using IDEF0* (illustrated edition.). Dorset House Publishing.

Forrester, J. W. (1991) System dynamics and the lessons of 35 years. *The systemic basis of policy making in the 1990s*;29:4224–4.

Gonzalez-Crussi, F. (2007). *A short history of medicine / by F. González-Crussi.* Modern Library chronicles. New York: Modern Library.

Jackson, M. C. (2000). *Systems Approaches to Management.* Springer.

Mintzberg, H. (1992). *Structure in Fives: Designing Effective Organizations* (Rep Sub.). Prentice Hall.

Morabito, J., Sack, I. and Bhate, A. (1999). *Organization Modeling: Innovative Architectures for the 21st Century.* Prentice Hall.

Onggo, B. S. S. (2009). Towards a unified conceptual model representation: a case study in healthcare. *Journal of Simulation, 3*(1), 40-49. doi:10.1057/jos.2008.14

Pahl, G, Beitz, G., Feldhusen, J. and Grote, K-H. (2006). *Engineering Design: A Systematic Approach.* In K. M. Wallace and L. T. M. Blessing (Eds.), Springer.

Paley, J. and Eva, G. (2010). Complexity theory as an approach to explanation in healthcare: A critical discussion. *International Journal of Nursing Studies.* doi:10.1016/j.ijnurstu.2010.09.012

Robinson, S. (2006). Issues in conceptual modelling for simulation: setting a research agenda. In *Proceedings of the 2006 OR Society Simulation Workshop.*

Robinson, S., Brooks, R., Kotiadis, K. and Der Zee, D. (Eds.). (2010). *Conceptual Modeling for Discrete-Event Simulation.* CRC Press. Retrieved from http://www.crcnetbase.com/doi/abs/10.1201/9781439810385-c7

Singer, C. J. (1957). *A short history of anatomy from the Greeks to Harvey : the evolution of anatomy / Charles Singer.* (2nd ed.). New York: Dover Publications.

Spencer, B. A. (1994). Models of Organization and Total Quality Management: A Comparison and Critical Evaluation. *The Academy of Management Review, 19*(3), 446-471. doi:10.2307/258935

Sterman, J. (1991). A skeptic's guide to computer models. In G. O. E. A. Barney, W. Kreutzer and M. J. Garrett (Eds.), *Managing a Nation: The Microcomputer Software Catalog* (pp. 209-229). Westview Press.

Szolovits, P. and Pauker, S. G. (1978). Categorical and probabilistic reasoning in medical diagnosis. *Artificial Intelligence, 11*(1-2), 115-144. doi:10.1016/0004-3702(78)90014-0

Thanes, S., Tricordant Limited, Whole Systems Organisation Development Consultants. (n.d.). . Retrieved January 14, 2011, from http://www.tricordant.com/index.html

Thibodeau, G. A. and Patton, K. T. (2009). *The Human Body in Health and Disease - Softcover (ANATOMY AND PHYSIOLOGY)* (5th ed.). Mosby.

Wynn, D. and Clarkson, J. (2005). Models of designing. *Design process improvement*, 34–59.

Young, T. and McClean, S. (2009). Some challenges facing Lean Thinking in healthcare. *International Journal for Quality in Health Care, 21*(5), 309 -310. doi:10.1093/intqhc/mzp038

Young, T., Brailsford, S., Connell, C., Davies, R., Harper, P. and Klein, J. H. (2004). Using industrial processes to improve patient care. *BMJ, 328*(7432), 162 -164. doi:10.1136/bmj.328.7432.162

INTERNATIONAL CONFERENCE ON ENGINEERING DESIGN, ICED11
15 - 18 AUGUST 2011, TECHNICAL UNIVERSITY OF DENMARK

PROPERTY RIGHTS THEORY AS A KEY ASPECT IN PRODUCT SERVICE ENGINEERING

Anna Katharina Dill[(1)], Herbert Birkhofer[(1)] and Andrea Bohn[(1)]
(1) Technische Universität Darmstadt

ABSTRACT
Product service systems (PSS) are a field of research which is supported by research in a large number of other areas. Product development and engineering design is the basis for most research projects but economic theory has a major influence too. The origin of the theory of property rights is the new institutional economy. Different types of rights concerning a property are described systematically and can be distributed separately. Although the distribution of property rights in general is a key aspect for the PSS design, the economic theory of property rights has not yet been introduced into PSS considerations in a broad and systematic way.
The aim of this paper is to close the gap and give a structured overview of the property rights theory and its potentials for PSS design. According to the procedure of the German VDI 2221 it is demonstrated how property rights considerations can support the different phases of a development process. Furthermore, it is demonstrated how property rights theory can support different goals in developing PSS and the authors present suggestions for a more differentiated look at the property rights distribution to improve the correlation with the requirements of PSS considerations.

Keywords: Product service systems, PSS, PSS development, property rights

1 INTRODUCTION
Not everybody is familiar with the research topic of product services systems (PSS) nowadays but the so called PSS are all around us. Concepts like renting and leasing of products are well known and widespread on the markets. In 1976 Obenberger and Brown promoted leasing and renting as a "consumption alternative in marketing" [1]. They suggested a change from the focus on ownership a focus on what they called *usership*, defined as "a broad term encompassing all types of consumption in which the consumer does not possess legal title of the product" [1]. This usership concept implies a rejection of the classic transfer of the legal title and a turn to renting and leasing concepts. The usership concept was not associated with the specific considerations of the economic property rights theory at that time.
The goals in PSS research range from a strong focus on ecological aspects, e.g. Mont [2] and Roy[3] to a focus on economic aspects, where most researchers consider a B2B context, e.g. Fuchs [4] and Schweitzer [5].
The original concept of PSS is relatively young and several authors set different focuses. The definition of Goedkoop et al., who define PSS as "a marketable set of products and services capable of jointly fulfilling a user's need" points out the quintessence of the PSS considerations [6].
Independent from the reason for developing something as a PSS – the difference to a classic product often is connected with the differentiated distribution of property rights concerning the product.
In the context of the New Institutional Economy, the theory of property rights provides a distinguished overview of rights and resulting duties towards a commodity that can be distributed in different combinations. This theory and the differentiation of rights will be explained later in section 3.
A brief overview of current PSS development approaches, problems of PSS development in general and the introduction to property rights theory is given in section 3 and the opportunities for PSS development are displayed. Thereafter the application of property rights theory along the different phases of the product development process is presented. Although the VDI 2221 is used to show the appropriateness of the concept in the development process, the presented approaches are applicable for other methods of product development and can support specific methods for PSS development too. Moreover, the consideration of property rights is the key to a change from product oriented development to result orientation, which sets focus on the customer needs.

2 CURRENT PROBLEMS AND STATE OF THE ART IN PSS-DESIGN

What appears at the market as some kind of PSS is often designed with traditional product development methods like VDI 2221 and not with specific methods that consider product and service as equally prioritized components in a system.

Today there is a number of approaches to support a systematic integrated PSS design (e.g. Botta 2007 [7] or Abdalla 2006 [8]). But the suggested approaches are quite special and base on a specific view on PSS. None of the approaches has yet become some kind of standard. And altough property rights theory is a core theory of new institutional economy, none of the approaches has considered it so far.

The consideration of property rights theory as a support for PSS development has only been published in two cases, until now: Hockerts suggested to change the focus on the service part of the PSS to a consideration of property rights distribution to enhance the eco-efficiency of PSS but without further advice for the integration of this findings in the development process [9]. Dill and Schendel refer to the property rights theory as a support for a systematic design of variants in the PSS development, but again without explicit advice how this could be integrated into existing PSS considerations [10]. Thus, the following description of types of property rights will be focusing on PSS considerations. Development process oriented examples of utilization of the theory are given thereafter.

3 PROPERTY RIGHTS AND THEIR SUPPORT FOR PSS DESIGN

3.1 The theory of property rights

The theory of property rights is one of the core theories of the new institutional economy – a rising subject of economic research since the 1970s. Next to property rights theory, the new institutional economy covers other important economic theories, e.g. the principal agent theory, the theory of transaction costs, assumptions of bounded rationality and asymmetric information [11]. The old picture of the *Homo oeconomicus* who acts rational appropriate to his preferences and aims for maximizing his utility is turned into a more realistic picture of human behavior and the situation in the economy. This implies consideration of opportunistic behavior and other characteristics that are in opposition to the old picture.

The following description of the types of property rights is based on Hockerts [9] and Furubotn/Pejovich [12]. It is important to point out, that the first three property rights include obligations which need to be considered for PSS design too. These obligations can lead to opportunities for the PSS design as well as the rights do.

1. The right to retain profits
   (The duty to cover losses)
2. The right to maintain and operate a product
   (The obligation to maintain a product)
3. The right to dispose of a product
   (the duty to pay for the disposal of a product)
4. The right to exclude others
5. The right to use a product

Additionally Furubotn and Pejovich add the concept of attenuation, which explains the existence of some degree of restriction an owner can have [12]. These restrictions can concern:

1. Changes in form, place of substance of an asset.
2. The transfer of all rights to an asset to others at a mutually agreed upon prince.

Such restrictions are traced back to the general legislation that can restrict the rights of an individual. In the first case it could be restricted by law. For example what types of buildings are allowed in a building area? What kinds of changes are allowed to be carried out when it comes to a landmarked house?

In the second case the reason could be fixed prices for special commodities in a market, e.g. drugs that are only available on prescription in a county.

This concept of attenuation can be extended for the consideration of property rights in PSS development by integrating the legal situation for the subject matter. Depending on what the PSS is supposed to contain, it could be restricted by different laws and provisions. This could be e.g. restrictions on chemicals or safety restrictions, as well as general environmental legislation.

3.2 Extended differentiation of property rights

The authors suggest a more detailed extension of the presented differentiation of property rights. To meet the requirements of PSS concepts and PSS development a separation of the rights to maintain and operate the product is suggested. In various PSS we have today, the service part offered by the provider is to maintain, while the customer operates the product/system.
Therefore the authors suggest the following:

2a: The right to maintain the product (Covering the obligation to maintain the product)
2b: The right to operate the product

It is important to point out, that this property right distribution only describes who is in charge of the process "maintenance" and therefore taking care of the product and fixing it if needed. The responsibility for the costs of the maintenance is not connected to the property right, but has to be defined in the legal contract of the PSS. These circumstances can be used in PSS design to handle problems of opportunistic behavior. E.g. in a car sharing scenario the provider is in charge for the maintenance. For the costs for repairing damages caused by accidents the users who were responsible are charged.

3.3 Potential of property rights in PSS design

The theory of property rights distribution can be applied to achieve different goals in PSS development. One of the popular goals of PSS is eco-efficiency. In 2008 Hockerts presented „property rights as a predictor for the eco-efficiency of product-service systems"[9]. He suggests turning the focus from a service centered development approach to a focus on property rights distribution. By explicit choice of the property rights constellation it is possible to directly focus on eco-efficiency topics such as material intensity or reduction of energy consumption. Moreover it is pointed out that problems like opportunistic behavior or asymmetric information can reduce the eco-efficiency of PSS, but can be solved or alleviated by selective property right distribution.
As a general approach for PSS design Dill and Schendel suggested to regard property rights distribution as an opportunity to conduct a systematic variation in PSS design with a well known pattern of variation.
These approaches are picked up again in section 4.1.

To set the property rights considerations into a well know product service systems context, the will be used to describe the eight types of product service systems by Tukker as they are presented in Figure 1.

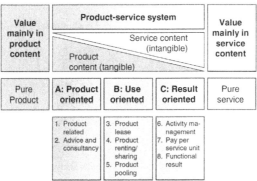

Figure 1. Types of PSS according to Tukker [15]

The following Table 1 provides an overview of correlations between the presented eight types of PSS according to Tukker in Figure1.

Table 1. Correlation of PSS type according to Tukker and property rights distribution

Type of PSS acording to Tukker	Property rights distributed to provider	Property rights distributed to customer
1. Product related	- Depending on the product related service, the provider get's the required rights to carry out the service	- In general all rights are distributed to the customer - Exceptions depend on the required rights to carry out the product related service
2. Advice and consultancy	- Usually no rights are distributed to the provider	- All rights are distributed to the customer
3. Product lease	- for a defined time span the provider gives all rights beside the right to dispose to the customer until the product is given back or the product is finally sold to the customer	- The right to exclude others
4. Product renting/sharing	- A right/duty to care for the maintenances	- The right to use the product - The right to exclude others while using
5. Product pooling	- A right/duty to care for the maintenances	- The right to use the product
6. Activity management	- Depending on the type of activity management	Depending on the type of activity management
7. Pay per service unit	- The provider holds all property rights but can't really use the right to exclude others	The rights to use
8. Functional result	All rights are distributed to the provider	No property rights on customer side

The distribution of the rights can be altering, depending on the point in time that is considered. The differences between various types of PSS are only visible if a longer period of time is regarded. This point will be explained with an example of car ownership, leasing and renting in the following:

By the time the driver is on her way in the car, driving from A to B all three variants look quite the same. At the End of the journey, the rented car will be returned to the car rental station, but the other cars stay with their driver.

When it comes to inspecting the car, the leased one usually is taken to the garage and paid for by the lessee, as well as in the case of owning the car. In contrast to this, the vehicle of a car rental service is taken there and paid for by the provider.

Figure 2 illustrates the difference between product lease and product sharing with an example of cars.

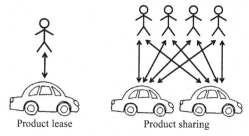

Product lease   Product sharing

Figure 2. Excamples of Types of PSS according to Tukker [15]

4 INTRODUCTION OF PROPERTY RIGHTS INTO SYSTEMATIC PSS DESIGN

4.1 Property rights in different phases of PSS design

In product development the German Guideline *VDI 2221 - Systematic approach to the development and design of technical systems and products* is well known in science and industry. It is known to be useful for developing material products as well as software [13].

The following Figure 3 shows the procedure of VDI 2221 with different phases of the development process. The phases of the development procedure are used to present different approaches to support the development via property right consideration.

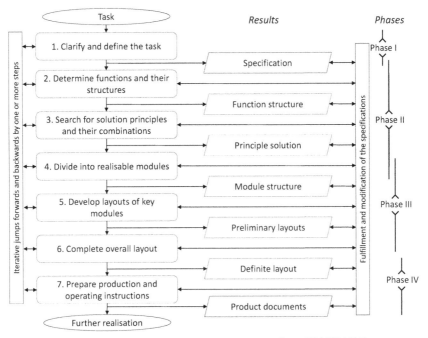

Figure 3. product development process according to VDI 2221 [13]

4.1.1 Phase I

For the clarification and definition of the task and the specification it is important to agree on the meaning of words and descriptions that are used. The usage of the property rights theory as a clarified concept of rights distribution can prevent misunderstandings and describes specific correlations with clear words.

Furthermore the consideration of property rights in the first phase provides inspiration for the participants in the process of clarifying and defining the task. This supports the turn from pure ownership thinking towards a more detailed look at different types of PSS and their feasibility to the task of development.

Additionally the consideration of property rights distribution can enhance the development process to reach a goal like "low material intensity" by specific choice of corresponding rights.

4.1.2 Phase II

For the second phase property rights distribution might have already been determined in the first phase. From this predetermined constellation further requirements can be derived and more detailed specifications of the property rights distribution can be established.

In the process of designing solution variants, property rights theory can be used to apply systematic variation in general, as well as to create special variants to reach a specific goal like eco-efficiency.

4.1.3 Phase III
For the steps in the third phase the property rights theory can be used to identify gaps in the solutions. A systematic survey of the different rights can assure that all aspects of the PSS are covered.

4.1.4 Phase IV
In the last phase a definite layout is available, the product documents are written and preparations for production and further realisation take place. With the support of the differentiated property rights considerations it is possible to describe the chosen constellation in a systematic way, that is easy to follow for everyone who is familiar with the basics of property rights theory.
The preparation of the legal contract is supported too. The requirements for the content of the contract can be derived to some instance from the allocation of the property rights.

4.2 Property rights theory as guideline for checklists in PSS design
Checklists are a common method to support the product development process [14]. The usage of checklists enables developers, independent from their specific knowledge. These property rights focused checklists help to find the right questions to cover important points during the development process as well as to support the necessary decisions. The property rights theory provides a structured basis for checklists due to the different opportunities they provide. Therefore it is possible to state different questions for the allocation of the property rights to enhance an efficient development.
The following Table 2 provides a checklist example to support an appropriate property right distribution in PSS development:

Table 2. Example checklist "Questions for property right distribution"

Type of right/duty	Distributed to provider	Distributed to customer
The right to retain profits (The duty to cover losses)	Is it possible to calculate the potential profits/losses? Can the profits or losses be influenced by the customers' behavior? Can the provider raise his profits or lower the losses by his behavior?	Is it possible to calculate the potential profits/losses? Has the customer influences on the profits or losses?
The right to maintain and operate a product (The obligation to maintain a product)	Is there the right infrastructure to maintain and operate the provided product(s)? Is there a reliable cost calculation?	Is the customer able to main-tain and operate the product? Are special tools or trainings needed to enable the customer to maintain and operate the product? Strengths and weaknesses in the areas of maintaining and operating the product compared with competitive products?
The right to dispose of a product (the duty to pay for the disposal of a product)	Is there a benefit for the provider? E.g. recycling of products parts of material needed?	Are there special requirements the customer has to take care of? Does the disposal cause extra costs/effort for the customers, which decrease the attractiveness of the PSS? Is there a law that prevents the distribution of the right to the customer?

The right to exclude others	Can the provider exclude others and still run the PSS? How can the others be excluded? Is there technical support?	If not given: take care of the parts that are responsible for personalizing the product for every new user if users change regularly How can the others be excluded? Is there technical support?
The right to use a product	Is it possible to only distribute this right to the provider? What bundle of rights is needed to be able to use the product?	Is it possible to only distribute this right to the user? What bundle of rights is needed to be able to use the product?

Checklists can also be used for the implementation of goals. In those types of checklists, certain allocations of the property right are proposed to achieve the desired effect, for example eco-efficiency.

5 DISCUSSION

Product service systems are a field of research that matches different research areas. Complex knowledge is necessary and existing procedures for PSS development have not yet been established as general standards. Methods and tools that support the development process of PSS – independent from the chosen design procedure – are required for further success and market penetration of the concept of PSS.

The presented theory of property rights is an established economic theory with a long tradition in research. It supports the view of different types of PSS in a structured way. Compared to earlier research like the eight types of PSS by Tukker [15], it is not a competing approach but more a complementation and extension of the concept.

The authors suggest dividing the existing consideration of "the right to maintain and operate the product" into separate rights to better meet the requirements of PSS design. Further changes to the traditional segmentation, towards a more PSS suitable segmentation of the rights, should be discussed and tested in development projects.

A weakness of the concept is the time aspect. In the classic segmentation it is not possible to distinguish e.g. a car sharing offer from a rented car without addition of information about the time span that is covered. Implementing a time aspect would support the usability for PSS design.

While talking about property rights distribution and usership arrangements it is important to point out that there are cases in which the core benefit is about having something somebody else can not have. Usually these needs are fulfilled by ownership arrangements, but even for such constellations it is possible to build usership arrangements regarding the adequate property rights distribution that fit.

REFERENCES

[1] Obenberger, R.W., Brown, S.W., *A Marketing Alternative: Consumer Leasing and Renting*, in *Business Horizons* 19 (5), 1976, pp82–86

[2] Mont, O., *Product-service systems. Panacea or myth?*, 2004 (Lund)

[3] Roy, R., Sustainable product-service systems, in *Futures Vol 32*, 2000, pp289–299

[4] Fuchs, C. H., *Life Cycle Management investiver Produkt-Service Systeme - Konzept zur lebenszyklusorientierten Gestaltung und Realisierung*, 2007 (Universität Kaiserslautern)

[5] Schweitzer, E., Mannweiler, C. and Aurich, J. C. (2009): Continuous Improvement of Industrial Product-Service Systems, in *Proc. 1st CIRP IPS2 Conference*, Cranfield, April 2009, pp16-24, (Cranfield Univ. Press, Cranfield)

[6] Goedkoop, M. J., van Halen, C.J.G., te Riele, H.R.M. and Rommes, P.J.M., *Product Service systems, Ecological and Economic Basics*, 1999

[7] Botta, C., *Rahmenkonzept zur Entwicklung von Product-Service - Systems Product-Service Systems Engineering*, 2007 (Eul, Lohmar)

[8] Abdalla, A.A., *TRIZ innovation management approach for problem definition and product service systems,* 2006 (Shaker, Aachen)

[9] Hockerts, K., *Property Rights as a Predictor for Eco-Efficiency of Product-Service Systems,* 2008. (CBS Working Paper Series, Frederiksberg)

[10] Dill, A. D. and Schendel, C., Implications of new institutional economy theory for PSS design in *Proc. 2nd CIRP IPS2 Conference,* Linköping, April 2010, pp43–49, (Linköping Univ. Press, Linköping)

[11] Richter, R., Furubotn, E. G. and Streissler, M. (2003), *Neue Institutionenökonomik - Eine Einführung und kritische Würdigung* 3., 2003 (Mohr Siebeck, Tübingen)

[12] Furubotn, E. G., Pejovich, S. (1972): Property Rights and Economic Theory: A Survey of Recent Literature, in *Journal of Economic Literature Vol 10*, pp1137-1162, (JSTOR)

[13] *VDI-Richtlinie 2221: Methodik zum Entwickeln und Konstruieren technischer Systeme und Produkte*, 1993 (VDI-Gesellschaft Entwicklung Konstruktion Vertrieb, Düsseldorf)

[14] Pahl, G., Beitz, W., Feldhusen, J., Grote, K.-H., *Engineering Design - A Systematic Approach*, 2007 (Springer, Berlin)

[15] Tukker, A., Eight types of product-service system: Eight ways to sustainability? Experiences from SusProNet, in *Business Strategy and the Environment* 13, 2004, pp246–260

Contact: Anna Katharina Dill
Technische Universität Darmstadt
Fachbereich Maschinenbau
Magdalenenstraße 4, 64289 Darmstadt
Germany
Tel: Int +49 6151 163383
Fax: Int +49 6151 163355
Email: dill@pmd.tu-darmstadt.de
URL: http://www.pmd.tu-darmstadt.de

Anna studied mechanical engineering and economics at the Technische Universität Darmstadt concluding in a diploma degree in September 2010. Since then she is a research associate at the Department of Mechanical Engineering. Her research interest is on product service systems (PSS) and eco design, in particular in increasing the eco-efficiency of PSS by systematic PSS-Design.

PRODUCT WITH SERVICE, TECHNOLOGY WITH BUSINESS MODEL: EXPANDING ENGINEERING DESIGN

Tomohiko Sakao[1], Tim McAloone[2]
(1) Linköping University, Sweden (2) Technical University of Denmark, Denmark

ABSTRACT
Looking back over the last decade, the importance of an expanded understanding of engineering design has been shared within the engineering design community. Presented concepts and methods to support such expansion include Functional Product Development, Service Engineering, and Product/Service-Systems (PSS) design. This paper first explains PSS design as an expansion of engineering design, away from merely the physical product. Secondly, it gives a review of PSS research and a projection of future research issues, also ranging out into untraditional fields of research. Finally, it presents a new promising concept beyond PSS design; via an integrated development of technology and business model. This can be of particular interest for further research, especially due to its high freedom for designers and thus high potential for innovation.

Keywords: Product/Service-System (PSS), Innovation, Servicizing

1 INTRODUCTION
Manufacturers in developed countries today face severe competition with hardware manufacturerd in low-wage countries. This competition is expected to become tougher as the quality of products by manufacturers in developing countries is ever-increasing. Thus, firms in developed countries need to find ways to distinguish themselves in terms of value for their customers. As regards value, product quality is just one component. Service is also an important element to create value for customers. Manufacturers in developed countries indeed regard service activities as increasingly important today. Increasing amounts of manufacturing firms are strategically shifting from a "product seller" towards a "service provider" paradigm [1]. Importantly, service activity is increasingly incorporated into the design space, an area which has been traditionally dominated by physical products in manufacturing industries (see conventional theories for mechanical design; e.g. [2]). To do so, companies need to expand the object of engineering design from a physical product to an offering consisting of products and services, so that the whole design is effective and efficient. Such an offering is often called a Product/Service-System (PSS) [3, 4]. PSS design has many commonalities to user-centred design (UCD), as the key stakeholders are brought to a central position in the design process. However, it is important to understand that PSS goes further than UCD, as PSS design incorporates service as a key design parameter, which is not necessarily the case with UCD.
Looking back over the last decade, the importance of such an expansion has been shared within the engineering design community [5-10]. Different groups have presented concepts and methods to support such expansion, e.g. Functional Product Development [8]; Functional Sales [11]; Integrated Product Service Engineering (IPSE) [12]; and Service Engineering [13]. In common with these concepts, the target of design comprises combinations of hardware and support services. Taking stock of the research that has been carried out in this area over the past decade, one could ask, whether we have been doing the right things and in which direction should we head?
To share these concerns within the engineering design community and in an attempt to answer these questions, this paper first aims at giving a review of PSS research. Thereafter, it presents a new promising concept beyond PSS design; integrated development of technology and business model. The remainder of the paper is as follows. Section 2 first explains PSS design as an expansion of the engineering design paradigm, and reviews PSS research. Section 3 introduces the idea of integrated development of technology and business model. Section 4 presents a concluding discussion.

2 ENGINEERING DESIGN OF PSS

2.1 Conditions for PSS Design: Differences from Traditional Engineering Design

Blessing, Chakrabarti and Wallace [14] state that "the aim of engineering design is to support industry by developing knowledge, methods and tools which can improve the chances of producing a successful product". In its generic nature, this statement provides a useful guide for engineers to consider their role and aims with design. However, this statement delimits itself to the domain of physical products, with services being excluded.

According to Tukker and Tischner [3] a PSS consists of "tangible products and services designed and combined to jointly fulfil specific customer needs", also regarded as a value proposition, including its network and infrastructure. As a design research field, PSS is a new and emerging area, where the definition and study of 'functional sales' [15], 'functional (total care) products' [16], 'servicizing' [17] and 'service engineering' [6] all have contributed to the foundation and our current understanding of PSS as a phenomenon.

In the meantime, the latter half of the first decade of the new millennium has seen a particularly increasing interest in PSS design methodology, from a broader and more multi-disciplinary group of researchers, representing engineering, technology management and economical disciplines (e.g. [18-20]). Thus the PSS arena described here is still in its formative stages, where definitions, understandings and approaches to the field are still fluid. The design object that PSS represents can therefore be seen as expanding in a series of directions, when compared to traditional product design. The following paragraphs take an excursion into basic differences in conditions for a PSS design to be possible. Six basic conditions have been identified.

Conditions of competencies and disciplines

According to Tan & McAloone [21], the underlying strategic principle of PSS is to shift from business based on the value of exchange of product ownership and responsibility, to business based on the value of utility of the product and services. This implies a fundamental reassessment of core business, ownership, transactions, development and delivery of the 'offering' (this term is chosen so as to avoid confusion about the nature of a product or service), and client-customer relationships.

Thus the object of value for the providing company transforms from merely the physical artefact, to any chosen and targeted transaction between the customer and the providing company. Compared to traditional product development, a new set of competencies must be present in the PSS design activity, to enable the design, development and maintenance of a satisfactory relationship with the customer, who is in a closer (and often contractual) relationship with the providing company.

Conditions defining new design objects

PSS has until now been regarded as the joint development of product and service, plus the providing company's subsequent delivery of services to the customer – when bundled together, dubbed "a system" [3]. Research in the field of PSS has so far established that the behaviour of services and products in the use phases of the product's life are identical [22]. We therefore see the need to arrive at more usable descriptions and definitions of product, service and PSS, linking to an integrated understanding of customer-oriented value and utility, thus freeing ourselves of the somewhat artificial distinction of {PSS = product + service}.

McAloone and Andreasen [22] take a domain-oriented view of PSS, where a PSS offering is described in terms of an artefact domain; a time domain; and a value domain. This view is closely inspired by Ropohl's system technical theory [23]. In each domain it is possible to describe the key distinctions and innovative developments that a company must undergo in order to create sustained value, customer lock-on and flexible solution-oriented business offerings to the customer.

New conditions regarding forms of production

The shift towards PSS for industrial companies can be described from many viewpoints, ranging from the desire to support a post-industrial society, the increased competition (and opportunity) to support an increasingly dematerialised world, to a necessary decoupling of competitive edge from cost, quality, time, etc. The current discussion of these reasons and observations of the augmentation of organisations' interests in usability, use and service is pointing towards the definition of new production forms [24].

There are a number of approaches towards implementing and integrating new production forms into the organisation, that are highly relevant to successful PSS design. By broadening the perspective from product life cycle to customer activity cycle [25], we expand the design object for PSS. And by placing the customer in focus and understanding their needs for functional, efficiency-based and/or social fulfilment (this weighting differs, dependent on Business to Consumer or Business to Business), it is possible to develop a competence- and network-based approach to supporting the customer's whole activity – and not merely providing a physical good.

New conditions and opportunities regarding choice
As previously implied, PSS design should be based upon new degrees of freedom in the design process, due to a more broadly defined design object, closer contact with the end-user and an extended service period, compared with traditional business. But what should a PSS give the user, seen from their perspective?
Traditional mass-produced products (anything from software to vacuum cleaners) come with in-built and implicitly regulated properties, that the user must reconcile him/herself with, or find out how to work around, if the properties limit the intended use. A large opportunity of PSS, on the other hand, is that the user is present in the specification of use and usability, leading to the creation of choice, as opposed to living with in-built regulation.

Conditions regarding interventions or touch points
Tan and McAloone [21] adopt a morphological approach to understanding PSS types and characteristics, based upon observations of a series of cases. In this morphological approach it is interesting to observe the varying types of *interventions* or *touch points* (exchanges between provider and user, product and user, product and provider, etc.), describing which party is active or responsible for certain key activities and elements of the PSS.
We feel it important to think in terms of interventions, as this gives useful insight into the key activity dimensions of a PSS; areas which normally are *not* up for discussion when designing a traditional artefact. This viewpoint ought to give the PSS designer the insight into how active or passive the user is in each element of the PSS concept and in which situations to choose whether to delegate or to keep responsibility for the good, the information, the service, and so on.

Conditions regarding value perceptions
The engineering community has focused for many years on effective approaches for ensuring high value products and systems. The challenge here has been in matching the customer's judgement of value (subjective evaluation of goodness vs. investment incurred) with the company's own ability to provide products of high quality. The very nature of PSS design – where the relationship with the customer is designed to be longer and more intense; where focus is given to functional provision and not merely sales of artefacts; and where the product life view is matched with a customer activity view – gives many opportunities for the development task to come much closer to an understanding of value perception than in a traditional product development situation.

2.2 Nature of PSS Design
Having the conditions for PSS design in Section 2.1, this section describes the nature of PSS design. PSS design addresses the customers while the functions of physical products and providers' activities are media for satisfaction delivery. It should be noted that providers' activities, such as maintenance services, are included in the usage process. The evaluation by customers has a premium. Approaches to PSS design involve changes in the traditional design procedures, delivering processes, and engineering mindsets. Therefore, it has much influence on the provider.
Sakao et al. [26] argue that the following three dimensions are necessary to create space used to map various types of elements for PSS-design research; the *offer*, the *provider*, and the *customer/user* dimensions. It should be noted that the first one refers to both "product" and "service" elements of PSS. In addition, the other two, i.e. the provider and the receiver, are indispensable to address PSS.
The *offer dimension* addresses the elements and activities in the offering's life cycle. This includes the lives of physical products being a part of the PSS, as well as service activities. Successful design of PSS depends on a thorough understanding of the solution life cycle and active design of beneficial linkages with involved heterogeneous systems.

The *provider dimension* addresses the evolvement of the product/service providers' organisation and operations. This covers such issues as the setup of development projects, organisational streamlining of the company towards service delivery and the identification of necessary partnerships for the successful operation of services.

The *customer/user dimension* addresses the evolving needs of service receivers. It is crucial for the provider of services and products to be able to anticipate receivers' reactions to new offerings. In principle, any PSS design is supposed to address at least some aspect in all three of these dimensions, since service includes activities of customers and providers and products. As such, these three dimensions are fundamental for PSS design. In addition, anticipating and utilising the dynamics along each dimension is crucial. This implies that the essence of PSS design, especially if compared to traditional engineering design, lies in the utilisation of dynamics of and among the three dimensions of offer, provider, and customer. Figure 1 illustrates the linkages of some of the research topics to the three identified dimensions.

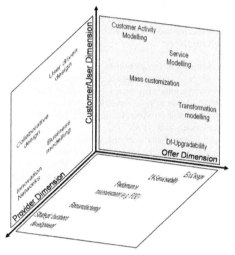

Figure 1. The three dimensions for PSS design [26].

In essence both products and services are just two modes in which companies attempt to deliver value to their customers. The 'product' versus 'service' discussion is not so much an issue of a new 'object' that has to be developed, but a new perspective on what kind of value is being created.

With PSS approaches we create a dependency between a (providing) company's operations and a (receiving) customer's activities. We have a close integration of operations, both tactically and strategically. PSS development models must inform us of the integration across the different levels of the company's development activities:

- Strategic business/product planning in cooperation with networks and service partners, i.e. development of PSS concepts.
- Product management and product development projects leading to new PSS 'offers', i.e. development of the product/service offer.
- PSS delivery system or function, which in steady relation to the customer delivers services, i.e. offer customisation and development of the service channel.

Fundamentally the difference in PSS in relation to traditional product development is that:
- the physical product is supported and enhanced throughout the customer's activities by the providing company (the business relationship with the customer may spread over several product upgrades and generations).

- the value creation is in the resulting activity where both the physical product, supporting services and the customer all play a vital role (the perception of value is beyond the physical product itself).
- the customer's activities are part of the value creation process and the providing company must interact closely with the customer throughout the life phases.

2.3 Existing Research on PSS

Researchers in the EU-funded SUSPRONET project [3] have contributed a great deal to the PSS research. From the perspective of engineering design, their contributions come mainly from an analytical-, as opposed to a synthesis/design viewpoint. For instance, the exhibition of *the product-service continuum* (from product-oriented, use-oriented, to result-oriented service) should be seen as an aid to *classification* as a part of the analytical phase of design (see Table 1). In addition, the work reported in [27] is exactly the result of *analysis*. Other literature such as Mont et al. [28] comes also from analytical viewpoint. This means that insight from this group of PSS research is limited to understanding the purpose of designing offerings, as opposed to methodology towards actually designing these.

Relatively recently, a group of research on PSS focusing on design has emerged. According to Sakao et al. [26], here we use the classification of the research targets into "PSS offer modelling", "PSS development process", and "PSS potential". The first two, i.e. offer modelling and development process, have been basic targets of engineering design research as presented by Finger and Dixon [29] and [30]. Table 1 illustrates the targets of the reviewed literature that were taken from international journals.

Table 1. Classification of PSS design literature (journal articles) into the three targets

Research target	2008 and before	2009 and onwards
PSS offer modelling	[13] [31] [32] [33] [16]	[34, 35] [36] [37] [38] [39] [40]
PSS development process	[41] [42] [31] [43]	[44] [37] [38] [45] [46] [47] [48]
PSS potential	[49]	[50]

Note: Some articles appear only in one target in this table, which should be interpreted to be the main target of the articles, although they may address two or three.

In an attempt to create normative aids to PSS development, McAloone et al. [22, 51] postulate that services and products are bound together by the use activity, that services are delivered or executed in a so-called 'transformation system' (TS), and that both products' and services' business aspects are based on value relations to the customer. The objective of this clarification is to allow the conceptualisation of new services to happen as a systematic pursuit of new solutions as it is known from engineering design. Based on a series of case studies and model-building endeavours with both industry and students, a series of interlinked PSS methodology proposals have been synthesised, supported by the domain insights into the technology, application area and necessary PSS competences.

Building on the above attempts, one can consider the expansion in nature of the PSS in relation to a product. By adopting whole (or at least extended) life responsibility for a product – which is inevitable in a PSS situation – the design task changes its nature. If seen from a PSS perspective, a manufactured product's destiny should be to be distributed, sold and domesticated, i.e. brought into the surroundings and context in which it is to serve for a period. In this situation we may focus upon the product itself, the man/machine interaction (learning, training, job-situation, working conditions etc.), the product's utilisation process (its productivity, reliability, yield, availability etc.) and the occurrence of failure, (repair, upgrading etc.). By adopting this stance the question of *system fit* arises, i.e. how well the product works together with other system elements and how well it contributes to the overall optimisation.

The product will be able to serve the user for a period, known as the *product service period*. After this period many different situations may occur with the product, from: returning to the manufacturer; being upgraded: re-used by a new owner; and finally subjected either to a planned and controlled disposal – leading to recycling, or a primitive disposal. So the total product life period from raw material allocation to this disposal situation may also be seen as a sequence of activities, all caused or disposed for by the designer.

A further significant consequence of PSS design is that the time domain is expanded, both by prolonging the period of time that the producing company has an active interest in and control over the product and also by creating the need to consider multiple product lives, where the product (artefact) can be subject to numerous users over longer periods. The consideration of multiple life phases poses a new and challenging set of criteria for the product planning and development activity to pay attention to, including: an extended stakeholder gallery; increased product liability; closer contact to the end-user; the risk of cannibalisation of existing/future products and markets; and a new opportunity (or necessity) to consider the meaning of core business [51].

2.4 Future Research Issues
A recent white paper on "Industrial PSS" [52] describes "... in 10 years the following statements will be relevant: Result-oriented business models evolve as an industry standard. Complex development processes are simplified by automatic configuration by Plug 'n' Play of product and service modules. Service will be provided globally by service supply chains based on modularised service processes." In order to realise this picture in ten years, there remains a lot to research about PSS design. Sakao et al. [26] further discuss future research needed in the PSS-design area: Design process, organisational structure, and mindset.

Design process
More research is needed for supporting companies in how they can be successful at integrated product and service development. Methods, tools, and procedures should support providers to develop services that are economically and environmentally beneficial and they need to be tested and validated in firms. Several concepts and suggestions (e.g. [34] and [10]) should be fed back for research; now we need the empirical testing.

Organisational structure
The organisational structure needs to shift in a company. How to organise the company in order to match the organisational arrangements according to the services offered is one area, where more research can be performed. Part of this is the competence profile of the company that needs to shift when moving into services (e.g. more service technicians or more business and service developers could be needed). A logistics system and a remanufacturing system may also need to be developed.

Mindset
Companies undergo major changes in mindset. Companies that have a strong culture and pride in their products must also build up a trust and belief in their services among their employees. Services also need to have a high status and be incorporated into the company. The importance of the mindset and how that can build, in line with new company values, will be an interesting research area.

3 INTEGRATED DEVELOPMENT OF TECHNOLOGY AND BUSINESS MODEL

3.1 Implication from Theoretical Reasoning
What would be an interesting issue if engineering design were expanded to PSS design and further beyond? What could be addressed in order to make a bigger impact or to be more fundamental? Ultimately, a technology adopted can be the most fundamental, in comparison to an employed system, physical product as a part of the system, and material used for the product in the hard, technical side. In other words from bottom up, a material is chosen based on the product, which depends on the system, assuming a given technology. Consider, for example, a hybrid personal motorcar, which adopts a new technology for its engine. This car performs far better than other cars designed differently with a traditional petrol-engine technology, in terms of green house gas emission. The point here is the increased performance by the new technology surpasses largely that by a certain "new design" such as a car with a new body.

Therefore, technology development would be an interesting issue to consider if engineering design were expanded and further developed. It has not been addressed as the focus of engineering design. This becomes more interesting if considering the principle; as discussed in Section 2.1, the earlier in the design process, the more freedom designers have. Technology development exists beforehand for the traditional engineering design to begin.

Having explored this further expansion above, let us now find what to learn from the achievement by PSS design. PS design addresses both 'hard' and 'soft' issues, while traditional product design deals with the physical product, including its materials. Having learnt that, it is interesting to ask what could be targeted as a part in the soft side of design object to increase freedom furthermore? The business model, rather than the organisational structure of a firm and service activity provided within the scheme, might have the biggest impact.

The result of this theoretical reasoning is depicted by Fig. 2 and has been named the "V-shape in techno-business". This allows the positioning of different disciplines. "Integrated development of technology and the business model" above PSS design has the highest impact. The length in the horizontal axis for each depicted area can be interpreted as the degree of freedom in design.

Figure 2. V-shaped relation in techno-business space (modified from [53]).

Longer time to market is characteristic of technology development (e.g. [54]) and has greater uncertainty compared to traditional product development. This poses a challenge to designers when expanding engineering design to include technology development. On the other hand, this is where services can be an effective way to decrease the impact of uncertainty. When a product with a new technology is launched, it could be combined effectively with a service as a package that takes care of technical risks.

Previous research has shown the business model to be an important factor in the PSS area (e.g. [28]). However, its integration with hard issues (i.e. product or technology) has not yet been discussed thoroughly. This is where the research need exists; in the integration of technology and business development. This integration is important, as they influence each other. For instance, a technology that was difficult or unfamiliar to users would require an intimate support service, which a provider would like to make money from.

Theoretical knowledge from the engineering perspective in this area is insufficient. Very little literature addresses this issue. For example, Efistathiades et al. [55] discuss an integrated process plan for implementing technologies, but do not focus on business models. This is no wonder, since developing such a theory of design/development aspects for PSS has just begun in the last decade. Indeed, utilising new technologies in developing PSS remains to be explored. However, this can be foreseen as a promising research issue following integration of products and services.

3.2 Industrial Needs

A driver for industry is pressure to decrease time-to-market generally. Emerging opportunities in the markets of developing countries is a particular driver. This is related to time-to-market, because current market opportunities may be lost without quick action. There is a need to implement new technologies for emerging markets, especially in the sector of environmental technologies. Developing countries, such as China, Russia and India, have a great need for solutions with environmental technologies to decrease their environmental impacts. In these situations, investigating alternative business models can be effective, because combining services in a different business model could decrease time-to-market. For instance, many Swedish firms have environmental technologies to potentially sell to the emerging markets, but building up a business model appropriate to the solution required is an issue [56].

3.3 Existing Knowledge and Research Opportunity

In previous research on integrated product development, Drejer highlights the need to integrate product and technology development that originates from the customers' requirement of shortened time to market [57]. Looking closely at technology development processes in the automotive sector, it is argued that technology development should happen before the requirement analysis for a product, because doing it the other way around takes more time and cost [58]. The technology developed is tuned after the requirement becomes available. An information processing model has been proposed to represent the process of developing products based on novel technical capabilities [59]. The process begins with exploration of the technological alternatives, and then moves to integration into a technological concept. Development of a detailed system and, then, production then occurs.

Drejer has argued for the need to integrate different disciplines, such as technology and sales [57]. Nyström demonstrates the need to address both marketing and R&D strategies within product development, and provided a framework for characterising and integrating marketing and technology strategies [60]. Another framework containing one line for business gates and another for technical decisions for a new product development process has been proposed, based on good practice in the chemical industry [61]. However, this framework does not address the design of a business model. Efstathiades et al. [55] discuss integrated process plans for implementing technologies, but do not focus on business models either.

Previous research in the PSS area has shown that the business model is an important factor (e.g. [28]). However, its integration with 'hard' issues (i.e. product or technology) has not yet been discussed thoroughly. This is where research is needed: the integration of technology and business development. This integration is important as they influence each other, for example, a technology difficult or unfamiliar to users requires an intimate support service, from which a provider would like to profit.

4 CONCLUDING DISCUSSION

By giving a review of PSS research in Section 2, we first highlighted some basic differences in conditions for a PSS design to be possible. This was followed by the nature of PSS design that the authors recognise. Finally, existing and future research was presented. To summarise our current insights, based on this exercise, we conclude with seven position statements:

- We believe that PSS development opens up for a greater arena of possibilities and therefore innovation practices than we have seen before.
- The engineering design object expands, implying that the product development task must handle more complex life cycle issues, multiple (and increased variance of) stakeholders, multiple product lives, societal issues, liability issues, etc.
- We have started to experiment and to attempt to develop a methodology for PSS development in the light of engineering design theory.
- We see the notional classification of the three elements: PSS offer modelling; PSS development process; and PSS potential, as a promising way to proceed with the support of PSS through an integrated approach.
- Considering the quantity and quality of the existing research on PSS design, there is a need to conduct the research further. Simply put, there are different types of elements of actions in PSS design that are not even well understood (see, for example, some recently initiated work, to understand more of PSS design [62]).
- We have presented a new promising concept beyond PSS design; integrated development of technology and business model.
- We are convinced that an integrated technology and business development model will shed great light on the task of PSS planning development and execution. However, little knowledge is available here also, and further research is demanded (an analysis of such a process can be studied in [63]).

ACKNOWLEDGEMENTS

This research is partly supported by the project Management of Innovation Processes for Business Driven Networks, funded by VINNOVA (The Swedish Governmental Agency for Innovation Systems) and partly by the Danish Agency for Science, Technology and Innovation.

REFERENCES

[1] Oliva, R. and Kallenberg, R., Managing the transition from products to services. *International Journal of Service Industry Management*, 2003, 14(2), pp160-172.

[2] Pahl, G. and Beitz, W., *Engineering Design: A Systematic Approach*. (Springer-Verlag, London, 1988).

[3] Tukker, A. and Tischner, U., *New Business for Old Europe*. (Greenleaf Publishing, Sheffield, 2006).

[4] Mont, O.K., Clarifying the concept of product–service system. *Journal of Cleaner Production*, 2002, 10(3), pp237-245.

[5] McAloone, T.C. and Andreason, M.M., Design for Utility, Sustainability and Social Virtues, Developing Product Service Systems. In *International Design Conference*. Dubrovnik, 2004. pp1545-1552.

[6] Tomiyama, T., Service Engineering to Intensify Service Contents in Product Life Cycles. In *Second International Symposium on Environmentally Conscious Design and Inverse Manufacturing*. Tokyo, 2001. pp613-618 (IEEE Computer Society)

[7] Sakao, T. and Shimomura, Y., A Method and a Computerized Tool for Service Design. In *International Design Conference*. Dubrovnik, 2004. pp497-502

[8] Brännström, O., Elfström, B.-O. and Thompson, G., Functional products create new demands on product development companies. In *International Conference on Engineering Design*. Glasgow, 2001.

[9] Roy, R. and Baxter, D., Special Issue - Product-Service Systems. *Journal of Engineering Design*, 2009, 20(4), pp327 – 431.

[10] Sakao, T. and Lindahl, M., eds. *Introduction to Product/Service-System Design*. (Springer, London, 2009).

[11] Lindahl, M. and Ölundh, G., The Meaning of Functional Sales. In *8th CIRP International Seminar on Life Cycle Engineering – Life Cycle Engineering: Challenges and Opportunities*. 2001. pp211-220.

[12] Lindahl, M., Sundin, E., Rönnbäck, A.Ö., Ölundh, G. and Östlin, J., Integrated Product and Service Engineering – the IPSE project. In *Changes to Sustainable Consumption, Workshop of the Sustainable Consumption Research Exchange (SCORE!) Network, supported by the EU's 6th Framework Programme*. Copenhagen, Denmark, 2006.

[13] Sakao, T. and Shimomura, Y., Service Engineering: A Novel Engineering Discipline for Producers to Increase Value Combining Service and Product. *Journal of Cleaner Production*, 2007, 15(6), pp590-604.

[14] Blessing, L., Chakrabatri, A. and Wallace, K., Designers - The Key to Successful Development. (Springer-Verlag, London, UK, 1998).

[15] Stahel, W.R., The Functional Economy: Cultural and Organizational Change. In Richards, D.J., ed. *The Industrial Green Game*, pp91-100 (National Academy Press, Washington, D.C., 1997).

[16] Alonso-Rasgado, T., Thompson, G. and Elfstrom, B., The design of functional (total care) products. *Journal of Engineering Design*, 2004, 15(6), pp515-540.

[17] White, A.L., Stoughton, M. and Feng, L., Servicizing: The Quiet Transition to Extended Product Responsibility. (Tellus Institute, Boston, 1999).

[18] Baines, T.S., Lightfoot, H.W., Evans, S., Neely, A., Greenough, R., Peppard, J., Roy, R., Shehab, E., Braganza, A., Tiwari, A., Alcock, J.R., Angus, J.P., Bastl, M., Cousens, A., Irving, P., Johnson, M., Kingston, J., Lockett, H., Martinez, V., Michele, P., Tranfield, D., Walton, I.M. and Wilson, H., State-of-the-art in product-service systems. *Proceedings of the Institution of Mechanical Engineers - B*, 2007, 221, pp1543-1552.

[19] Isaksson, O., Larsson, T.C. and Öhrwall Rönnbäck, A., Development of product-service systems: challenges and opportunities for the manufacturing firm. *Journal of Engineering Design*, 2009, 20(4), pp329 – 348.

[20] Rese, M., Strotmann, W.-C. and Karger, M., Which industrial product service system fits best?: Evaluating flexible alternatives based on customers' preference drivers. *Journal of Manufacturing Technology Management*, 2009, 20(5), pp640-653.

[21] Tan, A. and McAloone, T., Characteristics of Strategies in Product/Service-System Development. In *International Design Conference*. Dubrovnik, 2006. pp1435-1442

[22] McAloone, T.C. and Andreasen, M.M., Defining product service systems. In *Design for X*. TU Erlangen, 2002. pp51-60
[23] Ropohl, G., *Systemtechnik – Grundlagen und Anwendung*. (Carl Hanser Verlag, München, 1975).
[24] Andersen, P.D., Borup, M., Borch, K., Kaivo-oja, J., Eerola, A., Finnbjörnsson, T., Øverland, E., Eriksson, E.A., Malmér, T. and B.A., M., Foresight in Nordic Innovation Systems. (Nordic Innovation Centre, 2007).
[25] Vandermerwe, S., How increasing value to customers improves business results. *Sloan Management Review*, 2000, 42(1), pp27-37.
[26] Sakao, T., Sandström, G.Ö. and Matzen, D., Framing research for service orientation through PSS approaches. *Journal of Manufacturing Technology Management*, 2009, 20(5), pp754-778.
[27] Tukker, A., Eight Types of Product-Service System: Eight Ways to Sustainability? Experiences from Suspronet. *Business Strategy and the Environment*, 2004, 13, pp246 – 260.
[28] Mont, O., Singhal, P. and Fadeeva, Z., Chemical Management Services in Sweden and Europe: Lessons for the Future. *Journal of Industrial Ecology*, 2006, 10(1/2), pp279-292.
[29] Finger, S. and Dixon, J.R., A review of research in mechanical engineering design. Part I: Descriptive, prescriptive, and computer-based models of design processes *Research in Engineering Design*, 1989, 1, pp51-67.
[30] Finger, S. and Dixon, J.R., A review of research in mechanical engineering design. Part II: Representations, analysis, and design for the life cycle *Research in Engineering Design*, 1989, 1(2), pp121-137.
[31] Aurich, J.C., Fuchs, C. and Wagenknecht, C., Life cycle oriented design of technical Product-Service Systems. *Journal of Cleaner Production*, 2006, 14(17), pp1480-1494.
[32] Dausch, M. and Hsu, C., Engineering service products: the case of mass-customising service agreements for heavy equipment industry. *International Journal of Services Technology and Management*, 2006, 7(1), pp32 - 51
[33] Östlin, J., Sundin, E. and Björkman, M., Importance of closed-loop supply chain relationships for product remanufacturing. *International Journal of Production Economics*, 2008, 115(2), pp336-348.
[34] Sakao, T., Shimomura, Y., Sundin, E. and Comstock, M., Modeling Design Objects in CAD System for Service/Product Engineering. *Computer-Aided Design*, 2009, 41(3), pp197-213.
[35] Panshef, V., Dörsam, E., Sakao, T. and Birkhofer, H., Value-Chain-Oriented Service Management by Means of a 'Two-Channel Service Model'. *International Journal of Services Technology and Management*, 2009, 11(1), pp4-23.
[36] Moon, S.K., Simpson, T.W., Shu, J. and Kumara, S.R.T., Service representation for capturing and reusing design knowledge in product and service families using object-oriented concepts and an ontology. *Journal of Engineering Design*, 2009, 20(4), pp413 – 431.
[37] Doultsinou, A., Roy, R., Baxter, D., Gao, J. and Mann, A., Developing a service knowledge reuse framework for engineering design. *Journal of Engineering Design*, 2009, 20(4), pp389 – 411.
[38] Maussang, N., Zwolinski, P. and Brissaud, D., Product-service system design methodology: from the PSS architecture design to the products specifications. *Journal of Engineering Design*, 2009, 20(4), pp349-366.
[39] Hara, T., Arai, T. and Shimomura, Y., A CAD system for service innovation: integrated representation of function, service activity, and product behaviour. *Journal of Engineering Design*, 2009, 20(4), pp367-388.
[40] Aurich, J.C., Wolf, N., Siener, M. and Schweitzer, E., Configuration of product-service systems. *Journal of Manufacturing Technology Management*, 2009, 20(5), pp591-605.
[41] Alonso-Rasgado, T. and Thompson, G., A rapid design process for Total Care Product creation. *Journal of Engineering Design*, 2006, 17(6), pp509 - 531.
[42] Morelli, N., Product-service systems, a perspective shift for designers: A case study: the design of a telecentre. *Design Studies*, 2003, 24(1), pp73-99.
[43] McAloone, T.C., A Competence-Based Approach to Sustainable Innovation Teaching: Experiences within a New Engineering Programme. *Journal of Mechanical Design*, 2007, 129(6), pp769-778.

[44] Sakao, T., Birkhofer, H., Panshef, V. and Dörsam, E., An Effective and Efficient Method to Design Services: Empirical Study for Services by an Investment-machine Manufacturer. *International Journal of Internet Manufacturing and Services*, 2009, 2(1), pp95-110.
[45] Molloy, E.-M., Siemieniuch, C. and Sinclair, M., Decision-making systems and the product-to-service shift. *Journal of Manufacturing Technology Management*, 2009, 20(5), pp606-625.
[46] Kimita, K., Shimomura, Y. and Arai, T., Evaluation of customer satisfaction for PSS design. *Journal of Manufacturing Technology Management*, 2009, 20(5), pp654-673.
[47] Sundin, E., Lindahl, M. and Ijomah, W., Product design for product/service systems - design experiences from Swedish industry. *Journal of Manufacturing Technology Management*, 2009, 20(5), pp723-753.
[48] Tan, A.R., McAloone, T.C. and Lauridsen, E.H., Reflections on teaching product/service-system (PSS) design. *International Journal of Design Engineering*, 2010.
[49] Evans, S., Partidário, P.J. and Lambert, J., Industrialization as a key element of sustainable product-service solutions. *International Journal of Production Research*, 2007, 45(18 & 19), pp4225 - 4246.
[50] Azarenko, A., Roy, R., Shehab, E. and Tiwari, A., Technical product-service systems: some implications for the machine tool industry. *Journal of Manufacturing Technology Management*, 2009, 20(5), pp700-722.
[51] Matzen, D. and McAloone, T.C., From Product to Service Orientation in the Maritime Equipment Industry - A Case Study. In Mitsuishi, M., Ueda, K. and Kimura, F., eds. *Manufacturing Systems and Technologies for the New Frontier - Proceedings for The 41st CIRP Conference on Manufacturing Systems*, pp515-518 (Springer, Tokyo, 2008).
[52] Meier, H., Roy, R. and Seliger, G., Industrial Product-Service Systems - IPS^2. *CIRP Annals Manufacturing Technology*, 2010, 59(2), pp607-627.
[53] Sakao, T., Integrated Development of Technology and Business Model: A Discipline beyond Ecodesign and PSS Design toward Sustainability. *International Conference on Asia Pacific Business Innovation and Technology Management* (Cebu, Philippines, 2010).
[54] Tatikonda, M.V. and Rosenthal, S.R., Technology Novelty, Project Complexity, and Product Development Project Execution Success: A Deeper Look at Task Uncertainty in Product Innovation. *IEEE Transactions on Engineering Management*, 2000, 47(1), pp74-87.
[55] Efstathiades, A., Tassou, S. and Antoniou, A., Strategic planning, transfer and implementation of Advanced Manufacturing Technologies (AMT). Development of an integrated process plan. *Technovation*, 2002, 22, pp201-212.
[56] Swentec. Rapport till regeringen maj 2009 (2009).
[57] Drejer, A., Integrating product and technology development. *International Journal of Technology Management*, 2002, 24(2/3), pp124-142.
[58] Ueno, K., From Product-Oriented Development to Technology-Oriented Development. *IEEE Transactions on Reliability*, 1995, 44(2), pp220-224.
[59] Iansiti, M., Technology Development and Integration: An Empirical Study of the Interaction Between Applied Science and Product Development. *IEEE Transactions on Engineering Management*, 1995, 42, pp259-269.
[60] Nyström, H., Product Development Strategy: An Integration of Technology and Marketing. *Journal of Product Innovation Management*, 1985, 2, pp25-33.
[61] Shaw, N.E., Burgess, T.F., Hwarng, H.B. and Mattos, C.d., Revitalising new process development in the UK fine chemicals industry. *International Journal of Operations and Production Management*, 2001, 21(8), pp1133-1151.
[62] Sakao, T., Paulsson, S. and Mizuyama, H., Inside a PSS Design Process: Insights through Protocol Analysis. In *International Conference on Engineering Design*. Copenhagen, 15-18, August, 2011, in print.
[63] Sakao, T., Rönnbäck, A.Ö. and Sandström, G.Ö., Integrated Product Service Offerings for Facilitating Introduction of New Technologies to Emerging Markets. In *ISPIM Conference*. Hamburg, 12-15, June, 2011, to be submitted.

Contact: Tomohiko Sakao
Linköping University
Department of Management and Engineering
581 83 Linköping
Sweden
Tel: Int +46 13 282287
Email: tomohiko.sakao@liu.se
URL: www.iei.liu.se/envtech/om-oss/tomohiko_sakao

Tom is Professor of IPSE (Integrated Product Service Engineering) and Ecodesign (Environmentally conscious design). His focus at present is environmental and economic sustainability in manufacturing industries (from perspectives of both suppliers and purchasers) in the matured economy.

INTERNATIONAL CONFERENCE ON ENGINEERING DESIGN, ICED11
15 - 18 AUGUST 2011, TECHNICAL UNIVERSITY OF DENMARK

ANALYSING MODIFICATIONS IN THE SYNTHESIS OF MULTIPLE STATE MECHANICAL DEVICES USING CONFIGURATION SPACE AND TOPOLOGY GRAPHS

Somasekhara Rao Todeti[1] and Chakrabarti Amaresh[1]
(1) Indian Institute of Science, Bangalore, India

ABSTRACT
Automated synthesis of mechanical designs is an important step towards the development of an intelligent CAD system. Research into methods for supporting conceptual design using automated synthesis has attracted much attention in the past decades. In our research, ten experimental studies are conducted to understand how designers synthesize solution concepts for multi-state mechanical devices. The designers vnd that modification of kinematic pairs and mechanisms is the major activity carried out by all the designers. This paper presents an analysis of these synthesis processes, using configuration space and topology graph, to identify and classify the types of modifications that take place. Understanding these modification processes and the context in which they happened is crucial for developing a system for supporting design synthesis of multiple state mechanical devices that is capable of creating a comprehensive variety of solution alternatives.

Keywords: Automated synthesis, multiple state, conceptual design, mechanical device, analysis of synthesis processes, configuration space, topology graph

1 INTRODUCTION
The overall aim of this research is to develop a generic computational system to support designers during the conceptual phase of mechanical design to synthesize a wider variety of design alternatives than currently possible for multiple state mechanical devices. Conceptual design has the most significant influence on the overall product cost [1]. Conceptual design is a difficult task [2], which relies on the designer's intuition and experience to guide the process. A major issue within this task is that often not many potential solutions are explored by the designer during the design process [3, 4]. The major reasons for this are the tendency to delimit a design problem area too narrowly and thus not being able to diversify the possible set of design solutions, possible bias towards a limited set of ideas during the design process, and time constraints [5]. Evidence from earlier research suggests that a thorough exploration of the solution space is more likely to lead to designs of higher quality [6]. Therefore, a support system, automated or interactive, that can help generate a considerable variety of feasible design alternatives than currently possible at the conceptual design phase, is important to the development of intelligent CAD tools that can play a more active role in the mechanical design process, especially in its earlier phases.
Li [5] defined the operating state of a mechanical device by a set of relations between its input and output motions, which remain unchanged within an operating state. A multiple state device has more than one operating state. Other researchers [7, 8] defined state in various other ways. The definition of state used by Li [5] is adopted for use in this research work.

2 RESEARCH OBJECTIVES
The central question to be addressed in this research is – how to synthesize, automatically or interactively, a comprehensive set of possible device concepts that satisfy a given task comprising multiple states? In order to do this, we first wish to understand the process by which engineering designers synthesize multiple state devices. This, we hope, will throw light on how (and how well) multiple state synthesis tasks are currently handled, what can be learnt from these, and how this learning could be used in computational tools to help improve multi-state synthesis tasks. The research work presented in this paper is continuation of our previous work [9], where multi-state design tasks

are described in terms of a set of related functions, and this set of functions is given to the designers including the researcher to individually generate as many design alternatives as possible. All the designers are asked to think aloud while carrying out their synthesis processes. These synthesis processes are video recorded. The following practice has been observed in the case of each designer: an initial solution proposal is generated, which satisfies one of the functions of the design task. This initial proposal is modified for satisfying each of the remaining functions, taken one at a time. The initial solution proposal generated and the various types of *modifications,* carried on these solution proposals for satisfying previously unsatisfied functions, led to generation of various solutions. Modifications are found to be the major activity in synthesizing solutions for multi-state design tasks. To understand how these modifications are carried out and in what context, all the synthesis processes are analyzed using configuration space [10] and topology graph [10]. The objective of this paper is to analyze these synthesis processes using configuration space and topology graph in order to understand the modification processes. This understanding should help develop a support which can help generate a wider solution space for a given multistate design task than currently possible.

3 LITERATURE STUDY

Li [5] seems to be the only researcher who has directly addressed multiple state mechanical synthesis tasks. He has used the configuration space approach to represent and retrieve behaviors of kinematic pairs, and developed ADCS system for automatically generating solutions of mechanical devices that satisfy the given multiple state design tasks. However, ADCS is limited to generation of a single solution for a design task, rather than a comprehensive set of alternative designs that are possible to be generated for the task – a critical drawback if the goal is to support generation of a wider variety of concepts. Single state design synthesis [e.g. 11-16] is limited to either synthesizing a single kinematic pair, a mechanism or their combination, for single input and single output tasks using simulation based, configuration space or grammatical approaches. One approach promulgated earlier has been to generate solutions for each single state within a multi-state synthesis task, and find those solutions that are common across all these sets as solutions to the multi-state problem. However, if the intersection of these solution spaces leads to a null space, there would be no solution possible. An alternative approach is to modify, solutions to a single state task of a given multistate design problem until it satisfies all the other states constituting the problem. .

4 REPRESENTATION OF STATES OF MULTIPLE STATE MECHANICAL DEVICES

An existing multiple state device, a door attached with a latch, is a four state device. This device is analyzed for its functioning within each state and its state transitions.

4.1 Multi-state functioning of a door attached with a latch

The functions of a door are to allow and prevent the movement of energy or material. When a door is in the locked state, it completely prevents movement of these to and from the room. When it is in the opened state, it allows movement of these to and from the room. In between these two states, there are opening and closing states as shown in Figure 1. One way in which a door achieves these functions is

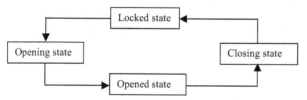

Figure 1. States and state transitions of door attached with a latch

when a latch is attached to it. The functions in each state depend on the type of latch attached to a door. For the type of latch shown in Figure 2(a), the functions performed in each state are:
- Locked state: Function1: the door is pushed, but it does not move.
                  Function2: the door is pulled, but it does not move.

- Opening state: Function3: the handle is rotated by applying an effort.
  Function4: keeping the effort in the function3 on the handle, another effort is applied to the door, and the door opens.
  Function5: as the effort on the handle is released, the handle rotates back.
- Opened state: Function 6: as the door is pulled, it opens further.
  Function 7: as the door is pushed, it closes further.
- Closing state: Function8: by applying further effort on the door, it comes to the locked state, where further push or pull does not move the door.

The door latch is disassembled to further study its components and interfaces among its components and how this latch when attached to the door attains the functioning of the door. The latch and its components (component H is the handle and component B is the wedge shaped block) are shown in Figures 2(b). From an understanding of these components and interfaces, the overall structure, and the functioning of the latch, the latch is modeled as shown in Figure 2(c). The latch has an L-shaped handle hinged at A, a torsion spring connected to the handle at A, a block, a rod attached to the block and a spring arrangement, where the spring is confined between the block and a support with a hole through which the rod can translate, a small pin attached to the rod protruding perpendicular to the plane of the paper, and a stop at C. This is a plane mechanism. The motion transformations between the handle (component H) and the block (component B) due to various efforts are described below as five functions:

- F1: Apply effort on the handle in anticlockwise direction, handle rotates (from $\theta = \theta_0$ to $\theta = \theta_1$), and simultaneously the block translates inside (from $x=x_0$ to $x=x_1$).
- F2: if effort is kept applied in the same direction when the handle is $\theta = \theta_1$, the handle does not rotate any further due to the obstruction C, and the block remains at $x=x_1$.
- F3: If the effort is released from the handle, the handle rotates back to $\theta = \theta_0$ from $\theta = \theta_1$ and simultaneously the block also translates back to $x=x_0$ from $x=x_1$.
- F4: Now if effort is applied on the block along negative x-axis, the block translates from $x=x_0$ to $x=x_1$, but there is no motion in the handle.
- F5: If the effort on the block is released, the block goes back to $x=x_0$ from $x=x_1$ but the handle does not move.

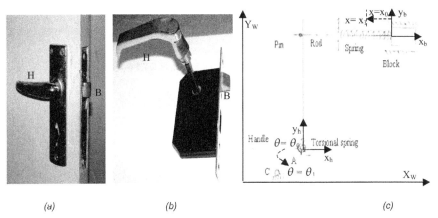

Figure 2. The door latch, its structure and components

4.2 Framing a door latch design task

By taking components L_1 and L_2 as shown in Figure 3(a) to act as the handle and the block respectively, a five-function design task can be devised for the above functions as follows:

1. F1: When torque is applied on L_1 in the counter clockwise direction around z-axis, L_1 rotates in counter-clockwise direction around z –axis from $\theta = \theta_0$ to $\theta = \theta_1$, and simultaneously L_2 translates
along x-axis in negative direction from $x = x_0$ to $x=x_1$.

2. F2: Even if torque is kept applied in the same direction on L_1, when L_1 is at $\theta = \theta_1$, L_1 does not rotate beyond $\theta = \theta_1$, and L_2 remains at $x = x_1$.
3. F3: If the torque is released on L_1, when L_1 is at $\theta = \theta_1$ then L_1 rotates back in the clockwise direction from $\theta = \theta_1$ to $\theta = \theta_0$ and L_2 simultaneously translates along x-axis from $x=x_1$ to $x=x_0$.
4. F4: Now if force is applied on L_2 along x-axis in the negative direction, L_2 translates along the axis in the negative direction from $x = x_0$ to $x = x_1$, but L_1 remains at $\theta = \theta_0$;
5. F5: If the force on L_2 is released, L_2 translates back to $x = x_0$ from $x = x_1$, but L_1 remains at $\theta = \theta_0$.

These five functions are given to the designers to generate as many solutions as possible.

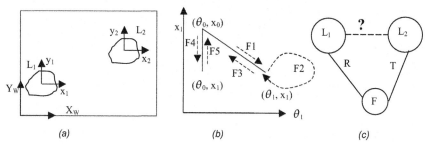

Figure 3. The required configuration space and topology graph for the door latch design task

A graph, shown in Figure 3(b), is drawn between the motions of L_1 and L_2. θ is the relative angular motion of the local coordinates system (x_1,y_1) attached to L_1 with the world coordinate system (X_W,Y_W), while x is the relative translatory motion of the local coordinates system (x_2,y_2) attached to L_2 with the world coordinate system (X_W,Y_W). The five functions (F1, F2, F3, F4 and F5) described above are shown in Figure 3(b). F2 involves no change in motions of L_1 and L_2 though effort is applied on L_1. So its starting and end configurations are the same and it is a point on the graph. The topology graph between L_1 and L_2 is shown in Figure 3(c). As L_1 is allowed only to rotate, it forms a revolute joint, R, with the frame (F), while L_2 forms a prismatic joint, T, with the frame. Now the task is to synthesize as many devices as possible, which are sets of components and interfaces, in each set we identify two components which can act as L_1 and L_2 such that they achieve the above five functions.

5 SYNTHESIS OF SOLUTIONS FOR THE LATCH DESIGN TASK

Some of the synthesis processes, carried out by the designers for generating solutions to the door latch design task, are analyzed below using configuration space and topology graph.

5.1 Solution1

After analyzing the given five functions of the door latch design task, first function F1 is focused on. As F1 requires rotary motion to be converted to translatory motion, a slider crank mechanism

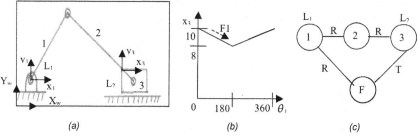

Figure 4. SP1, its configuration space and topology graph

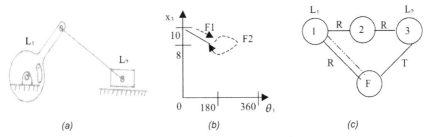

Figure 5. SP11, its configuration space and topology graph

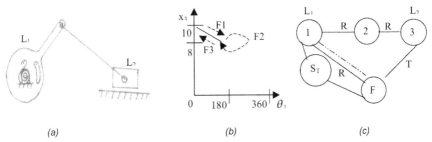

Figure 6. SP111, its configuration space and topology graph

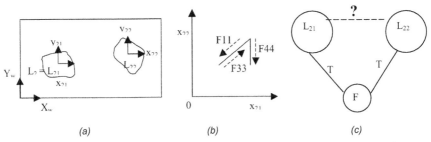

Figure 7. Newly derived design task, required configuration space and topology graph

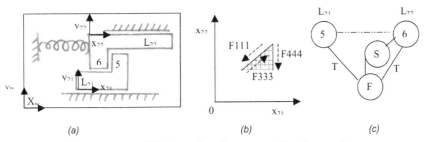

Figure 8. SP21, its configuration space and topology graph

is generated. We call this solution proposal1, or SP1. The slider-crank mechanism (SP1), its configuration space and topology graph are shown in Figures 4(a)-(c) respectively. Arbitrary numerical values are assigned for producing the configuration space in Figure 4(b). In the topology graph shown in Figure 4(c), the letters R and T stand for revolute pair and prismatic pair respectively. Here we can identify that the crank (Component1) acts as L_1, and the slider (Component3) as L_2

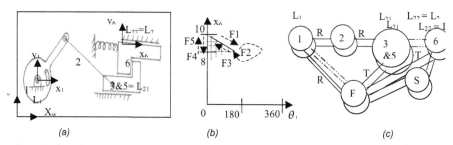

Figure 9. Combined solution from SP111 and SP21, its configuration space and topology graph

for function F1. Now SP1 is evaluated against F1, which it is found to satisfy. The same thing can be observed between the configuration spaces in Figures 4(b) and 3(b), i.e. the portion of configuration space of SP1 that is a line between the points (0, 10) and (180,8) in Figure 4(b) matches F1 of Figure 3(b). As F1 is satisfied, F2 is selected next. Keeping F1 and F2 in mind, SP1 is modified with a slot and pin arrangement, shown in Figure 5(a), producing SP11. SP11, its configuration space and topology graph are shown in Figures 5(a)-(c). The change in configuration space can be observed in terms of size reduction based on the slot and pin arrangement. The change in topology graph can be observed in terms of introduction of a line (chained line, which indicates a changing higher pair contact, i.e. the components are in contact at some configurations and not at other configurations.) between Component1 and the frame. As F1 and F2 are satisfied as shown in Figure 5(b), F3 is selected next. SP11 is modified by adding a torsional spring between the Component1 and the frame as shown in Figure 6(a), producing SP111. The configuration space and the topology graph for SP111 are shown in Figure 6(b) and 6(c) respectively. The configuration space remains the same, but the spring addition acts as a motion activator, which is required for F3. The corresponding change in topology graph can be observed by addition of a spring (S_T) between the frame and the Component1. As F1, F2, F3 are satisfied as shown in Figure 6(b), F4 is selected next. In F4, when force is applied on L_2, it translates along the negative x- direction, but without motion in L_1. But when force is applied on the Component3 (i.e. the slider, acting as L_2), Component3 moves and crank also rotates. So F4 is not satisfied, which can also be realized when configuration spaces shown in Figure 3(b) and Figure 6(b) are compared. i.e. the vertical line required in configuration space shown in Figure 3(b) does not exist in the existing configuration space of SP111 shown in Figure 6(b). As F4 is not satisfied, one more component (L_{22}) is chosen to act as the new L_2; this L_{22} is to be placed beside the slider (which is considered as L_2 till now; henceforth we call this L_{21}) of SP111. To satisfy F4, L_{21} and L_{22} have to be connected such that when L_{22} is pushed along the negative x- direction by a force, L_{21} should not move inwards. It must be kept in mind that the already satisfied functions F1,F2, and F3 should not be negated with the introduction of the new L_{22}. If the above four functions F1, F2, F3 and F4 are reconsidered with respect to SP111, they can be reframed as follows:
1. F11: if torque is applied on L_1 (the crank of SP111) in counter clockwise direction, then L_1 rotates from $\theta = \theta_0$ to $\theta = \theta_1$, L_{21} (the slider of SP111) translates inside along negative x-axis from
   x= x_0 to x=x_1 and L_{22} also has to translate inside along negative x-axis simultaneously.
2. F22: if torque is still kept applied on L_1 (the crank of SP111) in counter clockwise direction when L_1 is at $\theta = \theta_1$, L_1 does not rotate beyond $\theta = \theta_1$, L_{21} remains at x= x_1 and L_{22} also does not move.
3. F33: if torque is released from L_1, then L_1 rotates back to $\theta = \theta_0$ from $\theta = \theta_1$, L_{21} translates back along positive x-axis to x=x_1 from x= x_0 and L_{22} also translates along positive x-axis simultaneously.
4. F44: if force is applied on L_{22} along the negative x- direction, L_{22} translates inwards, L_{21} does not move and L_1 also does not move.

If only L_{21} and L_{22} are considered, the reformulated design task is as follows:
1. F111: If force is applied on L_{21} along the negative x-axis, L_{21} translates inward from $x=x_0$ to $x=x_1$, and L_{22} also translates inside simultaneously.
2. F222: If L_{21} stops, L_{22} also stops.
3. F333: If force is released from L_{21}, L_{21} translates along the positive x-axis from $x=x_0$ to $x=x_1$ and L_{22} also translates along positive x-axis simultaneously
4. F444: If force is applied on L_{22} along the negative x-axis, L_{22} translates along this axis, but L_{21} does not move.

So these functions (F111, F222, F333 and F444) together form a new design task as shown in Figures 7(a)-(c). The solution proposal (SP2) is a pair of two L- shaped blocks as shown in Figure 8(a). Component5 acts as L_{21} and Component6 as L_{22} with a changing contact higher pair interface between them that are constrained only to translate along the x- axis. SP2 is evaluated for F111, F222, F333 and F444. SP2 can not satisfy F333 as there is no motion actuating component. So SP2 is modified by adding a spring(S) between the frame and Component6, producing SP21 as shown in Figure 8(a). The configuration space and the topology graph for SP21 are shown in Figures 8(b) and 8(c) respectively. Now SP111 and SP21 are combined by attaching Component3 of SP111 with Component5 of SP21. The torsional spring between the crank is removed as it is redundant. The combined solution proposal, its configuration space and topology graph are shown in Figures 9(a) -(c) respectively. The combined solution proposal is evaluated for F5. It is found that F5 is satisfied as well as F1, F2, F3 and F4. This can be realized by comparing the configuration spaces shown in Figure 3(b) and Figure 9(b).

5.2. Solution2
A solution proposal (SP1) is generated as shown in Figure 10(a). Its configuration space and topology graph are shown in Figures 10(b) and 10(c) respectively. Component1 and Component2 act as L_1 and L_2 respectively. F1 can be realized as shown in Figure 10(b). As F1 is satisfied, F2 is selected next. SP1 is modified by adding a grounded obstruction, producing SP11 in Figure 11(a). The corresponding changes in the configuration space and the topology graph are shown in Figure 11(b)

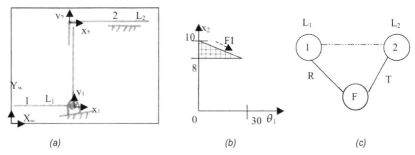

Figure 10. SP1, its configuration space and topology

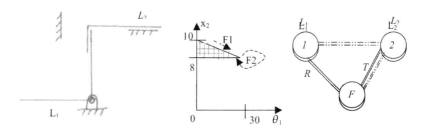

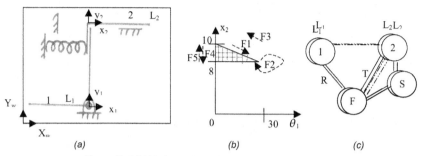

Figure 12. SP111, its configuration space and topology graph

and Figure 11(c) respectively. A chained dot line between the frame and Component2 indicates a changing contact higher pair between Component2 and the frame, as shown in Figure 11(c). F3 is selected next. In F3, as effort on L_1 is released, both L_1 and L_2 have to move back simultaneously. So a spring is connected between the frame and L_2, producing SP111 as shown in Figure 12(a). The configuration space and the topology graph for SP111 are shown in Figures 12(b) and 12(c) respectively. As F1, F2 and F3 are all satisfied as realized in Figure 12(c), F4 is selected next. SP111 is evaluated against F4 and is found to be satisfied, as can be observed in Figure 12(b). Next, SP111 is evaluated against F5, and this is also found to be satisfied. SP111 satisfies all five functions, as seen by comparing Figures 3(b) and 12(b). Component1 and Component2 act as L_1 and L_2 respectively.

5.3 Solution3

A gear pair (SP1) shown in Figure 13(a) is selected as the basis for developing a solution. Its configuration space and topology graph are shown in Figures 13(b) and 13(c) respectively. Component1 acts as L_1. Now F1 is selected, and is found that the translating component (L_2) does not

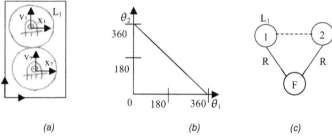

Figure 13. SP1, its configuration space and topology graph

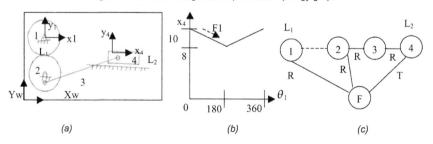

Figure 14. SP11, its configuration space and topology graph

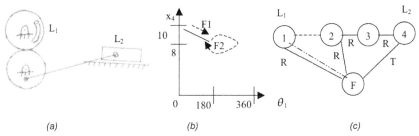

Figure 15. SP111, its configuration space and topology graph

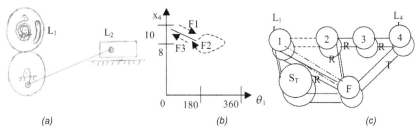

Figure 16. SP1111, its configuration space and topology graph

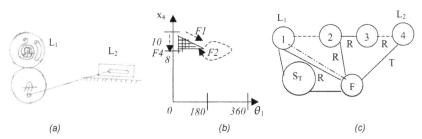

Figure 17. SP11111, its configuration space and topology graph

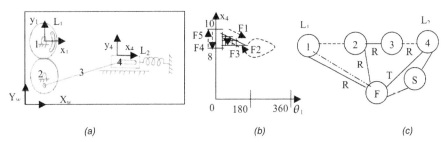

Figure 18. SP111111, its configuration space and topology graph

exist in SP1. So SP1 is modified by adding Component4 to act as L_2, which can only translate, and by adding a connecting rod (Component3) between Component2 and Component4 as shown in Figure 14(a), producing SP11. The configuration space (between Component1 and Component4) and the topology graph for SP11 are shown in Figures 14(b) and 14(c) respectively. SP11 is evaluated for F1, which satisfied F1 as seen in Figure 14(b). Next, F2 is selected, and SP11 is modified with a pin and slot arrangement, producing SP111 as shown in Figure 15(a). The configuration space and the topology graph for SP1111 are shown in Figures 15(b) and 15(c) respectively. Next F3 is selected, SP111 is modified by adding a torsional spring (S_T) between the frame and component1 producing SP1111 shown in Figure 16(a). The configuration space and the topology graph for SP1111 are shown in Figures 16(b) and 16(c) respectively. The realization of F1, F2 and F3 can be seen in Figure 16(b). Next, F4 is selected. SP1111 fails to satisfy F4, so it is modified by changing the interface between Comoponent3 and Component4 to a higher pair from revolute pair as shown in Figure 17(a) producing SP11111. The configuration space and the topology graph for SP11111 are shown in Figures 17(b) and 17(c) respectively. However, this modification fails F3, as Component4 can not translate back without any effort. So SP11111 is modified by adding a spring (S) between Component4 and the frame and removing the torsional spring (S_T), as it is redundant as shown in Figure 18(a), producing SP111111. The configuration space and the topology graph for SP111111 are shown in Figures 18(b) and 18(c) respectively. If configuration spaces shown in Figure 3(b) and Figure 18(b) are compared, it can be realized that F1,F2,F3,F4, and F5 are all satisfied.

6 FINDINGS AND PROJECTIONS

It is seen that components, their geometries and configurations, and interfaces between them play a crucial role in mechanical functions. It can be observed that in the design sessions studied, kinematic pairs are used as building blocks for constructing mechanical devices. A kinematic pair is retrieved first as it satisfies some of the functions of the multistate design task. It is then modified to satisfy the remaining functions. Let there be a kinematic pair that has two components I and O. 256 (i.e. 2*2*2*2*2*2*2*2) logical combinations can be made between possible existing and required situations of the effort states and motion states of these input and output components of the kinematic

Table 1. Possible Modification types with Observed Cases

Existing situation				Required situation			
Effort on I	Motion In I	Effort on O	Motion in O	Effort on I	Motion in I	Effort on O	Motion in O
Yes / No	Yes / No	Yes / No	Yes / No	Yes / No	Yes / No	Yes / No	Yes / No

pair, as shown in Table 1. On these 256 cases, the following conditions are now applied: the effort state on O is made to be 'No' in both the existing and required situations by applying the conditions that efforts are not applied on I and O at the same time and that there is virtual symmetry between I and O (i.e. similar situations would arise by exchanging I and O). So the number of cases of modification possible drops to 64 from 256. By applying another condition that the effort state of I between the existing and required situations remains the same, the number of cases drops further to 32. By applying yet another condition that at least one mismatch between the existing and the required situation must exist for a modification to be applied, the number of cases drops further to 24. These 24 are the distinct, realistic cases of modification that are theoretically possible, and are shown in Table 2.

Table 2. Possible Modification types with Observed Cases

	Existing situation				Required situation				Possible Modification
	Effort on I	Motion In I	Effort on O	Motion in O	Effort on I	Motion in I	Effort on O	Motion in O	
1	Yes	Yes	No	Yes	Yes	Yes	No	No	M1
2	Yes	Yes	No	No	Yes	Yes	No	Yes	M2
3	Yes	Yes	No	Yes	Yes	No	No	No	M3
4	Yes	Yes	No	Yes	Yes	No	No	Yes	M3 & M5
5	Yes	Yes	No	No	Yes	No	No	No	M3
6	Yes	Yes	No	No	Yes	No	No	Yes	M3 & M5
7	Yes	No	No	Yes	Yes	Yes	No	No	M4 & M6
8	Yes	No	No	Yes	Yes	Yes	No	Yes	M4

9	Yes	No	No	No	Yes	Yes	No	Yes	M4
10	Yes	No	No	No	Yes	Yes	No	No	M1 & M4
11	Yes	No	No	No	Yes	No	No	Yes	M1, M4 & M5
12	Yes	No	No	Yes	Yes	No	No	No	M6
13	No	Yes	No	Yes	No	No	No	No	M6
14	No	Yes	No	Yes	No	No	No	Yes	M1
15	No	Yes	No	No	No	No	No	Yes	M5 & M6
16	No	Yes	No	No	No	No	No	No	M6
17	No	Yes	No	No	No	Yes	No	Yes	M5
18	No	Yes	No	Yes	No	Yes	No	No	M1
19	No	No	No	Yes	No	Yes	No	Yes	M2
20	No	No	No	Yes	No	Yes	No	No	M5 & M6
21	No	No	No	No	No	Yes	No	Yes	M5
22	No	No	No	No	No	Yes	No	No	M5
23	No	No	No	No	No	No	No	Yes	M5
24	No	No	No	Yes	No	No	No	No	M6

Possible modifications on a pair can be classified into 6 types: M1: Introduce Relative Degree of Freedom (RDOF) between the input and output components; M2: Constrain RDOF between the input and output components; M3: Constrain RDOF between the frame and a component; M4: Provide RDOF between the frame and a component; M5: Introduce a spring between a component and the frame; and M6: Remove a spring between a component and the frame.

To obtain the required situation from the existing situation, at least one of these six modifications need to be performed. For example, in Case1 in Table 2, the existing situation is that when an external effort is applied on the input component, it moves as well as the output component, but the required situation demands that when external effort is applied on the input component, it should move without the output component moving. A possible modification to achieve this is M1, which is to introduce a RDOF between the input and the output components, as observed in the case in Figure 17(a), where a slot and pin arrangement is introduced between the slider and the connecting rod to release the RDOF.

7 CONCLUSIONS AND FUTURE WORK

The behavior and structure of a multi state device can be represented using a configuration space and a topology graph respectively. A series of synthesis processes are analyzed using these, and an exhaustive set of plausible, generic modification processes and their contexts are identified. These processes will be utilized in devising the synthesis rules for developing a multi-state design task synthesis support to help generate a wider solution space for greater novelty and quality.

REFERENCES
[1] Dixon J.R. and Welch R.V. Guiding conceptual design through behavioral reasoning. *Research in Engineering Design*, 1994, 6, 169–188.
[2] Seering W.P. and Ulrich K.T. Synthesis of schematic descriptions in mechanical design. *Research in Engineering Design*, 1989, 1, 3–18.
[3] Bligh T.P. and Chakrabarti A. An approach to functional synthesis of solutions in mechanical conceptual design. Part I: Introduction and knowledge representation. *Research in Engineering Design*, 1994, 6, 127–141.
[4] Jansson D.G. and Smith S.M. Design fixation. *Design Studies*, 1991, 22(1), 3 11.
[5] Li C.L. *Conceptual design of single and multiple state mechanical devices: an intelligent CAD approach*, PhD thesis, 1998 (Department of Mechanical Engineering, University of Hong Kong)
[6] Langdon P.M. and Chakrabarti A. Browsing a large solution space in breadth and depth in. *In The 12th International Conference in Engineering Design (ICED'99)*, Vol. 3, 24-26 Aug 1999.
[7] Forbus K.D. *Qualitative reasoning, CRC Hand-book of Computer Science and Engineering*, 1996 (CRC Press).
[8] Brown J.S. and de Kleer J. A qualitative physics based on confluences. *Artificial Intelligence*, 1984, 24 (1-3), 7-83.
[9] Somasekhara Rao Todeti and Amresh Chakrabarti. An empirical model of process of synthesis of multiple state mechanical devices, *In Proceedings of the 17th International Conference on Engineering Design (ICED'09)*, Vol.4, Pages. 239-250, Stanford, CA, August 24-27, 2009.

[10] Elisha Sacks and Leo Joskowicz, *The Configuration Space Method for Kinematic Design of Mechanisms,* 2010(The MIT Press).
[11] Shean Juinn Chiou and Sridhar kota. Automated conceptual design of mechanisms. *Mechanism and machine theory,* 1999, 34,467-495.
[12] Murakami T. and Nakajima N. Mechanism concept retrieval using configuration space. *Research in Engineering Design,* 1997, 9, 99- 111.
[13] Chakrabarti A. and Bligh T. P. Functional Synthesis of Solution-Concepts in Mechanical Conceptual Design. Part II: Kind Synthesis, *Research in Engineering Design,* Vol. 8, No. 1, pp-52-62, 1996.
[14] Chakrabarti A. and Bligh T. P. An approach to functional synthesis of solutions in mechanical conceptual design. Part III: Spatial configuration. *Research in Engineering Design,* 1996, 2, 116-124.
[15] Sun K. and Faltings B. FAMING: Supporting innovative mechanism shape design. *Computer Aided Design,* 1996, 28(3), 207 – 216.
[16] Subramanian D. and Wang C. S. Kinematic synthesis with configuration spaces. *Research in Engineering Design,* 1995, 7, 193-213.

Contact: Prof. Amaresh Chakrabarti
Indian Institute of Science (IISc), Bangalore
Centre for Product Design and Manufacturing (CPDM)
Bangalore
India
Tel: +91 80 2293 2922
Fax: +91 80 2360 1975
E-mail: ac123@cpdm.iisc.ernet.in
URL: http://www.cpdm.iisc.ernet.in/people/ac/ac.htm

Amaresh Chakrabarti is Professor in the Department of Centre for Product Design and Manufacturing (CPDM) at Indian Institute of Science, Bangalore, India. His research interests are functional synthesis, design Creativity, design Methodology, collaborative design, eco-design, engineering design, design synthesis, requirements management, knowledge management, computer aided design, and design for variety.

Somasekhara Rao Todeti is Research Scholar in the Department of Centre for Product Design and Manufacturing (CPDM) at Indian Institute of Science, Bangalore, India. His research interests are engineering design, functional synthesis, mechanisms and machines, geometric modeling, and computer-aided design.

Index of Authors

A
Abdullah, Z — 8-1, 8-13
Abi Akle, A — 6-370
Abramovici, M — 9-1
Achiche, S — 10-102, 3-40, 7-44, 7-74
Agarwal, A — 2-120
Agogino, A M — 8-255
Agogué, M — 2-266
Ahmad, N — 6-500, 1-538
Ahmed-Kristensen, S — 10-200, 3-11, 3-377, 1-266
Alber-Laukant, B — 10-341
Albers, A — 9-196, 4-268, 2-256, 1-256
Albino, M — 9-335
Aleksandar, S — 10-456
Amaral, D C — 10-290
Ameri, F — 6-121
Anacker, H — 4-329
Andersen, C L — 7-74, 7-44
Andersson Schaeffer, J — 10-331
Andersson, K — 4-288
Andreasen, M M — 4-133
Annamalai Vasantha, G V — 4-67
Antonsson, E K — 10-402
Aoyama, K — 3-253
Appio, F P — 10-102
Apreda, R — 2-304, 2-1
Aquino Shluzas, L M — 10-159
Armstrong, G — 1-134
Aryana, B — 7-254
Aschehoug, S H — 5-145
Atkinson, S R — 7-152
Auriol, G — 10-228
Aurisicchio, M — 10-443, 7-437, 7-182, 6-468
Austin, S — 10-209
Aasland, K E — 10-47

B
Badin, J — 6-161
Badke-Schaub, P — 7-414, 7-404, 6-382
Baer, T — 5-220
Baeriswyl, M C — 8-140
Balachandran, L K — 1-113
Ball, A — 6-65
Balmelli, L — 6-282
Banerjee, B — 7-384
Baron, C — 10-228
Baronio, G — 8-161
Barth, A — 2-41
Barthès, J — 2-366
Bassler, D — 3-306
Bauer, F — 4-329
Baxter, T — 10-456
Becker, I — 9-91
Beckman, S L — 3-21, 8-255
Beckmann, G — 8-130, 5-249
Behncke, F H — 1-344
Behrisch, J C — 5-1
Bénabès, J — 9-324
Benassi, J L — 10-290
Bendaya, M — 3-306
Bennis, F — 9-324
Berdillon, V — 9-399
Bergema, K — 3-211
Bergsjö, D — 6-249
Bernardes, M M — 3-52, 1-147
Bernstein, W Z — 8-55
Bertolotti, F — 3-233
Bertoluci, G — 3-264, 1-211
Bertoni, A — 9-226
Bertoni, M — 9-226
Bertsche, B — 10-270
Bey, N — 4-77
Beyer, C — 3-274
Biancuzzo, E — 9-335
Bidermann, W — 3-317
Biedermann, W — 4-205, 4-369, 4-11
Bierer, A — 5-176
Binz, H — 10-178
Birch, D — 5-346
Birkhofer, H — 6-342, 9-257, 9-11, 9-314, 5-324, 5-302, 5-293, 4-441, 4-337, 2-236, 10-433, 1-299
Bitzer, M — 3-123
Bjärnemo, R — 2-356
Björklund, T A — 3-325

Blanco, E	6-231	Cartes, J	8-236
Blessing, L	1-393, 2-344	Casakin, H	7-107, 7-22
Bocquet, J	1-321, 3-93, 1-211	Cascini, G	9-246, 1-509
		Cash, P	2-151
Boelskifte, P	7-44, 7-74, 8-275	Cassotti, M	2-266
Bohn, A	1-299, 9-11, 9-257, 10-433, 9-275, 6-342, 5-324, 5-302, 5-293, 4-441, 4-337, 2-236, 9-314	Castro, J	3-245
		Catic, A	1-157
		Chakrabarti, A	1-334, 4-461, 9-111
		Chami, M	4-100
		Chamoret, D	6-161
		Chan, J	7-85
Bohn, M	9-275	Chapotot, E	6-370
Bojčetić, N	6-131	Chen, H	6-153
Boks, C	5-145, 2-13	Chen, W	10-35
Boland, R	10-426	Chen, Y	2-51
Bolognini, F	9-367	Chevalier, B	5-312
Bonaccorsi, A	2-304	Childs, P R	10-443
Bonnemaire, G	2-130	Cho, C K	2-314
Bont. de, C	3-211	Cho, Y C	5-282
Borg, J	10-139, 8-45	Choi, Y	4-230
Borgianni, Y	9-246	Choi, Y M	3-116
Boujut, J	1-12, 7-342	Choulier, D	8-173, 2-323, 7-285
Bracewell, R H	6-303, 10-443		
Brannemo, A	1-429	Christensen, B	7-265
Breitsprecher, T	6-108, 10-78	Christian, J L	10-91
Briede Westermeyer, J C	8-236	Christian, Z	9-196
Brix, T	6-272	Christiaans, H	7-458
Broberg, O	7-64	Christoph, L	6-192
Brodersen, S	8-275	Cicconi, P	5-198
Brombacher, A	10-1	Claesson, A	4-55
Brosch, M	5-249	Clarkson, P J	6-500, 1-538, 4-430, 1-475, 5-60, 7-152, 1-134, 7-1, 4-369
Bruch, J	6-21		
Buda, A	9-267		
Buijs, J	1-79		
Burvill, C R	6-414, 8-1, 8-13, 5-103	Coatanéa, E	2-323, 9-267
		Cogez, P	2-214
Bustamante, A	8-236	Čok, V	4-122
		Conway, A	8-66
C		Corin Stig, D	6-249
Cagan, J	6-360, 7-85	Cortimiglia, M N	3-354
Caillaud, E	1-441, 2-41	Cox, M F	8-55
Caldwell, N H	7-152, 5-60	Csernak, S	9-275
Calvi, R	3-176	Culley, S	6-424, 6-65, 2-151
Camilleri, K	8-45		
Campana, E	6-392	Currano, R M	7-374
Cardillo, A	9-246, 1-509		
Cardoso, C	7-404	D	
Carlsson, S	3-153	Da Cunha, C	9-305
		Da Silva Vieira, S L	6-382

Daly, S R	10-91	Elstner, S	5-271
Daniilidis, C	4-400	Eng, N L	6-468, 10-443
Darlington, M	6-65	Engelhardt, R A	9-314, 9-257,
de Paula, I C	3-354		5-324, 10-433
de Vere, I	8-226, 8-216	Ensici, A	7-414
de Weck, O L	1-355, 10-126,	Ensslin, V	4-400
	10-278	Eppinger, S D	8-140
Deamer, J	10-209	Erbe, T	4-222
Dekoninck, E	7-140	Erden, A	10-466
Del Frate, L	2-204	Erden, Z	10-466
Dentsoras, A	10-311	Ericson, Å	6-480, 3-284,
Deschinkel, K	5-236		3-143
Dewberry, E L	5-165	Eriksen, K R	4-167
Dewulf, S	6-210	Eriksson, J	1-429
Di Minin, A	10-102	Eriksson, Y	10-331, 7-194
Dias, M	10-248	Estrada, G	6-490
Dienst, S	9-1	Evans, S	4-67
Dill, A K	4-441	Eynard, B	1-321
Dinar, M	6-392		
Dodds, S	4-430	F	
Doenmez, D	7-1	Fain, N	1-366, 8-120
Dong, A	7-54, 7-117	Falconer, M	4-238
Dong, H	6-153	Fantoni, G	2-1, 2-304
Dorst, K	2-142, 10-1	Farndon, R	1-417
Dowlen, C	2-183	Farrokhzad, B	1-122
Doyle, K	5-265	Farrugia, P	8-45, 10-139
Drémont, N	9-176	Fathi, M	9-1
Du Bois, E	10-258	Feldhusen, J	4-157
Duffy, A	9-176	Felk, Y	2-214
Duflou, J R	6-210	Feng, X	4-1
Duhovnik, J	4-122, 1-366	Fernandes, A A	6-382
Dunn, J	4-238	Ferreira Junior, L D	10-290
During, C	4-145	Feustel, F	9-275
Döring, U	6-272	Field, B W	8-35, 8-23, 8-13,
			8-1
E		Figge, A	4-21
Ebel, B	2-256	Fixson, S	10-370
Eben, K G	4-400, 1-101,	Florin, U	7-194, 10-331
	3-199	Follmer, M	4-258, 9-122
Eckert, C M	5-60	Fonseca, T	6-382
Eifler, T	5-324, 10-433,	Forest, J	2-323
	9-257, 9-314	Fortin, C	6-183
Eilmus, S	4-299	Franić, L	6-131
Eisenbart, B	2-344	Fraslin, M	6-231
Eisto, T	7-32	Frillici, F S	1-509
Ekman, A	3-153, 2-120	Fritz, O	10-270
Ekman, S	3-153, 2-120	Fu, K	6-360, 7-85
Elezi, F	3-317	Fujita, K	9-50
Elgh, F	6-86	Fukushige, S	5-135

Furuhjelm, J	1-180	Gudem, M	10-167, 2-13
Förg, A	1-233	Gudmundsson, H P	7-74, 7-44
		Guenand, A	9-399
G		Guenov, M D	1-113
Gabelloni, D	2-304, 2-1	Gumienny, R	6-446
Galafassi, A	3-52	Gumpinger, T	10-349, 4-175
Galea, A	10-139	Gutiérrez, E	1-373
Garcia-Noriega, S	5-336	Gutu, D	6-456
Gardoni, M	7-285	Gwilliam, J	5-19
Garner, S	10-66	Gylling, M	7-480
Gaukstern, T	4-329	Götze, U	5-176
Gausemeier, J	10-413, 4-329		
Gebauer, K J	6-402	H	
Gedell, S	4-55	Hahn, A	9-61
Georgiev, G V	7-234	Hales, C	2-163
Geraci, D	9-335	Halfmann, N	5-271
Gerber, E	7-468	Han, C	7-384
Gerhard, D	6-192	Han, D M	1-288
Gericke, K	1-393, 2-344	Hansen, C L	4-133
Germani, M	5-198	Hansen, C T	2-142
Gero, J S	2-194, 7-117, 2-294	Hansen, M S	10-190
		Hansen, Z N	3-11
Giacobbe, F	9-335	Harkema, C	10-1
Gidel, T	2-366, 3-187	Harrison, T	7-437
Giess, M D	3-1	Hatcher, G D	5-259
Gish, L	3-83	Hatchuel, A	2-87, 2-214
Giurco, D	5-1	He, L	10-35
Goehner, P	10-178	Heebøll, J	3-31
Goh, Y M	6-96	Hehenberger, P	9-122, 1-122, 4-258, 4-309
Gomes, S	5-236, 6-161		
Gonçalves, M G	7-404	Hellenbrand, D	9-215, 1-521
Gondhalekar, A C	1-113	Hellström, D	10-392
Gonzalez, R	10-91	Helten, K	3-199, 3-133, 1-101
Gooch, S	4-238, 4-185		
Gopsill, J A	6-141	Hendriks, L	2-275
Gorbea, C E	10-278	Henriques, E	1-417, 5-28
Gorissen, D	6-436	Henze, L	1-79
Graessler, I	4-45	Hepperle, C	7-1, 1-233
Graignic, P	9-176	Herberg, A	1-499
Gramlich, S	1-299, 4-337	Herbert, B	5-324, 1-299, 10-433, 9-314, 2-236, 4-337, 9-11, 6-342, 5-302, 4-441, 5-293, 9-257
Grech, A	10-139		
Greene, R T	5-113		
Grierson, H J	7-12		
Gries, B	1-531		
Große Austing, S	9-61		
Grote, K	3-274	Herman, B	9-101
Gruber, G	10-78	Hermann, K	9-314
Grubisic, V V	3-187	Heß, C	10-270
Gual, J	5-155	Hicks, B J	6-141, 6-31, 2-151
		Hirosaki, M	5-135

Hisarciklilar, O	3-40	Jorgensen, U	8-275, 1-453
Hoffenson, S	5-81	Jowers, I	10-66
Hofmann, D	10-270	Juhl, J	7-480
Hong, Y K	7-448	Jun, T	4-430
Hong, Y S	3-221, 4-88, 9-165	Jung, M F	7-244, 7-384
		Junk, S	5-12
Horvath, I	10-258, 2-108	Jupp, J R	9-285
Hosnedl, S	8-287	Jursch, S	4-381
Howard, T J	3-1, 3-387, 6-65, 7-140	Just, V	4-329
		Juuti, T	1-405
Huang, S	10-456	Järvenpää, Jo	4-319
Huet, G	6-183	Järvenpää, Ju	4-248
Hunt, G R	6-392	Jørgensen, U	1-453, 8-275
Hussain, R	4-67		
Husung, S	2-226	**K**	
Hvam, L	4-133	Kain, A S	1-487, 9-295, 3-346
Hwang, H	7-214		
Häusler, S	9-61	Kallenborn, O	4-357
Högman, U	3-62	Kan, J W	2-194
Hölttä, V	7-32	Kang, C	4-88
		Kapoor, A	8-226, 8-216
I		Karaka_i_, M	4-122
Iacob, C	6-1	Karlsson, A	6-322
Ijomah, W L	5-259	Karniel, A	1-243
Ion, W	8-66	Kastensson, Å	3-284
Isaksson, O	4-347	Kauling, G B	1-147
		Kayani, O K	4-195
J		Kazakçi, A	2-266, 2-275
Jablokow, K W	2-377	Kazamia, K I	5-19
Jacobsen, P	10-190	Keane, A	6-436
Jang, M	4-230	Keller, R	1-355
Jankovic, M	1-199, 1-321	Kendira, A	2-366
Janus, A	6-292	Kerley, W P	1-134
Jauregui-Becker, J M	6-402, 9-389	Khan, S	4-288
Jee, H	2-314	Kihlander, I	8-100, 10-360
Jensen, N	3-377	Kim, C	7-458
Jensen, O K	3-377	Kim, H	6-43, 5-70
Jiménez-Narvaez, L	7-285	Kim, J H	7-448
Johannes, M	5-324, 10-433, 9-314	Kim, K	7-214
		Kim, K J	4-88
Johannesson, H	4-55, 3-62	Kim, S	10-302
Johansson, C	6-332, 9-226	Kim, S R	5-282
Johansson, G	6-21	Kim, Y S	4-230, 7-297, 7-448, 5-282, 4-390, 3-221, 2-314, 1-288
Johansson, P	6-332		
Johnson, A L	6-303		
Jomaa, I	9-305		
Jonas, H	10-349, 4-112	Kim, Y M	7-448
Jones, A	2-366	Kirihara, K	4-215
Jong, F d	1-79	Kirisci, P T	9-80

Kirjavainen, S	3-325	L	
Kirschner, R	9-295, 3-346, 1-487	Ladeby, K R	10-190
		Lande, M	7-384
Kishita, Y	5-135	Lang, A	1-487, 3-346, 9-295
Kissel, M	9-215		
Kitajima, K	3-253	Langbein, S	9-186
Kittel, K	4-309	Langer, S	1-499
Klein, P	9-80	Langley, P	6-392
Kleinsmann, M	3-211	Lanz, M	4-248
Klingender, M	4-381	Laratte, B	5-312
Kljajin, M	4-122	Larroude, V	5-49
Kloberdanz, H	10-433, 6-342, 9-257, 5-324, 2-236	Larsson, A	3-143
		Larsson, T	3-143
		Le Dain, M	3-176
Klokkehaug, J A	9-71	Le Masson, P	2-214, 2-87
Knoblinger, C	3-133	Leary, M	10-456, 5-103, 6-414
Ko, Y D	2-314		
Koehler, S	5-176	Lee, H K	1-288
Koga, T	3-253	Lee, J	1-288
Koh, D C	4-390	Lee, J H	9-165, 3-221
Kohn, A	6-350	Lee, S W	4-390, 1-288
Kokkolaras, M	4-347	Lee, S	4-230
Komashie, A	4-430	Legadeur, J	7-285
Komoto, H	2-334	Lehtonen, T	1-405, 4-248
Konez Ero_lu, A	10-466	Leidich, E	5-176
Kopp, M	10-270	Leifer, L J	7-244, 7-374, 10-167, 7-384, 10-159
Koppe, R	9-61		
Kortler, S	4-410, 3-317		
Korzenowski, A	3-354	Leino, S S	9-91, 4-319
Kotera, Y	9-50	Lenau, T	1-310
Kotovsky, K	7-85, 6-360	Lenne, D	2-366
Kratzer, M	10-178	Leroy, Y	1-199
Krause, D	4-299, 8-130, 5-271, 5-249, 4-112, 4-175, 10-349	Lewis, W P	8-23
		Li, Y	8-88
		Liang, H	5-346
		Liem, A	9-357, 2-130, 6-53, 8-194, 7-254
Krebber, S	6-342		
Krehmer, H	1-1, 6-292		
Kreimeyer, M	7-1, 8-184	Lindberg, T	6-446
Kreitler, S	7-107	Lindegaard, H	1-453, 8-275
Kress, G	7-353	Lindemann, U	4-11, 1-57, 1-101, 1-344, 1-499, 1-521, 10-278, 3-199, 3-346, 1-233, 4-205, 4-369, 4-400, 4-410, 6-350, 7-1, 9-30, 9-215, 9-295, 3-317, 1-487
Kreye, M E	6-96		
Kroll, E	6-76, 1-169		
Krollmann, J	1-57		
Krüger, D	10-380		
Kubisch, C	3-274		
Kwak, M	5-70		
Königs, S F	4-357		
Köppen, E	7-162		
Körber, K	1-499		

Lindh, H	3-264	Matthews, P C	1-69
Lindner, A	9-1	Matthiesen, S	9-236
Lindow, K	8-205	Maurer, C	3-74
Linsey, J	7-309, 7-394	Maurer, M	1-24
Liu, Y	6-262	Mazur, M	10-456, 6-414
Liu, Z	2-51	McAloone, T C	10-102, 3-387, 4-77, 4-449
Lloveras, J	6-490, 5-155, 8-78	McAlpine, H C	6-141, 6-31
Lodgaard, E	10-47, 9-71	McArthur, C	6-121
Lohmeyer, Q	2-256	McKay, A	10-66
Lommatzsch, N	4-337	McMahon, C A	2-173, 3-1, 6-65
Lopes, J C	10-248	McSorley, G	6-183
Lorentz, B	4-268	Medland, A J	4-185, 4-238, 8-245
Lovallo, D	7-54		
Lowe, D	6-241	Meerkamm, H	1-1
LU, W	7-170	Mehta, P P	8-265
Lund, K	7-275, 8-100	Meijer, S	7-424
Luyk - de Visser, I	10-1	Meinel, C	6-446, 7-162
Lygin, K	9-186	Mejborn, C O	1-310
Löfqvist, L G	3-164	Melles, G	8-226, 8-216
		Metraglia, R	8-161
M		Metzler, T	7-330
Maclachlan, R	7-12	Miller, E	8-88
Macredie, R	6-153	Miller, M	7-492
Macrì, D M	3-233	Millet, D	5-49
Maeno, T	10-302	Minel, S	6-370
Maier, A M	1-188, 7-152, 7-1	Mirson, A	3-317
Maier, M	9-132	Mitchell-Baker, A	4-430
Makino, Ya	10-302	Mizuno, Y	5-135
Makino, Yu	10-302	Mizuyama, H	3-365
Mandolini, M	5-198	Moatari Kazerouni, A	3-40
Marco, A	10-443, 7-437, 7-182, 6-468	Mocko, G M	9-377, 7-204
		Modzelewski, M	9-80
Marfisi, E	6-468	Monceaux, A	10-228
Marini, V K	1-266	Montecchi, T	7-362
Marinov, M	6-456	Monticolo, D	6-161
Marion, T J	10-370	Morgan, T	7-320
Marion, W	5-324, 9-314, 9-257	Mortensen, N H	1-276, 4-133
		Moss, M A	1-417, 1-134
Marjanović, D	9-345, 6-173, 1-383, 6-131	Motte, D	2-356
		Mougaard, K	4-77, 3-387
Marle, F	3-104, 3-93	Moulin, C	2-366
Martin, P	4-268	Mounarath, R	7-54
Masko, M L	5-60	Mullineux, G	9-40
Mathias, J	10-433, 5-324, 9-314	Murakami, T	7-224
		Mutoh, Y	4-278
Mathieson, J	7-492	Müller, P	8-205
Matsuka, Y	10-11	Mäkinen, H	4-319
Matthews, J	8-245	Möhringer, S	1-221
		Mörtl, M	1-57, 1-233

N

Nagai, Y	7-234
Nagarajah, A	4-157
Nakagawa, S	7-224
Naumann, T	4-357
Nehuis, F	10-321
Neugebauer, L M	4-77
Newnes, L B	6-96
Nezhadali, V	4-195
Nguyen Van, T	1-321
Nicquevert, B	1-12
Nielsen, T A	4-77
Nielsen, T H	4-167
Nilsson, S	8-100
Nishimura, H	6-282
Nomaguchi, Y	9-50
Norell Bergendahl, M	8-100, 6-200
Nunez, M	1-113

O

Oduncuoglu, A	10-114
Oehmen, J	3-133, 3-306
Ogliari, A	3-187
Ognjanovic, M	2-23
Ohashi, S	2-61
Oizumi, K	3-253
Oja, H	1-405
Olivetti, E	10-248
Olsson, A	3-264
Onarheim, B	7-265
Opiyo, E Z	10-56
Orawski, R	1-57
Oriakhi, E	7-394
Orsborn, S	1-465
Ortiz Nicolas, J C	10-443, 7-182
Osborn, J	7-204
Osorio Gómez, G	8-151
Otsuka, Y	4-278
Ovtcharova, J	5-220, 6-456
Öberg, Å	10-331
Ölvander, J	4-1

P

Paetzold, K	9-143, 5-228, 4-222
Pakkanen, J	4-248
Palmer, G	8-110
Panarotto, M	5-187
Panchal, J	4-33
Papalambros, P Y	10-149, 5-81
Park, J	10-426
Park, K T	4-88
Park, Y	2-61
Parvan, M I	9-30
Pasqual, M C	10-126
Paulsson, S	3-365
Pavković, N	6-131
Pearson, G	6-414
Peças, P	5-28
Pechuan, A	3-317
Pei, E	6-153
Pellegård, Ø	9-71
Peng, X	7-394
Pepe, C	1-134, 1-417
Perez, M	8-236
Personnier, H	3-176
Petersen, S I	3-21, 3-31
Petetin, F	1-211
Petiot, J	7-170, 9-305
Pettersson, M	4-1
Philip, A D	1-69
Phleps, U	5-91
Plaumann, B	10-349
Poirson, E	9-305, 9-324
Poltschak, F	1-122
Poppen, F	9-61
Poulet, A	1-441
Pourmohamadi, M	2-294
Prats, M	10-66
Prieto, P A	9-208
Pulkkinen, A	1-405
Pulm, U	1-221
Punz, S	9-122, 4-258
Puyuelo, M	5-155

Q

Qamar, A	4-145

R

Radkowski, R	10-413
Ramani, K	8-55
Ramanujan, D	8-55
Ramirez, M	5-39, 5-1
Randmaa, M	3-387
Rasmussen, L B	10-190
Rath, K	5-293
Raucent, B	9-101
Raudberget, D S	1-45
Rauscher, M	10-178

Rauth, I	7-162	Sakaguchi, S	2-73
Ravaut, Y	9-324	Sakao, T	4-449, 3-365
Rayner, H	4-430	Salehi, V	2-173
Razzaq, H	4-195	Salunkhe, U	2-120
Rebentisch, E	3-133	Salustri, F	2-151
Redon, R	3-294	Samuel, P	2-377
Reeßing, M	6-272	Sandberg, M	4-347
Rehman, F	8-88	Santos, R	10-248
Reich, Y	1-243, 10-24, 6-220, 7-424, 2-87	Sanya, I	6-241
		Sapin, J	9-101
		Sato, K	10-11
Reichel, T	10-238	Savšek, T	6-173
Reitmeier, J	9-143	Scales, D	10-221
Ren, Y	10-149	Scanlan, J	6-436
Restrepo, J	1-531, 1-266	Schar, M	7-353
Reyes, T	5-312, 5-124	Schindler, C	3-335
Rianantsoa, N	3-294	Schmidt III, R	10-209
Ribeiro, I	5-28	Schmitz, S	4-369, 1-538, 6-500, 1-475
Rieg, F	10-341		
Rihtarsic, J	8-120	Schnjakin, M	7-162
Riitahuhta, A	1-405	Schotborgh, W O	9-389
Ringen, G	9-71	Schubert, S	4-157
Rio, M	5-124	Schunn, C	7-85
Ritzén, S	8-100	Schwalmberger, A	9-30
Robert, A	5-236	Seemüller, H	4-100
Robinson, B	1-355	Seering, W	3-133, 3-306, 3-245
Roland, E	9-314, 9-257, 10-433, 5-324		
		Seifert, C M	10-91
Rosa, F	9-21	Seki, K	6-282
Roschuni, C	8-255	Sen, C	9-377
Rose, B	2-41, 1-441	Seppälä, M	9-267
Rosen, R	4-258	Seshia, A	9-367
Rosenqvist, T S	1-453	Shah, J J	6-392, 7-127
Rosenstein, D	10-24	Shai, O	6-220
Roth, R	10-248	Shapiro, A A	10-402
Roth, S	5-236	Shea, K	9-345, 9-367, 7-330
Rothwell, A	4-238		
Rotini, F	1-509, 9-246	Shehab, E	6-241
Roucoules, L	5-124	Shihmanter, A	6-76
Rovida, E	9-21	Shimizu, H	4-278
Roy, R	4-67	Shin, C	7-97
Ruiz Arenas, S	8-151	Shin, D I	3-221
Russo, D	7-362	Shin, J	7-297
Rünger, G	10-238	Shirasaka, S	10-302
Röder, B	9-11	Siddiqi, A	1-355
Röhner, S	6-108	Sigurjonsson, J B	8-194
		Simonyi, A	6-456
S		Singh, V	7-117
Sadek, T	5-210, 9-186	Siyam, G	1-475
Saga, N	4-215	Smith, S	7-127

Smulders, F	1-79, 7-424	Tingström, J	7-275, 1-180
Snider, C	7-140	Tiryakioglu, C	6-31
Sohrabpour, V	10-392	Tiwari, A	4-67
Sonalkar, N	7-384	Toche, B	6-183
Sop Njindam, T	5-228	Todeti, S R	9-111, 4-461
Srp, Z	8-287	Toivonen, V	9-91
Stankovic, T	9-345	Tomiyama, T	2-334
Stappers, P J	1-79	Torcato, R	10-248
Stark, R	4-21, 8-205	Torlind, P	5-187, 6-322
Stechert, C	10-321	Torry-Smith, J M	1-276
Steger, D	10-238	Tran Duy, K	9-101
Steinert, M	7-374, 3-21, 10-159, 10-167, 3-245	Trevelyan, J	10-221
		Trimingham, R	5-336
		Troussier, N	9-176
Stephan, N K	3-335	Tryfonas, T	7-320
Stetter, R	5-91, 1-221, 4-100	Tränkle, M	5-12
		Tsumaya, A	2-73
Stevanović, M	1-383	Tucker, C	6-43
Stockinger, A	9-153, 10-380	Tuokko, R	4-248
Stoll, T	9-153	Tuttass, I	4-357
Stompff, G	1-79	Tyapin, I	4-347
Strong, D	7-12	Törlind, P	6-322, 5-187
Störrle, H	1-188		
Subic, A	6-414	U	
Subrahmanian, E	7-424	Umeda, Y	5-135
Sudin, M N	10-200	Uuttu, O	4-319
Suistoranta, S	1-405		
Summers, J	8-110, 7-204, 9-377, 7-492	V	
		Vajna, S	4-309
Sun, G	7-501	Valkenburg, R	3-211, 3-74
Sutherland, J W	8-55	van Houten, F J	6-402
Swan, A H	1-180	van Schaik, J R	6-436
Szots, D M	6-456	Vandevenne, D	6-210
Štorga, M	9-345, 6-173	Vareille, J	2-246
		Venkataraman, S	1-334
		Verhaegen, P	6-210
T		Vermaas, P	2-98
Takeda, K	6-436	Vidal, L	3-104, 3-93
Takiguchi, S	4-278	Vielhaber, M	3-123, 1-157
Talecki, H	1-465	Vietor, T	10-321
Tan, X	8-88	Viganò, R	9-21
Tarkian, M	4-1, 4-195	Vignoli, M	3-233
Tartler, D	6-292	Villa, V	8-161
Taura, T	2-73, 2-61	Viswanathan, V	7-309
Tegel, O	3-274	Vliembergen, E V	1-79
Thane, S	4-430	Voos, H	4-100
Thoben, K	9-80	Vosgien, T	1-321
Thomson, A I	6-313	Vukšinović, N	1-366
Thomson, V	3-40, 10-114		
Thor, P	6-480		

W

Wada, H	5-135
Wadell, C	6-200
Walde, F	9-1
Walla, W P	5-220
Wallace, K	2-163
Walter, M	10-78
Wang, H	6-303
Warell, A	9-357
Wartzack, S	10-78, 9-153, 6-292, 10-380, 1-1, 1-91, 6-108
Watty, R	8-184
Weber, C	4-222, 2-226
Weber, F	6-424
Weil, B	2-214, 2-266, 2-87
Weir, J G	5-103, 8-23
Weite, P	8-173
Welo, T	2-13, 10-167
Wendland, M	5-210
Wenngren, J	6-480
Wenzel, H	1-113
Westphal, C	1-91
Whitfield, R I	9-176
Whitney, D	1-417
Wiebel, M	5-324, 9-257, 9-314
Wiedner, A	1-256
Wikander, J	4-145
Wikström, A	10-331
Windmill, J F	5-259
Winkelman, P	2-287
Wodehouse, A J	7-12
Wood, K	7-85
Woodfine, N	8-88
Woodward, J	7-127
Wren, G	8-66
Wuttke, F	9-275
Wynn, D C	1-538, 4-369, 6-500, 1-475

X

Xie, Yi	6-424
Xie, Yo	2-51
Xu, H	10-238

Y

Yamamoto, E	2-73, 2-61
Yan, X	5-236, 8-88
Yanagisawa, H	7-224
Yang, S	3-354
Yannou, B	6-370, 3-294, 1-199
Yao, S	7-501
Yilmaz, S	10-91
Yogev, O	10-402
Yosie, N	3-253
Yue, H	7-140
Yvars, P	4-420, 5-49

Z

Zadnik, Ž	4-122
Zainal Abidin, S B	9-357
Zapaniotis, A	10-311
Zapf, J	10-341
Zeiler, W	6-11, 2-31, 1-35
Zeman, K	4-258, 4-309, 9-122
Zeng, Y	2-246
Zhang, X	10-228
Zhang, Z	2-51
Zhao, F	8-55
Zhao, S	5-302
Zhu, S	6-282
Ziebart, J R	10-321
Zier, S	2-236
Žavbi, R	8-120

CPSIA information can be obtained
at www.ICGtesting.com
Printed in the USA
BVOW06s2027050317
477783BV00021B/145/P